geometry and complex variables

PURE AND APPLIED MATHEMATICS

A Program of Monographs, Textbooks, and Lecture Notes

EXECUTIVE EDITORS

Earl J. Taft
Rutgers University
New Brunswick, New Jersey

Zuhair Nashed
University of Delaware
Newark, Delaware

CHAIRMEN OF THE EDITORIAL BOARD

S. Kobayashi
University of California, Berkeley
Berkeley, California

Edwin Hewitt
University of Washington
Seattle, Washington

EDITORIAL BOARD

M. S. Baouendi
University of California, San Diego

Donald Passman
University of Wisconsin-Madison

Jack K. Hale
Georgia Institute of Technology

Fred S. Roberts
Rutgers University

Marvin Marcus
University of California, Santa Barbara

Gian-Carlo Rota
Massachusetts Institute of Technology

W. S. Massey
Yale University

David L. Russell
Virginia Polytechnic Institute
and State University

Leopoldo Nachbin
Centro Brasileiro de Pesquisas Fisicas
and University of Rochester

Jane Cronin Scanlon
Rutgers University

Anil Nerode
Cornell University

Walter Schempp
Universität Siegen

Mark Teply
University of Wisconsin-Milwaukee

LECTURE NOTES

IN PURE AND APPLIED MATHEMATICS

1. *N. Jacobson*, Exceptional Lie Algebras
2. *L. -Å. Lindahl and F. Poulsen*, Thin Sets in Harmonic Analysis
3. *I. Satake*, Classification Theory of Semi-Simple Algebraic Groups
4. *F. Hirzebruch, W. D. Newmann, and S. S. Koh*, Differentiable Manifolds and Quadratic Forms (out of print)
5. *I. Chavel*, Riemannian Symmetric Spaces of Rank One (out of print)
6. *R. B. Burckel*, Characterization of C(X) Among Its Subalgebras
7. *B. R. McDonald, A. R. Magid, and K. C. Smith*, Ring Theory: Proceedings of the Oklahoma Conference
8. *Y.-T. Siu*, Techniques of Extension on Analytic Objects
9. *S. R. Caradus, W. E. Pfaffenberger, and B. Yood*, Calkin Algebras and Algebras of Operators on Banach Spaces
10. *E. O. Roxin, P.-T. Liu, and R. L. Sternberg*, Differential Games and Control Theory
11. *M. Orzech and C. Small*, The Brauer Group of Commutative Rings
12. *S. Thomeier*, Topology and Its Applications
13. *J. M. Lopez and K. A. Ross*, Sidon Sets
14. *W. W. Comfort and S. Negrepontis*, Continuous Pseudometrics
15. *K. McKennon and J. M. Robertson*, Locally Convex Spaces
16. *M. Carmeli and S. Malin*, Representations of the Rotation and Lorentz Groups: An Introduction
17. *G. B. Seligman*, Rational Methods in Lie Algebras
18. *D. G. de Figueiredo*, Functional Analysis: Proceedings of the Brazilian Mathematical Society Symposium
19. *L. Cesari, R. Kannan, and J. D. Schuur*, Nonlinear Functional Analysis and Differential Equations: Proceedings of the Michigan State University Conference
20. *J. J. Schäffer*, Geometry of Spheres in Normed Spaces
21. *K. Yano and M. Kon*, Anti-Invariant Submanifolds
22. *W. V. Vasconcelos*, The Rings of Dimension Two
23. *R. E. Chandler*, Hausdorff Compactifications
24. *S. P. Franklin and B. V. S. Thomas*, Topology: Proceedings of the Memphis State University Conference
25. *S. K. Jain*, Ring Theory: Proceedings of the Ohio University Conference
26. *B. R. McDonald and R. A. Morris*, Ring Theory II: Proceedings of the Second Oklahoma Conference
27. *R. B. Mura and A. Rhemtulla*, Orderable Groups
28. *J. R. Graef*, Stability of Dynamical Systems: Theory and Applications
29. *H.-C. Wang*, Homogeneous Branch Algebras
30. *E. O. Roxin, P.-T. Liu, and R. L. Sternberg*, Differential Games and Control Theory II
31. *R. D. Porter*, Introduction to Fibre Bundles
32. *M. Altman*, Contractors and Contractor Directions Theory and Applications
33. *J. S. Golan*, Decomposition and Dimension in Module Categories
34. *G. Fairweather*, Finite Element Galerkin Methods for Differential Equations
35. *J. D. Sally*, Numbers of Generators of Ideals in Local Rings
36. *S S. Miller*, Complex Analysis: Proceedings of the S.U.N.Y. Brockport Conference
37. *R. Gordon*, Representation Theory of Algebras: Proceedings of the Philadelphia Conference
38. *M. Goto and F. D. Grosshans*, Semisimple Lie Algebras
39. *A. I. Arruda, N. C. A. da Costa, and R. Chuaqui*, Mathematical Logic: Proceedings of the First Brazilian Conference

40. F. Van Oystaeyen, Ring Theory: Proceedings of the 1977 Antwerp Conference
41. F. Van Oystaeyen and A. Verschoren, Reflectors and Localization: Application to Sheaf Theory
42. M. Satyanarayana, Positively Ordered Semigroups
43. D. L. Russell, Mathematics of Finite-Dimensional Control Systems
44. P.-T. Liu and E. Roxin, Differential Games and Control Theory III: Proceedings of the Third Kingston Conference, Part A
45. A. Geramita and J. Seberry, Orthogonal Designs: Quadratic Forms and Hadamard Matrices
46. J. Cigler, V. Losert, and P. Michor, Banach Modules and Functors on Categories of Banach Spaces
47. P.-T. Liu and J. G. Sutinen, Control Theory in Mathematical Economics: Proceedings of the Third Kingston Conference, Part B
48. C. Byrnes, Partial Differential Equations and Geometry
49. G. Klambauer, Problems and Propositions in Analysis
50. J. Knopfmacher, Analytic Arithmetic of Algebraic Function Fields
51. F. Van Oystaeyen, Ring Theory: Proceedings of the 1978 Antwerp Conference
52. B. Kedem, Binary Time Series
53. J. Barros-Neto and R. A. Artino, Hypoelliptic Boundary-Value Problems
54. R. L. Sternberg, A. J. Kalinowski, and J. S. Papadakis, Nonlinear Partial Differential Equations in Engineering and Applied Science
55. B. R. McDonald, Ring Theory and Algebra III: Proceedings of the Third Oklahoma Conference
56. J. S. Golan, Structure Sheaves over a Noncommutative Ring
57. T. V. Narayana, J. G. Williams, and R. M. Mathsen, Combinatorics, Representation Theory and Statistical Methods in Groups: YOUNG DAY Proceedings
58. T. A. Burton, Modeling and Differential Equations in Biology
59. K. H. Kim and F. W. Roush, Introduction to Mathematical Consensus Theory
60. J. Banas and K. Goebel, Measures of Noncompactness in Banach Spaces
61. O. A. Nielson, Direct Integral Theory
62. J. E. Smith, G. O. Kenny, and R. N. Ball, Ordered Groups: Proceedings of the Boise State Conference
63. J. Cronin, Mathematics of Cell Electrophysiology
64. J. W. Brewer, Power Series Over Commutative Rings
65. P. K. Kamthan and M. Gupta, Sequence Spaces and Series
66. T. G. McLaughlin, Regressive Sets and the Theory of Isols
67. T. L. Herdman, S. M. Rankin, III, and H. W. Stech, Integral and Functional Differential Equations
68. R. Draper, Commutative Algebra: Analytic Methods
69. W. G. McKay and J. Patera, Tables of Dimensions, Indices, and Branching Rules for Representations of Simple Lie Algebras
70. R. L. Devaney and Z. H. Nitecki, Classical Mechanics and Dynamical Systems
71. J. Van Geel, Places and Valuations in Noncommutative Ring Theory
72. C. Faith, Injective Modules and Injective Quotient Rings
73. A. Fiacco, Mathematical Programming with Data Perturbations I
74. P. Schultz, C. Praeger, and R. Sullivan, Algebraic Structures and Applications Proceedings of the First Western Australian Conference on Algebra
75. L. Bican, T. Kepka, and P. Nemec, Rings, Modules, and Preradicals
76. D. C. Kay and M. Breen, Convexity and Related Combinatorial Geometry: Proceedings of the Second University of Oklahoma Conference
77. P. Fletcher and W. F. Lindgren, Quasi-Uniform Spaces
78. C.-C. Yang, Factorization Theory of Meromorphic Functions
79. O. Taussky, Ternary Quadratic Forms and Norms
80. S. P. Singh and J. H. Burry, Nonlinear Analysis and Applications
81. K. B. Hannsgen, T. L. Herdman, H. W. Stech, and R. L. Wheeler, Volterra and Functional Differential Equations

82. *N. L. Johnson, M. J. Kallaher, and C. T. Long,* Finite Geometries: Proceedings of a Conference in Honor of T. G. Ostrom
83. *G. I. Zapata,* Functional Analysis, Holomorphy, and Approximation Theory
84. *S. Greco and G. Valla,* Commutative Algebra: Proceedings of the Trento Conference
85. *A. V. Fiacco,* Mathematical Programming with Data Perturbations II
86. *J.-B. Hiriart-Urruty, W. Oettli, and J. Stoer,* Optimization: Theory and Algorithms
87. *A. Figa Talamanca and M. A. Picardello,* Harmonic Analysis on Free Groups
88. *M. Harada,* Factor Categories with Applications to Direct Decomposition of Modules
89. *V. I. Istrățescu,* Strict Convexity and Complex Strict Convexity: Theory and Applications
90. *V. Lakshmikantham,* Trends in Theory and Practice of Nonlinear Differential Equations
91. *H. L. Manocha and J. B. Srivastava,* Algebra and Its Applications
92. *D. V. Chudnovsky and G. V. Chudnovsky,* Classical and Quantum Models and Arithmetic Problems
93. *J. W. Longley,* Least Squares Computations Using Orthogonalization Methods
94. *L. P. de Alcantara,* Mathematical Logic and Formal Systems
95. *C. E. Aull,* Rings of Continuous Functions
96. *R. Chuaqui,* Analysis, Geometry, and Probability
97. *L. Fuchs and L. Salce,* Modules Over Valuation Domains
98. *P. Fischer and W. R. Smith,* Chaos, Fractals, and Dynamics
99. *W. B. Powell and C. Tsinakis,* Ordered Algebraic Structures
100. *G. M. Rassias and T. M. Rassias,* Differential Geometry, Calculus of Variations, and Their Applications
101. *R.-E. Hoffmann and K. H. Hofmann,* Continuous Lattices and Their Applications
102. *J. H. Lightbourne, III, and S. M. Rankin, III,* Physical Mathematics and Nonlinear Partial Differential Equations
103. *C. A. Baker and L. M. Batten,* Finite Geometries
104. *J. W. Brewer, J. W. Bunce, and F. S. Van Vleck,* Linear Systems Over Commutative Rings
105. *C. McCrory and T. Shifrin,* Geometry and Topology: Manifolds, Varieties, and Knots
106. *D. W. Kueker, E. G. K. Lopez-Escobar, and C. H. Smith,* Mathematical Logic and Theoretical Computer Science
107. *B.-L. Lin and S. Simons,* Nonlinear and Convex Analysis: Proceedings in Honor of Ky Fan
108. *S. J. Lee,* Operator Methods for Optimal Control Problems
109. *V. Lakshmikantham,* Nonlinear Analysis and Applications
110. *S. F. McCormick,* Multigrid Methods: Theory, Applications, and Supercomputing
111. *M. C. Tangora,* Computers in Algebra
112. *D. V. Chudnovsky and G. V. Chudnovsky,* Search Theory: Some Recent Developments
113. *D. V. Chudnovsky and R. D. Jenks,* Computer Algebra
114. *M. C. Tangora,* Computers in Geometry and Topology
115. *P. Nelson, V. Faber, T. A. Manteuffel, D. L. Seth, and A. B. White, Jr.* Transport Theory, Invariant Imbedding, and Integral Equations: Proceedings in Honor of G. M. Wing's 65th Birthday
116. *P. Clément, S. Invernizzi, E. Mitidieri, and I. I. Vrabie,* Semigroup Theory and Applications
117. *J. Vinuesa,* Orthogonal Polynomials and Their Applications: Proceedings of the International Congress
118. *C. M. Dafermos, G. Ladas, and G. Papanicolaou,* Differential Equations: Proceedings of the EQUADIFF Conference
119. *E. O. Roxin,* Modern Optimal Control: A Conference in Honor of Solomon Lefschetz and Joseph P. Lasalle
120. *J. C. Díaz,* Mathematics for Large Scale Computing
121. *P. S. Milojević,* Nonlinear Functional Analysis

122. *C. Sadosky*, Analysis and Partial Differential Equations: A Collection of Papers Dedicated to Mischa Cotlar
123. *R. M. Shortt*, General Topology and Applications: Proceedings of the 1988 Northeast Conference
124. *R. Wong*, Asymptotic and Computational Analysis: Conference in Honor of Frank W. J. Olver's 65th Birthday
125. *D. V. Chudnovsky and R. D. Jenks*, Computers in Mathematics
126. *W. D. Wallis, H. Shen, W. Wei, and L. Zhu*, Combinatorial Designs and Applications
127. *S. Elaydi*, Differential Equations: Stability and Control
128. *G. Chen, E. B. Lee, W. Littman, and L. Markus*, Distributed Parameter Control Systems: New Trends and Applications
129. *W. N. Everitt*, Inequalities: Fifty Years On from Hardy, Littlewood and Pólya
130. *H. G. Kaper and M. Garbey*, Asymptotic Analysis and the Numerical Solution of Partial Differential Equations
131. *O. Arino, D.E. Axelrod, and M. Kimmel*, Mathematical Population Dynamics: Proceedings of the Second International Conference
132. *S. Coen*, Geometry and Complex Variables
133. *J. A. Goldstein, F. Kappel, and W. Schappacher*, Differential Equations with Applications in Biology, Physics, and Engineering
134. *S. J. Andima, R. Kopperman, P. R. Misra, J. Z. Reichman, and A. R. Todd*, General Topology and Applications

Other Volumes in Preparation

Girolamo Cardano (1501- 1576)

Luigi Cremona (1830 - 1903)

Federigo Enriques (1871 - 1946)

Luigi Fantappiè (1901 - 1956)

Photographs courtesy of the Istituto Nazionale di Alta Matematica, Rome

geometry and complex variables

proceedings of an international meeting on the occasion
of the IX centennial of the University of Bologna

edited by

Salvatore Coen
University of Bologna
Bologna, Italy

Alma Mater Studiorum
Sæcularia Nona

Marcel Dekker, Inc.   New York • Basel • Hong Kong

Library of Congress Cataloging--in--Publication Data

Geometry and complex variables: proceedings of an international meeting on the occasion of the IX centennial of the University of Bologna/edited by Salvatore Coen.
      p. cm. -- -- (Lecture notes in pure and applied mathematics; v.)
  "The meeting was for the most part carried on within a period of approximately twelve months from July 1988 to June of 1989 and consisted of a series of 38 conferences with a more intensive series in February 1989 and an historical session in March of the same year"-- --Pref.
  Includes bibliographical references and index.
  ISBN 0-8247-8445-6
  1. Geometry, Differential-- --Congresses. 2. Functions of complex variables-- --Congresses. I. Coen, S. II. Series.
QA641.G444 1991
516.3'6-- --dc20   91-12514
                                                                                            CIP

This book is printed on acid-free paper

Copyright © 1991 by MARCEL DEKKER, INC. All Rights Reserved

Neither this book nor any part may be reproduced or transmitted in any form or by any means, electronic or mechanical, including photocopying, microfilming, and recording, or by any information storage and retrieval system, without permission in writing from the publisher.

MARCEL DEKKER, INC.
270 Madison Avenue, New York, New York 10016

Current printing (last digit):
10 9 8 7 6 5 4 3 2 1

PRINTED IN THE UNITED STATES OF AMERICA

Preface

In the year 1988, the University of Bologna celebrated its 900 years of existence dating back to 1088, the acknowledged founding date of this prestigious university known to its students and professors as *"Alma Mater Studiorum."*

The commemoration lasted approximately two years and climaxed in September of 1988 when in Bologna's Piazza Maggiore, the Magna Charta delle Università Europee was solemnly signed in the presence of more than 350 deans from some of the most important universities in the world. The festivities were marked with a series of events: scientific meetings, exhibits, artistic performances, conferring of honorary degrees, restoration of museums, grants, and awards.

It was, therefore, in this commemorative spirit that a committee* was established for the purpose of organizing this meeting. The scientific committee consisted of professors M. Heins (University of Maryland, College Park), G.-C. Rota (Massachusetts Institute of Technology, Cambridge), E. Vesentini (Scuola Normale Superiore, Pisa), and S. Coen (Università di Bologna). The meeting was for the most part carried on within a period of approximately 12 months, from July 1988 to June 1989 and consisted of a series of 38 conferences, with a more intensive series in February 1989 and a historical session in March of the same year.

At the happy conclusion of the meeting on May 15, 1990, the Science Faculty of the University of Bologna conferred an honorary degree *honoris causa* in mathematics on Prof. J. J. Kohn of Princeton University for his outstanding scientific work.

Purpose of the meeting

It was the aim of the meeting to bring together experts to present and discuss research based on the works of great mathematicians who have taught at the University of Bologna in the field of geometry and complex variables.

---

*The organizing committee consists of S. Coen, President (Università di Bologna), M. Manaresi, Vice President (Università di Salerno), G. Bolondi, Secretary (Università di Camerino), G. Forni, Assistant Secretary (Università di Bologna), U. Bottazzini (Università di Bologna), G. Menichetti (Università di Bologna), R. Musti (Università di Bologna), P. Salmon (Università di Bologna), C. Tinaglia (Università di Bologna), and A. Vaz Ferreira (Università di Bologna).

For a better understanding of the meeting's theme, it seems opportune to briefly recall some of the professors who have over the centuries been associated with the subject of mathematics in Bologna.

Historical notes on mathematics in Bologna

Among the earliest names of importance, we find that of Luca Pacioli (Borgo San Sepolcro 1445, Roma 1514), lecturer *ad Mathematicam* in Bologna 1501-1502, and author of the famous work *De divina Proportione* and (in his time) the widely circulated *Summa de Arithmetica, Geometria, Proportioni et Proportionalità*. We must also mention Domenico Maria Novara (Ferrara 1454, Bologna 1504), astronomy lecturer in the period 1483-1504, famous for his accurate astronomical observations and his ability to interpret them, as well as for his role in guiding Copernicus during his Bologna period.

We must also bear in mind that for centuries (from the twelfth century) mathematics was being taught in Bologna by abacus teachers (*Maestri d'Abbaco*) and astrology lecturers.

One of the best periods for the study of mathematics in Bologna is that of the *Grandi Algebristi*, who worked in Bologna during the sixteenth century: Scipione dal Ferro (Bologna 1465, Bologna 1526), mathematics lecturer from 1496 to 1526, the fascinating personality Girolamo Cardano (Pavia 1501, Roma 1576), mathematician, philosopher, and lecturer in medicine at Bologna from 1561 to 1570, Ludovico Ferrari (Bologna 1522, Bologna 1565) lecturer in 1564-1565, and, finally, Rafael Bombelli, about whom we know very little and who died around 1575. The above mentioned, together with Tartaglia and others, had the courage and ability to look for and find solution formulas for third- and fourth-degree algebraic equations. In line with our interest in complex variables, it is worthwhile to take a moment to consider an important problem that came up at that time in the solution of the cubic equations $x^3 = px + q$. Let $\Delta = q^2/4 - p^3/27$, the discriminant of this equation. Scipione del Ferro and Tartaglia's solution formula

$$\sqrt[3]{q/2 + \sqrt{\Delta}} + \sqrt[3]{q/2 - \sqrt{\Delta}}$$

was without meaning if $\Delta < 0$. This situation, called the "irreducible case," is particularly important since it produces real distinct solutions. Cardano made some important observations on the subject, but it is to Bombelli that credit is generally assigned for the definitive solution of the difficult problem obtained by introducing imaginary numbers.

In the seventeenth century, the University of Bologna would once more boast of its great mathematicians: these, of course, dedicated much of their work to developing infinitesimal analysis and geometry. Pietro Antonio Cataldi (Bologna 1552, Bologna 1626), lecturer in mathematics from 1583 to 1626, is famous for having introduced and studied continued fractions. The most representative figure, however, is Bonaventura Cavalieri (Milano 1598, Bologna 1647), a follower and student of Galileo since his student days in Pisa, who was lecturer in mathematics at Bologna from 1629 to 1647. Recent important studies on Cavalieri, who is most famous for his work *Geometria indivisibilibus continuorum nova quadam ratione promota*, have come up with a complete picture of all his work. Pietro Mengoli (Bologna 1625 or 1626, Bologna 1686), a student of Cavalieri's and lecturer from 1650 to 1686, is usually remembered for his work *Geometria speciosa* and for contributions to series theory, to the computation of special integrals, and to the theory of logarithms. E. Giusti [G] wrote in reference to *Novae Quadraturae Arithmeticae* that Mengoli pushes the preciseness of concepts to and often above the limits of pedantry, defining each term, even the most commonly known and used, and never leaving the safe terrain of geometrically demonstrated propositions.

Preface

Even at the risk of seeming to deviate somewhat from our principal subject area, we must note that in the seventeenth century some great astronomers were *ad Mathematicam* lecturers in Bologna. Among these we find: Giovanni Antonio Magini (Padova 1555, Bologna 1618), lecturer from 1588 to 1618, about whom it is interesting to remember that he was chosen for Bologna over Galileo, whose theories he always implacably opposed, and Giandomenico Cassini (Pirinaldo, Imperia, 1625, Paris 1712), head of a well-known family of astronomers and geodesicists, lecturer from 1652 to 1669, when he transferred to Paris. Let us also remember Domenico Guglielmini (Bologna 1655, Padova 1710), the hydraulic engineering expert who from 1689 to 1699 was lecturer *ad Mathematicam*, which title was later changed to *ad Mathematicam hydrometricam*.

Among the mathematicians of the 1700s, we find Gabriele Manfredi (Bologna 1681, Bologna 1761) and Vincenzo Riccati (Castelfranco Veneto 1707, Castelfranco Veneto 1775), the first being well known for his studies on homogeneous differential equations of differential geometry and hydraulic problems, and the second belonging to an illustrious family of Italian mathematicians and recognized for his research on hyperbolic functions as well as for a famous mathematics treatise (on which he collaborated). The name of Maria Gaetana Agnesi (Milano 1718, Milano 1799) should also be mentioned since Pope Benedict XIV, secularly the Prospero Lambertini who did so much for studies in Bologna, offered her a mathematics lectureship in Bologna after she published her *Instituzioni analitiche*. There is no record, however, of her ever having taught at the university. While discussing the development of science in Bologna during the eighteenth century, it is important to mention the founding of the Istituto di Bologna through the work and patronage of Luigi Ferdinando Marsigli (1658-1730), a man of great intelligence, a soldier, and a scholar.

At this point, we find ourselves in a period considered critical for mathematics in Bologna. While we must await publication of recent research on the history of mathematics during the 1800s to give us more information, we can proceed to the situation of mathematical studies in Bologna immediately after the unification of Italy.

In 1860, Luigi Cremona (Pavia 1830, Roma 1903) was named Professor of Higher Geometry (Geometria Superiore) at the University of Bologna. We can, in fact, point out that Cremona was nominated personally by the then Minister of Education, Terenzio Mamiani della Rovere (Pesaro 1799, Roma 1885), who, with a view to improving studies in Bologna, at the same time nominated as professors the geologist Giovanni Capellini (Spezia 1833, Bologna 1922), Giosuè Carducci (Val di Castello, Lucca 1835, Bologna 1907; Nobel prize winner for literature in 1906), and the philosopher Bertrando Spaventa (Bomba, Chieti 1817, Napoli 1882). Then in 1862, we have Eugenio Beltrami (Cremona 1835, Roma 1900) as complementary algebra professor. The two young, well-qualified, and intelligent professors, Cremona and Beltrami, brought mathematical studies in Bologna to a turning point. Unfortunately, their time in Bologna was short. Cremona in 1867 went to the Istituto Tecnico Superiore (presently the Politecnico) in Milan and soon afterward to Rome, while Beltrami remained in Bologna for only a year, after which he went to Pisa for three years, returning to Bologna until 1873, with a professorship in *Meccanica Razionale*. Their time in Bologna, however, was scientifically productive for both. Cremona published the monograph *Preliminari di una teoria geometrica delle superficie* in 1866, following the *Introduzione ad una teoria geometrica delle curve piane* (1862), while Beltrami produced the work *Teoria degli spazi a curvatura costante* in 1868. From 1888 until their deaths, the University of Bologna included them as *Professori Onorari*. A rather difficult period followed for the study of mathematics in Bologna owing to the scarcity of professorships in this field; we must nevertheless remember Riccardo De Paolis (Roma 1854, Roma 1892), who taught algebra and analytic geometry in the years 1878-1880.

It is only toward 1880-1881, during what we can call the mathematics renaissance at the university, that three young, bright, and enthusiastic mathematicians in their thirties, namely, Salvatore

Pincherle (Trieste 1853, Bologna 1936), Cesare Arzelà (Santo Stefano di Magra 1847, Bologna 1912), and Luigi Donati (Fossombrone 1846, Bologna 1932) came to Bologna from the school of Pisa. None of these three transferred elsewhere. In 1896, Federigo Enriques (Livorno 1871, Roma 1946) was assigned the geometry professorship; in 1908, Pietro Burgatti (Cento, Ferrara 1868, Bologna 1938) was called to Meccanica Razionale and in 1922, Leonida Tonelli (Gallipoli 1885, Pisa 1946) was called to Analisi Superiore. In 1923 Enrico Bompiani (Roma 1889, Roma 1975) succeeded Enriques in Bologna until 1927. We can say that at the beginning of the nineteenth century, Bologna finally realized an esteemed position in mathematics. In 1924, at the International Congress of Mathematicians in Toronto, Salvatore Pincherle was named president of the International Mathematical Union. Later, Pincherle was assigned the task of organizing in Bologna the first international mathematicians' congress after World War I including the participation of German mathematics scholars. Pincherle, who had already founded the Istituto di Matematica a few years earlier as well as the Unione Matematica Italiana in 1923, becoming its first president, carried out his work for the international meeting very well. His work was faced with serious difficulties, but the Bologna International Congress was a success.

It seems noteworthy that years earlier, in 1911, the mathematician/philosopher Federigo Enriques had been assigned the organizing task for the Fourth International Philosophy Congress. We might state that by then Bologna had reached a position of high acclaim. Successors were all of high caliber. To replace the retiring Pincherle in 1928, we find Beppo Levi (Torino 1875, Rosario, Argentina 1961). For Tonelli, in 1930, Giuseppe Vitali (Ravenna 1875, Bologna 1932) arrived but stayed a very short time, passing away unexpectedly two years later in 1932; and in 1932, to replace Vitali we find Luigi Fantappiè (Viterbo 1901, Bagnaia, Viterbo 1956). In 1931 Beniamino Segre (Torino 1903, Roma 1977) was called. In this brief listing we have omitted assistants and assigned professors, some of whom had noteworthy names. Among the many we can mention for their work in geometry and complex variables: Ugo Amaldi (Verona 1875, Roma 1957), a student of Pincherle, assistant professor of algebra and analytic geometry in 1899-1900 and then of projective and descriptive geometry in 1900-1901, Francesco Severi (Arezzo 1879, Roma 1961), assistant professor of projective and descriptive geometry in 1902-1903, Annibale Comessatti (Udine 1866, Padova 1945), assigned professor for several years starting 1936, Oscar Chisini (Bergamo 1889, Milano 1967), being graduated from the University of Bologna and assigned projective geometry professor in the years 1921-1922. Concluding this brief study, we note that we must not undervalue the difficulties encountered periodically by mathematics professors in Bologna owing to the, at times dramatic, scarcity of teachers.

In the academic year 1937-1938, the Professori Ordinari of the Istituto di Matematica of Bologna were Pietro Burgatti, Luigi Fantappiè, Beppo Levi, and Beniamino Segre. Only an external malefic event could destroy such a solid institution. And it is with the year 1938, the eight hundred and fiftieth anniversary of the university's founding, a year of trauma and change for the Istituto di Matematica, that we end this brief account.

Of the splendid period 1880-1938, we have only very briefly traced the university's development in mathematics, since it is amply and very well described in the various papers of this meeting.

For the history of mathematics at the University of Bologna up to the scientific activity in Bologna by Pincherle, Arzelà, and Donati, one can refer to the work of Ettore Bortolotti [B].

Outline of the contents of this volume

Many of the contributions in this book are works of pure mathematical research, while others represent research on the history of mathematics, and some can be considered substantial survey papers.

The contents of this volume are necessarily multidisciplinary. In addition, the special multidisciplinary nature of this meeting can even be found in some individual papers. It would be difficult for the editor to classify some of these articles in one special branch of mathematics rather than in another. For this reason we have chosen to present them in alphabetical order.

Almost half the articles are concerned with algebraic geometry, differential geometry, and the fundamentals of geometry. This could be expected when we recall that Luigi Cremona and Beniamino Segre taught in Bologna and that Federigo Enriques conducted a school of mathematical and philosophical thinking there.

Many contributors come under the theory of analytic functions of one or several complex variables, and even this is to be expected when we remember the scientific work carried out by Salvatore Pincherle and Luigi Fantappiè.

Various works are related to the history of mathematics, justifiably when we consider the contributions of research in Bologna as outlined above.

The scientific personages most studied in this volume are: F. Enriques, B. Segre, and L. Fantappiè. In fact, as regards the last mentioned, this volume probably contains the most thorough and up-to-date study of his scientific work in the field of complex analysis.

Please note that this work contains a listing of the mathematicians mentioned throughout.

We might mention that in asking the contributors to speak on mathematics research in Bologna, we specified that it was not absolutely necessary to deal with research carried out only in the period when these mathematicians were actually working in Bologna. In addition, we would also like to state that we did not strive to be comprehensive. And finally, we must point out that we realize only too well that much of the interesting research carried out in the limited period 1880-1938 (the most extensively discussed at this meeting) has not even been mentioned.

In conclusion, we would like to thank CNR (the Consiglio Nazionale delle Ricerche) for their role in making this publication possible. The editor appreciates the work of the conference participants and their contributions which, delving into the rich history of mathematics in Bologna, have produced interesting results that have in some cases brought to light important possibilities for further study. We would be most pleased to learn that the meeting has served to inspire new research in the history of mathematics or some new development in geometry or complex variables.

Acknowledgments. C.N.R. and M.U.R.S.T. 60% funds have been gratefully used for the preparation and final publication of this scientific effort.

Salvatore Coen

References

[B] Ettore Bortolotti, *La Storia della Matematica nella Università di Bologna*, N. Zanichelli Editore, Bologna, 1947.

[G] Enrico Giusti, Le prime ricerche di Pietro Mengoli: la somma della serie *These Proceedings*.

Contents

Preface		v
Contributors		xix
Vincenzo Ancona	A Vanishing Theorem for the Cohomology of a Vector Bundle whose Curvature Can Change Its Sign	1
A. Barlotti	Incidence Structures and Galois Geometries	5
E. Beltrami	Beltrami's Operator and the Tail of the Cat	11
Carlos A. Berenstein	From Salvatore Pincherle's «Operazioni Distributive» to J. Ecalle's Resurgent Functions	19
Umberto Bottazzini	Pincherle e la teoria delle funzioni analitiche	25
Maria Virginia Catalisano, Silvio Greco	Linear Systems: Developments of Some Results by F. Enriques and B. Segre	41
Fabrizio Catanese	Recent Results on Irregular Surfaces and Irregular Kaehler Manifolds	59
Ciro Ciliberto	A Few Comments on Some Aspects of the Mathematical Work of F. Enriques	89
Salvatore Coen	Geometry and Complex Variables in the Work of Beppo Levi	111
Alberto Conte	When is a Conic Bundle Rational?	141
Mischa Cotlar	Quadratic Inequalities for Hilbert Transforms and Hankel Forms in the Spaces $\mathcal{L}^2(\mathcal{F})$ and $\mathcal{L}^2(B)$	147
Paolo de Bartolomeis	Some New Global Results in Twistor Geometry	155
Igor V. Dolgachev	Enriques Surfaces: Old and New	165
Philippe Ellia, Monica Idà	Some Connections Between Equations and Geometric Properties of Curves in $\mathbf{P}^3$	177
F. Gherardelli	Biholomorphic Invariants of Real Submanifolds of $\mathbb{C}^n$	189

Enrico Giusti	Le prime ricerche di Pietro Mengoli: la somma delle serie	195
Maurice Heins	Composition and Fatou Limits	215
G.M. Henkin, A.A. Shananin	The Bernstein Theorems for the Fantappiè Indicatrix and Their Applications to Mathematical Economics	221
J.J. Kohn	Microlocal Analysis on Three Dimensional CR Manifolds	229
D. Loeb, G.C. Rota	Recent Contributions to the Calculus of Finite Differences: A Survey	239
Luigi Muracchini	Local Differential Properties of Algebraic Surfaces	277
V.P. Palamodov	The Method of Holomorphic Waves and Its Applications to Partial Differential Equations	281
Luigi Pepe	Ricerca matematica e vita accademica a Bologna nel secolo XVIII	291
R. Michael Range	Cauchy-Fantappiè Formulas in Multidimensional Complex Analysis	307
J.H. Sampson	Maria Gaetana Agnesi	323
Daniele C. Struppa	An Extension to Fantappiè's Theory of Analytic Functionals	329
Bernard Teissier	On B. Segre and the Theory of Polar Varieties	357
Giuseppe Tomassini	Pseudoconformal Invariants and Differential Geometry	369
A. Vaz Ferreira	Giuseppe Vitali and the Mathematical Research at Bologna	375
Alessandro Verra	Contact Curves of Two Kummer Surfaces	397
Edoardo Vesentini	Kreĭn Spaces and Holomorphic Isometries of Cartan Domains	409
Stefano Francesconi	L'insegnamento della matematica nell'Università di Bologna dal 1860 al 1940	415

Index 475

Contributors

Vincenzo ANCONA	Dipartimento di Matematico "U. Dini", Università di Firenze, Viale Morgagni 67/A, I-50134 Firenze, Italia
Adriano BARLOTTI	Dipartimento di Matematica "U. Dini", Università di Firenze, Viale Morgagni 67/A, I-50134 Firenze, Italia
Edward BELTRAMI	Department of Applied Mathematics and Statistics, State University of New York, Stony Brook, New York 11794,U.S.A.
Carlos A. BERENSTEIN	Department of Mathematics, University of Maryland, College Park, MD 20742 USA
Umberto BOTTAZZINI	Dipartimento di Matematica, Università di Bologna, Piazza di Porta S. Donato 5, I-40127 Bologna, Italia
Maria Virginia CATALISANO	Dipartimento di Matematica, Università di Genova, Via L.B.Alberti 4, I-16132 Genova, Italia
Fabrizio CATANESE	Dipartimento di Matematica, Università di Pisa, Via Filippo Buonarroti 2, 56100 Pisa, Italia
Ciro CILIBERTO	Dipartimento di Matematica, Università di Roma II, Via O.Raimondo I-00173 Roma, Italia
Salvatore COEN	Dipartimento di Matematica, Università di Bologna, Piazza di Porta S.Donato 5, I-40127 Bologna, Italia
Alberto CONTE	Dipartimento di Matematica, Università di Torino, Via Carlo Alberto 10, I-10123 Torino, Italia
Mischa COTLAR	Central University of Venezuela, Facultad de Ciencias, U.C.V., Caracas A.P. 20513 Venezuela
Paolo DE BARTOLOMEIS	Istituto di Matematica Applicata "G. Sansone", Via di S.Marta 3 I-50139 Firenze, Italia
Igor V. DOLGACHEV	Department of Mathematics, University of Michigan, Ann Arbor, Michigan, 48109, USA
Philippe ELLIA	Département des Mathématiques, Université de Nice, Parc Valrose, F-06034 Nice Cedex, France
Stefano FRANCESCONI	Via Stradella 1, I-42100 Reggio Emilia, Italia

Francesco GHERARDELLI	Dipartimento di Matematica "U.Dini", Università di Firenze, Viale Morgagni 67/a, I-50134 Firenze, Italia
Enrico GIUSTI	Dipartimento di Matematica, Università di Firenze, Viale Morgagni 67/a, I-50134 Firenze, Italia
Silvio GRECO	Dipartimento di Matematica, Politecnico di Torino, Corso Duca degli Abruzzi 24, I-10124 Torino, Italia
Maurice HEINS	Department of Mathematics, University of Maryland, College Park, MD 20742 USA
Gennadi M. HENKIN	Academy of Sciences, Central Economic and Mathematical Institute, ul. Krasikova 32, 117418 Moscow, U.S.S.R.
Monica IDÀ	Dipartimento Scienze Matematiche, Università di Trieste, Piazzale Europa I-34100 Trieste, Italia
Josep J. KOHN	Department of Mathematics, Fine Hall, Princeton University, Princeton N.J. 08544, U.S.A.
Daniel LOEB	Départment des Mathématiques et de l'Informatique, Université de Bordeaux 1, 351 cours de la Liberation, 33405 Talence Cedex, France
Luigi MURACCHINI	Dipartimento di Matematica, Università di Bologna, Piazza di Porta S. Donato, 5, I-40127 Bologna, Italia
Viktor P. PALAMODOV	University of Moscow, MGU, Department of Mechanics and Mathematics Leninskije gori, 119899 Moscow, USSR.
Luigi PEPE	Dipartimento di Matematica, Università di Ferrara, Via Machiavelli 35 I-44100 Ferrara, Italia
R. Michael RANGE	Department of Mathematics, State University of New York at Albany, Albany, New York 12222, USA.
Gian-Carlo ROTA	MIT Department of Mathematics, Cambridge, MA 02139 U.S.A.
Joseph H. SAMPSON	Department of Mathematics, The Johns Hopkins University, Baltimore, Maryland 21218, USA
A.A. SHANANIN	Academy of Sciences U.S.S.R.
Daniele C. STRUPPA	Department of Mathematics, Univ. of Calabria, 87036 Arcavacata di Rende(Cs), Italia and Department of Mathematical Sciences, George Mason Univ. Fairfax, VA 22030, USA
Bernard TEISSIER	Département de Mathématiques et d'Informatique, Ecole Normale Supérieure 45 rue d'Ulm, 75830 Paris, France
Giuseppe TOMASSINI	Scuola Normale Superiore; Piazzale dei Cavalieri 7, I-56126 Pisa, Italia

Arturo VAZ FERREIRA — Dipartimento di Matematica, Università di Bologna, Piazza di Porta S. Donato, 5, I-40127 Bologna, Italia

Alessandro VERRA — Dipartimento di Matematica, Università di Genova, Via L.B.Alberti 4, I-16132, Genova, Italia

Edoardo VESENTINI — Scuola Normale Superiore, Piazza dei Cavalieri 7, I-56126 Pisa, Italia

geometry and complex variables

A Vanishing Theorem for the Cohomology of a Vector Bundle Whose Curvature Can Change Its Sign

VINCENZO ANCONA Istituto Matematico «U. Dini», Università di Firenze, Viale Morgagni, 67/A, 50134 Firenze Italy

Let X be a Kähler compact manifold, E an hermitian holomorphic vector bundle on X. The famous Kodaira-Nakano theorem states the vanishing of the cohomology groups $H^q(X, K_X \otimes E)$ for $q > 0$ under a strict positivity assumption for the curvature of E [3], [5]. Later Grauert and Riemenschneider [2] showed the same result under the weaker assumption that the curvature form is nonnegative everywhere and strictly positive at least at one point.

Here we prove the above vanishing when the curvature form can change its sign, provided the set of points where it is positive prevails over that of points where it is negative. The complete proof has appeared in [1]; it uses techniques similar to those in [4].

Let X be a compact connected n-dimensional Kähler manifold. In local coordinates $z = (z_1, \ldots, z_n)$ let us denote by $(g_{i\bar{j}})$ the matrix of the metric on X, and:

$$g = \det(g_{i\bar{j}})$$

$$\Gamma^i_{jk} = g^{\bar{l}i} \partial_j g_{k\bar{l}}$$

$$R_{\bar{i}j} = \partial_{\bar{i}} \partial_j \log g$$

$$R^{\bar{k}}_{\bar{j}} = g^{\bar{k}l} R_{\bar{j}l}$$

Lecture given on February 8, 1989, at the Università degli Studi di Bologna, on the occasion of the 900th anniversary of its founding.

Let us consider an holomorphic vector bundle E on X, a local holomorphic frame $(s_1, \ldots, s_r)$ of E, and an hermitian metric $a = (a_{\rho\bar{\sigma}})$ on X, given by $a_{\rho\bar{\sigma}} = \langle s_\rho, s_{\bar{\sigma}}\rangle$.

The matrix of the connection compatible with the hermitian metric is given by

$$\Theta^\mu_{i\nu} = a^{\bar{\lambda}\mu} \partial_i a_{\nu\bar{\lambda}}$$

and the corresponding curvature matrix by

$$\Theta^\mu_{\nu i \bar{j}} = -\partial_{\bar{j}} \Theta^\mu_{i\nu}$$

We denote by $\mathcal{R}(E^*)$ the differentiable bundle of complex frames of E^*, and by $\mathcal{U}(E^*)$ the subbundle of unitary frames of E^*, and we put

$$\mathcal{R} = \mathcal{R}_E = \mathcal{R}(T_\mathbb{C} X) \times_X \mathcal{R}(E^*)$$
$$\mathcal{U} = \mathcal{U}_E = \mathcal{U}(T_\mathbb{C} X) \times_X \mathcal{U}(E^*)$$

A point in $\mathcal{R}_E$ is given by

$$r_x = (e_1, \ldots, e_n, \epsilon^{*1}, \ldots, \epsilon^{*r}) \tag{1}$$

where $x \in X$ and $(e_1, \ldots, e_n)$ is a frame of $(1,0)$-tangent vectors, $(\epsilon^{*1}, \ldots, \epsilon^{*r})$ a complex frame of E^* at x.

A $(0,q)$-form ϕ with values in E can be written locally

$$\phi = \phi^\alpha \otimes s_\alpha \tag{2}$$

where

$$\phi^\alpha = \phi^\alpha_{\bar{i}_1 \ldots \bar{i}_q} d\bar{z}^{i_1} \wedge \ldots \wedge d\bar{z}^{i_q} \tag{3}$$

The $(0,q)$-form ϕ defines a function

$$\mathcal{R}(\phi) : \mathcal{R}_E \to (\wedge^q \mathbb{C}^n)^r$$

in the following way: if $x \in X$ and r_x is a frame as in (1), $\mathcal{R}(\phi)(r_x)$ is the expression of ϕ in the frame r_x, whose components are

$$\mathcal{R}(\phi)^\alpha_{\bar{i}_1 \ldots \bar{i}_q}(r_x) = \langle \phi, \bar{e}_{i_1} \wedge \ldots \wedge \bar{e}_{i_q} \otimes \epsilon^{*\alpha}\rangle$$

We define

$$A^{\rho\bar{i}}_{\sigma\bar{j}}(r_x) = g^{\bar{i}k}\Theta^\rho_{\sigma k\bar{j}}$$

thus obtaining a square matrix $(A^{\rho\bar{i}}_{\sigma\bar{j}})$ of order rn whose entries are functions on $\mathcal{R}_E$ with values in $\mathbb{C}$. On $\mathcal{U}_E$, it induces for every integer q $(0 < q \le n)$ a bounded operator

$$A^{(q)} : L^2(\mathcal{U}, (\wedge^q \mathbb{C}^n)^r) \to L^2(\mathcal{U}, (\wedge^q \mathbb{C}^n)^r)$$

by the formula

$$(A^{(q)}\xi)^\rho_{\bar{j}_1 \ldots \bar{j}_q} = \sum_{k=1}^n A^{\rho\bar{i}}_{\sigma\bar{j}_k} \xi^\alpha_{\bar{j}_1 \ldots (\bar{i})_k \ldots \bar{j}_q}$$

for $\xi = (\xi^\alpha_{\bar{j}_1...\bar{j}_q})$ in $L^2(\mathcal{U}, (\wedge^q \mathbb{C}^n)^r)$.

In the same way we define the operator

$$R^{(q)} : L^2(\mathcal{U}, (\wedge^q \mathbb{C}^n)^r) \to L^2(\mathcal{U}, (\wedge^q \mathbb{C}^n)^r)$$

by

$$(R^{(q)}\xi)^\rho_{\bar{j}_1...\bar{j}_q} = \sum_{k=1}^n R^{\bar{i}}_{\bar{j}_k} \xi^\rho_{\bar{j}_1...(\bar{i})_k...\bar{j}_q}$$

For a fixed q and $x \in X$ let

$$\delta^{(q)}(x) = \inf_{r_x \in \mathcal{U}} \{\text{eigenvalues of } A^{(q)}(r_x)\}$$

and write

$$\delta^{(q)}(x) = \delta^{(q)^+}(x) - \delta^{(q)^-}(x)$$

Let λ be the least nonzero eigenvalue of the scalar Laplace operator on X.
Finally let K_X be the canonical line bundle on X.
Our result, due to B. Gaveau and the author [1], is:

THEOREM. *Let us suppose* $\text{vol}(X) = 1$. *If*

$$\|\delta^{(q)^-}\|_\infty \leq \max_{0 \leq t \leq \gamma_q} \left(t\|\delta^{(q)^+}\|_{L^1} - \frac{4t^2}{\lambda} \|\delta^{(q)^+}\|_\infty^2 \right)$$

where

$$\gamma_q = \inf(1, \lambda/4\|\delta^{(q)^+}\|_\infty)$$

then the following vanishing holds:

$$H^q(X, K_X \otimes E) = 0$$

We give now a sketch of the proof of the theorem.

We define horizontal vector fields $L_k, L_{\bar{k}}(k = 1,\ldots,n)$ on $\mathcal{R}_E$. If $x \in X$ and $r_x \in \mathcal{R}_E$ as in (1), we write $e_k = \frac{1}{2}(\tilde{e}_k - i\tilde{e}_{k+n})$ for real and imaginary parts. If $\tilde{e}_k(t)$ is a path on X starting at x and tangent to $\tilde{e}_k$, $\gamma_k(t)$ its horizontal lift to $\mathcal{R}_E$ and f a complex C^1 function in a neighbourhood of r_x, we put

$$L'_k f(r_x) = \frac{d f(\gamma_k(t))}{dt}\bigg|_{t=0}$$

and

$$L_k = \frac{1}{2}(L'_k - iL'_{k+n}), \quad \overline{L}_k = \frac{1}{2}(L'_k + iL'_{k+n}) \quad (k = 1,\ldots,n)$$

A tedious computation shows that the so-called Weitzenböck formula translates to

PROPOSITION. *Let $\Box$ be the Laplace-Beltrami operator acting on $(0,q)$-forms ϕ with values in E. The following identity holds on $\mathcal{U}_E$:*

$$\mathcal{R}(\Box \phi) = -g^{\bar{k}s} L_s \bar{L}_k \mathcal{R}(\phi) + (A^{(q)} - R^{(q)})\mathcal{R}(\phi)$$

If we replace now E by the bundle $\tilde{E} = k_X \otimes E$ the above formula reads on $\mathcal{U}_{\tilde{E}}$

$$\tilde{\mathcal{R}}(\tilde{\Box}\phi) = -\tilde{\mathcal{L}}\tilde{\mathcal{R}}(\phi) + \tilde{A}^{(q)}\tilde{\mathcal{R}}(\phi)$$

with $\tilde{\mathcal{L}} = \sum_{k=1}^{n} L_k \bar{L}_k$, since $\tilde{A}^{(q)} = A^{(q)} + R^{(q)}$.

We want to show that $\tilde{\Box}\phi = 0$ implies $\phi = 0$.

Let us denote by H the closed subspace of $L^2(\mathcal{U}_{\tilde{E}}, (\wedge^q \mathbb{C}^n)^r)$ consisting of functions $\mathcal{R}(\phi)$, where ϕ is a $(0,q)$-form with values in $\tilde{E}$, $H_1 = L^2(\mathcal{U}(T_\mathbb{C} X), \mathbb{C})$, $H_2 = L^2(X, \mathbb{C})$.

We only need to prove that for $f \in H$, $-\tilde{\mathcal{L}}f + A^{(q)}f = 0$ implies $f = 0$.

The operator $\tilde{\mathcal{L}}$ is self-adjoint and positive. The matrix $A^{(q)}$ can be written as $A^{(q)^+} - A^{(q)^-}$, $A^{(q)^+}$ and $A^{(q)^-}$ being hermitian, positive semi-definite matrices, defining two bounded operators in H.

Define:

$$\mu = \inf_H \operatorname{Spec}(-\tilde{\mathcal{L}} + A^{(q)^+})$$

$$\mu_1 = \inf_{H_1} \operatorname{Spec}(-\tilde{\mathcal{L}} + \delta^{(q)^+})$$

$$\mu_2 = \inf_{H_2} \operatorname{Spec}(-\Delta + \delta^{(q)^+})$$

where Δ is the scalar Laplace-Beltrami operator on X. Then it is not hard to prove: $\mu \geq \mu_1 = \mu_2$.

By Malliavin [4] the assumption in theorem gives $||\delta^{(q)^-}||_\infty < \mu_2$, which implies $||\delta^{(q)^-}||_\infty < \mu$. It is now clear that it exists no $f \in H$, $f \neq 0$ such that $-\tilde{\mathcal{L}}f + A^{(q)}f = 0$.

REFERENCES

[1] V. Ancona and B. Gaveau, Annulation de la cohomologie pour des fibrés a courbure de signe variable sur une variété Kählérienne, *Annali di Mat. Pura e Appl.* 151, 97-107, (1988).
[2] H. Grauert and O. Riemenschneider, Verschwindungssätze für analytische kohomologiegruppen auf komplexen, Raumen, *Inv. Math.* 11, 263-292, (1970).
[3] K. Kodaira, On a differential geometric method in the theory of analytic stacks, *Proc. Nat. Ac. USA* 39, 274-291, (1953).
[4] P. Malliavin, Annulation de cohomologie et calcul des perturbations dans L^2, *Bull. Sc. Math.* 100, 331-336, (1976).
[5] S. Nakano, On complex analytic vector bundles, *J. Math. Soc. Japan* 7, 1-12, (1955).

Incidence Structures and Galois Geometrics

A. BARLOTTI Istituto Matematico «U. Dini», Università di Firenze, Florence, Italy

1. INTRODUCTION

First I wish to express my sincere thanks to the University of Bologna and in particular to Professor Salvatore Coen for the Honour I received with the invitation to present a lecture at this Conference. According to the leading idea fixed by the organizers I shall consider questions originated by the work of Beniamino Segre who was one of the many eminent mathematicians teaching at this University. B. Segre was active in many areas of Mathematics. Among these I choose the field in which He did pioneering work in the last part of His life, obtaining a large amount of important results.

I feel that we have to remember that when He was Professor of Geometry here in Bologna He was dismissed as a consequence of the racial persecution by the fascist government, had to go to England, and was able to come back to His position only at the end of World War II.

2. SEGRE'S THEOREM ON OVALS IN FINITE DESARGUESIAN PLANES OF ODD ORDER

In 1954 B. Segre presented the following theorem in which he proved a conjecture formulated in 1949 by Paul Kustaanheimo (cf. [22], [23]):

THEOREM. A $(q+1)$-arc (called also an oval) of $PG(2,q)$, q odd, is an irreducible conic.

This graphic characterization of the conics, beautiful and unespected [1], gave rise to intense researches in several directions. We shall recall here briefly the following ones:

a) To determine other graphic characterizations for algebraic varieties.
b) To study the theory of k-arcs and (k,n)-arcs not only in Galois planes but also in non-Desarguesian planes, and in more general geometric structures.
c) To make an accurate study of the «Packing problem».
d) To develop a systematic investigation of the various ways in which a geometric structure may be represented within a different one.

It is certainly impossible to present here all the noteworthy results obtained in the above areas. We shall therefore restrict our exposition only to few questions connected in a particular way to Segre's work. For all the results concerning $(q+2)$-arcs in $PG(2,q)$, q even, and $(q+1)$-caps in $PG(3,q)$, we simply refer to [17], [18]. For these and many other questions see also [2].

3. ON THREE FUNDAMENTAL PROBLEMS OF SEGRE IN $PG(r,q)$

In 1955, in [24], B. Segre posed the following problems:

i) PROBLEM $I(r,q)$: For given r and q what is the maximum value of k for which there exist k-arcs in $PG(r,q)$? And how is it possible to characterize geometrically such k-arcs ?

ii) PROBLEM $II(r,q)$: For what values of r and p, with $q > r+1$, is every $(q+1)$-arc of $PG(r,q)$ the point set of a normal rational curve?

iii) PROBLEM $III(r,q)$: For given r and p, with $q > r+1$, what are the values of $k(\leq q)$ for which every k-arc of $PG(r,q)$ is contained in a normal rational curve of this space? And how many are the curves of this type containing the k-arc ?

Segre was considering in [24] these problems in the case q odd and proved that for any k-arc of $PG(r,q)$ with $r = 2,3$ or 4 is $k \leq q+1$. He found also that for any r the point set of a normal rational curve is a $(q+1)$-arc. For $r = 2,3$ the converse is true (cfr. [24], [27]). Only recently D.G. Glynn [13] has given an example of a 10-arc in $PG(4,9)$ which is not the point set of a normal rational curve. J.A. Thas (cfr. [28], [29]) proved various results on Problem $III(r,q)$.

A substantial progress in connection with the above three problems has been obtained in 1988 by A. Bruen, J.A. Thas and A. Blokhuis (cfr. [6]) in the case q even. The leading idea is the following: to each arc K in $PG(3,q)$ (respectively $PG(4,q)$) it is associated a surface (respectively, hypersurface) $S(K)$. From properties of $S(K)$ there follows an embedding theorem for arcs in $PG(3,q)$ (respectively, $PG(4,q)$). These embedding theorems led to fundamental results answering in the even case and for large q the problems of B. Segre. We quote as an example that in $PG(n,q)$, $q = 2^s$, $n \geq 4$ and $q \geq (n-1)^3$, every $(q+1)$-arc is a normal rational curve.

---

[1] The conjecture by Kustaanheimo was considered unplausible by several mathematicians; even the author of the conjecture, later (cf. [24] p. 359) on the basis of approximate counting arguments considered the fact almost impossible.

The interest on the above three problems is enhanced from the fact that they are connected with classical combinatoric structures, as the orthogonal arrays and with the theory of error correcting codes, in particular for what concernes the construction of «maximum distance separable codes» (cfr. [6] [16]).

4. THE PACKING PROBLEM

The main linear coding theory problem, i.e. the problem of finding the largest value of M for which there exists a linear (n, M, d)-code over $GF(q)$ is strictly connected with the «packing problem». This consists in finding the maximum number, $m(r,q)$, of points in $PG(r,q)$ that belong to a k-cap if $r \geq 3$, or to a k-arc if $r = 2$. The known values of $m(r,q)$ are listed below:

$$m(r,2) = 2^r \qquad \text{(Bose 1947)}$$

$$m(2,q) = \begin{cases} q+1 & q \text{ odd} \quad \text{(Bose 1947)} \\ q+2 & q \text{ even} \quad \text{(Bose 1947)} \end{cases}$$

$$m(3,q) = q^2 + 1 \qquad \begin{cases} q \text{ odd} & \text{(Bose 1947)} \\ q \text{ even} & \text{(Qvist 1952)} \end{cases}$$

$$m(4,2) = 20 \qquad \text{(G. Pellegrino 1970)}$$

$$m(5,3) = 56 \qquad \text{(Hill 1973)}$$

To determine the value of $m(r,q)$ in more general cases seems to be a very difficult problem. A considerable amount of work has been done providing upper bounds for $m(r,q)$ (see, e.g. [27], p. 166).

5. INCIDENCE STRUCTURES FROM GALOIS SPACES

The idea of studying a geometric structure by using a representation of this in another one which appears more convenient for our purpose has been very fruitful. A very nice application of this fact is given by the representation of translation planes (in particular the non-Desarguesian ones) in high dimensional (Desarguesian) projective spaces (cfr. [1], [7], [8], [19], [26]). Representations for other kinds of non-Desarguesian planes were also found (cfr. [4], [5]) and the possibility to discover new classes of projective planes using similar kind of constructions is still open.

The above constructions allow also generalizations which lead to classes of incidence structures different from planes. Well known examples were given by Heft (cfr. [14]) who constructed various classes of divisible designs and partial designs.

Various other ways based on properties of finite geometric structures are leading to the construction of classes of designs. Some of these may be based on the existence of a covering or of a partition of a geometric object obtained using substructures of some type. We may choose two particular families, A and B, of elements appearing in the construction as, respectively, the set of blocks and the set of points of the design, the incidence in the design being given by some particular relation.

Several nice examples of this kind of construction were obtained by I. Vecchi in his «Tesi di laurea» discussed here in Bologna in 1981 (cf. [32]). I wish to point out explicitely that B. Segre's idea of arcs and caps plays a fundamental role in the various constructions. Other examples, resulting from a careful analysis of the structure of non-degenerate quadrics in $PG(4,q)$ and $PG(5,q)$, are given in [9].

REFERENCES

[1] J. Andrè, Über nicht-Deserguessche Ebenen mit transitiver Translationsgruppe, *Math. Z.* 60, pp. 156-186 (1954).

[2] A. Beutelspacher, Geometry from B. Segre's point of view. Variations on a classical theme. To appear.

[3] R.C. Bose, Mathematical theory of the symmetrical factorial design, *Sankhyā* 8, 107-166 (1947).

[4] R.C. Bose, «On a representation of Hughes planes». In *Proc. International Conference on projective planes*, Edited by M.J. Kallaher and T.G. Ostrom, Wash. State Univ. Press, pp. 27-57 (1973).

[5] R.C. Bose and A. Barlotti, Linear representation of a class of projective planes in a four dimensional space, *Annali di Mat. Pura e Appl.* (4) 88, pp. 9-32, (1971).

[6] A.A. Bruen, J.A. Thas and A. Blokhuis, On M.D.S. codes, arcs in $PG(n,q)$ with q even and a solution of three fundamental problems of B. Segre, *Invent. Math.* 92, 441-456 (1988).

[7] R.H. Bruck and R.C. Bose, The construction of translation planes from projective spaces. *J. Algebra* 1, pp. 85-102 (1964).

[8] R.H. Bruck and R.C. Bose, Linear representation of projective planes in projective spaces. *J. Algebra* 4, pp. 117-172 (1966).

[9] I. Casaglia, Some block designs associated with quadrics of $PG(4,q)$ and $PG(5,q)$, *J. Statist. Plann. Inference*, 21, 265-272 (1989).

[10] L.R.A. Casse, «A solution to B. Segre's problem $I_{r,q}$.» *VII Österreichischer Mathematikerkongress*, Linz, 1968.

[11] L.R.A. Casse and D.G. Glynn, The solution to Beniamino Segre's problem $I_{r,q}, r = 3, q = 2^h$. *Geom. Dedicata* 13, 157-163 (1982).

[12] L.R.A. Casse and D.G. Glynn, On the uniqueness of $(q+1)_4$-arcs of $PG(4,q), q = 2^h, h \geq 3$. *Discrete Math.* 48, 173-186 (1984).

[13] D.G. Glynn, The non-classical 10-arc of $PG(4,9)$. *Discrete Math.* 59, 43-51 (1986).

[14] S.M. Heft, *Spreads in projective geometry and associated designs*. Ph. D. Thesis, Chapel Hill, N.C. (1971).

[15] R. Hill: On the largest size of cap in $S_{5,3}$, *Atti Accad. Naz. Lincei Rend.* 54, 378-384 (1973).

[16] R. Hill: *A first course in coding theory*. Oxford: Clarendon Press 1986.

[17] J.W.P. Hirschfeld, *Projective geometries over finite fields*. Oxford: Clarendon Press 1979.

[18] J.W.P. Hirschfeld, *Finite projective spaces of three dimensions*. Oxford: Clarendon Press 1985.

[19] G. Panella, Isomorfismo tra piani di traslazione di Marshall Hall. *Annali di Mat. Pura e App.* (4) 47, pp.169-181 (1959).

[20] G. Pellegrino, Sulle calotte massime dello spazio $S_{4,3}$, *Atti Acc. Sci. Lett. Palermo* (4) 34, 297-328 (1976).
[21] B. Qvist, Some remarks concerning curves of the second degree in a finite plane, *Ann. Acad. Sci. Fenn.* n. 134, 1-27 (1952).
[22] B. Segre, Sulle ovali nei piani lineari finiti, *Atti Accad. Naz. Lincei Rend.* 17, 141-142 (1954).
[23] B. Segre, Ovals in a finite projective plane, *Canad. J. Math.* 7, 414-416 (1955).
[24] B. Segre, Curve razionali normali e k-archi negli spazi finiti, *Annali Mat. Pura e Appl.*, (4) 39, 357-379 (1955).
[25] B. Segre, *Lectures on modern geometry*, Ed. Roma: Cremonese 1961.
[26] B. Segre, Teoria di Galois, fibrazioni proiettive e geometrie non desarguesiane, *Annali di mat. Pura e Appl.* (4) 64, pp. 1-76 (1964).
[27] B. Segre, Introduction to Galois geometries, *Mem. Accad. Naz. Lincei* 8, 133-236 (1967).
[28] J.A. Thas, Normal rational curves and k-arcs in Galois spaces, *Rend. Mat.* (6) 1, 331-334 (1968).
[29] J.A. Thas, Connection between the Grasmannian $G_{k-1;n}$ and the set of the k-arcs of the Galois space $S_{n,q}$, *Rend. Mat.* (6) 2, 121-134 (1969).
[30] J.A. Thas, Normal rational curves and $(q+2)$-arcs in Galois space $S_{q-2,q}(q=2^h)$. *Atti Accad. Naz. Lincei Rend.* 47, 249-252 (1969).
[31] J.A. Thas, Complete arcs and algebraic curves in $PG(2,q)$, *J. Algebra* (in press).
[32] I. Vecchi, Some results on coverings of Galois spaces with ovoids and related BIB-designs, *J. Statist. Plann. Inference* 10, 219-225 (1984).

Beltram's Operator and the Tail of the Cat

E. BELTRAMI Department of Applied Mathematics, State University of New York, Stony Brook, New York 11794.

INTRODUCTION

I was asked to comment on the work of Eugenio Beltrami (no relation to me, as far as I know) during his stay at the University of Bologna. Being an applied mathematician I thought it best to avoid a discussion of his influence in non-euclidean geometry or Hodge theory, for example, and to restrict myself to some elementary applications of the Beltrami operator which originated in his seminal paper «Delle Variabili Complesse Sopra Una Superficie Qualunque», published in 1867 in the Annali di Matematica Pura ed Applicata, serie II, Tomo 1, 329-366.

Although the usual undergraduate applied mathematics or engineering student rarely, if ever, encounters Beltrami's name there are some simple but instructive examples of surface Laplacian's that can be taught at a reasonably introductory level. I would like to give two examples below. One occurs in connection with the notion of curvature on a surface, an idea which students of today are not unfamiliar, and the other is an amusing exercise in mathematical biology. There are, of course, less elementary examples of the Beltrami operator in applied mathematics, such as the work of Lax and Phillips in scattering theory [1] and Huygen's principle [2], but the two I discuss are more accessible to the average student prepared in classical applied mathematical methods.

The second section below justifies the whimsical title of this talk, as we shall see.

CURVATURE AND SURFACE LAPLACIANS

Our discussion follows Reilly [3]. Let S be a smooth surface in $\mathbb{R}^3$ defined parametrically by

$$r(u_1, u_2) = (x_1(u_1, u_2), x_2(u_1, u_2), x_3(u_1, u_2)) \tag{1}$$

The metric coefficients g_{ij} are obtained in the usual way as $g_{ij} = r_{u_i} \cdot r_{u_j}$ using the ordinary dot product of vectors in $\mathbb{R}^3$. Then Beltrami's operator on S is defined by

$$\Delta_S = \sum_{i=1}^{2} \sum_{j=1}^{2} \frac{1}{\sqrt{g}} \frac{\partial}{\partial u_i} \left(\sqrt{g} g^{ij} \frac{\partial}{\partial u_j} \right) \qquad (2)$$

where $g = \text{Det}(g_{ij})$ and $(g^{ij}) = (g_{ij})^{-1}$.

The problem with expression (2) for the beginning student is that it is too abstract. Even though it reduces to the ordinary Laplace operator

$$\sum_{i=1}^{2} \frac{\partial^2}{\partial u_i^2} \qquad (3)$$

on a planar surface, it would be useful to show directly that (2) is indeed a generalization of (3) to curved surfaces. This can be done locally about any point p of the surface S whenever it admits a parametric representation of the form

$$(u_1, u_2, h(u_1, u_2)) \qquad (4)$$

for some smooth function h. Letting $x_i = u_i$, Let us choose p to be at the origin of the x_1, x_2 plane. The outward normal n is in the x_3 direction, perpendicular to the tangent plane T at p. Any plane Γ passing through p and orthogonal to T defines a smooth curve $\gamma = \Gamma \cap S$. By choosing planes along the x_1 and x_2 coordinate axis, respectively, we generate curves γ_1 and γ_2 through p whose tangents there are orthogonal.

Now pick a C^2 function F on $\mathbb{R}^3$ whose restriction to S is defined locally by $F(x_1, x_2, h(x_1, x_2)) = f(x_1, x_2)$. Denote the arclength along γ_1 and γ_2 by s_1 and s_2, respectively. Then the Laplace operator Δ at p on S is defined to be

$$\Delta = \sum_{i=1}^{2} \frac{\partial^2}{\partial s_i^2} \qquad (5)$$

in which $\frac{\partial^2}{\partial s_i^2}$ signifies the second derivative of h with respect to arclength along γ_i. It is clear that (5) is a direct generalization of the Laplacian

$$\sum_{i=1}^{2} \frac{\partial^2}{\partial x_i^2}$$

on the x_1, x_2 plane to an operator on the curved surface over this plane. At this point we attempt to enlighten the student by showing that in fact Δ and Δ_S, namely (2) and (5), are the same. An added benefit of this demonstration is that it allows one to introduce the notion of mean curvature and to see how this concept is tied in to that of the Beltrami operator. Since the typical student will already have been exposed to some elementary differential geometry of curves and surfaces, such an approach can only deepen his appreciation of the connection between geometry and potential theory.

Curvature of the surface S as defined by (1) is given in terms of the coefficients L_{ij} of the second fundamental form:

$$L_{ij} = r_{u_i u_j} \cdot n$$

The trace of the matrix (L_{ij}) defines the mean curvature H of S at p by

$$H(p) = \frac{1}{2} \sum_{i=1}^{2} L_{ii}$$

In terms of the parametrization (4) this evidently reduces to

$$H(p) = \frac{1}{2} \sum_{i=1}^{2} \frac{\partial^2 h(0,0)}{\partial x_i^2} \tag{6}$$

Moreover, since $g_{ij} = \delta_{ij}$ in our case, a direct evaluation of Δ_S from (2) shows that

$$\Delta_S f = \sum_{i=1}^{2} \left(\frac{\partial^2 f}{\partial x_i^2} + \frac{\partial f}{\partial x_3} \frac{\partial^2 h}{\partial x_i^2} + \frac{\partial^2 f}{\partial x_3^2} \left(\frac{\partial h}{\partial x_i} \right)^2 \right) \tag{7}$$

with all derivatives taken at p. But, since $\nabla h = 0$ there, a combination of (6) and (7) shows that

$$\Delta_S = \sum_{i=1}^{2} \frac{\partial^2}{\partial x_i^2} + 2 H(p) \frac{\partial}{\partial n} \tag{8}$$

where $\frac{\partial}{\partial n}$ denotes the directional derivative given by $\frac{\partial f}{\partial n} = \nabla f \cdot n$ (in our case $n = (0,0,1)$). A similar straightfoward computation also establishes that the right side of (8) is identical to the operator Δ. It follows that the Beltrami operator $\Delta_S = \Delta$ is given in terms of the ordinary Laplace operator on a plane plus a term involving the mean curvature. At this point the student can be led in several directions including a study of harmonic functions on S and minimal surfaces but instead we will apply it to elucidate a boundary value problem in biology. Before doing so, however, observe that if $\frac{\partial^2}{\partial n^2}$ denotes the quadratic form $n \cdot \nabla^2 f n$, in which $\nabla^2 f$ is the Hessian of f at p then this reduces to $\frac{\partial^2 f}{\partial x_3^2}$ in the present case. We therefore obtain from (7) that the ordinary Laplacian in $\mathbb{R}^3$ can be expressed as

$$\sum_{i=1}^{3} \frac{\partial^2 f}{\partial x_i^2} = \Delta_S f - 2 H(p) \frac{\partial f}{\partial n} + \frac{\partial^2 f}{\partial n^2}. \tag{9}$$

An example of (9) is provided by considering the sphere $\sum_{i=1}^{3} x_i^2 = r^2$ having the parametric representation $(r \cos \theta \sin \phi, r \sin \theta \sin \phi, r \cos \phi)$. Then the well known formula for $\sum \frac{\partial^2}{\partial x_i^2}$ in spherical coordinates is

$$\frac{1}{r^2 \sin^2 \phi} \frac{\partial^2}{\partial \theta^2} + \frac{1}{r^2 \sin \phi} \frac{\partial}{\partial \phi} \left(\sin \phi \frac{\partial}{\partial \phi} \right) + \frac{2}{r} \frac{\partial}{\partial r} + \frac{\partial^2}{\partial r^2}. \tag{10}$$

We may compute Δ_S directly from (2) letting $u_1 = \theta, u_2 = \phi$ to obtain the Beltrami operator as

$$\Delta_S = \frac{1}{r^2 \sin^2 \phi} \frac{\partial^2}{\partial \theta^2} + \frac{1}{r^2 \sin \phi} \frac{\partial}{\partial \phi} \left(\sin \phi \frac{\partial}{\partial \phi} \right). \tag{11}$$

Moreover $\frac{\partial^2}{\partial n^2}$ becomes $\frac{\partial^2}{\partial r^2}$ in spherical coordinates and $\frac{\partial}{\partial n} = \frac{\partial}{\partial r}$, and so (10) reveals that $H(p) = -1/r$. This result for the sphere can also be verified directly from (6), of course.

A BOUNDARY VALUE PROBLEM IN BIOLOGY

Even though the exact genetic mechanism of animal coat markings is not known it is possible to speculate that a reaction-diffusion process controls the patterns which appear on the skins of some mammal species. The biological and mathematical details of such a model are presented by Murray [4] and what we propose to do here is to highlight the essential ideas of this model as an illustration of the role of Beltrami's operator. The precursor of this morphogenic model is due to Turing [5], a variant of which in the context of a marine predator-prey system is described in Beltrami [6].

The coloring of skin is due to specialized pigment cells called meloncytes. We assume that the production of these cells is regulated by the concentration c of a chemical C. A high enough concentration of C activates the growth of the cells. Another chemical D, of concentration d, tends to inhibit the growth of C. These chemicals react to each other in the presence of an enzyme, as well as diffuse, on the two dimensional surface of animals such as giraffes and zebras. More specifically I have in mind my cat, Jeffrey, whose striped tail resembles that of a racoon.

If c is large enough at some portion of the surface then pigmentation is triggered, as we said, and distinctive patterns emerge such as stripes or spots. Otherwise no markings appear there.

The actual spatio-temporal model governing the reaction and diffusion of C and D can be written in non-dimensional form as a pair of non-linear equations

$$\begin{aligned} \frac{\partial c}{\partial t} &= f_c(c,d) + \mu_c \Delta_S c \\ \frac{\partial d}{\partial t} &= f_d(c,d) + \mu_d \Delta_S d \end{aligned} \tag{12}$$

in which f_c, f_d are suitable reaction rate functions, which we need not specify here, μ_c, μ_d are diffusion rate constants, and is Beltrami operator on the body surface. A wealth of patterns may be revealed by (12) depending on the geometry involved. Here we consider only the tapering cylindrical tail of an animal, my cat say, which we approximate by a conical surface. The molecular weight of C is considered to be higher than that of D and so the diffusion constant μ_c is smaller than μ_d.

We wish to show that if the thickness of the tail is small enough near its tip, then the only patterns which can emerge are stripes. However, at the other end of the tail near the rump, where the tail is thicker, it is possible for irregular spots to appear. This conclusion is consonant with the observation of real animal tails (Murray [4]).

Beltrami's Operator

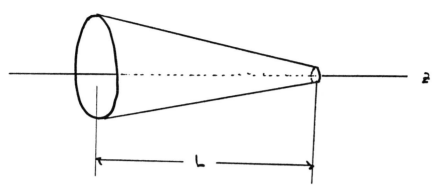

Figure 1

The conical surface S has axial coordinate $z, 0 \leq z \leq L$, a radius r at location z, and azimuthal coordinate θ (Figure 1). It's parametric representation is

$$(z, r\cos\theta, r\sin\theta)$$

in which $u_1 = z, u_2 = \theta$. Since r is fixed at any given point p of S, the corresponding Beltrami operator is, from (2),

$$\Delta_S = \frac{\partial^2}{\partial z^2} + \frac{1}{r^2}\frac{\partial^2}{\partial \theta^2}. \tag{13}$$

It is required that the solutions $c(z,\theta), d(z,\theta)$ to (12) be periodic functions of θ. Moreover no flux conditions are imposed at the endpoints $0, L$ of the form

$$\frac{\partial c}{\partial z} = \frac{\partial d}{\partial z} = 0.$$

The necessity of these conditions is obvious at the narrow end of the cone where the tail terminates (at $z = L$) while at $z = 0$ this is due to the fact that the rump end of the tail, near the animal's underbelly, is where an unpigmented regime begins.

Let $\bar{c}, \bar{d}$ be spatially uniform and time independent (viz. constant) solutions to (12). Linearize about $\bar{c}, \bar{d}$ by writing

$$w = \begin{pmatrix} c - \bar{c} \\ d - \bar{d} \end{pmatrix}$$

Then

$$\frac{\partial w}{\partial t} = Mw + D\Delta_S w \tag{14}$$

where $M = (m_{ij})$ is the Jacobian of the transformation $\begin{pmatrix} f_c \\ f_d \end{pmatrix}$ at $\bar{c}, \bar{d}$, and

$$D = \begin{pmatrix} \mu_c & 0 \\ 0 & \mu_d \end{pmatrix}$$

Let $W(z, \theta)$ be a solution to the eigenvalue problem

$$\Delta_S W = -k^2 W \tag{15}$$

wich satisfies the given boundary conditions. By separating variables in (15) it is easy to see that each of the two components of W is a sum of terms of the form

$$B_{m,n} \cos n\theta \cos m\pi z/L \qquad (16)$$

in which $B_{m,n}$ is a constant vector for each m, n and for which the corresponding eigenvalues are

$$k_{m,n}^2 = m^2\pi^2/L + n^2/r^2. \qquad (17)$$

We see that

$$w(z,\theta,t) = e^{\lambda t}W(z,t) \qquad (18)$$

is a solution to (14) provided that λ satisfies the quadratic equation

$$\text{Det}(M - k^2 D - \lambda I) = 0 \qquad (19)$$

We now require that the real part of the roots λ be negative in the absence of diffusion. This ensures that $w(z,\theta,t)$ decays to zero as $t \to \infty$ for all initial perturbations which are spatially homogeneous. Put in another way, the uniform constant state is stable to all uniform disturbances and so patterns can emerge only from perturbation which are initially non-homogeneus in space. In effect we hypothesize that pattern formation must depend on diffusion of the substances C and D as well as on the reaction between them.

To translate this into mathematical terms let $k^2 = 0$ so that the diffusion term drops out in (19). Then λ is an eigenvalue of M and so Re $\gamma < 0$ if and only if Det $M > 0$ and Trace $M < 0$. These conditions are assumed to be true from now on. Next, suppose that $k^2 \neq 0$. Then λ is an eigenvalue of $M - k^2 D$ and we require that Re $\lambda > 0$ to quarantee that a perturbation about the homogeneous state be unstable. This means that a small non-homogeneous initial disturbance will grow in time, which is the onset of pattern formation.

Since Trace M is negative evidently the same is true for Trace $(M - k^2 D)$ because D has positive elements. Therefore Re $\lambda > 0$ is assured for some given interger k only if $g(k^2) = \text{Det}(m - k^2 D)$ is negative. Now $g(k^2)$ is a quadratic in k^2 given by

$$\gamma k^4 - (m_{11}\gamma + m_{22})k^2 + \text{Det } M, \qquad (20)$$

where $\gamma = \mu_c/\mu_d$. Since $m_{11} + m_{22} = \text{Trace } M < 0$ it is clear that γ must not equal unity otherwise $m_{11}\gamma + m_{22} < 0$ and (20) then shows that the quadratic is positive. Therefore one necessarily must have that $\gamma \neq 1$, a condition that we have explicitly assumed to be true from the very beginning.

For k^2 zero or k^2 very large the quadratic $g(k^2)$ is positive and it becomes negative only if it crosses the k^2 axis. Therefore the negativity of $g(k^2)$ requires that $g(k^2) < 0$ in the range

$$\alpha < k^2 < \beta \qquad (21)$$

for two distinct points $0 < \alpha < \beta$.

It can be shown using specific functions f_c, f_d which are written out in Murray [4], that condition (21) can indeed be satisfied. Using (17) this is restated as

$$\alpha < m^2\pi^2/L^2 + n^2/r^2 < \beta \qquad (22)$$

If r is below a certain critical value r_{min} (at the thin end of the tail) then (22) is satisfied only for $n = 0$. From (16) above this means that the eigenfunctions are periodic in z, uniformly

Figure 2

in θ, for certain finite range of integers m. This means that the concentration c alternates between peaks and throughs along the length of the tail with coat markings which may be expected near every peak of c. This translates physically into a regular pattern of stripes.

Now suppose that $r > r_{min}$ so that (22) can be satisfied for a finite range of both m and n. This means that we are now along a thicker part of the tail. The eigenfunctions (16) are doubly periodic in this case and genuinely two dimensional patterns of spots will appear wherever the function peaks. Thus there is a transition from a two dimensional coat marking to a one dimensional set of stripes as we move along the tapered tail from rump to tip (Figure 2).

This is what the linear theory indicates but of course it cannot tell us how the nonlinear system actually evolves. What we have described here is only the onset of pattern formation. However a numerical investigation of the nonlinear reaction-diffusion system (12) vindicates the conclusions drawn from the linear perturbation analysis (Murray [4]). In any event the model is only a suggestive metaphor for what actually takes place in certain mammal hides. Moreover these conclusions are very much dependent on the particular geometry of the animal's body surface and it could be, for example, that no pattern appears anywhere.

Thus we have seen another connection, not very deep perhaps but hopefully amusing, between geometry and the Beltrami operator.

REFERENCES

[1] P. Lax and R. Phillips, *Scattering Theory For Automorphic Functions*, Annals of Math Studies 87, Princeton Univ. Press, 1976.
[2] P. Lax and R. Phillips, An Example Of Huygen's Principle, *Comm. Pure & Applied Math., 31* (1978), 415-421.
[3] R. Reilly, Mean Curvature, The Laplacian, And Soap Bubbles, *Am. Math. Monthly* (1982), 180-198.
[4] J. Murray, A Pre-pattern Formation Mechanism For Animal Coat Markings, *J. Theor. Biology 88* (1981), 161-199.
[5] A. Turing, The Chemical Basis Of Morphogenesis, *Phil. Trans. Royal Soc. London B 237* (1952), 37-72.
[6] E. Beltrami, *Mathematics For Dynamic Modeling*, Academic Press, 1987.

From Salvatore Pincherle's "Operazioni Distributive" to J. Ecalle's Resurgent Functions

CARLOS A. BERENSTEIN Department of Mathematics, University of Maryland, College Park, MD 20742, USA

Abstract. In his monograph «Le Operazioni Distributive e le loro Applicazioni all'Analisi» (1901), Pincherle considered in a systematic way different linear operators in spaces of analytic functions. Among them he emphasized the consideration of the process of analytic continuation as an operator, inducing a representation of the homotopy group of the domain. In particular, it becomes natural to consider functional equations involving this operator. J. Ecalle's theory of the resurgent functions (1974) could be considered as an attempt to develop an *algebraic* theory of the operators of analytic continuation. In between lie a number of very interesting (from the analytic point of view) particular cases of these very general approaches. In this talk we will mention one of them, my joint work with Ahmed Sebbar on an old problem of A. Hurwitz (Math. Werke, vol. 2, p. 752 (1933)). Namely, let $\mathcal{F}$ be the space of all holomorphic functions in the disk of center 1 and radius 1 such that they admit an analytic continuation along any path in $\mathbb{C}\setminus\{0\}$. The operator Γ of analytic continuation along the unit circle in the counterclockwise direction is well defined in $\mathcal{F}$. In modern notation the problem of Hurwitz was the description of $K = \ker\{\Gamma - \frac{d}{dz} : \mathcal{F} \to \mathcal{F}\}$. H. Lewy gave an example showing that e^z was not the only function in K; later Naftalevich (Mich. Math. J. 22 (1975)) showed that $\dim K = \infty$. We construct a biorthogonal system determining K.

From very early on, the work of Salvatore Pincherle shows his interest in considering

---

Lecture given on February 9, 1989, at the Università degli Studi di Bologna on the occasion of the 900th anniversary of its founding.

systematically large classes of operators acting on spaces of analytic functions. The «operazioni funzionali» of the 1886 memoir [8(a)] includes both linear integral transforms and non-linear transformations, but very quickly he concentrates on the linear case (later called «operazioni distributive»). In particular, he consideres infinite order differential operators acting on analytic functions, and tries to solve the corresponding equations by introducing a sort of functional calculus. Concerned with problems of analytic continuation and different representations of analytic functions, he considers the analytic continuation of functions given by integrals [8(b)], and eventually studies systematically the linear operator of analytic continuation, pointing out that it gives a representation of the homotopy group of the plane with punctures in the case the operator is restricted to the finite dimensional space of solutions of an ordinary differential equation with analytic coefficients [8(c)].

It is this thoroughly modern point of view, as well as concrete applications (hypergeometric functions and their properties, for instance), that it is explained in detail in his book of lectures given at the University of Bologna, «Le Operazioni Distributive e le loro Applicazioni all'Analisi», published in 1901 [9]. He recapitulates there his work on the functional calculus of linear operators acting on spaces of analytic functions, including a chapter on the geometry of function spaces.

Pincherle was not unique in attempting to systematize the operation of analytic continuation, certainly Weierstrass, Mittag-Leffler and Poincaré thought about the same lines. Some of this accumulated knowledge, though not lost, is not any longer part of the basic bag of tricks of every working complex analyst, much less every mathematician. In fact, if at all, only a «negative» result is mentioned in our courses on Complex Analysis, the Fabry-Hadamard gap theorem.

Let $f(z) = \sum_{n=0}^{\infty} a_n z^n$ be a power series with radius of convergence exactly 1. Clearly there are many entire functions g which interpolate the coefficients of f, that is, $g(n) = a_n$. If g has many zeros, then $|z| = 1$ is the natural boundary of f, more precisely a sufficient condition is that if $\{\lambda_k\} = \{n \in \mathbb{N} : g(n) \neq 0\}$, then $\lambda_{k+1} - \lambda_k > \theta \lambda_k$ for some fixed $\theta > 0$ (Hadamard, 1892).

«Positive» results were also considered at the end of the last century and the beginning of this one. For instance, Leau (1899), Wigert (1900), Faber (1903), obtained that the power series f has an analytic continuation as an analytic function in $\mathbb{C} \setminus [1, \infty)$ if g is a function of exponential type zero. If g is a polynomial of degree m, the point 1 is a pole of order $m + 1$. In fact, f has an analytic continuation to the Riemann surface of $\text{Log}(1 - z)$ with at most singularities at points lying above $z = 1$ (cf. [3], p. 337).

I mentioned this theorem to show the richness of the old results in this and other related areas of complex analysis, like differential equations in the complex plane. Also, it plays a role in what follows.

Jean Ecalle started well before 1974 to consider systematically the operators of analytic continuation acting on analytic functions defined on the universal covering space of $\mathbb{C} \setminus \mathbb{Z}$, more generally, $\mathbb{C} \setminus$ lattice. The point being that one can consider functional equations that involve both, classical operators, differentiation and multiplication, and the operators of analytic continuation. His objective was the complete understanding of local objects, like the infinite dimensional Lie group

$$G = \{f \in \mathcal{O}_0(\mathbb{C}) : f(z) = z + \ldots\}$$

of local biholomorphic maps tangent to the identity. Over the years this study has brought together and developed further an *algebraic* approach to asymptotic developments, analytic continuation, etc. There are at least three thick volumes of prepublications by Ecalle (many other smaller), a forthcoming book by Pham et al., on the resurgent functions, namely those that reappear after continuation in another guise (see the example below)[†]. A major triumph of the theory is the still unpublished work of Ecalle, Martinet and Ramis solving the Poincaré Conjecture on the finitude of the number of limit cycles for autonomous algebraic planar systems of ordinary differential equations [4].

In an attempt to understand a tiny bit of the problems of resurgence we considered with Ahmed Sebbar the following simplest example of an equation of resurgence.

Let $\mathcal{F}$ be the space of power series in $z-1$ with radius of convergence ≥ 1 and such that they have an analytic continuation along *any* path starting at $z = 1$ which avoids the origin. (In other words, this is the space of functions holomorphic everywhere in the Riemann surface of Log z).

In the space $\mathcal{F}$ it makes sense to consider the operator Γ or analytic continuation along the unit circle traversed counterclockwise starting at $z = 1$.

In [6] Hurwitz asked whether the equation (of resurgence)

$$f' = \Gamma f \qquad (*)$$

has any solution in $\mathcal{F}$ beyond the obvious ones, $f(z) = \text{const} \times e^z$. Lewy found the explicit solution

$$f(z) = \int_{-\infty}^{0} \exp\left(zt + \frac{(\log|t| - \pi i)^2}{4\pi i}\right) dt.$$

Later, A. Naftalevich [7] used his previous work on difference-diferential equations to show that if P is a polynomial and H an entire function, there are infinitely many affinely independent solutions F of the equation

$$F(\zeta + 1) = \exp[P(\zeta)]F'(\zeta) + H(\zeta). \qquad (\dagger)$$

This result of Naftalevich shows that the Hurwitz equation has infinitely many linearly independent solutions since a change of variable $z = \exp(2\pi i \zeta)$ brings $(*)$ to the form

$$F(\zeta + 1) = \left(\frac{-i}{2\pi}\right) \exp[-2\pi i \zeta] F'(\zeta). \qquad (**)$$

The very remarkable fact about Naftalevich's paper was that there was no way to find out whether Hans Lewy's solution was one of the solutions found by Naftalevich's iteration method. It was also clear that the simplicity of Lewy's solution ougth to hide some deeper properties of the solutions of $(*)$. For instance, could it be that all solutions are some sort of Laplace transforms? Could we find all solutions of $(*)$ (or equivalently $(**)$)? We have been able to answer both questions on the affirmative [1]. Before proceeding, let us say that equations of the type $(\dagger)$ have been studied for a long time; the most notable results are due to N.G. de Bruijn who, motivated by problems on Analytic Number Theory, considered the equation

$$F'(\xi) = \exp(\alpha \xi + \beta) F(\xi - 1) \qquad (\dagger\dagger)$$

for $\xi \in \mathbb{R}$, where $\alpha > 0$ and $\beta \in \mathbb{C}$. He showed in [2] that for every $B \in \mathbb{R}$ there is a family $(F_n)_{n \geq 1}$ of solutions of $(\dagger\dagger)$ for $\xi \geq B$ such that any other solution of $(\dagger\dagger)$ in $\xi \geq B$

[†] Part of the difficulty in understanding the work of Ecalle stems from the fact that he prefers to state «mandalas», i.e., general principles, instead of going over all the details. On the other hand, these mandalas are very profound indeed.

can be written in the form $F = \sum_{n\geq 1} c_n F_n$, the series being uniformly convergent on any interval of the form $[B+1, C]$.

Regretfully, de Bruijn's method doesn't adapt well to our case $(**)$, because $-2\pi i \notin \mathbb{R}$. We reduce the problem to Wiener-Hopf equation and with the help of theorems of the type of the Leau-Wigert-Faber theorem mentioned above, and the Avanissian-Gay transform, we construct explicitly a duality bracket, and a biorthonormal system of functions $\{F_n\}_{n\in\mathbb{Z}}$, $\{G_n\}_{n\in\mathbb{Z}}$, F_n solutions of $(**)$, such that any entire solution of $(**)$ is of the form

$$F(\zeta) = \sum_{-\infty}^{\infty} c_n F_n(\zeta) + c \exp[e^{2\pi i \zeta}]. \qquad (***)$$

In this system, F_0 coresponds to Hans Lewy's solution. Furthermore, one can give precise estimates for the coefficients c_n, which establish necessary and sufficient conditions on the sequence $\{c_n\}_{n\in\mathbb{Z}}$, to define a solution of $(**)$ via the development $(***)$.

The collection of all continuous linear operators in $\mathcal{F}$ which commute with Γ is precisely those of the form

$$Tf(z) = \sum_{n=0}^{\infty} A_n(z) \frac{d^n}{dz^n} f(z),$$

where the A_n are (single-valued) analytic functions in $\mathbb{C}\setminus\{0\}$ the «symbol»

$$\zeta \mapsto \sum_{n=0}^{\infty} A_n(z)\zeta^n$$

is a entire function of exponential type zero, uniformly for $z \in K (K \subset\subset \mathbb{C}\setminus\{0\})$. It will be natural to study the equations of resurgence

$$Tf = \Gamma f$$

and find whether they have non-trivial, i.e., multiple-valued, analytic solutions. Some simple cases show one needs to admit at least polar singularities. The MS thesis of Paul Gilden [5] considered a few simple and strange properties of this last equation. In particular, he pointed out that if we replace Γf by the condition that f becomes Tf after continuation along *some* path, then we do not have a linear space of solutions, e.g., if $Tf := -f$, $f = \sqrt{z}$ and $\sqrt[4]{z}$ will satisfy $Tf = \Gamma f$ for two *different* analytic continuation operators.

Acknowledgement. I would like to thank Professor Salvatore Coen for the invitation to deliver this lecture, the National Science Foundation, and the Consiglio Nazionale delle Ricerche, who made possible my participation in this colloquium.

REFERENCES

[1] C.A. Berenstein and A. Sebbar, *Sur un problème de Hurwitz*, to appear.
[2] N.G. de Bruijn, The difference-differential equation $F'(x) = e^{\alpha x + \beta} F(x-1)$, *Indagationes Math.* 15 (1953), 449-464.
[3] P. Dienes, *The Taylor series*, Dover Publ., NY (1957).

[4] J. Ecalle, *Les fonctions resurgentes*, vol. 3, Prepubl. Osay, 1985 (and references therein).
[5] P. Gilden, M.S. thesis, Univ. of Maryland, 1984.
[6] A. Hurwitz, *Mathematische Werke*, Band II, Birkhäuser Verlag, Basel (1933), p. 752.
[7] A. Naftalevich, On a differential-difference equation, *Mich. Math. J.* 22 (1975), 205-223.
[8] S. Pincherle, *Opere Scelte*, Edizione Cremonese, Roma (1954):
   (a) Studi sopra alcune operazioni funzionali, Bologna (1886), (Section 3);
   (b) Sur certaines opérations fonctionelles représentées par des intégrales définies, Acta math. (1887) (Section 4);
   (c) Contributo alla integrazione delle equazioni differenziali lineari mediante integrali definiti, Bologna (1892) (Section 10).
[9] S. Pincherle, *Le Operazioni Distributive e le loro Applicazioni all'Analisi*, Nicola Zanichelli, Bologna (1901).

Pincherle e la Teoria delle Funzioni Analitiche

UMBERTO BOTTAZZINI, Dipartimento di Matematica, Università di Bologna, Piazza di Porta S. Donato 5, 40127, Bologna, Italia.

1. Nei fascicoli del *Giornale di Matematiche* di Battaglini apparvero nel corso del 1880 le varie parti di un «Saggio di una introduzione alla teoria delle funzioni analitiche secondo i princìpi del prof. C. Weierstrass compilato dal Dott. S. Pincherle».

L'occasione di quello scritto del giovane matematico italiano era stata il «conseguimento di un posto di studi all'estero», che gli aveva permesso di seguire nell'anno 1877-78 i corsi del celebre analista di Berlino. Tornato in patria, scrive Pincherle presentando ai lettori del *Giornale* il suo lavoro, «mi credeva quasi in obbligo di far conoscere almeno in parte, ai miei compagni di studio, le nuove vedute e i concetti nuovi che il prof. Weierstrass va introducendo nella scienza» (Pincherle 1880b, 178). Concetti che allora cominciavano a diffondersi in Germania, «ma che rimangono ancora quasi sconosciuti agli studenti italiani per la nota avversione di quel maestro per la stampa».

Weirstrass del resto non vedeva nemmeno di buon occhio la pubblicazione di lavori come quelli cui si accingeva Pincherle. Non mancavano dunque le ragioni per legittimare le esitazioni che egli sentiva di fronte alla «difficoltà di una conveniente esposizione di argomenti delicati e per la loro novità soggetti a controversia», dove «una parola impropriamente adoperata basta a sviare il concetto» (Ibid., 178).

Quel «Saggio», compilato sulla base degli appunti presi a lezione dallo stesso Pincherle e «di fascicoli di corsi antecedenti messi alla [sua] disposizione dalla gentilezza di alcuni suoi scolari» doveva comunque restare a lungo, e non solo in Italia, una delle poche fonti disponibili a chi volesse familiarizzarsi con le teorie che Weierstrass esponeva nei suoi corsi.

Il «Saggio» di Pincherle si apre con una sezione dedicata ai «Princìpi fondamentali dell'aritmetica»: «L'analisi fondandosi senza alcun postulato sul solo concetto di numero, conviene stabilire anzitutto la definizione delle varie specie di numeri e delle operazioni che su di essi si possono eseguire» (Pincherle 1880b, 179). Vengono così introdotti i numeri interi e i razionali, e poi presentata la teoria weierstrassiana dei numeri reali (cioè dei «numeri composti d'infiniti elementi positivi e negativi», nella sua terminologia) e la teoria dei «numeri formati con due unità principali», cioè i numeri complessi.

Dopo aver trattato le proprietà di serie e prodotti infiniti (numerici), la Seconda Sezione del «Saggio» è dedicata alla dimostrazione di teoremi «sulle grandezze in generale», tra cui la «Proposizione fondamentale» seguente: «Se in una varietà ad una dimensione si hanno infiniti posti soddisfacenti ad una definizione comune si troverà in quella varietà per lo meno un posto avente le proprietà che in qualunque suo intorno, per piccolo che si voglia prendere, esisteranno sempre infiniti posti soddisfacenti a quella definizione» (Pincherle 1880b, 237). E' in questo modo che si trova enunciato il celebre teorema di Bolzano-Weierstrass, a cui fanno seguito alcuni altri teoremi fondamentali, come quello sui limiti superiore e inferiore per «infiniti numeri reali definiti da una legge qualunque».

Solo nella Terza Sezione viene introdotto il concetto di funzione, quello di continuità e quello di derivata, a commento dei quali, riportando le parole di Weirstrass, Pincherle scrive: «Si è creduto fino a questi ultimi tempi che l'essere continua bastasse ad una funzione per ammettere derivata e in molti trattati si trova «dimostrato» il «teorema»: ogni funzione continua ammette una derivata. Ma queste dimostrazioni ammettono tutte implicitamente qualche proprietà che non è contenuta nel concetto generale di funzione» (Pincherle 1880b, 247). Un esempio particolarmente significativo era dato dal *Traité de calcul différentiel* di Bertrand, apparso nel 1864, che si apriva proprio con la dimostrazione di quel «teorema».

Nell'ultima Sezione quarta dedicata a «Le funzioni razionali e le serie di potenze» sono infine introdotti i concetti e i teoremi classici sulla convergenza uniforme delle serie, sulla derivabilità termine a termine delle serie uniformemente convergenti e il metodo del prolungamento analitico; in altre parole, gli elementi essenziali della *Funktionenlehre* secondo Weierstrass.

2. Il «Saggio» di Pincherle si affiancava alla *Teorica* di Casorati e per molti aspetti ne rappresentava un naturale complemento. La *Teorica*, di cui era apparso il primo volume (e il solo pubblicato) nel 1868 era infatti ispirato alle vedute di Cauchy e Riemann e anzi, ancora alla fine del secolo costituiva a parere di Klein la migliore esposizione della teoria delle funzioni di una variabile complessa secondo il punto di vista riemanniano.

Quando Casorati preparava la redazione del suo trattato, delle radicali novità che gli analisti di Berlino andavano introducendo nella «scienza» si aveva in Italia una vaga notizia. Casorati si limitava a dare un rapido cenno dei lavori di Weierstrass nell'ampia sezione di «Notizie storiche» con cui si apre la *Teorica*, discutendo due memorie (Weierstrass 1854 e 1856) relative al problema dell'inversione degli integrali iperellittici e alla teoria degli integrali abeliani.

«Il sig. Weierstrass deve aver compiuto ricerche generalissime sull'argomento» scriveva Casorati (1868, 43). «Ma di esse non siamo in grado di dare alcuna precisa informazione, attesochè finora non abbiamo potuto averne alcuna esatta contezza. L'illustre matematico, forse a cagione delle gravi malattie ma, del resto, non unico fra i sommi nel rifuggire dalle noje gravose della stampa, non diede finora alle medesime vera pubblicità; limitandone la comunicazione alla stretta cerchia degli amici o dei colleghi nell'Accademia». Anche la consultazione dei *Monatsberichte* dell'Accademia di Berlino non offriva maggiori lumi, giacchè la gran parte delle comunicazioni di Weiestrass era ridotta al semplice titolo. Lo stesso, o peggio, avveniva per le lezioni all'Università.

Per avere notizie più precise e dirette sui risultati ottenuti dagli analisti di Berlino, lo stesso Casorati nell'autunno del 1864, quando Riemann soggiornava a Pisa, si era recato nella capitale prussiana per incontrare Weierstrass, Kronecker e i loro allievi.

Da Kronecker egli aveva appreso dell'esistenza di funzioni che «non ammettono coefficiente differenziale, che non possono rappresentare linee», come si legge negli appunti presi da Casorati durante quegli incontri. E ancora, che non è sempre possibile prolungare analiticamente una funzione in ogni parte del piano complesso, contrariamente a quanto si era in generale sempre supposto «anche dal Riemann». Così avveniva per esempio per la funzione definita dalla serie

$$\theta(q) = 1 + 2\sum_{n\geq 1} q^{n^2}$$

che ammette la circonferenza $|q| = 1$ come luogo di punti singolari e non è prolungabile fuori dal disco unitario.

Weierstrass raccontò a Casorati di aver trovato il teorema «che ogni funzione (monodroma s'intende) la quale non abbia punti in cui cessa di essere definita è necessariamente razionale (la funzione $e^{1/x}$ nel punto $x = 0$ non è definita poiché può avervi qualunque valore). Egli credeva che siffatti punti non potessero costituire una continuità e che quindi almeno per un punto P si potesse passare sempre da una porzione A chiusa del piano ad un punto M qualunque di esso». Ma le cose non stavano affatto così, come mostrava la funzione propostagli da Kronecker. Fu «cercando appunto la dimostrazione della possibilità generale ch'egli si accorse della non generale possibilità» (in: Bottazzini 1986, 264).

La figura di Riemann appare continuamente sullo sfondo di questi incontri. Casorati viene così a sapere che «in Berlino le cose del Riemann fecero difficoltà» mentre Weierstrass non esitava a dichiarare che «egli capì il Riemann perchè possedeva già i risultati delle ricerche» e che «i discepoli del Riemann hanno il torto di attribuire tutto al loro maestro, mentre molte cose erano già fatte e dovute a Cauchy etc. ed il Riemann non fece che vestirle alla maniera sua per suo comodo» (in: Bottazzini 1986, 263).

La critica si fa più esplicita e puntuale nelle parole di Kronecker: «I matematici (...) sono un po' *hochmütig* nell'uso del concetto di funzione. Anche Riemann generalmente molto esatto, non è irreprensibile sotto questo rapporto. Se una funzione cresce e poi diminuisce o viceversa dice Riemann dovervi essere un minimo od un massimo (vedi la dimostrazione del così detto *Dirichlet'sche Princip*; mentre dovrebbesi restringere la conclusione alla sfera delle funzioni per così dire ragionevoli».

Delle conversazioni con i matematici berlinesi non dovevano restare molte tracce nella *Teorica*, se si esclude la dimostrazione del teorema di «Casorati-Weierstrass» sul comportamento di una funzione nell'intorno di una singolarità essenziale, che si può intravedere nelle parole di Weierstrass. Ma certo i dubbi sollevati da Kronecker e Weierstrass, rimasti irrisolti, dovevano avere un peso decisivo nel convincere Casorati a rinunciare al portare a termine il suo trattato. Il secondo volume della *Teorica* avrebbe dovuto infatti trattare in maniera dettagliata la teoria delle «funzioni abeliane» secondo i metodi di Riemann, dove il «principio di Dirichlet» gioca un ruolo essenziale. E fu proprio l'incapacità di venire a capo delle difficoltà sollevate intorno all'ammissibilità di tale principio a rappresentare per Casorati uno scoglio insuperabile. «Anche il principio di Dirichlet contribuisce non poco al mio indugio» egli confessava a Battaglini nel novembre del 1869. «Molto mi peserebbe doverne far senza; e non meno mi peserebbe di presentarlo ancora così incompleto come Riemann l'ha dato» (in: Neuenschwander 1978, 28).

Quello stesso anno durante l'estate Casorati aveva incontrato Kronecker in vacanza sul lago di Como e insieme avevano discusso dell'argomento. «Kronecker mi sembra non avesse creduto possibile uno stabilimento rigoroso di esso principio in tutta la generalità riemanniana», continua la lettera di Casorati. «Avendone discorso alquanto, ne ricavai le più ampie conferme de' miei dubbi, ma non i mezzi per porvi rimedio». Anche i lavori di Schwarz (1869a, 1869b) «riferisconsi a casi particolari» mentre «quello recentissimo» di Weber (1869) «non mi pare inappuntabile, benchè finora non feci che leggerlo in gran fretta».

Una breve comunicazione di Weierstrass (1870) all'Accademia di Berlino rendeva poi di dominio pubblico un controesempio, che stabiliva in maniera definitiva l'inammissibilità di quel principio, nella formulazione datane da Riemann.

Negli anni seguenti Casorati continuò senza successo le ricerche intorno alla difficile questione, nella speranza di poter completare il proprio trattato col secondo volume previsto. Egli si rivolse a Schläfli, a Prym e poi a Schwarz nel 1872, per chiedergli copia dei suoi lavori (Schwarz 1870a, b, c, 1872). Ancora il discusso principio era oggetto della corrispondenza intercorsa nel 1875 tra Casorati e H. Weber, il curatore delle opere di Riemann (Neuenschwander 1978). Quello stesso anno giungeva a Pavia Pincherle, per cominciare la sua attività di insegnamento in quella città.

3. Conclusi gli studi alla Scuola Normale Superiore, Salvatore Pincherle (11 marzo 1853 - 10 luglio 1936) a poco più di ventun anni si era laureato in matematica a Pisa. Per le sue doti egli si era già segnalato giovanissimo, ricevendo il *1° prix d'excellence* al termine della *Classe de Mathématiques préparatoires* a Marsiglia, dove la famiglia si era trasferita dalla nativa Trieste.

Nel 1870 Pincherle aveva ottenuto un posto come alunno interno alla Normale, dove seguì i corsi di Enrico Betti e Ulisse Dini. Mentre quest'ultimo teneva lezioni di analisi superiore, da diversi anni Betti aveva fatto della fisica matematica l'oggetto delle sue ricerche e del suo insegnamento (Bottazzini 1982).

Agli studi sulla capillarità, che Betti aveva pubblicato nel 1866 in una lunga memoria «in guisa che possa servire come una monografia del soggetto», si affiancavano ora quelli sulla teoria dell'elasticità, apparsi nel corso del 1872 e 1873 nel *Nuovo Cimento*. Betti vi raccoglieva il contenuto essenziale delle lezioni tenute per il corso di Fisica matematica e forniva numerose applicazioni del cosiddetto «teorema di reciprocità» che «nella teoria delle forze elastiche dei corpi solidi, tiene il luogo che il teorema di Green ha nella teorica delle forze che agiscono secondo le leggi di Newton» (in: Bottazzini 1982, 269).

La sostanziale adesione al modello newtoniano e il ruolo giocatovi dalla funzione potenziale portò Betti in maniera naturale a far i conti con il «principio di Dirichlet». E' ragionevole supporre che egli esortasse Dini, che dagli iniziali interessi in geometria differenziale si era ormai definitivamente indirizzato nel campo dell'analisi, a lavorare intorno a quel complesso di problemi che la critica di Weierstrass aveva reso di grande attualità. Di fatto Dini, che allora intrattenne una fitta corrispondenza scientifica con Schwarz su questi argomenti, pubblicò diversi lavori dedicati al problema di Neumann di determinare una funzione armonica in un dominio, quando siano dati i valori assunti sul contorno dalla derivata normale (Dini 1870-71, 1871-73).

Queste ricerche di Dini non dovettero apparentemente interessare molto il giovane Pincherle, se egli preferì laurearsi sotto la guida di Betti, con una tesi sulle superfici di capillarità e le relative costanti caratteristiche, il cui contenuto fu riassunto in due articoli apparsi nel *Nuovo Cimento* (Pincherle 1874, 1875).

Quei primi lavori erano tuttavia destinati a restare senza seguito nella produzione scientifica di Pincherle, così come occasionale era il carattere delle ricerche portate a termine nei primi tempi trascorsi a Pavia e presentate da Casorati per la pubblicazione nei *Rendiconti* dell'Istituto Lombardo (Pincherle 1876, 1877).

Nel professore pavese Pincherle trovò un interlocutore autorevole, che influì non poco nell'orientare verso l'analisi l'indirizzo delle sue ricerche. D'altra parte, il soggiorno a Berlino di Pincherle fornì a Casorati la possibilità di avere notizie dettagliate e di prima mano su quanto Weierstrass andava esponendo a lezione. Di fatto, al ritorno a Pavia Pincherle venne invitato a tenere un corso all'università, il cui contenuto essenziale fu poi riassunto nel «Saggio» (Pincherle 1880b).

Per molti aspetti il «Saggio» assumeva un carattere emblematico: dopo che Riemann aveva trovato in Italia, presso Betti e Casorati in particolare, interlocutori che ne avevano apprezzato «l'immensa generalità» dei metodi e delle teorie, «un magnifico lavoro di puro pensiero» – ebbe a dire Betti – che «soddisfa completamente le principali tendenze della analisi moderna» (in: Bottazzini 1986, 280), a poco a poco le critiche di Weierstrass e Kronecker dal punto di vista del rigore avevano convinto i matematici italiani della necessità di abbandonare i metodi riemanniani. In fondo il «Saggio» si presentava come un manifesto del «nuovo» modo di fare analisi, di cui Pincherle si annunciava come il più avvertito rappresentante.

Il periodo di studi trascorso a Berlino aveva segnato infatti una svolta anche nella produzione scientifica di Pincherle, annunciata da una breve nota che nel maggio 1878 era apparsa nei *Rendiconti* dell'Istituto Lombardo, presentata ancora una volta da Casorati.

«Il sig. Weierstrass ha dimostrato – esordiva Pincherle (1878, 382) – che qualunque funzione intera avente un numero finito od infinito di posti degli zeri si può scomporre in prodotto di funzioni della stessa specie avente ciascuna un solo posto-zero tutt'al più; e questa scomposizione essendo analoga a quella ordinaria di una funzione intera razione nei suoi fattori lineari», ecco che si può porre la questione «se e con quali restrizioni, si conservino per le funzioni intere trascendenti le relazioni che passano fra i coefficienti e le radici delle razionali».

In analogia con quanto avviene per i polinomi, la determinazione di quelle relazioni non è molto di più di un esercizio di manipolazione di serie. Ma quella nota di Pincherle è interessante, dal nostro punto di vista, sia per quello che dice che per quello che non dice.

Pincherle si ispira infatti dichiaratamente alla grande memoria di Weierstrass sulle funzioni monodrome (Weierstrass 1876), ne adotta il linguaggio e i risultati, ma significativamente (come Weierstrass del resto) tace sul fatto che il teorema di fattorizzazione delle funzioni intere era già stato ottenuto una quindicina d'anni prima da Betti. In un corso sulla teoria delle funzioni ellittiche, tenuto a Pisa e poi pubblicato negli *Annali di matematica pura e applicata*, Betti (1860-61) aveva mostrato che se il prodotto infinito

$$\prod_i (1 - \frac{z}{\alpha_i})$$

dove $\alpha_i (i \in \mathbb{N})$ assume un'infinità di valori che sono gli zeri di una funzione intera W, è convergente per ogni z finito, allora esso rappresenta una funzione intera che ha come zeri tutti e soli gli α_i e può essere data nella forma

$$e^w \prod_i (1 - \frac{z}{\alpha_i})$$

dove w è una funzione intera.

Il problema successivo affrontato da Betti era quello di determinare una funzione intera quando è dato un sistema infinito di numeri complessi α_i che sono gli zeri della funzione. Nell'ipotesi che gli zeri siano tutti a distanza finita tra loro, Betti affermava che «la funzione intera potrà decomporsi in un numero infinito di fattori di primo grado ed esponenziali» (1860-61, 246).

Quell'articolo di Betti non fu apparentemente mai letto da Weierstrass, se questi scriveva ancora nel giugno del 1881 a Sonja Kowalewski: «Vengo a sapere una curiosità. *Herr* Betti afferma che il teorema che una funzione trascendente intera con zeri prefissati – di cui solo un numero finito si trova in un dominio finito – è stato dato da lui stesso. E' un peccato che le idee giuste non si affermino sempre al momento giusto!».

Ancora ispirate a un corso inedito di Weierstrass sulle funzioni ellittiche (come ci informa lo stesso Pincherle), erano due sue note sulle funzioni «monodrome» (Pincherle 1879, 1880a).

L'intento era quello di presentare «un saggio di una nuova trattazione delle funzioni ellittiche» che «forse non sembrerà del tutto oziosa» quando si consideri che «essa riduce lo studio delle funzioni ellittiche a quello di nuove funzioni che godono di proprietà analoghe ma più semplici per molti riguardi» (Pincherle 1880a, 92). La proprietà cui faceva allusione Pincherle era di soddisfare ad una «importante relazione di natura algebrica».

Dopo aver definito, seguendo Casorati (1868) e Weierstrass (1876) che una funzione ha «il carattere di funzione razionale» nell'intorno di un punto x_o se ivi è sviluppabile in serie della forma

$$\sum_{k=0} c_{m+k}(x - x_o)^{m+k} \quad (m \in \mathbb{Z})$$

Pincherle si pone il problema se esistono funzioni analitiche, monodrome, razionali nell'intorno di ogni valore della variabile e tali che

$$\varphi(x) = \varphi(\omega x)$$

con ω costante.

Supponendo ammessa l'esistenza di funzioni razionali dappertutto tranne che in un numero finito di punti singolari, Pincherle dimostra che tali punti possono essere solo l'origine e il punto $x = \infty$, e che il «moltiplicatore» ω non può avere la forma $e^{2\pi i r}$ con $r \in \mathbb{R} - \mathbb{Q}$, ma è dato dalle potenze intere (positive o negative) di un numero $\rho e^{i\theta}$ con $\rho > 1$.

Inoltre, suddiviso il piano complesso in corone circolari mediante circonferenze i cui raggi sono in progressione geometrica di ragione ρ, una funzione φ riprende gli stessi valori in ogni corona; di più, diventa nulla, infinita e assume qualunque valore lo stesso numero (finito) di volte in ogni corona, ciò che consente una classificazione in «ordini» delle funzioni φ. Infine, Pincherle prova che non esistono funzioni di ordine inferiore al secondo e che $x\varphi'(x)$ ha «la stessa natura» di $\varphi(x)$.

Dimostrata poi l'esistenza di tali funzioni con la costruzione di una di esse mediante un'opportuna funzione ausiliaria, e trovatene le principali proprietà, Pincherle passa a studiare le funzioni $E(x)$ che si ottengono dalle $\varphi(x)$ con la sostituzione $z = e^{ax}$. Sono queste le funzioni ellittiche, intendendo «col prof. Weierstrass *ellittiche* tutte le funzioni doppiamente periodiche che hanno il solo posto $z = \infty$, per posto singolare» (Pincherle 1880a, 118).

Nella seconda parte di questa memoria Pincherle presenta uno studio più dettagliato delle funzioni monodrome $f(x)$ aventi «un'equazione caratteristica». Come aveva mostrato in (Pincherle 1879), tali funzioni «soddisfano tutte ad equazioni funzionali della forma

$$f(x) = f\left(\frac{ax + b}{cx + d}\right)\text{»} \quad (\text{Pincherle 1880a, 121}).$$

A coronamento della sua attività di ricerca, nella primavera del 1880 Pincherle vinse il concorso per la cattedra di Algebra complementare e geometria analitica nell'università di Palermo. Dal primo gennaio dell'anno seguente egli era tuttavia nominato per lo stesso insegnamento all'università di Bologna, dove svolse ininterrottamente per oltre quarant'anni la propria attività di docente.

4. La nuova posizione accademica segnò per Pincherle la fine della consuetudine di rapporti e di discussioni scientifiche con Casorati, che aveva accompagnato gli anni trascorsi a Pavia.

Nonostante la presenza di Arzelà, a Bologna infatti Pincherle lavorò in condizioni di relativo isolamento, affidandosi alle riviste per tenersi al corrente dei risultati della ricerca e, al tempo stesso, cercando di stabilire regolari contatti epistolari con matematici italiani e stranieri. Di quella corrispondenza non è restata traccia, se si escludono le lettere a Casorati e alcune lettere scambiate con Poincaré.

La lettura nei *Comptes Rendus* dell'Accademia di Parigi di due brevi note dell'analista francese (Poincaré 1881a, b) sulle funzioni «fuchsiane» convinse Pincherle dell'opportunità di segnalargli i propri lavori (Pincherle 1879, 1880a) «sur des sujéts qui me semblent avoir quelque analogie avec ceux que vous traitez d'une façon si remarquable» (in: Dugac 1989, 210).

Rivendicando l'interesse dei propri risultati, Pincherle aggiungeva: «Je crois que la determination des fonctions ayant la propriété

$$f(x) = f\left(\frac{ax+b}{cx+d}\right)$$

(et sans autres points singuliers que ceux qui sont exigés par l'equation précedente) résulte immédiatement de mes deux Notes».

La lettera di risposta di Poincaré non ci è pervenuta, ma è sufficiente una rapida lettura dei lavori di quest'ultimo per rendersi conto che il rapporto tra le ricerche dei due non va oltre «quelque analogie», essenzialmente dovuta al fatto che la proprietà delle funzioni studiate è la stessa. Infatti, mentre Poincaré è interessato allo studio geometrico delle proprietà del disco unitario (il *cerchio fondamentale* di centro nell'origine e raggio 1, nella sua terminologia) invariante per le trasformazioni del gruppo iperbolico $z \to \frac{az+b}{cz+d}$ e alla classificazione di tutti i «gruppi fuchsiani» (sottogruppi del gruppo iperbolico), nulla di tutto ciò, come s'è visto, si trova in Pincherle.

Egli infatti seguiva una propria linea di ricerca, sulla via tracciata da Weierstrass nella sua grande memoria «Zur Funktionenlehre» (1880). Lo stesso Pincherle, riguardando molti anni dopo ai suoi primi esordi, così motivava l'orientamento allora preso: «Tandis que l'étude des développements d'une fonction arbitraire, au sens général de Dirichlet, ou, comme on dit aussi, d'une fonction de variable réelle, constituait déjà, vers 1880, un chapitre considérable de l'analyse à la suite des travaux de Dirichlet, de Riemann, de Du Bois-Reymond, de Dini, etc., on n'avait guère étudié que dans des cas particuliers les développements de fonctions analytiques en séries ordonnés par C. Neumann, Heine, Thomé, Frobenius suivant les polinômes de Legendre, les fonctions de Bessel, les produits spéciaux de la forme $(x-a_1)(x-a_2)\ldots(x-a_n)$ » (Pincherle 1925-341).

I teoremi di Weierstrass sulle serie uniformemente convergenti di funzioni analitiche permettevano ora di affrontare «d'une façon générale l'étude des développements à caractère analytique». Ed era a questo studio che Pincherle si era dedicato fin dai primi mesi del 1882, come testimoniano di quaderni di appunti manoscritti che, raccolti in volumi, permettono di seguire passo passo l'evolversi delle sue ricerche dal 1882 fino all'anno della sua morte (Bottazzini-Francesconi 1989).

Pincherle inaugura il primo «Quaderno» di appunti (Gennaio 1882-Agosto 1882) con alcuni studi sulle funzioni sferiche. Data la generica funzione sferica

$$P_n(x) = \frac{1}{2^n n!} \frac{d^n(x^2-1)^n}{dx^n}$$

Pincherle dimostra che esiste un opportuno dominio in cui $P_n(x)$ è analitica e inoltre che, assegnata una arbitraria funzione analitica $f(x)$, esiste un dominio del piano complesso in cui $f(x)$ può essere sviluppata in serie di funzioni sferiche. In particolare, si può sviluppare in questo modo qualunque funzione regolare in un cerchio di centro nell'origine e raggio qualunque.

Considerata poi la serie

$$\frac{1}{y-x} = \sum_0^\infty x^n \frac{1}{y^{n+1}}$$

convergente in un cerchio di raggio opportuno, egli pone

$$\frac{1}{y-x} = \sum_0^\infty Q_n(y) P_n(x)$$

e mostra che i coefficienti $Q_n(y)$ – funzioni sferiche di II° specie – sono funzioni regolari fuori dal cerchio di centro nell'origine e raggio 1.

Questi risultati si ottenevano come applicazioni particolari di alcuni teoremi sugli sviluppi in serie di funzioni analitiche, che Pincherle pubblicava (senza dimostrazione) in una breve nota di una pagina, presentata dall'Istituto Lombardo da Casorati il 23 marzo 1882 (Pincherle 1882a).

Se si ha un sistema di funzioni analitiche $p_n(x)$ regolari in un cerchio $|x| < R$, affermava Pincherle, allora le potenze intere x^m sono ivi sviluppabili in serie della forma

$$x^m = \sum_{n=0}^\infty \alpha_{m,n} p_n(x)$$

e se inoltre si possono assegnare due numeri positivi K e σ, tali che entro lo stesso cerchio si abbia

$$\sum_{n=0} |\alpha_{m,n} p_n(x)| < \sigma^m K$$

allora qualunque funzione $F(x)$ regolare in un cerchio $|x| < R', R' > \sigma$ è sviluppabile in serie di funzioni $p_n(x)$:

$$F(x) = \sum_{n=0} c_n p_n(x) \ .$$

«Sotto le stesse ipotesi», continuava Pincherle, «esiste un secondo sistema di funzioni» $q_n(x)$ – coniugato del sistema $p_n(x)$ – tale che «in un campo assegnabile» vale

$$\frac{1}{1-x'x} = \sum_{n=0} q_n(x') p_n(x) \ .$$

Infine, «se alle ipotesi già fatte aggiungiamo che le funzioni del sistema $p_n(x)$ siano polinomi razionali interi in x e di grado n, avremo che le funzioni $q_n(x)$ del sistema conjugato saranno serie di potenze intere di x principianti colla potenza x^n, ed inversamente, se le funzioni $p_n(x)$ sono tali serie di potenze, le $q_n(x)$ sono polinomi di grado n in x» (Pincherle 1882a, 225).

Come applicazioni di questi teoremi si potevano considerare le funzioni sferiche di 1° e 2° specie, le funzioni di Bessel con le loro coniugate (Neumann 1867) o le funzioni di Lamé.

La generalizzazione di questi risultati portava Pincherle a studiare le condizioni di convergenza di serie della forma

$$\sum_{n=0} \varphi_n(x) f_n(x)$$

dove le $\varphi_n(x)$ costituiscono un sistema di funzioni analitiche tali che la serie $\sum \varphi_n(x)$ converge incondizionatamente e uniformemente in un dominio J e le $f_n(x)$ sono funzioni analitiche definite in un dominio connesso A del piano complesso e poi, «le condizioni sotto cui date funzioni analitiche si possono sviluppare in serie di tale forma».

Questi studi prendevano corpo in una memoria (Pincherle 1882b), che costituisce il suo primo contributo di rilievo alla teoria delle funzioni analitiche.

Pincherle cominciava col considerare una successione infinita di funzioni $f_n(x)$ meromorfe (a «carattere razionale») in un dominio connesso A del piano complesso. Se $i_n, i'_n, \ldots,$ $i_n^{(r)}$ erano i poli delle funzioni $f_n(x), n \in \mathbb{N}$ e j i loro «punti limite», Pincherle considerava il dominio J (non necessariamente connesso) ottenuto togliendo da A i punti j mediante cerchi di raggio piccolo a piacere oppure striscie sottili quanto si vuole (a seconda che i punti j fossero rispettivamente isolati o costituenti linee). Tolti con una tecnica analoga da J i restanti punti i_n (isolati e in numero finito) Pincherle otteneva un dominio K nel quale le funzioni «hanno carattere razionale intero».

A questo punto egli intraprendeva sulle funzioni $f_n(x)$ «una ricerca importante» che presentava «difficoltà non lievi» e per la quale mancavano «considerazioni generali»: che cosa si poteva dire del limite

$$(x \in K) \quad \lim_{n \to \infty} f_n(x) \ ?$$

Per cercare di aggirare la difficoltà Pincherle assoggettava la successione delle $f_n(x)$ a delle condizioni, introducendo «peut-être pour la première fois le concept de "système de fonctions limitées" dans "leur ensemble", depuis si communu» (Pincherle 1925, 341).

Le funzioni $f_n(x)$ – affermava Pincherle (1882b, 66) – «rimangono finite» per il valore x_o di x se si può assegnare un $N > 0$ tale che, per ogni $n, |f_n(x_o)| < N$. Chiamato $m =$ inf N, si poteva dire in maniera analoga che le funzioni $f_n(x)$ rimangono finite nell'intorno di un punto x_o «se i numeri m corrispondenti ai vari punti dell'intorno hanno un limite superiore finito» $L(x_o)$. Infine, le funzioni rimangono finite entro tutto un campo C se si può assegnare un $N > 0$ tale che, per ogni $x \in C$ e per ogni $n, |f_n(x)| < N$.

L'esser finite nell'intorno di ogni punto di un campo connesso C e del contorno comportava per le funzioni $f_n(x)$ di essere finite in tutto C; un'osservazione, faceva notare Pincherle, che discendeva come caso particolare dalla «proposizione generale» seguente, «non inutile da enunciare benchè ovvia»: «Se ad ogni punto x_o di un campo connesso C e del contorno, corrisponde un valore ed uno solo di una quantità X (funzione di x nel senso più generale della parola) e se si può assegnare un tale intorno di x_o che il limite superiore dei valori assoluti di X corrispondenti ai punti dell'intorno sia un numero finito $L(x_o)$, esisterà un numero N tale che sia in tutto il campo $|X| < N$» (Pincherle 1882b, 67).

Pincherle si limitava ad osservare in nota che il teorema di Dini «sulla continuità uniforme di una funzione continua in tutti punti di un campo ad una dimensione» e quello di Weierstrass che una serie uniformemente convergente nell'intorno di tutti i punti di un dato campo è uniformemente convergente in quel campo, «sono casi particolari», di quella proposizione generale. Solo nella «Notice sur les travaux» scritta nel 1925 per gli *Acta Mathematica* rivendicava l'originalità e l'importanza di quel suo teorema «qui correspond, pour les aires planes, à la célèbre proposition de Heine-Borel» (Pincherle 1925, 341).

Il risultato principale stabilito da Pincherle nella sua memoria era la generalizzazione di un teorema dato da Weierstrass (1880, 221).

Nell'ipotesi che le $f_n(x)$ fossero limitate nel dominio J (escluso al più un numero finito di esse) egli considerava infatti una seconda successione di funzioni $\varphi_n(x)$ regolari in tutto J e tali che la serie

$$\sum_{n=0} \varphi_n(x)$$

fosse uniformemente e incondizionatamente convergente nell'intorno di tutti i punti di J. Allora si poteva dimostrare che: «La serie

$$S(x) = \sum_{n=0} \varphi_n(x) f_n(x)$$

rappresenta entro tutto il campo J un ramo ad un valore di funzione analitica monogena, se J è connesso, e se J è sconnesso, tanti rami di funzioni analitiche (anche diverse) quanti sono i pezzi separati di cui è composto J : le quali funzioni non hanno nei punti di J singolarità essenziali, ma al più un numero finito d'infiniti ordinari nei punti dove divengono infinite le funzioni $f_n(x)$ » (Pincherle 1882b, 68-69).

Nel resto della memoria Pincherle studiava alcune applicazioni del suo teorema, facendo prima delle ipotesi opportune sulla distribuzione dei poli delle $f_n(x)$, e poi considerando tra gli altri il caso in cui $\varphi_n(x) = c_n x^n$ e le $f_n(x)$ sono delle funzioni regolari in un intorno dell'origine, ottenendo così gli sviluppi in serie di funzioni di Bessel e le serie di Lambert generalizzate.

5. Insieme agli studi preparatori dei propri lavori originali, nei quaderni di appunti Pincherle annotava le osservazioni che gli venivano dalla lettura di articoli di altri matematici. Tutto ciò consente di avere un'idea molto precisa dei successivi momenti del suo lavoro di ricerca.

Così, un articolo di Lindemann (1882) sullo sviluppo delle funzioni di variabile complessa in funzioni di Lamè suggerisce a Pincherle che a tale sviluppo si può associare un sistema di curve del 4° ordine dal quale si possono ottenere le curve di Cassini «mediante un processo di passaggi al limite che spero mostrare in una prossima occasione. Dalla presenza di queste ovali segue che lo sviluppo in serie di funzioni di Lamè per funzioni di variabile complessa si ottiene col teorema di Cauchy, ma dà luogo ad una disciplina più estesa che per le serie di funzioni sferiche». Di queste «considerazioni» si troverà traccia in (Pincherle 1883-84b).

Tra le più «recenti memorie d'analisi» studiate in quel periodo da Pincherle figura anche una di Hermite (1880) sulla determinazione di una funzione i cui punti singolari coincidano con gli elementi di una data successione – una questione affrontata da Weierstrass e Mittag-Leffler (Bottazzini 1986, 281-285) – e una di Poincaré «sur les fonctions à éspaces lacunaires», un argomento anch'esso sollevato da Weierstrass. E' stato questi infatti, scrive Pincherle, a richiamare l'attenzione «sulle funzioni aventi uno spazio lacunare, cioè che non si possono *continuare* in un certo campo».

A Poincaré comunque si rivolgeva Pincherle con una lunga lettera (10 giugno 1882) che merita di essere letta per esteso. Pincherle infatti delineava una sorta di programma degli argomenti da trattare nel corso di analisi superiore di cui era incaricato, e al tempo stesso accennava ad alcuni temi delle sue future ricerche.

Elencando i «principaux problèmes dont s'occupent actuellement les spécialistes de la théorie générale des fonctions» Pincherle ci fornisce inoltre una rassegna dei problemi aperti in quel periodo.

«Il me semble – esordiva Pincherle (in: Dugac 1989, 211) – que les principaux problèmes qui forment l'objet de la théorie générale des fonctions puissent se partager en quatre grandes classes:

A. "Etant donné un élément de fonction analytique, reconnaître les propriétés de la fonction qu'il définit".

On sait (Weierstrass) qu'un élément de fonction sert à définir la fonction dans tout le champ de sa validité; et l'on obtient la valeur, ou les valeurs, de la fonction pour tous les points de ce champ au moyen de la continuation (*Fortsetzung*) de proche en proche. Mais cette méthode est peu pratique pour faire connaître:

1° Les limites du champ de validité.

2° Si la fonction est uniforme ou multiforme.

3° Si elle satisfait à équation algébrique, ou à une équation algébrico-différentielle, ou si elle appartient à une classe connue de fonctions.

Il semble donc que l'un des principaux problèms de la théorie des fonctions devrait être le suivant: «Reconnaître, par la loi des coefficients de l'élément, les trois caractères sus-énoncés dans la fonction».

J'ignore si ce problème a été résolu, hors le cas des *séries récurrentes*, et des *séries hypergéométriques*.

B. "Quelles sont les fonctions qui peuvent s'exprimer au moyen de *formes arithmétiques* déterminées"?

Une forme arithmétique, qui ne contient que les 4 opérations en nombre fini, représente une fonction rationnelle et ne donne lieu à aucune observation. Mais si la forme contient des opérations en nombre infini (séries, produits infinis ou fractions continues), on se trouve en présence des problèmes de cette seconde classe: "Convergence ou divergence de la forme arithmétique. Champ de convergence, à une ou deux dimensions. Si le champ est à deux dimensions, la forme représente-t-elle une fonction analytique, et dans quel cas"? Ces questions sont résolues en partie pour les séries; *j'ignore* si elles ont été traitées pour les produits infinis ou les fractions continues (sauf la première).

Pour les séries, la convergence *au même degré (gleichmässig)* donne un critérium pour reconnaître si la série représente une fonction analytique, quand en outre le champ de convergence est à deux dimension (Weierstrass: *Zur Functionenlehre*), mais c'est une condition suffisante, et non nécessaire. Je ne crois pas que cette condition ait été étendue à d'autres formes arithmétiques que les séries.

A cette classe se rattachent encore les problèmes ayant pour but la construction de fonctions avec les zéros ou des infinis déterminés: résolus par Betti, Weierstrass, Mittag-Leffler, et récemment étendus aux fonctions multiformes par Picard.

Un des résultats les plus importants obtenus dans cette classe de problèmes est le fait analytique qu'une seule et même forme arithmétique peut représenter deux fonctions analytiques différentes: c'est-à-dire qui ne peuvent se déduire l'une de l'autre par *continuation*.

Enfin à cette même classe de problèmes se rattache l'étude des séries de la forme $\sum a_n P_n(x)$, où les fonctions $P_n(x)$ constituent un système donné (Séries de Fourier, de fonctions sphériques, de Frobenius, de Lindemann, etc.). En particulier, en quels cas ces séries peuvent-elles être des développements de zéro.

C. "Recherche des fonctions qui satisfont à une propriété donnée, dans tout le champ de validité".

Cette classe de problèmes comprend la résolution des équations finies (fonctions implicites), des équations différentielles, et des équations fonctionnelles; et, d'abord, la démonstration de la possibilité de la solution au moyen des fonctions analytiques (Cauchy et Weierstrass).

D. Enfin, les problèmes de la dernière classe sont ceux qui regardent la manière d'être de la fonction aux limites de son champ de validité; ainsi:

"Manière d'être d'une fonction uniforme dans le voisinage d'un point singulier (Weierstrass).

Manière d'être d'une fonction multiforme dans le voisinage d'un point de diramation (Riemann).

Détermination d'une fonction uniforme ayant un nombre donné de points singuliers (Weierstrass).

Fonctions à espaces lacunaires (Weierstrass, Poincaré).

Manière d'être d'une fonction le long d'une ligne, soit un contour du champ de validité, soit la circonférence de convergence de l'un de ses éléments; d'où les fonctions de variable réelle (Dirichlet, Dini Cantor)"».

La risposta di Poincaré (conservata tra le pagine del «Quaderno» di Pincherle) non si faceva attendere.

«Je vous remercie beaucoup de votre interessante lettre et des aperçus nouveaux et ingénieux qu'elle renferme» – scriveva Poincaré a Pincherle il 15 giugno 1882. «C'est bien ainsi, ce me semble, qu'il convient d'exposer la théorie générale des fonctions si l'on veut bien faire comprendre le véritable sens del problèmes qu'on a à traiter.

Le problème qui consiste à reconnaître, d'après les coefficients d'un développement en série de puissances, quelles sont les propriétés essentielles de la fonction représentée par ce développement est loin d'être résolu, comme vous le faites fort bien remarquer et il y a encore beaucoup à faire dans ce sens.

Vous citez le cas des séries récurrentes et celui des séries hypergéométriques; je pense que vous comprenez sous ce dernier nom, non seulement la série de Gauss, mais toutes les séries represéntant des intégrales d'équations différentielles linéaires à coefficients rationnels; il y a en effet entre p coefficients consécutifs d'une pareille série (tout à fait analogue à la série de Gauss) une relation linéaire de récurrence dans laquelle entre le rang n du premier de ces p coefficients. Voilà donc une condition qui permet de reconnaître d'après la loi des coefficients si la série satisfait à une équation linéaire; et par conséquent *si elle représente une fonction algébrique.*

Il y a aussi des cas où la loi des coefficients montre immédiatement quel est le champ de validité de la fonction; je ne parle pas seulement ici du cas simple des séries convergentes dans tout le plan; mais des séries telles que celles-ci

$$\sum \frac{x^{3^n}}{2^n} \quad \text{ou} \quad \sum \varphi_p(n) x^n,$$

où $\varphi_p(n)$ représente la somme des puissances $p^{\text{ièmes}}$ des diviseurs de n. On voit immédiatement en effet que le champ de validité est le cercle de rayon 1 et de centre 0.

Je passe au second de vos problèmes: étant donné un développement en série, ou un produit infini, ou une fraction continue, reconnaître si ce développement représente une fonction analytique. Le cas du produit infini se ramène aisément à celui de la série traité par Weierstrass; il suffit de passer aux logarithmes. Quant au cas des fractions continues, je ne crois pas qu'il ait été approfondi comme il mériterait de l'être.

Il est encore une autre classe de problèmes qui sont un peu différents de ceux dont vous parlez, ce sont ceux qui se rattachent à l'*aenliche Abbildung* d'une contour sur un autre et au principe de Dirichlet.

Malheureusement beaucoup de ces problèmes ont été longtemps traités sans une rigueur suffisante, mais on en trouve une solution rigoureuse dans les *Monatsberichte* de l'Académie de Berlin, octobre 1870, page 767 et suivantes, dans un mémoire de M. Schwarz».

6. Nell'agosto del 1882 Pincherle cominciava la stesura di una memoria sui sistemi di funzioni analitiche (1883-84a), intimamente connessa ad una delle questioni che egli aveva sollevato nella sua lettera a Poincaré. Egli prendeva le mosse dal teorema di Neumann che ogni funzione analitica regolare entro un'ellisse di fuochi ± 1 è ivi sviluppabile (escluso il contorno) in serie di funzioni sferiche incondizionatamente e uniformemente convergente.

«Ce théorème – osservava Pincherle (1925, 342) – doit dépendre d'une propriété générale qui met en rapport les courbes de convergence des séries ordonnées suivant les fonctions d'une suite donnée $f_n(x)$ avec la nature de la fonction génératrice des $f_n(x)$». Questa era la proprietà che egli cercava di mettere in luce nel suo lavoro, generalizzando il risultato di Neumann.

Egli cominciava col considerare una funzione di due variabili

$$T(x,y) = \sum_m \sum_n a_{m,n}(x-x_o)^m y^n$$

le cui singolarità soddisfano un'equazione $f(x,y) = 0$, con f intera. Egli mostrava poi che le curve entro cui converge la serie

$$\sum c_n p_n(x) \quad \text{con } p_n(x) = \sum a_{m,n}(x-x_o)^m$$

dipendono dai moduli delle radici di $f(x,y) = 0$. Come caso particolare si otteneva poi il teorema di Neumann, quando $f(x,y) = 1 - 2xy + y^2$.

Riprendendo poi i propri risultati (Pincherle 1882a) nell'ultima parte della memoria egli considerava quei sistemi di funzioni $p_n(x)$ «di speciale importanza» per i quali si poteva determinare, entro un opportuno dominio, un sistema associato $P_n(x)$ tale che:

$$\sum p_n(x) P_n(x) = \frac{1}{y-x}. \tag{1}$$

In questo caso infatti il teorema integrale di Cauchy consente di determinare lo sviluppo in serie di funzioni $p_n(x)$ o $P_n(x)$ per qualunque funzione regolare entro il dominio di validità della (1).

Pincherle risolveva il problema dello sviluppo con un proprio metodo originale, ricorrendo a dei sistemi di relazioni «qui représentent une première extension à l'infini de la théorie des déterminants» (Pincherle 1925, 342).

Pubblicata questa memoria nell'aprile dell'anno successivo, egli si affrettava a inviarne a Poincaré un estratto, annunciandogli al tempo stesso le linee di sviluppo delle sue ricerche (e di fatto il contenuto essenziale di (Pincherle 1883-84b)):

«Comme je le dis dans ce travail – scrive Pincherle in una lettera dell'11 novembre 1883 – je me propose d'examiner dans des Mémoires à suivre:

1° Ce qu'on peut dire en général sur le développement de zéro (*Nullentwickelung*)

$$\sum c_n p_n(x) = 0.$$

2° Si, étant donné un systéme $p_n(x)$, on peut énoncer les conditions sous les quelles une fonction $f(x)$, (holomorphe si les $p_n(x)$ le sont dans un certain champ, ou ayant des singularités dépendant de celles des $p_n(x)$) peut se développer en série $\sum c_n p_n(x)$.

Á la premiére demande je donne une réponse qui me semble satisfaisante, et qui fait dépendre le nombre et la nature des *Nullentwickelung* de certains points singuliers: les dévelopements de ce genre, trouvés par M. Frobenius (Crelle, t. 73) y rentrent parfaitement comme cas particuliers. Quant à la seconde, elle me semble difficile en général, toutefois j'espère arriver à une réponse au moyen d'une généralisation de la formule de Cauchy

$$f(x) = \frac{1}{2\pi i} \int_c \frac{f(y)}{y-x} dy$$

qui consiste à substituer à $\frac{1}{y-x}$ une fonct [ion] convenable de x et y. (...) Ces résultats paraissent prochainement, j'éspère, dans les *Annali di Matematica*».

7. La pubblicazione delle due memorie sui sistemi di funzioni analitiche conclude il primo periodo di ricerche di Pincherle. Nei «Quaderni» dei suoi appunti si affacciano nuove problematiche, sollecitate dallo studio dei lavori di Cantor sugli insiemi di punti, pubblicati nel 1883 negli *Acta Mathematica*, della memoria di Dedekind e Weber (1882) sulle funzioni algebriche o di contemporanei articoli di Kronecker, Hurwitz e Mittag-Leffler.

La pubblicazione dell'opuscolo di Klein (1882) sulla teoria riemanniana delle funzioni algebriche fa scrivere a Pincherle che «per acquistare sode cognizioni in analisi, credo necessario uno studio approfondito sull'argomento della superficie di Riemann (che ha per applicazione la teoria delle funzioni algebriche, abeliane, fuchsiane, ecc.) coi metodi del Klein».

Un'ammissione notevole, se si pensa alla sua precedente produzione, improntata ai metodi di Weierstrass. In realtà, l'attenzione alla letteratura scientifica contemporanea sgombra da pregiudizi di «scuole» è la condizione che permette a Pincherle di elaborare una propria linea di ricerca, i cui motivi ispiratori egli aveva delineato nella lettera «programmatica» a Poincaré.

«Gli studi sulle funzioni – si legge infatti in un appunto dell'estate 1884 – possono essere di due specie: di valore o di forma. Gli studi di valore sono quelli d'indole più elementare, che si presentano per primi e nei quali si studiano le variazioni che avvengono nel valore di una variabile, e le conseguenti proprietà delle funzioni stesse.

Riscontrando proprietà più o meno analoghe in varie funzioni queste si riuniscono in classi aventi proprietà comuni, e lo studio di queste classi si può chiamare studio di *forma*. (...) In ultima analisi lo studio di forma sarà quello in cui, in luogo di entrare in giuoco i vari studi di una funzione, figura la funzione come un tutto, come un elemento della questione, e non figurano, almeno essenzialmente, i valori della variabile indipendente.

Richiamo la lettera al sig. Poincaré. Ogni problema di valore sulle funzioni, conduce a problemi di forma. Per esempio, dato un elemento di funzione analitica, studiare le proprietà della funzione rappresentata è un problema di valore che conduce al problema: quali funzioni si possono rappresentare mediante serie di una determinata natura? I problemi delle classi B e C di quella lettera sono di *forma*, D è invece una classi di problemi di valore, come pure l'integrazione di una certa equazione in un campo con date condizioni al contorno».

Questo criptico accenno annunciava l'esordio di Pincherle in un nuovo campo di ricerche che, condotte in diverse riprese per più di un anno, lo portavano nel novembre del 1885 alla stesura definitiva della sua prima memoria sulle operazioni funzionali (Pincherle 1886).

BIBLIOGRAFIA

E. BETTI, (1860-61), La teorica delle funzioni ellittiche, *Annali di Matematica pura e appl. 3*, 65-159; 298-310; *4*, 26-45; 57-70; 297-336 (= *Opere Matematiche*, vol. I, Milano, 1903, 228-412).

U. BOTTAZZINI, (1982), Enrico Betti e la formazione della scuola matematica pisana, in:*La storia delle matematiche in Italia. Atti del convegno*, a cura di O. Montaldo e L. Grugnetti, Università di Cagliari, 1982, 229-276.

U. BOTTAZZINI, (1986), *The higher calculus. A history of real and complex analysis from Euler to Weierstrass*, New York, Springer Verlag.

U. BOTTAZZINI, S. FRANCESCONI, (1989), Manuscript volumes and lecture notes of Salvatore Pincherle, *Historia Mathematica 16*, 379-380.

F. CASORATI, (1868), *Teorica delle funzioni di variabili complesse*, vol. I, Pavia.

U. DINI, (1870-71), Sopra le funzioni di una variabile complessa, *Annali di matematica pura e appl. (2), 4*, 159-174 (= *Opere*, a cura dell'UMI, vol. II, Roma 1954, 245-263).

U. DINI, (1871-73), Sull'integrazione dell'equazione $\Delta^2 u = 0$, *Annali di Matematica pura e appl. (2) 5*, 305-345 (= *Opere*, a cura dell'UMI, vol. II, Roma 1954, 264-310).

P. DUGAC, (1989), Henri Poincaré. La correspondance avec des matématiciens de J à Z, *Cahiers du Séminaire d'Histoire des Mathématiques, 10*, 83-230.

G. FROBENIUS, (1871), Über die Entwicklung analytischer Funktionen in Reihen, die nach gegebenen Funktionen fortschreiten, *Journ. reine angew. Mathematik 73*, 1-30.

F. KLEIN, (1882), *Über Riemanns Theorie der algebraischen Funktionen und ihrer Integrale*, Leipzig.

E. NEUENSCHWANDER, (1978), Der Nachlass von Casorati (1835-1890) in Pavia, *Archive Hist. Ex. Sciences 19*, 1-90.

C. NEUMANN, (1867), Über die Entwickelung beliebig gegebener Funktionen nach den Besselschen Funktionen, *Journ. reine angew. Mathematik 67*, 310-314.

S. PINCHERLE, (1874), Sulle superficie di capillarità, *Il Nuovo Cimento (2) 12*, 19-64.

S. PINCHERLE, (1875), Sulle costanti di capillarità, *Il Nuovo Cimento (2) 14*, 17-25.

S. PINCHERLE, (1876), Sopra alcuni problemi relativi alle superfici d'area minima, *Rend. Ist. Lombardo (2) 9*, 444-456.

S. PINCHERLE, (1877), Sulle equazioni algebrico-differenziali di primo ordine di primo grado a primitiva generale algebrica, *Rend. Ist. Lombardo (2) 10*, 143-152.

S. PINCHERLE, (1878), Relazioni fra i coefficienti e le radici di una funzione intera trascendente, *Rend. Ist. Lombardo (2) 11*, 381-398.

S. PINCHERLE, (1879), Sulle funzioni monodrome aventi un'equazione caratteristica, *Rend. Ist. Lombardo (2) 12*, 536-542.

S. PINCHERLE, (1880a), Ricerche sopra una classe importante di funzioni monodrome, *Giornale di matematiche 18*, 92-136.

S. PINCHERLE, (1880b), Saggio di una introduzione alla teoria delle funzioni analitiche secondo i princìpi del prof. C. Weierstrass, *Giornale di matematiche 18*, 178-254; 347-357.

S. PINCHERLE, (1882a), Alcuni teoremi sopra gli sviluppi in serie per funzioni analitiche, *Rend. Ist. Lombardo (2) 15*, 224-225.

S. PINCHERLE, (1882b), Sopra alcuni sviluppi in serie per funzioni analitiche, *Mem. Accad. Sci. Ist. Bologna (4) 3*, 149-180 (= *Opere scelte*, a cura dell'UMI, vol. I, Roma 1954, 64-91).

S. Pincherle, (1883-84a), Sui sistemi di funzioni analitiche e le serie formate coi medesimi. Memoria I, *Annali di Matematica pura e appl. (2) 12*, 11-41.

S. Pincherle, (1883-84b), Sui sistemi di funzioni analitiche e gli sviluppi in serie formati coi medesimi. Memoria II, *Annali di Matematica pura e appl. (2) 12*, 107-133.

S. Pincherle, (1886), Studi sopra alcune operazioni funzionali, *Mem. Accad. Sci. Ist. Bologna (4) 7*, 393-442 (= *Opere scelte*, a cura dell'UMI, vol. I, Roma 1954, 91-141).

S. Pincherle, (1925), Notice sur les travaux, *Acta Mathematica 46*, 341-362 (= *Opere scelte*, a cura dell'UMI, vol. I, Roma 1954, 45-63).

H. Poincaré, (1881a), Sur les fonctions fuchsiennes, *Compt. Rend. Acad. Sci. Paris 92*, 333-335 (= *Œuvres*, a cura di G. Darboux, vol. II, Paris 1916, 1-4).

H. Poincaré, (1881b), Sur les fonctions fuchsiennes, *Compt. Rend. Acad. Sci. Paris 92*, 395-398 (= *Œuvres*, a cura di G. Darboux, vol. II, Paris 1916, 5-7).

H.A. Schwarz, (1869a), Über einige Abbildungsaufgaben, *Journ. reine angew. Mathematik 70*, 105-120 (= *Ges. Math. Abh.*, vol. II, 65-83).

H.A. Schwarz, (1869b), Notizia sulla rappresentazione conforme di un'area ellittica sopra un'area circolare, *Annali di Matematica pura e appl. (2) 3*, 166-170 (= *Ges. Math. Abh.*, vol. II, 102-107).

H.A. Schwarz, (1870a), Über die Integration der partiellen Differentialgleichung $\frac{\partial^2 u}{\partial x^2} + \frac{\partial^2 u}{\partial y^2} = 0$ für die Fläche eines Kreises, *Vierteljahrschrift der Naturforschenden Gesellschaft in Zürich 15*, 113-128 (= *Ges. Math. Abh.*, vol. II, 175-190).

H.A. Schwarz, (1870b), Über einen Grenzübergang durch alternirendes Verfähren, *Vierteljahrschrift der Naturforschenden Gesellschaft in Zürich 15*, 272-286 (= *Ges. Math. Abh.*, vol. II, 133-143).

H.A. Schwarz, (1870c), Über die Integration der partiellen Differentialgleichung $\frac{\partial^2 u}{\partial x^2} + \frac{\partial^2 u}{\partial y^2} = 0$ unter vorgeschriebenen Grenz- und Unstetigkeitsbedingungen, *Monatsberichte der K. Akad. der Wiss. Berlin*, 767-795 (= *Ges. Math. Abh.*, vol. II, 144-171).

H.A. Schwarz, (1872), Zur Integration der partiellen Differentialgleichung $\frac{\partial^2 u}{\partial x^2} + \frac{\partial^2 u}{\partial y^2} = 0$ *Journ. reine angew. Mathematik 74*, 21-253 (= *Ges. Math. Abh.*, vol. II, 175-210).

H. Weber, (1869), Note zu Riemanns Beweis des Dirichletschen Princips, *Journ. reine angew. Mathematik 71*, 29-39.

K. Weierstrass, (1854), Zur Theorie der Abel'schen Funktionen, *Journ. reine angew. Mathematik 47*, 289-306 (= *Math. Werke*, vol. I, 133-152).

K. Weierstrass, (1856), Theorie der Abel'schen Funktionen, *Journ. reine angew. Mathematik 52*, 285-339 (= *Math. Werke*, vol. I, 297-355).

K. Weierstrass, (1870), Über das sogenannte Dirichlet'sche Princip, in: *Math. Werke*, vol. II, 49-54.

K. Weierstrass, (1876), Zur Theorie der eindeutigen analytischen Funktionen, *Abh. der K. Akad. der Wiss. Berlin*, 11-66 (= *Math. Werke*, vol. II, 77-124).

K. Weierstrass, (1880), Zur Funktionenlehre, *Monatsberichte der K. Akad. der Wiss. Berlin*, 719-743 (= *Math. Werke*, vol. II, 201-230).

Linear Systems: Developments of Some Results by F. Enriques and B. Segre

MARIA VIRGINIA CATALISANO   Dipartimento di Matematica, Università di Genova, Genova, Italy.

SILVIO GRECO   Dipartimento di Matematica, Politecnico di Torino, Torino, Italy.

INTRODUCTION

This paper is intended as a tribute to the outstanding algebraic geometers F. Enriques and B.Segre, who have been working for many years at the University of Bologna; it consists in a survey report on some recent results on linear systems of plane curves inspired by or related with a minor but interesting part of their work, and shows how one can find plenty of ideas also in little things.

Our starting points are two theorems of B. Segre, one relating complete intersections and postulation, and the other dealing with regularity of linear systems; and one by Enriques on projections of space curves. Fortunately these results, besides some pretty ideas, contain some obscure points, and thus, as we shall see, have given a lot of inspiration for further developements.

Work done under the auspices of "Progetto Nazionale Geometria Algebrica e Algebra Commutativa", supported by the Italian M.P.I.

We shall dicuss here the above mentioned results, and give a few related results taken from some papers either in print, or in preparation, trying to give some ideas of the techniques used in the proofs.

1 PRELIMINARIES

Let $\mathbb{P}^2$ be the projective plane over an algebraically closed field k and let Z be a zero-dimensional subscheme of $\mathbb{P}^2$, corresponding to the homogeneous ideal $\alpha \subset R = k[X_0, X_1, X_2]$, and to the ideal sheaf $I \subset \mathcal{O}_{\mathbb{P}^2}$. Let S_t be the linear system of all plane curves of degree t containing Z as a subscheme (so that S_t corresponds to the subspace $H^0(\mathbb{P}^2, I(t))$ of $H^0(\mathbb{P}^2, \mathcal{O}_{\mathbb{P}^2}(t))$ and to the homogeneous part α_t of α).

A classical problem is to study the properties of S_t in terms of the geometrical properties of Z, see for example the outstanding paper by Castelnuovo [Ca].

The first thing one wants to know is the *dimension* of S_t defined as

$$\dim S_t := h^0(I(t)) - 1 = \dim_k \alpha_t - 1.$$

From the exact sequence $0 \to I \to \mathcal{O}_{\mathbb{P}^2} \to \mathcal{O}_Z \to 0$ twisting by t and taking cohomology one gets

$$h^0(I(t)) = (t+1)(t+2)/2 - \delta + h^1(I(t))$$

where $\delta := h^0(\mathcal{O}_Z)$ is the *degree* of Z. It follows:

$$\dim S_t = t(t+3)/2 - \delta + h^1(I(t)).$$

The number

$$\text{vir.dim } S_t := t(t+3)/2 - \delta$$

is called *virtual dimension* of S_t, while $h^1(I(t))$ is called *superabundance* of S_t. Finally the function

$$H(Z,t) := \dim_k R_t/\alpha_t = (t+1)(t+2)/2 - h^0(I(t))$$

is called the *Hilbert function* of Z. Note that $H(Z,t)$ is just "the number of conditions that Z imposes to the curves of degree t". So to compute the dimension of S_t is equivalent to compute either its superabundance or $H(Z,t)$.

The linear system S_t is said to be *regular* if its dimension is equal to its virtual dimension, i.e.

if $h^1(I(t)) = 0$; and this happens for all sufficiently large t's. Then the first thing to try in order to get control on $\dim S_t$ is to find upper bounds for the integer

$$\tau = \tau(Z) := \min \{t : S_t \text{ is regular}\}.$$

The number $\tau + 1$ (often denoted by σ in the literature) is the least integer for which the difference function of $H(Z,t)$ vanishes.

An interesting particular case of the above situation is the linear system consisting of all curves of degree t having at certain fixed distinct points $P_1, \ldots, P_s$ multiplicities $m_1, \ldots, m_s$ at least. This corresponds to the homogeneous ideal $\cap \wp_i^{m_i}$, where $\wp_i$ is the homogeneous prime ideal corresponding to P_i $(i = 1, \ldots, s)$. The corresponding subscheme Z is also called the subscheme of *fat points* $(P_1, \ldots, P_s; m_1, \ldots, m_s)$, and its degree is $\delta = \sum_1^s m_i(m_i+1)/2$.

Another interesting case is the linear system of the adjoints of degree t to an irreducible plane curve Γ: it consists of all curves of degree t verifying the linear conditions of "passing throuh each e-fold point of Γ (either actual or infinitely near) with multiplicity $e - 1$ at least". The corresponding ideal is the pull-back to $\mathbb{P}^2$ of the conductor of Γ. These linear systems provide a powerful tool to study the linear series on the normalization of Γ.

2 POSTULATION AND COMPLETE INTERSECTIONS

In this section we discuss the first result of B.Segre mentioned in the introduction.

Assume Z is the complete (scheme theoretical) intersection of two curves of degrees e and d with no common components. Then the following hold:

(a) (Bezout) $\delta = ed$;
(b) (Jacobi) S_{e+d-3} is non-regular.

A natural question is to see to which extent (a) and (b) imply that Z be a complete intersection of two curves of degree e and d (easy examples show that this is not always the case, e.g., Z consisting of 8 distinct points, 5 of which on a line and the remaining 3 on a different line). Segre's answer [S1] is:

THEOREM A Assume Z is a closed subscheme of a non-singular curve C of degree $d \geq 3$, and assume (a) and (b) hold. Then there exists a curve C' of degree e such that $Z = C \cap C'$.
Proof (Segre). Consider the linear series σ cut out on C by S_{e+d-3} outside of Z (σ corresponds to $\mathcal{O}_C((e+d-3)H - Z)$, where H is a hyperplane section). By a direct and easy calculation one shows that σ is a g_{2p-2}^{p-1} where p is the genus of C. Then σ is the canonical series and hence any divisor D of σ is the complete intersection of C with a curve of degree $d - 3$. Since $D + Z$ is the complete intersection of C with a curve of degree $e+d-3$, the conclusion follows by Noether's Restsatz. ∎

This theorem and its proof have caused a lot of further developements expecially in view of the following remarks.

REMARKS 2.1 (i) The original statement of Theorem A does not contain the assumption "C non-singular" but only "C integral", and it is clear that the proof does not work in this generality; the best one can get is to assume that Z be a Cartier divisor on C and use the theory of Cartier divisors. So it is natural to ask whether Segre's original statement is correct, and, in particular if (a) and (b) imply that Z is a Cartier divisor.

(ii) The theorem points out a connection between the geometry of Z as a subscheme of C (i.e., being a complete intersection on C) and its postulation (non-regularity of S_{e+d-3}). So it is natural to ask if similar connections do hold in more general settings.

We shall mention a few developements.

REMARK 2.2 The first complete proof of Theorem A is due to E.Davis [D1], and is based on purely algebraic properties of the Hilbert function, developed by various authors (see, e.g., [GMR]). From this one gets "a posteriori" that Z is indeed a Cartier divisor on C.

The proof given in [D1] has been generalized to the case when Z is an Arithmetically Cohen Macaulay one-codimensional subscheme of a (not necessarily irreducible) projective hypersurface, see [D2]; for a better explanation of the methods of proof see also [D3].

In [G3] (see also [G1] for a preliminary version) it is shown how to fill-up the gap in Segre's proof. The main result is the following

THEOREM 2.3 Let C be a complete integral curve in $\mathbb{P}^r$, of degree d and arithmetic genus g, and let Z be a zero-dimensional subscheme of C, corresponding to the ideal sheaf $J \subset \mathcal{O}_C$. Put

$$\lambda = \min \{h^0(\text{coker } f); \quad f: J \to \omega_C \text{ injective}\},$$

and assume $n > (\delta - \lambda + 2g - 2)/d$. Then $h^1(J(n)) = 0$.

Proof. If $h^1(J(n)) \neq 0$ by duality on C we get $\text{Hom}(J, \omega_C) \neq 0$, whence an exact sequence $0 \to J \to \omega_C \to F \to 0$; then by standard facts involving the Euler characteristic and Riemann-Roch one gets a contradiction. ∎

REMARK 2.4 The above theorem can be used to check the vanishing of $h^1(I_Z(n))$, where $I_Z \subset \mathcal{O}_{\mathbb{P}^r}$ is the ideal sheaf of Z, via the exact sequence $0 \to I_C \to I_Z \to J \to 0$; of course one needs assumptions on the vanishing of $h^1(I_C(n))$, and methods to compute λ. The first assumption is verified in a number of interesting situations (Arithmetically Cohen-Macaulay curves, in particular complete intersections, e.g., plane curves; or in general for $n \geq d - r + 1$, by the Castelnuovo Theorem as generalized in [GLP]). The evaluation of λ is possible in some cases. For example, if ω_C is invertible, we have (l.cit.):

(1) $\lambda = 0$ if and only if Z is a Cartier divisor;

(2) if Z is reduced, then λ is the number of singular points of C belonging to Z;

Linear Systems

(3) if Z is a scheme of fat points $(P_1, \ldots, P_s; m_1, \ldots, m_s)$ in the plane, and if C is an integral curve having at P_i an ordinary m_i-fold point, then $\delta - \lambda = \sum_1^s m_i$.

Now from the example (1) above and Theorem 2.3 it follows that the assumptions (a) and (b) of Theorem A do imply that Z is indeed a Cartier divisor. Then Segre's proof gives the following

THEOREM A' Let $C \subset \mathbb{P}^r$ be an integral curve of degree d which is the complete intersection of $r-1$ hypersurfaces of degrees $a_1, \ldots, a_{r-1}$, and put $a = \sum a_i$. Let Z be a closed subscheme of C having degree δ. Then the following are equivalent:

(i) Z is the (scheme-theoretic) intersection of C with a hypersurface of degree e;

(ii) $\delta = de$, and $h^1(I_Z(e+a-r-1)) \neq 0$. ∎

Another application of Theorem 2.3 (using 2.4.(3)) is the following

THEOREM 2.5 ([G3],Th.4.2) Consider in $\mathbb{P}^2$ the scheme of fat points $(P_1, \ldots, P_s; m_1, \ldots, m_s)$, and assume that there is an integral curve of degree d having at each P_i an ordinary point of multiplicity m_i. Then $\tau \leq [\sum m_i/d] + d - 2$. ∎

REMARK 2.6 Theorems 2.3 and A' can be generalized to one-codimensional subschemes of integral complete intersections in $\mathbb{P}^r$, see [G2]. These results are somehow complementary to the ones mentioned in Remark 2.2.

3 FAT POINTS NO THREE ON A LINE AND SEGRE'S BOUND FOR τ

Let us consider now a scheme Z of fat points $(P_1, \ldots, P_s; m_1, \ldots, m_s)$ in $\mathbb{P}^2$ (see preliminaries). The second theorem of B.Segre we want to discuss gives a bound for the regularity of S_t.

THEOREM B ([S2],VIII) Assume the P_i's are "generic", and $m_1 \geq m_2 \geq \ldots \geq m_s > 0$ $(s \geq 2)$. Then

$$\tau \leq \max(m_1 + m_2 - 1, [\sum_1^s m_i / 2]).$$

Segre's proof is only sketched and consists of two steps.

Step 1. By letting the points "move into" an irreducible conic the bound for regularity cannot decrease: so it is sufficient to prove the theorem when the P_i's lie on an irreducible conic.

Step 2. One proves the theorem when the points lie on an irreducible conic by induction, comparing the given system "with the new one (either effective or virtual) obtained by taking away the conic". ∎

Now the proof of step 2 can be achieved without much trouble (see Prop. 3.1 below), but step 1 does not seem so obvious. However Theorem B is correct: we shall indeed describe four different proofs of it, and several further developements. We begin with

PROPOSITION 3.1 Let $m_1 \geq \ldots \geq m_s \geq 0$, $s \geq 2$. If the P_i's are distinct points of a non-singular conic C, then $\tau \leq t := \max(m_1 + m_2 - 1, [\sum_1^s m_i /2])$.

Proof. It suffices to prove that S_t is regular. We proceed by induction on $\sum_1^s m_i$.

Since the conclusion is obvious for $\sum_1^s m_i = 0$, we assume $\sum_1^s m_i \neq 0$. If either (a) $\sum_1^s m_i \leq 2m_1 + 2m_2 - 2$, or (b) $\sum_1^s m_i \geq 2m_1 + 2m_2 - 1$ and $\sum_1^s m_i$ is odd, then the line $P_1 P_2$ or the conic C, respectively, are fixed components for all curves of S_t. Let Z' denote the scheme $(P_1, \ldots, P_s; \max(m_1-1, 0), \max(m_2-1, 0), m_3, \ldots, m_s)$ in case (a), and the scheme $(P_1, \ldots, P_s; \max(m_1-1, 0), \ldots, \max(m_s-1, 0))$ in case (b). Let I' and δ' denote the sheaf associated to Z' and the degree of Z', respectively. In case (a) we have:

$h^1(I(t)) = h^0(I(t)) - (t+1)(t+2)/2 + \delta = h^0(I'(t-1)) - (t+1)(t+2)/2 + \delta = t(t+1)/2 - \delta' + h^1(I'(t-1)) - (t+1)(t+2)/2 + \delta = h^1(I'(t-1))$.

The conclusion follows from the inductive hypothesis. An analogous calculation works for case (b).

If (c) $\sum_1^s m_i$ is even and $\sum_1^s m_i \geq 2m_1 + 2m_2$ (so $m_4 > 0$), let Z' and I' be defined as in the previous case (b). By a straightforward calculation for $m_5 = 0$, and using the inductive hypothesis for $m_5 > 0$, it is easy to prove that $H(Z, t-1) = t(t+1)/2 - h^0(I(t-1)) = t(t+1)/2 - h^0(I'(t-3)) = \delta - 1$. Since the Hilbert function is strictly increasing up to the point at which it stabilizes, we get $H(Z, t) = \delta$, hence the conclusion. ∎

Now we describe the first two methods to complete the proof of Theorem B. They are based on the following theorem which can be of independent interest.

THEOREM 3.2 ([C1], Th.4.3) Assume that no three of the P_i's are collinear, and that they do not lie on a conic. Let $m_1 \geq \ldots \geq m_s > 0$. Let $q \geq \max(m_1 + m_2, [\sum_1^s m_i /2])$.

Then the generic curve of S_q is integral, and has at each P_i an ordinary singularity of multiplicity m_i. Moreover the characteristic series (see [Ca]) of S_q is non-special.

Proof. One can construct "many" reducible curves of S_q (starting with conics and lines, and using induction on $\sum_1^s m_i$), having the prescribed behaviour at the base points. Moreover these curves allow to apply Bertini's Theorem to show that the generic curve is integral (this is easy in characteristic zero, as the classical Bertini works; while in positive characteristic one can apply [J], 6.11.3, to show that the generic curve is irreducible, and observe that, because of the reducible curves described above, the generic curve is also reduced, being such at all the base points). Finally, if σ is the characteristic series, a direct calculation shows that $\deg \sigma \geq 2p-1$, where p is the genus of C. ∎

Now we restate Theorem B in a slightly more general form and we give the first two proofs of it.

THEOREM 3.3 Assume $s \geq 2$, no three of the P_i's are collinear, and $m_1 \geq \ldots \geq m_s \geq 0$. If $t := \max(m_1 + m_2 - 1, [\sum_1^s m_i /2])$, then S_n is regular for every $n \geq t$, i.e., $\tau \leq t$.

First proof of 3.3 (see [C1], Th.4.5) The proof is by induction on $\sum_1^s m_i$, and splits into three cases:

(a) if the P_i's lie on a conic, apply 3.1;

(b) if the P_i's do not lie on a conic, and $\sum_1^s m_i \geq 2m_1 + 2m_2$, the conclusion follows from Theorem 3.2, and a classical theorem of Castelnuovo ([Ca],p.162), stating that an irreducible linear system is regular iff its characteristic series is "non-special";

(c) if the P_i's do not lie on a conic, and $\sum_1^s m_i \leq 2m_1 + 2m_2 - 1$, then observe that the line $P_1 P_2$ is a fixed component for all curves of S_t. The conclusion follows by the inductive hypothesis. ∎

Second proof of 3.3 If the points lie on a conic use again 3.1. Otherwise use 3.2 and 2.5. ∎

Our third proof of Theorem B consists in proving step 1 of Segre's proof. This has been settled by G.Paxia [P]. The main result is:

THEOREM 3.4 For any s-tuple of fixed positive integers $m_1, \ldots, m_s$ there is an irreducible projective flat family containing all schemes of fat points $(P_1, \ldots, P_s; m_1, \ldots, m_s)$.
Proof. Consider the closed set $F \subset (\mathbb{P}^2)^s$ whose closed points are $(P_1, \ldots, P_s)$ with $P_i = P_j$ for some $i \neq j$, and let T be the complement of F. Let then V_i, $i = 1, \ldots, s$, be the closed set of $T \times \mathbb{P}^2$ whose closed points are $(P_1, \ldots, P_s, P)$ with $P = P_i$. Let J_i be the ideal sheaf of V_i considered as a reduced closed subscheme of $T \times \mathbb{P}^2$. Finally let V be the subscheme of $T \times \mathbb{P}^2$ corresponding to the ideal sheaf $\cap J_i^{m_i}$ and let $p: V \to T$ be the projection: one can show that this is the required family. (Remark: this family may contain each scheme more than once; to avoid this it is sufficient to "symmetryze" the factors of T corresponding to equal m_i's). ∎

COROLLARY 3.5 The integer valued function $\tau: T \to \mathbb{N}$ defined by $(P_1, \ldots, P_s) \to \tau[(P_1, \ldots, P_s; m_1, \ldots, m_s)]$ is upper semicontinuous.
Proof. It follows from 3.4 and the lower semicontinuity of the Hilbert function (which is, in turn, a consequence of the semicontinuity for cohomology, see, e.g., [BG] for details). ∎

Third proof of 3.3. It is clear that Corollary 3.5 settles step 1 of Segre's proof of Theorem B. Hence, using again Proposition 3.1, we have the conclusion. ∎

The fourth proof of Theorem B will be given in the next section.

Theorem B leads naturally to the following problems:

(a) find upper bounds for τ when the P_i's lie on an integral curve of degree $d > 2$;

(b) find the Hilbert function of fat points, when the P_i's lie on an integral curve of degree $d \geq 1$;

(c) find upper bounds for τ when the P_i's have a different kind of genericity.

In the next sections, we give some results about the problems above. More precisely, in section 4 we give bounds for τ when the P_i's are non-singular points of an integral curve of degree d (Th.4.3); in section 5 we discuss the case of points on a conic; in section 6 we find bounds for τ when the P_i's are in uniform position (Th.6.1), or they are generic points of $\mathbb{P}^2$ (Th.6.2).

4 FAT POINTS ON AN INTEGRAL CURVE OF DEGREE d

Under the assumption that $P_1, \ldots, P_k$ ($1 \leq k \leq s$) lie on an integral curve of degree d, the next theorem relates the superabundance $h^1(I(t))$ of S_t to $h^1(I'(t-d))$, where I' is the sheaf associated to a suitable subscheme of Z. So we get an induction algorithm for checking regularity.

THEOREM 4.1 ([C2],Th.3.1) Assume $P_1, \ldots, P_k$ ($k \leq s$) are distinct, non-singular points of an integral curve C of degree d, and P_i, $k < i \leq s$, are distinct points not on C.

Let $m_i > 0$ for every i ($1 \leq i \leq s$), and let I' denote the sheaf associated to the scheme $(P_1, \ldots, P_s; m_1 - 1, \ldots, m_k - 1, m_{k+1}, \ldots, m_s)$. Let D be a Cartier divisor of the linear series $tH \cdot C$, where tH is the linear system of all curves of degree t. Let E denote the Cartier divisor on C defined by $E = \sum_1^k m_i P_i$, and $i(D-E)$ the index of speciality of $D-E$. Then:

for $t \geq d$, $\quad i(D-E) \leq h^1(I(t)) \leq i(D-E) + h^1(I'(t-d));$

for $t < d$, $\quad i(D-E) \leq h^1(I(t)) + \frac{(d-t-1)(d-t-2)}{2} \leq i(D-E) + h^1(I'(t-d)).$

Sketch of proof. Let f be a polynomial defining C. Multiplication by f gives a short exact sequence $0 \to I'(-d) \to I \to J \to 0$. By a straightforward calculation, we find that J is canonically isomorphic to $\mathcal{O}_C(-E)$. Twisting by t, taking cohomology, and using the well-known values of cohomology of $\mathbb{P}^2$, as well as $i(D-E) = h^1(\mathcal{O}_C(D-E))$, we get the conclusion. ∎

As an obvious consequence, we have the next Corollary, that provides an inductive process to lower multiplicities.

COROLLARY 4.2 Notation and hypotheses as in 4.1, let g denote the arithmetic genus of C. If (i) $t \geq d$; (ii) $td - \sum_{i \leq k} m_i \geq 2g - 1$; (iii) $h^1(I'(t-d)) = 0$, then S_t is regular. ∎

As a first application, we give the fourth proof of Theorem 3.3.

Fourth proof of Theorem 3.3. By induction on $\sum_1^s m_i$. Obvious for $\sum_1^s m_i \leq 3$, assume $\sum_1^s m_i > 3$. For s=2 the conclusion is well-known. For $s > 2$, use 4.2 with C=L, where L is the line $P_1 P_2$, and the inductive hypothesis. ∎

Now we give bounds for τ when all the P_i's lie on C. This result ([C2],Th.4.1), is proved by induction on m_1, using 4.1 with k=s, and 4.2.

THEOREM 4.3 Let $P_1, \ldots, P_s$ be distinct non-singular points of an integral curve C of degree $d \leq s$, and let $m_1 \geq m_2 \geq \ldots \geq m_s > 0$. Set:

$v = \max\{ n \in \mathbb{N} : m_d = \ldots = m_n \};$

$x_1 = \sum_1^d m_i - 1;$ $\qquad\qquad\qquad\qquad\qquad t_1 = x_1;$

$x_2 = (\sum_1^s m_i + 1)/d + d - 3;$ $\qquad\quad t_2 = \min\{n \in \mathbb{N} : n \geq x_2\};$

$x_3 = (\sum_1^v m_i + (m_d - 1)(d^2 - v) + 1)/d + d - 3;$ $\quad t_3 = \min\{n \in \mathbb{N} : n \geq x_3\}.$

Then $\tau \leq t := \max(t_1, t_2, t_3)$. ∎

Linear Systems

In studying the sharpness of this bound, we find that it is closely related to the following problem of complete intersection.

PROBLEM 4.4 Let d, s, $m_1, \ldots, m_s$ be positive integers, let e, r be integers such that $\sum_1^s m_i = ed+r$ ($0 \leq r \leq d-1$). Do there exist an integral curve C of degree d, a curve C_e of degree e, simple distinct points $P_1, \ldots, P_s$ of C, and integers $m_1', \ldots, m_s'$ such that $0 \leq m_i' \leq m_i$ for every i, $\sum_1^s m_i' = ed$, and $C_d \cdot C_e = \sum_1^s m_i' P_i$?

We have the following result ([C2],Th.5.4), whose proof mainly relies upon finding suitable P_i's such that $i(D-E) > 0$, and applying 4.1.

THEOREM 4.5 Notation as in Theorem 4.3, let (*) denote the following case

(*) $\qquad s > d \geq 3 \qquad m_d > 1 \qquad t_2 > t_1 \qquad t_2 > t_3$.

Then:

(i) if we are not in case (*), then for any integral curve C of degree d, there exist simple distinct points $P_1, \ldots P_s$ on C, so that $\tau = t$, i.e., the bound of 4.3 is sharp.

(ii) Assume we are in case (*). If Problem 4.4 has an affirmative answer, then there exists an integral curve C of degree d, and simple distinct points $P_1, \ldots P_s$ on C, so that $\tau = t$. If moreover d divides $\sum_1^s m_i$, then also the converse is true.∎

REMARKS 4.6 (i) For $d=3$, in char $k = 0$, Problem 4.4 has an affirmative answer (see [C2],Rem.5.6). So the bound of Theorem 4.3 is sharp for $d=3$ (in char $k = 0$).

(ii) In case $s \leq d(d+3)/2$, $d \geq 3$, and $\sum_1^d m_i > 2d$, Gimigliano ([G],Th.1.1,Rem.1.8) shows that the bound $\tau \leq \sum_1^d m_i$ holds for simple distinct points of an integral curve C of degree d, unless $s = d(d+3)/2$, and $m_1 = \ldots = m_s \geq 2$. In this latter case, he finds $\tau \leq \sum_1^d m_i + 1$. Theorem 4.3 generalizes and improves Gimigliano's result. His method of proof consists in studying the geometry of the rational surface X obtained by blowing-up $\mathbb{P}^2$ at the points $P_1, \ldots, P_s$. The motivation for studying X is as follows. If E_i is the exceptional divisor in X over P_i, and E_0 is the strict transform of a generic line in $\mathbb{P}^2$, then the linear system $|D| = |tE_0 - m_1 E_1 - \ldots - m_s E_s|$ on X has the same dimension as $H^0(I(t))$. More precisely $h^i(I(t)) = h^i(\mathcal{O}_X(D))$, for every $i \geq 0$. So it is possible to study problems about linear systems of plane curves by studying the divisors $|D_t| = |tE_0 - m_1 E_1 - \ldots - m_s E_s|$ on X.

5 FAT POINTS ON A CONIC

We know (Prop.3.1) that if the P_i's lie on a non-singular conic C, then $\tau \leq t := \max(m_1 + m_2 - 1, [\sum_1^s m_i / 2])$. As a consequence of Proposition 3.1, we give (Th.5.1) an algorithm for computing the Hilbert function $H(Z, n)$ when the P_i's lie on C, from the Hilbert function of a suitable subscheme Z' of Z. Then, observing that for points on a conic not only $\tau \leq t$, but infact $\tau = t$ holds, we investigate when, if no three of the P_i's are collinear, $\tau = t$ implies that the P_i's lie on a conic. Finally we give an algorithm for constructing a minimal set of homogeneous generators of α.

PROPOSITION 5.1 Let $m_1 \geq \ldots \geq m_s \geq 0$, $s \geq 2$, and let the P_i's be distinct points of a non-singular conic C. Define Z' as follows:

(a) for $\sum_1^s m_i \leq 2m_1 + 2m_2 - 2$,
$$Z' = (P_1, \ldots, P_s \, ; \, \max(m_1-1, 0), \max(m_2-1, 0), m_3, \ldots, m_s);$$

(b) for $\sum_1^s m_i > 2m_1 + 2m_2 - 2$,
$$Z' = (P_1, \ldots, P_s \, ; \, \max(m_1-1, 0), \ldots, \max(m_s-1, 0)).$$

Set $t := \max(m_1 + m_2 - 1, [\sum_1^s m_i / 2])$. Then:

$$\begin{aligned}
H(Z, n) &= \delta & &\text{if } n \geq t; \\
&= n + 1 + H(Z', n-1) & &\text{if } 0 \leq n < t \text{ and } \sum_1^s m_i \leq 2m_1 + 2m_2 - 2; \\
&= 2n + 1 + H(Z', n-2) & &\text{if } 0 \leq n < t \text{ and } \sum_1^s m_i > 2m_1 + 2m_2 - 2.
\end{aligned}$$

(By convention, $H(Z, n) = 0$ for $n < 0$).

Proof. Obvious for $n \leq 1$, and by 3.1 for $n \geq t$. For $2 \leq n < t$ the conclusion follows by a straightforward calculation, by observing that either the line $P_1 P_2$, or the conic C is a fixed component for the curves of S_n. ∎

REMARKS 5.2 (i) It is worth noting that $H(Z, n)$ is completely determined by the m_i's, i.e., the actual disposition of the P_i's on C is unimportant. This is also true for points on a line, as it is proven by E.Davis and A.Geramita in [DG]. There they give a complete analysis of Hilbert functions of fat points on a line.

(ii) S.Giuffrida in [Gf], assuming no three of the P_i's lie on a line, and no six lie on a conic, computes the Hilbert function of $s \leq 6$ fat points. He also finds the least integer t such that the generic curve of S_t is integral.

(iii) By studying the geometry of suitable divisors on the blowing-up $X \to \mathbb{P}^2$ of $\mathbb{P}^2$ at the P_i's, as done by Gimigliano (Rem.4.6), B.Harbourne gives algorithms for computing the Hilbert function of fat points on an integral curve of degree $d \leq 3$. His algorithm is much more complicated if the points lie on an integral cubic [H1], than if the points lie on a line, or on a conic [H2]. Actually for points on a cubic $H(Z, n)$ does not depend only on the m_i's. In [H3], Harbourne is concerned with the Hilbert function of $s \leq 8$ fat points. Under further restrictions, i.e., assuming no three of the P_i's lie on a line, no six lie on a conic, and, in case $s=8$, there is no cubic through all eight points having a singularity at one of them (so the P_i's are smooth points of some irreducible reduced cubic), he proves that the Hilbert function is completely determined by the m_i's.

From 5.1, by an easy calculation, we get $H(Z, t-1) \leq \delta - 1$. It follows that

COROLLARY 5.3 Let $m_1 \geq \ldots \geq m_s \geq 0$, $s \geq 2$, and let the P_i's be distinct points of a non-singular conic C.

Then $\tau = t := \max(m_1 + m_2 - 1, [\sum_1^s m_i / 2])$. ∎

Linear Systems

It is classically known that if the points are on a line, then $\tau = \sum_1^s m_i - 1$, and that this implication can be reversed, i.e., if $\tau = \sum_1^s m_i - 1$, then the P_i 's are collinear (see, e.g., [DG]).

It is natural to ask if an analogous theorem holds for points on a conic. The answer is in the negative, but we have the following results ([C2],Th.6.3,Th.6.4).

THEOREM 5.4 Let $m_1 \geq \ldots \geq m_s > 0$, $s \geq 6$. Assume:
  (i) $\sum_1^s m_i$ is even and $t = \sum_1^s m_i /2$;
  (ii) no three of $P_1, \ldots, P_s$ are collinear;
  (iii) $\sum_1^s m_i \geq 2m_1 + 2m_2$;
  (iv) vir.dim $S_{t-1} \geq -1$;
  (v) $h^1(I(t-1)) > 0$.
Then $P_1, \ldots, P_s$ lie on a non-singular conic.

THEOREM 5.5 Let $m_1 \geq \ldots \geq m_s > 0$, $s \geq 6$. Assume:
  (i) $\sum_1^s m_i$ is odd and $t = (\sum_1^s m_i - 1)/2$;
  (ii) no three of $P_1, \ldots, P_s$ are collinear;
  (iii) $\sum_1^s m_i \geq 2m_1 + 2m_2 + 1$;
  (iv) vir.dim $S_{t-1} \geq -1$;
  (v) $h^1(I(t-1)) > 0$;
  (vi) $\sum_1^s m_i \geq 2m_1 + 7$;
  (vii) $m_i \geq 2$ for every i.
Then $P_1, \ldots, P_s$ lie on a non-singular conic.

Sketch of proofs. With regard to Theorem 5.4, in case $\sum_1^s m_i \geq 2m_1 + 2m_2 + 2$, if not all the P_i's lie on a conic, then by 4.2 we have $h^1(I(t-1)) = 0$, and the conclusion follows by contradiction. The proof continues now by induction on $\sum_1^s m_i$: observe that we may assume $m_2 > 1$, and apply 5.4 to $Z' = (P_1, \ldots, P_s ; m_1 - 1, m_2 - 1, m_3, \ldots, m_s)$. The conclusion follows by the inductive hypothesis.

The proof of 5.5 is similar. We only remark that in case $\sum_1^s m_i \geq 2m_1 + 2m_2 + 3$, to force $h^1(I(t-1)) = 0$, we have to use also 5.4.∎

REMARK 5.6 The hypotheses of Theorems 5.4 and 5.5 cannot be weakened. For each of the hypotheses (ii) through (v), (respectively (ii) through (vii)) it is possible to construct a set of points not on a non-singular conic for which all hypotheses, but that one, hold.

Now we are concerned with the construction of a minimal set of homogeneous generators of the homogeneous ideal $\alpha = \oplus \alpha_i \subset R$ associated to Z, say a *minimal homogeneous basis*, m.h.b. for short. Since the case $m_2 = 0$ is well-known, we state next theorem only in case $m_2 \geq 1$.

THEOREM 5.7 ([C3],Prop.3.3) Let $m_1 \geq \ldots \geq m_s \geq 0$, $s \geq 2$, $m_2 \geq 1$, and let the P_i 's be distinct points of a non-singular conic C. Set $q = \max(m_1 + m_2, [(\sum_1^s m_i +1)/2])$. Define Z' as in 5.1. Let α' be the homogeneous ideal associated to Z', and $\{g_1, \ldots, g_\mu\}$ a m.h.b. of α'.

If f and r are polynomials defining C and the line $P_1 P_2$, respectively, then:

(i) if $\sum_1^s m_i \leq 2 m_1 + 2 m_2 - 2$, then there exists $v \in \alpha_q$ such that $\{rg_1, \ldots, rg_\mu, v\}$ is a m.h.b. of α;

(ii) if $\sum_1^s m_i > 2 m_1 + 2 m_2 - 2$, and $\sum_1^s m_i$ is odd, then there exist $h_1, h_2 \in \alpha_q$ such that $\{fg_1, \ldots, fg_\mu, h_1, h_2\}$ is a m.h.b. of α;

(iii) if $\sum_1^s m_i > 2 m_1 + 2 m_2 - 2$, and $\sum_1^s m_i$ is even, then there exists $h \in \alpha_q$ such that $\{fg_1, \ldots, fg_\mu, h\}$ is a m.h.b. of α.

Sketch of proof. The proof consists of three steps.

Step 1. Let $\delta = \oplus \delta_i$ denote the ideal generated by α_q. Then $\alpha_i = \delta_i$ for every $i \geq q$; equivalently the degrees of the elements of a m.h.b. of α are not greater than q.

Step 2. (i) Let δ and δ' denote the ideals generated by α_q and α'_{q-1}, respectively. Since $\dim_k \alpha_q / r\alpha'_{q-1} = 1$, let $\bar{v} \neq 0$ be an element of $\alpha_q / r\alpha'_{q-1}$. We have $\delta = r\delta' + (v) \subset r\alpha' + (v) \subset \alpha$. If g is a homogeneous element of α, and $\deg g \geq q$, then by step 1 we have $g \in \delta \subset r\alpha' + (v)$. If $\deg g < q$, then $g \in r\alpha'$. It follows that $\alpha = r\alpha' + (v) = (rg_1, \ldots, rg_\mu, v)$.

Step 3. (i) $\{rg_1, \ldots, rg_\mu, v\}$ is a minimal set of generators for α, i.e., by the homogeneous Nakayama Lemma, none of the rg_i's, nor v is in the ideal generated by the remaining ones. The proof of this statement is by contradiction, and straightforward calculations. (Analogously for (ii) and (iii)). ∎

6 FAT POINTS IN UNIFORM POSITION AND GENERIC FAT POINTS

In this section we give a bound for τ, when the P_i's are distinct points in uniform position. The bound we find is not sharp, but it is the best known to the authors for points with that kind of genericity.

For points in uniform position, we use the definition given by A.Geramita and F.Orecchia in [GO], i.e., a set $\mathcal{S}$ of s points of $\mathbb{P}^2$ is said to be *in uniform position*, if for any $s' \leq s$ points $P_1, \ldots, P_{s'}$ of $\mathcal{S}$, and any n, we have $h^0(I'(n)) = \max(0, (n+1)(n+2)/2 - s')$, where I' denotes the sheaf associated to the ideal $\cap_{1 \leq i \leq s'} \wp_i$.

Observe, on setting $s' = \binom{f+1}{2} + r$ ($0 \leq r \leq f$), that the s' points impose independent conditions to the curves of degree f, and, for $r = 0$, even to the curves of degree $f - 1$.

We have the following result ([C2],Th.7.3):

THEOREM 6.1 Let $P_1, \ldots, P_s$ be distinct points in uniform position, and $m_1 \geq \ldots \geq m_s \geq 0$, $s \geq 5$. Set $s_i = \max \{n \in \mathbb{N}: m_n \geq i\}$ ($1 \leq i \leq m_1$). Let r_i, f_i be the integers defined by $s_i = \binom{f_i + 1}{2} + r_i$, $0 \leq r_i \leq f_i$. Set:

$d_1 = f_1 - 1$ for $r_1 = 0$; $d_1 = f_1$ for $r_1 > 0$;
$d_i = f_i$ for $i > 1$ and $0 \leq r_i \leq 2$; $d_i = f_i + 1$ for $i > 1$ and $r_i > 2$.

Linear Systems

Then
$$\tau \leq t := \max(m_1 + m_2 - 1, [\textstyle\sum_1^5 m_i/2], \textstyle\sum_1^{m_1} d_i).$$

Sketch of proof. By induction on $\sum_1^s m_i$. Apply 4.2 to S_t, with suitable curves C. To choose the curves C, it is useful the following result of Maggioni and Ragusa [MR]: if $s = \binom{f+1}{2} + r$, through the points $P_1, \ldots, P_s$ there exists a non-singular curve C of degree d, with $d=f$ if $0 \leq r \leq 2$, while $f \leq d \leq f+1$ if $r > 2$. ∎

Let U be the subset of $(\mathbb{P}^2)^s$ consisting of the s-tuples $(P_1, \ldots, P_s)$ of distinct points of $\mathbb{P}^2$ in uniform position and let $s = \binom{f+1}{2} + r$, $0 \leq r \leq f$. By [GO], U is a nonempty open subset of $(\mathbb{P}^2)^s$. For any s-tuple $(P_1, \ldots, P_s) \in U$, we know that there exists a non-singular curve C of degree d, $f \leq d \leq f+1$, through $P_1, \ldots, P_s$; moreover, by [MR], there is a nonempty open subset $U' \subseteq U$, for which C can be found of degree $d=f$. By this and the techniques of Theorem 6.1, one can prove the following result ([C2],Th.7.7).

THEOREM 6.2 Let $(P_1, \ldots, P_s)$ be a generic s-tuple of points of $\mathbb{P}^2$, let s_i, f_i, r_i be as in Theorem 6.1. Set:

$$d_1 = f_1 - 1 \quad \text{for } r_1 = 0; \qquad d_1 = f_1 \quad \text{for } r_1 > 0;$$
$$d_i = f_i \quad \text{for } i > 1.$$

Define t as follows:

$$t = m_1 d_1 + 1 \quad \text{either for } m_1 = m_s > 1, r_1 = f_1, s \geq 9$$
$$\text{or for } m_1 = m_{s-1} > 1, m_s = 1, r_1 = 0, s \geq 10;$$
$$t = \max(m_1 + m_2 - 1, [\textstyle\sum_1^5 m_i/2], \textstyle\sum_1^{m_1} d_i) \quad \text{otherwise.}$$

Then $\tau \leq t$. ∎

REMARKS 6.3 (i) The bound given by Theorem 6.1 is not sharp for all possible sequences $m_1, \ldots, m_s$, but we can produce cases for which S_{t-1} is not regular. As an example, let C_4 be an integral quartic with three nodes P_1, P_2, P_3 and let $P_4, \ldots, P_{14}$ lie on C_4, so that $P_1, \ldots, P_{14}$ are distinct points in uniform position.

Let $Z = (P_1, \ldots, P_{14}; 2, \ldots, 2)$. By Theorem 6.1, we get $t=9$. Now we prove that S_8 is non-regular. Let I' be the sheaf associated to the scheme $Z' = (P_4, \ldots, P_{14}; 1, \ldots, 1)$. By Bézout's Theorem, C_4 is a fixed component for all curves in S_8, so $h^0(I(Z)) = h^0(I'(Z')) = 4$, while $\text{vir.dim } S_8 = 2$.

(ii) Gimigliano proves that the bound he gives for τ in case of points on a curve of degree d (see Rem.4.6) holds also for generic points of $\mathbb{P}^2$. The bound of Theorem 6.2 is an improvement.

(iii) It is worth noting that for generic points of $\mathbb{P}^2$, in char $\Bbbk = 0$, Gimigliano ([G],Prop.2.2) proves that, if $m_1 = \ldots = m_s \geq 2$, $d \geq 3$, $s = d(d+3)/2$, then $\tau \leq m_1 d$. By his result, and by 6.2 and 4.1, it is easy to deduce that (hypotheses and notation as in 6.2), for char $\Bbbk = 0$, we have:

$$\tau \leq \max(m_1 + m_2 - 1, [\textstyle\sum_1^5 m_i/2], \textstyle\sum_1^{m_1} d_i).$$

7 PLANE CURVES AS PROJECTIONS OF NONSINGULAR SPACE CURVES

The fact that every integral plane curve can be obtained as a projection of a non-singular curve in some projective space from a linear space was first proved by Veronese [V], and is a non-difficult fact, as pointed out, e.g., by Severi [S]. Indeed let Γ be an integral plane curve and let C be its normalization: then Γ is a projection of the projective embedding C' of C corresponding to any very ample linear series of the form $|L+D|$, where L is the pull back of a linear section and D is an effective divisor of sufficiently large degree. We observe that neither Veronese, nor Severi seem to be very interested in the degree of C', or in the dimension of the projective space containing it.

On the other hand there is the following theorem by Enriques.

THEOREM C ([E], vol. II, p.549) Every plane integral curve can be obtained as a projection of a non-singular curve in $\mathbb{P}^4$ from a straight line. ∎

Although the proof of this theorem does not seem complete, it points out an interesting method based on linear systems of adjoints. We shall discuss here this method and some further developements, due to L.Caire [C].

Let Γ and C be as above and let ψ be an adjoint of degree $m-1$. Let x_0, x_1, x_2 be homogeneous coordinates in $\mathbb{P}^2$ and consider a linear system S_m generated by $\psi x_0, \psi x_1, \psi x_2, \psi_3, \ldots, \psi_r$, where the ψ_i's are adjoints of degree m. Let σ be the linear series cut out on C by S_m, after forgetting the base points. Assume $\dim \sigma = r > 2$, and denote by $\rho: C \to C' \subset \mathbb{P}^r$ the morphism corresponding to σ. By construction we can choose coordinates $y_0, \ldots, y_r$ in $\mathbb{P}^r$ such that ρ is the unique lifting to C of the rational map $\Gamma \dashrightarrow \mathbb{P}^r$ defined by the homogeneous equations

$$t y_0 = x_0 \psi(x_0, x_1, x_2)$$
$$\cdots\cdots\cdots\cdots\cdots$$
$$t y_r = \psi_r(x_0, x_1, x_2)$$

From this one gets:

PROPOSITION 7.1 Γ is (homographic to) the projection of C' from the linear space $L: y_0 = y_1 = y_2 = 0$ onto the plane $y_3 = \ldots = y_r = 0$; and $L \cap C'$ corresponds to the points of $\psi \cap \Gamma$ which are not base points for S_m. ∎

Let us denote by $\pi: C' \dashrightarrow \Gamma$ the projection, and by F' the set of points of C' where π is not defined. Put also $E = \text{supp}(\Delta)$, where Δ is the "double points" divisor (i.e. the divisor corresponding to the conductor) and $E' = \rho(E)$. Then from 7.1 one gets:

PROPOSITION 7.2 The set $C' \setminus (E' \cup F')$ consists of simple points. ∎

The main result in [C] is the following improvement of Theorem C:

THEOREM 7.3 Let the notation be as above. Then:

(i) if $r > 3$ and σ separates points and tangent vectors in the set $E \cup \rho^{-1}(L \cap C')$, then C' is non-singular;

(ii) if (i) holds, then one can assume $r = 4$ (by a suitable choice of $\psi_3, \psi_4 \in S_m$);

(iii) if $r = 3$ and σ separates points and tangent vectors in E, then $C' \setminus F'$ is non-singular; moreover C' is non-singular if and only if the intersection multiplicity of ψ and Γ outside all base points of S_m is ≤ 1.

Proof. (i) is proved by using 7.2 and a direct calculation. (ii) is achieved by showing that a general pencil in S_m separates points and tangent vectors in $E \cup \rho^{-1}(L \cap C')$; and the proof of (iii) is of the same nature. ∎

REMARKS 7.4 (i) Theorem C is contained in Theorem 7.3 (ii). We observe here that in [E] it is claimed that C' turns out to be non-singular just because ψ is assumed to cut Γ tranversally far from the base points: but this only says, in our notation, that F' is empty: it is not so clear why this implies the conclusion.

(ii) Theorem 7.3 allows, at least in principle, to get control of the degree of C'; indeed if σ' is the linear series cut out on C' by S_m then the base divisor of σ' is $\Delta + D$, where D is effective. So one gets $\deg(C') = \deg(\sigma) = m \deg(\Gamma) - \deg(\Delta + D)$. So one can try to operate on D and m to lower the degree. Note that D is avoided in [E].

(iii) The fact of considering D as above allows also to get a converse of Theorem 7.3, namely:

THEOREM 7.5 Every non-singular C' which projects onto Γ can be obtained with a suitable S_m.

Proof. Let ψ be an adjoint of degree $m - 1$ which cuts out $L \cap C'$ (plus something else). Then one can recover the linear series cut out by the hyperplanes by a suitable S_m (using Noether's Restsatz). ∎

As an illustration of the above we have the following

EXAMPLE 7.6 Let Γ be a plane non-singular curve of degree $d \geq 4$. Then Γ is the projection of a non-singular $C' \subset \mathbb{P}^4$ of degree $2d - 1$, and no lower degree. In this case S_m is obtained with a line ψ and two conics through a point of $\psi \cap \Gamma$.

REFERENCES

[BG] Boratynski, M., and Greco, S. (1985). *Hilbert functions and Betti numbers in a flat family*, Ann. Mat. Pura Appl., 142, pp. 277-292.

[C] Caire, L. *Plane curves as projections of non-singular space curves,* in preparation.

[Ca] Castelnuovo, G. (1891). *Ricerche generali sopra i sistemi lineari di curve piane*, Memorie della Reale Accademia delle Scienze di Torino, s^e II, t. XLII , in *Memorie Scelte* , N.Zanichelli, Bologna (1937-XV), pp. 137-188.

[C1] Catalisano, M.V. (1989). *Regolarità e irriducibilità dei sistemi lineari di curve piane passanti per punti di* $\mathbb{P}^2$ *in posizione generale con assegnata molteplicità* , Atti Accad. Sci. Torino, to appear.

[C2] Catalisano, M.V. (1989). *Linear systems of plane curves through fixed "fat" points of* $\mathbb{P}^2$, Queen's Papers in Pure and Appl. Math., VI, to appear.

[C3] Catalisano, M.V. (1989). *Standard bases of some ideals defining "fat" points in* $\mathbb{P}^2$, Dip. Mat. Univ. Genova, preprint.

[D1] Davis, E. (1984). *On a theorem of Beniamino Segre* , Queen's Papers in Pure and Appl. Math., 67, pp. 1 D - 10 D.

[D2] Davis, E. (1985). *Complete intersections of codimension 2 in* $\mathbb{P}^r$ *: the Bezout-Jacobi-Segre Theorem revisited* , Rend. Sem. Mat. Torino, 43, pp. 333-353.

[D3] Davis, E. (1986). *0-dimensional subschemes of* $\mathbb{P}^2$ *: new application of Castelnuovo's function* , Ann. Univ. Ferrara, 32 , pp. 93-107.

[DG] Davis, E., and Geramita, A. (1984). *The Hilbert function of a special class of 1-dimensional Cohen-Macaulay graded algebras* , Queen's Papers in Pure and Appl. Math., 67 , pp. 1 H - 29 H.

[E] Enriques, F., and Chisini, O. (1918). *Lezioni sulla teoria geometrica delle equazioni e delle funzioni algebriche* , vol.II, Libro IV, N.Zanichelli, Bologna, (Ristampa 1985).

[GMR] Geramita, A., Maroscia, P., and Roberts, L.G. (1983). *The Hilbert function of a reduced K-Algebra* , J. London Math. Soc. (2), 28 , pp. 443-452.

[GO] Geramita, A., and Orecchia, F. (1981). *On the Cohen-Macaulay type of s-lines in* $\mathbb{A}^{n+1}$, J. Algebra, 70 , pp. 116-140.

[G] Gimigliano, A. *Regularity of linear systems of plane curves* , J. Algebra, to appear.

[Gf] Giuffrida, S. (1985). *Hilbert function of a 0-cycle in* $\mathbb{P}^2$, Le Matematiche, 40, pp. 252-266.

[G1] Greco, S. (1984). *On the postulation of finite subschemes of projective integral curves* , Queen's Papers in Pure and Appl. Math., 67 , pp. 1 I - 30 I.

[G2] Greco, S. (1985). *Alcune osservazioni sui sottoschemi di codimensione 1 di una varietà proiettiva* , Sem. Geom., Univ. Bologna, pp. 89-100.

[G3] Greco, S. *Remarks on the postulation of zero-dimensional subschemes of projective space* , Math. Ann., to appear.

[GLP] Gruson, L., Lazarsfeld, R., and Peskine, C. (1983). *On a theorem of Castelnuovo and the equations defining space curves* , Invent. Math., 72 , pp. 491-506.

[H1] Harbourne, B. (1985). *Complete linear systems on rational surfaces* , Trans. Amer. Math. Soc., 289, pp. 213-226.

[H2] Harbourne, B. (1986). *The Geometry of rational surfaces and Hilbert functions of points in the plane* , CMS Conf. Proc., 6 , pp. 95-111.

[H3] Harbourne, B. *Hilbert functions of points in good position in* $\mathbb{P}^2$, Queen's Papers in Pure and Appl. Math., to appear.

[J] Jouanolou, J.P. (1983). *Théorèmes de Bertini et applications* , Birkhäuser, 42.

[MR] Maggioni, R., and Ragusa, A. (1985). *Nonsingular curves passing through points of $\mathbb{P}^2$ in generic position, I* , J. Algebra, 92, pp. 176-193.
[P] Paxia, G. *On flat families of fat points* , in preparation.
[S1] Segre, B. (1947). *Sui teoremi di Bézout, Jacobi e Reiss* , Ann. Mat. Pura Appl. (4), 26, pp. 1-26.
[S2] Segre, B. (1961). *Alcune questioni su insiemi finiti di punti in Geometria Algebrica* , Atti Convegno Internaz. Geometria Algebrica, Torino, pp. 15-33.
[S] Severi, F. (1926). *Trattato di Geometria Algebrica* , vol.I, Parte I, N.Zanichelli, Bologna.
[V] Veronese, G. (1882). *Behandlung der projectivischen Verhältnisse der Räume von verschiedenen Dimensionen durch das Prinzip des Projicirens und Schneidens* , Math.Ann., 19, pp. 161-234.

Recent Results on Irregular Surfaces and Irregular Kaehler Manifolds

FABRIZIO CATANESE, Dipartimento di Matematica, Università di Pisa, via Buonarroti 2, 56127 Pisa, Italia

Abstract. In this lecture note, after a commemorative foreword which the reader interested in mathematical statements may entirely skip, we focus on some recent results obtained by several authors, and which originate in an attempt to understand some questions posed by Enriques in his book «Le Superficie Algebriche».

Notably, after describing the basic properties of irregular manifolds, and some history of these results, we shall concentrate on work of the present author concerning fibrations between irregular manifolds, generalizing early work of Castelnuovo-De Franchis, and giving topological characterizations of those fibrations.

After surveying results on surfaces and manifolds with irrational pencils, due to Beauville, Xiao, Siu and ourselves, we shall discuss some very recent and interesting progress (due to Green, Lazarsfeld and ourselves) in the direction of classification of irregular Kaehler manifolds.

Discussing results on the moduli of algebraic surfaces, especially results on moduli of irregular surfaces due to the present author and Reider, will naturally then lead to consider the theory of paracanonical systems on surfaces, introduced by Enriques and carried through by Green-Lazarsfeld and Beauville.

Contents

§1 Irregular surfaces and the Fundamental theorem.
§2 Kaehler-Hodge theory and irregular Kaehler manifolds.
§3 Moduli of algebraic surfaces and irrational pencils.

§4 The paracanonical system and generic vanishing theorems.
§5 Classification of irregular manifolds, and char p analogues.

Foreword

I am very happy to have been participating in the celebration of the 900th anniversary of the University of Bologna: on the one hand my mind was once again impressed by the firm roots upon which the tree of scientific knowledge relies, but on the other hand this static view was completely wiped out by the remembrance of the peregrinations undertaken by the medieval «Clerici vagantes». And we find out that international communication and free exchange of ideas has always been the vital lymph which has nourished every scientific and intellectual achievement.

Thus the purpose of my talk is to outline a thread of continuity between the work done by one of the great mathematicians who taught at the University of Bologna and some very active current research being done in China, France, Italy and United States: but, oddly enough, some ideas which originated in Bologna some 85 years ago did not reach the young Italian generation by direct tradition, they spread around the globe and were revitalized in Russia, United States and Japan.

I am of course referring to Federigo Enriques, who in 1904 had a chair of Geometry in Bologna, and to the mathematical heritage of the most important part of his work, his book «Le Superficie Algebriche», which was printed here again in Bologna by Nicola Zanichelli in 1949, shortly after Enriques' departure. This treatise is «a report of the exploration of a territory where many gems have been collected and many others await a person worth of discovering them *» and as such I have sometimes glimpsed at some page of it: pretty much with an attitude similar to the one of the Clerici vagantes, who would seek inspiration out of some page of the Holy writings.

While I procrastinate to a near future the undertaking of a thorough and critical reading of the treatise, I would like here, also in view of Ciliberto's and Dolgachev's contributions to this volume, to restrict my consideration to some topics treated in chapter IX, with a particular emphasis on paragraphs 6 and 8 (pages 339-347, respectively pages 354-357).

I should warn the reader that I will make no attempt to discuss the interesting historical problem of evaluating the contributions of the italian geometers to the theory exposed in chapter IX, entitled «Irregular surfaces and continuous systems of inequivalent curves». According to G. Castelnuovo, chapter IX is the hardest and most interesting, even if the *Fundamental Theorem on irregular surfaces* is not proven there, neither does the proposed algebro-geometric approach (using continuity arguments and the principle of degeneration) seem (even to Castelnuovo) feasible.

But I would like to emphasize the importance of Enriques' *Heuristic Method*: his main mathematical attitude was oriented towards the discovery of interesting mathematical truths, and he would achieve this goal by asking problems, verifying many examples, giving incomplete but convincing «proofs».

Of course Enriques was well aware of the meaning of a rigourous proof, but his main concern was not to avoid errors (this is rather easy, as he points out, whenever one deals with

---

* G. Castelnuovo, foreword to «Le Superficie Algebriche».

subtle and intricate classifications), but to provide the reader with an historical picture of the (often difficult and dirty) road that ideas had to follow before crystallizing in beautiful abstract patterns.

G. Castelnuovo wondered, in the foreword to the book, whether would there soon come a continuator of Enriques' work, a mathematician who could bring the theory of surfaces to a level of perfection similar to the one of the theory of curves.

We are lucky that pretty early his scepticism turned out to be ill founded: Enriques' heritage was revived in the 60's by three very important schools, the Russian school of I.R. Shafarevich, Zariski's Seminar at Harvard and the Japanese school of K. Kodaira.

The Steklov Institute Seminar Book ([Sh]) starts with a famous quotation: «Eschylus said that his tragedies were but fragments of Homerus' banquet», and indeed Shafarevich and his coworkers were particularly fond of the algebro-geometric methods, and worked out many ideas of Enriques, especially concerning the general classification of algebraic surfaces according to their plurigenera.

At Harvard, great clarification was done in the algebraic direction, and indeed a great achievement was the extension of Enriques' classification to characteristic p, envisaged by Mumford and completed with the collaboration of Bombieri (whose Seminar at Pisa was my introduction to the subject).

Kodaira however, pushing through the transcendental methods, was able to extend Enriques' classification to the case of compact complex surfaces (especially in the beautiful second series [Kod2] written when the signature theorem had become available). By looking at the borrower's card at the I.A.S. in Princeton you can verify that he spent a few years meditating on Enriques's book, and indeed he gave finally a very simple and general proof of the main theorem of surface classification.

The heritage which these schools took up again, while profiting from the new and powerful machinery of sheaves and cohomology, was not only related to the mathematical contents of Enriques's work, but had indeed the same scholarly attitude of discussing new problems, posing daring conjectures, dreaming about mathematics. In a way I regard the recent successes of the Japanese school (through work of Iitaka, Ueno, Kawamata, Mori, Miyaoka and many others) concerning the classification of higher dimensional varieties, as the most ripe fruit grown out of Enriques' seeds.

1. Irregular surfaces and the fundamental theorem

For the benefit of graduate students or non experts, we stick to modern terminology and concepts, explaining the classical definitions whenever convenient.

Throughout the paper we let X be either a compact Kaehler manifold of $\mathbb{C}$-dimension equal to n, or a complete projective variety (of dimension n and) defined over an algebraically closed ground field k.

In case when $n = 2$, we shall call X a surface.

Already since Max Noether the existence of irregular surfaces was recognized: these are the surfaces for which the **numerical irregularity** $\mathbf{p_g - p_a}$ is strictly positive.

In modern terms, the **geometric genus** $\mathbf{p_g}$ equals $h^0(\Omega_X^2)$, which, by Serre duality, is the same as $h^2(\mathcal{O}_X)$, while the numerical irregularity is simply $h^1(\mathcal{O}_X)$. In classical terms, if X is (the normalization of) a surface of degree d in $\mathbb{P}^3$ with ordinary singularities, $\mathbf{p_g}$ is the number of independent surfaces of degree $d - 4$ passing through the singular curve of

X, whereas **the arithmetic genus** p_a is a coefficient appearing in the **postulation formula**, which gives, for m at least $d-3$, the number $r(m)$ of independent surfaces of degree m passing through the singular curve of X.

One way to write the postulation formula, in terms of the degree δ of the singular curve of X, is the following:

$$r(m) = C(m+3-d, 3) + 1/2(m+4-d)[dm - 2\delta] + 1 + p_a \quad (1.1)$$

(where $C(a, b)$ stands for the binomial coefficient).

On the other hand, there is the other notion, of the so called **geometrical irregularity**, denoted by **q**.

The latter can be defined in terms of algebraic systems of divisors on X (these are curves if X is a surface), as we are going to indicate.

Firstly, two divisors D and D' are said to be **linearly equivalent** if there is a meromorphic function f such that $D - D'$ is the divisor of f (divisor of zeros minus divisor of poles).

An algebraic (or continuous) system of divisors parametrized by T is a divisor $\mathcal{D} \subset X \times T$ such that for all $t \in T$, $D_t = \mathcal{D} \cap (X \times \{t\})$ is a divisor in X. If T is connected, all the divisors D_t are said to be **directly algebraically equivalent**, and this notion generates the relation of algebraic equivalence in the additive group of all divisors. Finally, an algebraic system of divisors with irreducible parameter space T is said to be **effectively parametrized** if for all $t \in T$, the set $\{t' \in T | D_{t'}$ is algebraically equivalent to $D_t\}$ is finite.

DEFINITION 1.2. *The geometric irregularity q of X equals the maximal dimension (i.e., dim T) of an effectively parametrized continuous system of divisors on X.*

(1.3) **The fundamental theorem on irregular surfaces** *asserts that the geometric irregularity* **q** *equals the arithmetic irregularity* $p_g - p_a$.

This theorem, valid in characteristic 0, but false in characteristic p (as it was shown by Igusa in 1953 ([Igu])), was conjectured (over $\mathbb{C}$) by Enriques in Bologna in 1904, and proven, using transcendental methods, by H. Poincare' in 1909-10 ([Poi]). At that time F. Severi was an assistant to Enriques. When Severi, shortly after his graduation, arrived from Torino in 1902, Enriques got him interested in the problem of continuous systems, suggesting the investigation of the surfaces which are the symmetric square of a curve.

For a long time Severi and Enriques tried to give a geometric proof of the theorem, but eventually realized that differential geometric or topological arguments were indeed needed.

The first breakthrough, due to Castelnuovo in 1905, consists in introducing an Abelian variety, called Picard variety in honour of E. Picard and denoted by $\text{Pic}^\circ(X)$.

Letting $\text{Div}_\ell(X)$ be the group of divisors linearly equivalent to zero, resp. $\text{Div}_a(X)$ the group of divisors algebraically equivalent to zero, $\text{Pic}^\circ(X)$ fits into an exact sequence

$$0 \to \text{Div}_\ell(X) \to \text{Div}_a(X) \to \text{Pic}^\circ(X) \to 0. \quad (1.4)$$

Moreover, $\text{Pic}^\circ(X)$ has dimension equal to q.

For convenience of the modern reader, we state the theorem in modern terminology (we shall further dwell upon this in the subsequent paragraph).

1.5. Fundamental theorem on irregular surfaces

Let X be a complex algebraic compact surface, then

Irregular Surfaces   63

i) $\text{Pic}^\circ(X)$ is the group of isomorphism classes of topologically trivial line bundles on X, hence $\cong H^1(\mathcal{O}_X)/H^1(X,\mathbb{Z})$, and in particular its dimension equals the numerical irregularity.

ii) There exists a line bundle $\mathscr{P}$ on $X \times \text{Pic}^\circ(X)$, called Poincare' line bundle, such that the restriction of $\mathscr{P}$ to $X \times \{t\}$ is a bundle on X whose isomorphism class is t. Moreover, if H is a sufficiently ample divisor on X, and we denote by H' its pull back to $X \times \text{Pic}^\circ(X)$, then $\mathscr{P}(H')$ has a section whose divisors of zeros $\mathcal{D}$ yields an effectively parametrized algebraic system of divisors D_t on X, and of maximal dimension $= q$.

iii) The complex torus $\text{Pic}^\circ(X)$ is an Abelian variety and admits as a dual Abelian variety the following complex torus, called Albanese variety in honour of Giacomo Albanese, and denoted by $\text{Alb}(X) = A$:

$$A = (H^0(\Omega_X^1))^\vee / j(H_1(X,\mathbb{Z})),$$
$$(j : H_1(X,\mathbb{Z}) \to (H^0(\Omega_X^1))^\vee \text{ being given by integration}).$$

REMARK: note that i), ii) above yield (1.3), i.e., equality of the numerical and geometrical irregularity.

As we just saw, this important result has a corollary, also obtained independently by Severi in 1905, namely the equality between the geometric irregularity and the dimension of the space of holomorphic 1-forms: i.e., over $\mathbb{C}$, $q = h^0(\Omega_X^1)$.

As we mentioned earlier, not only was the first proof given by Poincare' using transcendental methods, but the theorem fails over an algebraically closed field of positive characteristic: its failure was completely analyzed by A. Grothendieck in 1960, who found necessary and sufficient conditions for its validity.

We defer the reader to D. Mumford's beautiful book «Lectures on curves on an algebraic surface» ([Mu1]) for a lively and detailed exposition of the whole theory: we simply recall the main result

1.6. Fundamental theorem on irregular surfaces in positive characteristic

i) $\text{Pic}^\circ(X)$ is a not necessarily reduced group scheme, whose tangent space at the origin is canonically identified with $H^1(\mathcal{O}_X)$. Via this identification the reduced subscheme $\text{Pic}^\circ(X)_{\text{red}}$ has as tangent space at the origin the intersection of the kernels of certain Bockstein homomorphisms β_r defined inductively on $\ker(\beta_{r-1})$ and with values in $H^2(\mathcal{O}_X)/\text{Im}(\beta_{r-1})$. In particular, the geometrical irregularity q is smaller than the arithmetic irregularity $h^1(\mathcal{O}_X)$, and indeed equal if $p_g = h^2(\mathcal{O}_X) = 0$.

ii) Again there exists a Poincare' line bundle $\mathscr{P}$ on $X \times \text{Pic}^\circ(X)$, and a divisor $\mathcal{D}$ in $X \times \text{Pic}^\circ(X)$, yielding an effectively parametrized algebraic system of divisors D_t on X, of maximal dimension $= q$.

It is interesting to remark that Enriques himself had realized that the main difficulties came from the case $p_g > 0$, and that essentially the problem could be solved with differential geometric methods (here, the Taylor series is gotten upon dividing iterated derivatives by suitable factorials, and the Bockstein operations «measure» the obstruction to carry out this procedure in positive characteristic).

Somehow, I feel the picture is not complete in characteristic p regarding the relations occurring between the geometric irregularity and the dimension of certain subspaces of the space $H^0(\Omega_{X^1})$.

First of all, we have to remark that not all the regular 1-forms are closed in char $= p$, i.e., $H^0(d(\mathcal{O}_X)) \neq H^0(\Omega_X^1)$ (cf. [Mu2], [Ny], [Fos]), the simplest example, due to Mumford, being the differential form (xdy) on the affine plane, which, although having a pole on the line at infinity in the projective plane, becomes regular on a separable Artin-Schreyer covering of the projective plane.

Now, the Albanese variety $A = \text{Alb}(X)$ exists also for varieties in characteristic p (cf. [Lg]), and clearly the regular 1-forms on A, being translation invariant, are d-closed; henceforth also their pull backs to X under the Albanese morphism satisfy this property, and then the dimension of A, which equals the geometric irregularity, and $1/2$ of the ℓ-adic first Betti number, is less than $h^0(d(\mathcal{O}_X))$.

On the other hand, Bombieri and Mumford ([BM3]) proved the existence, on supersingular Enriques surfaces in characteristic 2, of a nonzero form in $H^0(d(\mathcal{O}_X))$, whereas the arithmetic irregularity equals 1 but $\text{Pic}^\circ(X)$ is the nonreduced group scheme α_2, which has dimension zero.

In fact, (cf. [Mi1] prop. 4.14) this form is also closed under the Cartier operator, and classifies the α_2-principal homogeneous space giving the purely inseparable K3 double cover of X.

Thus the problem here is whether one can describe the 1-forms coming from the Albanese variety (they form a q-dimensional space) in terms of kernels of suitable differential operators: the answer is apparently not yet known.

On the other hand, after trying to give an idea of why the long sought for algebraic proofs waited long before coming, I want to show Hodge's theory of harmonic integrals which extended to higher dimensions the early results of Picard, Castelnuovo and Poincare' can still give very interesting geometric applications.

2. Kaehler-Hodge theory and irregular Kaehler manifolds

Let X be a compact Kaehler manifold of complex dimension n, then (cf. [K-M]) the 3 Laplace operators associated in the given Hermitian metric to the operators $d, d^{1,0}, d^{0,1}$ (where $d = d^{1,0} + d^{0,1}$ is the type decomposition of the operator of exterior differentiation) coincide up to a factor of 2, hence,

(2.1) if $\varphi = \sum_{p+q=k} \varphi_{p,q}$ is the type decompostion of a k-form, then φ is harmonic if and only if each $\varphi_{p,q}$ is harmonic.

The following is nowadays the usual form in which it is stated the

(2.2) Hodge - De Rham theorem:
0) Letting $H^k(X, \mathbb{C})$ be the De Rham cohomology group $= Z_k/B_k$, where, as usual, Z_k is the space of closed differential k-forms (the φ's such that $d\varphi = 0$), B_k is the subspace of exact forms, i.e. the φ's which can be written as $\varphi = d\psi$, then
  1) $H^k(X, \mathbb{C}) = \oplus_{p+q=k} H^{p,q}$, where $H^{p,q}$ is the space of harmonic forms of type p, q
  2) $H^{p,q} \cong H^q(\Omega_X^p)$ (in particular, holomorphic p-forms are d-closed)
  3) $\overline{H^{p,q}}$ si the complex conjugate subspace $(H^{p,q})^*$ of $H^{p,q}$, in particular

$$H^1(X, \mathbb{C}) = H^0(\Omega_X^1) \oplus (H^0(\Omega_X^1))^*.$$

Irregular Surfaces

We shall assume throughout X to be irregular, i.e., such that

$$q = h^0(\Omega_X^1) = h^1(\mathcal{O}_X) > 0 .$$

A fundamental geometric object for the study of X is given then by the Albanese morphism of X, $\alpha : X \to A$, where $A = \mathrm{Alb}(X)$ is the Albanese variety of X, the q-dimensional complex torus

(2.3) $A = \mathrm{Alb}(X) = (H^0(\Omega_X^1))^V/j(H_1(X,\mathbb{Z}))$, and, once fixed a base point x_0, $\alpha(x)$ is given, up to translation, by integration along any path from x_0 to x.

(2.4) The Picard variety $\mathrm{Pic}^\circ(X) \cong H^1(\mathcal{O}_X)/H^1(X,\mathbb{Z})$ is the dual complex torus since, by the Hodge theorem identification $H^1(\mathcal{O}_X) \cong (H^0(\Omega_X^1))^*$, $H^1(\mathcal{O}_X) \times (H^0(\Omega_X^1))^V$ carries a natural Hermitian form whose imaginary part is the bilinear form yelding the natural duality between $H^1(X,\mathbb{Z})$ and $j(H_1(X,\mathbb{Z})) \cong H_1(X,\mathbb{Z})/$ torsion.

Usually one sees that $\mathrm{Pic}^\circ(X) = H^1(\mathcal{O}_X^*)^\circ \cong H^1(\mathcal{O}_X)/H^1(X,\mathbb{Z})$ is a complex torus by looking at the commutative diagram of exact sequences

$$0 \to \mathbb{Z} \to \mathcal{O}_X \to \mathcal{O}_X^* \to 0$$
$$0 \to \mathbb{C} \to \mathcal{O}_X \to d\mathcal{O}_X \to 0$$

yielding, again since X is Kaehler,

$$0 \to H^0(d\mathcal{O}_X) = H^0(\Omega_X^1) \to H^1(X,\mathbb{C}) \to H^1(\mathcal{O}_X) \to 0 \text{ and}$$
$$H^1(\mathcal{O}_X)/H^1(X,\mathbb{Z}) \cong H^1(X,\mathbb{C})/(H^0(\Omega_X^1) \oplus H^1(X,\mathbb{Z})) \cong$$
$$\cong H^1(X,\mathbb{R})/H^1(X,\mathbb{Z})$$

Finally, via the exact sequence of locally constant sheaves

$$0 \to \mathbb{Z} \to \mathbb{R} \to U(1) \to 0$$

we identify

$$\mathrm{Pic}^\circ(X) \cong H^1(X,U(1)) , \qquad (2.5)$$

and the correspondence runs explicitly as follows: we let $\mathbb{L}$ be a locally constant sheaf determined by a cocycle $g_{\alpha\beta}$ in $H^1(X,U(1))$ (i.e., $\mathbb{L}$ is locally isomorphic to $\mathbb{C}$ and the transition functions $g_{\alpha\beta}$ are unitary complex numbers), and then to $\mathbb{L}$ we associate the holomorphic invertible sheaf $\mathcal{L} = \mathbb{L} \otimes \mathcal{O}_X$.

It is important for later developments that the Hodge - De Rham theory can be extended to $\mathbb{L}$ valued differential forms, yielding the following

(2.6) Twisted version of the Hodge - De Rham theorem:

0) Letting $H^k(X,\mathbb{L})$ be the De Rham cohomology group $= Z_k/B_k$, where, as usual, Z_k is the space of $\mathbb{L}$-valued closed differential k-forms (the φ's such that $d\varphi = 0$), B_k is the subspace of exact forms, i.e. the φ's which can be written as $\varphi = d\psi$, then

1) $H^k(X,\mathbb{L}) = \oplus_{p+q=k} H^{p,q}(\mathbb{L})$ where $H^{p,q}(\mathbb{L})$ is the space of harmonic forms of type p,q with values in $\mathbb{L}$
2) $H^{p,q}(\mathbb{L}) \cong H^q(\Omega_X^p \otimes \mathbb{L})$
3) $H^{q,p}(\mathbb{L}^*)$ is the complex-conjugate subspace $H^{p,q}(\mathbb{L})^*$ of $H^{p,q}(\mathbb{L})$, in particular $H^1(X,\mathbb{L}) = H^0(\Omega_X^1 \otimes \mathbb{L}) \oplus (H^0(\Omega_X^1 \otimes \mathbb{L}^*))^*$.

We remark here that if we set $\mathcal{L} = \mathbb{L} \otimes \mathcal{O}_X$, then $\mathcal{L}^{-1} = (\mathbb{L}^*) \otimes \mathcal{O}_X$.

As we shall see in the sequel, one can study the geometry of irregular varieties by considering the point of view related to one of the two dual varieties, Alb and Pic : studying points or divisors on X.

The first point of view was the one in which the Italian school was most successful, and, not for mere nationalistic reasons, we shall devote the first and greater part of our discussion to this point of view.

So, let's first recall some elementary properties of the Albanese variety and of the Albanese morphism

(2.7) Universal property of the Albanese variety.

$\alpha : X \to A$ enjoys the property that for every holomorphic map $f : X \to T$ of X to a complex torus T, there exists a holomorphic map $\beta : A \to T$, such that f is the composition of β with α (moreover β is a homomorphism for a proper choice of the origin in T).

As a consequence of the universal property, it follows immediately that

   i) $Y = \alpha(X)$ generates A, i.e. there is an integer m such that every $a \in A$ can be written as the sum of m points in Y.

   ii) If Y is a curve, then Y is smooth and α has connected fibres.

i) implies the following

COROLLARY 2.8. *If X is algebraic, then A is dominated by an algebraic variety, hence A is an Abelian variety.*

Conversely, if A is an Abelian variety, A is projective, hence Y is projective, and in particular X is projective if $\dim X = \dim Y$ (this follows from a theorem of Moishezon, cf. [Moi], asserting that a Kaehler manifold bimeromorphic to a projective variety is indeed projective).

REMARK 2.9. *By a theorem of Ueno ([Ueno] thm. 10.9), given a proper subvariety Y of a complex torus A, letting A_1 be the complex subtorus $A_1 = \{x \in A | x + Y = Y\}$, and $u : A \to A_2 = A/A_1$ be the quotient map, then A_2 is an Abelian variety, $Z = u(Y)$ is an algebraic variety of general type, and u makes Y an analytic bundle over Z with fibre A_1.*

Kawamata ([Kaw]) extended Ueno's result obtaining an analogous structure theorem for X under the assumption $\dim X = \dim Y$.

We now want to make an important observation which, although not difficult, we have never encountered before: clearly the dimension a of the image Y of the Albanese map is invariant by biholomorphisms, but indeed this is also a topological invariant of X (and also a bimeromorphic invariant, of course).

It is therefore meaningful to set the following

DEFINITION 2.10. *Let $Y = \alpha(X)$: then $\dim(Y)$ is called the **Albanese dimension** of X, and shall be denoted by $\mathbf{a} = \mathrm{alb}(X)$.*

PROPOSITION 2.11. *The Albanese dimension a of X is a topological invariant of X. More precisely, let $\Lambda(X) \subset H^*(X, \mathbb{C})$ be the exterior graded subalgebra generated by $H^1(X, \mathbb{C})$, respectively $\Lambda^{\mathrm{hol}}(X)$ be the subalgebra generated by $H^0(\Omega_X^1) = H^{1,0}(X)$. Then $\Lambda(X)_{2a} \neq 0$, while $\Lambda(X)_{2a+1} = 0$ (resp.: $\Lambda^{\mathrm{hol}}(X)_a \neq 0$, while $\Lambda^{\mathrm{hol}}(X)_{a+1} = 0$).*

Proof. By the Hodge decomposition we have

$H^1(X, \mathbb{C}) = H^0(\Omega_X^1) \oplus (H^0(\Omega_X^1))^*$, where $*$ denotes complex conjugation. Hence if $\Lambda^{hol}(X)_{a+1} = 0$, certainly $\Lambda(X)_{2a+1} = 0$.

Since $H^0(\Omega_X^1) = \alpha^*(H^0(\Omega_A^1))$, clearly $a = \dim(\alpha(X))$ is the maximal integer m with $\Lambda^{hol}(X)_m \neq 0$.

To show that $\Lambda(X)_{2a} \neq 0$, we apply the following

LEMMA 2.12. *Let $\omega_1, \omega_2, \ldots, \omega_r$ be holomorphic forms in $H^0(\Omega_X^1)$ such that $\omega = \omega_1 \wedge \omega_2 \wedge \ldots \wedge \omega_r$ is $\neq 0$ in $H^0(\Omega_X^r)$. Then $\omega \wedge \omega^* \neq 0$ in $H^{2r}(X, \mathbb{C})$, and $\eta = \eta_1 \wedge \ldots \wedge \eta_r \neq 0$ in $H^r(X, \mathbb{C})$, for every choice of $\eta_i = \omega_i$ or ω_i^*.*

Proof. Let ξ be the Kaehler closed $(1,1)$ form on X.

If $\omega \wedge \omega^* = 0$ in $H^{2r}(X, \mathbb{C})$, then we would have

$$\int_X \omega \wedge \omega^* \wedge \xi^{n-2r} = 0.$$

But a constant times the integrand is positive, and strictly positive at each point where $\omega \neq 0$, a contradiction.

The second assertion follows immediately from the first, since $\eta \wedge \eta^* = \pm \omega \wedge \omega^* \neq 0$ in $H^{2r}(X, \mathbb{C})$.

Q.E.D. for 2.11 and 2.12

It is also worthwhile to make a second remark, namely that the differentiable structure of the fibres of the Albanese map only depends upon the differentiable structure of X.

To see this, consider again the Albanese map $\alpha : X \to A = \text{Alb}(X) = (H^0(\Omega_X^1))^\vee / j(H_1(X, \mathbb{Z}))$, $j : H_1(X, \mathbb{Z}) \to (H^0(\Omega_X^1))^\vee$ being the homomorphism given by integration.

Clearly α depends upon the complex structure, but only to a certain extent. The complex space $H^0(\Omega_X^1)$, via the real and imaginary parts of forms belonging to it, determines a subspace V of the space of d-closed 1-forms which maps isomorphically to the De Rham cohomology group $H^1_{DR}(X, \mathbb{R})$; viewing A as a differentiable torus, a change of choice for V alters α by adding $[g]$, the projection onto A of a differentiable vector valued function $g : X \to \mathbb{R}^{2q}$.

The map $\beta = (\alpha + [tg], id_{[0,1]}) : X \times [0, 1] \to A \times [0, 1]$ is proper and of maximal rank (equal to $2a + 1$) over a locally closed submanifold M of dimension $2a + 1$, and with $M \cap (A \times \{t\}) \neq \emptyset$ for each t in view of 1.4: hence $\beta^{-1}(M)$ is a differentiable fibre bundle over M, and we have proven the following

PROPOSITION 2.13. *The differentiable structure of the general fibre of the Albanese map of X is completely determined by the differentiable structure of X.*

We now want to point out the important role played, in the classification of irregular manifolds, by the manifolds for which $\dim X = \dim Y$, and Y is a proper subvariety of A.

To do this, let's go back to the times around 1905.

At that time a very important result was obtained by De Franchis and Castelnuovo (cf. [D-F], [Cas 1], and [Bo], [B-P-V] for a modern exposition and applications), who stated the following theorem only for algebraic varieties (cf. also [D-F-2]).

(2.14.) CdF theorem (Theorem of Castelnuovo-De Franchis):
Let X be a compact Kaehler manifold, and let $w_1, w_2 \in H^0(\Omega_X^1)$ be two linearly independent forms such that $w_1 \wedge w_2 = 0$: then there exists a holomorphic map $f : X \to C$ to a curve such that $w_1, w_2 \in f^*(H^0(\Omega_C^1))$ (hence C has genus $g \geq 2$). Moreover, if we are given a form w, such that $w_1 \wedge w = 0$, then also $w \in f^*(H^0(\Omega_C^1))$.

Sketch of proof. Since $w_1 \wedge w_2 = 0$, there is a meromorphic function φ such that $w_1 = \varphi w_2$.
Now, applying exterior differentiation, we get $dw_1 = 0 = d\varphi \wedge w_2$, which implies that the form w_2, at the points where $d\varphi$ is not zero, is a holomorphic multiple of $d\varphi$, $w_2 = \lambda d\varphi$. But X is compact, hence λ is constant on the connected components of the fibres of φ. Hence, if C is the Riemann surface associated to the monodromy of the components of φ (C is called nowadays the Stein factorization of φ, by invoking a much more general theorem), φ_2 pulls back from C, and the same argument applies also to w_1, and w. Finally, since C has positive genus, the meromorphic map $f : X \to C$ is indeed holomorphic.

Q.E.D.

Castelnuovo used the above result ([Cas1]) to imply that surfaces with negative arithmetic genus are ruled.

We want now to give a new application of the above classical theorem (cf. [Cat4]).

THEOREM 2.15. *Let X be a compact Kaehler manifold: then there exists a non constant holomorphic map with connected fibres $f : X \to C$ to a curve C of genus $g \geq 2$, if and only if there is a g-dimensional subspace V of $H^1(X, \mathbb{C})$ which is a maximal isotropic subspace, where a subspace W is said to be isotropic (or 1-wedge) if $\Lambda^2 W$, the natural image of $\Lambda^2(W)$ into $H^2(X, \mathbb{C})$, is zero.*

Moreover, any maximal isotropic subspace V as above occurs as a pull back $f^*(V')$ of a maximal isotropic subspace V' of $H^1(C, \mathbb{C})$ for some $f : X \to C$ as above.

Proof. Assume that $\varphi_1, \varphi_2, \ldots, \varphi_g$ are a basis of V.

Since $H^1(X, \mathbb{C}) = H^0(\Omega_X^1) \oplus (H^0(\Omega_X^1))^*$, we can write $\varphi_1 = w_1 + \eta_1^*, \varphi_2 = w_2 + \eta_2^*, \ldots, \varphi_g = w_g + \eta_g^*$, and we let U be the span of $w_1, w_2, \ldots, w_g$, respectively W be the span of $\eta_1, \eta_2, \ldots, \eta_g$.

We know that, for $i, j = 1, \ldots g$, the following form gives a zero cohomology class:

$$(w_i + \eta_i^*) \wedge (w_j + \eta_j^*) = 0 \text{ in } H^2(X, \mathbb{C}) . \tag{2.16}$$

But, on a Kaehler manifold, the type decomposition passes to cohomology, hence (2.16) is equivalent to

$$w_i \wedge w_j = 0, \ \eta_i \wedge \eta_j = 0, \ (w_i \wedge \eta_j^*) + (\eta_i^* \wedge w_j) = 0 . \tag{2.17}$$

Clearly (2.17) imply $\Lambda^2 U = \Lambda^2 W = 0$, and we can apply CdF to U and W, provided their dimension is at least 2.

Assume $\dim U = 1$: then we may assume $w_2 = \ldots = w_g = 0$, and again (2.17) plus lemma 2.12 imply $w_1 \wedge \eta_j = 0$ for each j (for $j = 1, w_1 \wedge \eta_1 = 0$ follows then from $\eta_1 \wedge \eta_j = 0$ for each j). Thus CdF can be applied to the subspace $U + W$, whose dimension is ≥ 2, hence V pulls back from a curve of genus $\geq g$, but indeed of genus g by the maximality of V. The same argument applies if $\dim W = 1$.

If on the contrary, $\dim U, \dim W$ are both ≥ 2, we get maps $f : X \to C, f' : X \to C'$, by applying CdF to U, respectively to W.

Consider then the product map $\varphi = (f \times f') : X \to C \times C'$.

If the image of φ is a curve C'', we are done as before, since V pulls back from $H^1(C'', \mathbb{C})$.

If φ is surjective, we derive a contradiction to (2.17), since by the Kuenneth formula $H^*(C \times C', \mathbb{C}) = H^*(C, \mathbb{C}) \otimes H^*(C', \mathbb{C})$, and φ^* is injective.

Q.E.D.

After I had some partial results in this direction, the above result was inspired by the following result proven by Siu using the theory of harmonic maps ([Siu]) and independently discovered by Beauville who derives it (cf. [Be4]), as we shall see in §4, from the generic vanishing theorems for the cohomology of topologically trivial line bundles.

THEOREM 2.18. (Siu-Beauville) *Let X be a compact Kaehler manifold and C be a curve of genus $g \geq 2$: then there is a non constant holomorphic map $f : X \to C'$, to some curve C' of genus $g' \geq g$, if and only if there is a surjective homomorphism $\pi_1(X) \to \pi_1(C)$.*

DEFINITION 2.19. *An irrational pencil of X is a morphism of the given variety X to a curve C of genus ≥ 1.*

On the other hand, a morphism to a curve of genus 1 is not necessarily stable by deformation, as it is shown by the example of Abelian surfaces with a fixed polarizatzion type. In fact these surfaces form an irreducible family, but the Abelian surfaces admitting a (non constant) holomorphic map to a curve of genus 1 are exactly the ones isogenous to a product of elliptic curves, hence they form a countable union of proper algebraic subsets of the parameter space of the family.

For this reason, since instead we have seen that the existence of a (non constant) holomorphic map to a curve of genus ≥ 2 is a topological property of X, we introduce the class of Albanese general type manifolds which, in the realm of irregular manifolds, is a good generalization of the class of curves of genus at least 2.

DEFINITION 2.20. *An irregular Kaehler manifold X (or a normal complex space X bimeromorphic to a Kaehler manifold) of dimension n is said to be of* **Albanese general type** *if $q > n$, and, for its Albanese dimension a, we have $a = n$ (that is, its Albanese image Y has dimension n and is a proper subvariety of $A = \mathrm{Alb}(X)$).*

What in fact occurs is that, again via Hodge theory, the existence of a fibration f between X and an irregular normal Kaehler space Y of Albanese general type is merely a multilinear algebra property of the real cohomology algebra of X.

In fact, if Y is of dimension $k < n$, f is completely determined by the pull-back F of the first De Rham cohomology group of Y.

F enjoys certain properties which we shall summarize by saying that F is a saturated real $2k$-wedge subspaces of $H^1(X, \mathbb{C})$, and conversely such a subspace determines a fibration f (by definition, a fibration is a morphism with connected fibres).

Before we even get to a precise statement of the theorem, we need to discuss a generalization of the classical Castelnuovo-De Franchis theorem, which was obtained independently by several authors, all of them apparently unaware of prior work by Z. Ran ([Ran1]). Around the same time, Green-Lazarsfeld ([G-L 2]), Peters, and ourselves, ([Cat4]), reproved it with different proofs. Of course I prefer to give mine, which is of a rather differential geometric nature and allows a further generalization.

THEOREM 2.21. (GCdF) *Let X be a compact Kaehler manifold, and let $U \subset H^0(\Omega_X^1)$ be a $(k+1)$-dimensional strict k-wedge subspace, i.e., a subspace with a basis given by forms $\omega_1, \omega_2, \ldots, \omega_{k+1}$, and such that $\omega_1 \wedge \omega_2 \wedge \ldots \wedge \omega_{k+1} = 0$, while $\Lambda^k(U)$ embeds into $H^0(\Omega_X^k)$ under the natural homomorphism.*

Then there is a holomorphic map $f : X \to Y$ with connected fibres to a k-dimensional normal variety such that $U \subset f^*(H^0(\Omega_Y^1))$ (hence Y is of Albanese general type).

Proof. U determines a rank k saturated subsheaf $\mathscr{H}$ of Ω_X^1, inducing a subbundle of the cotangent bundle in an open set whose complement $\sum$ has codimension at least 2.

Let $\mathscr{F}$ be the foliation defined by $\mathscr{H}$ in the holomorphic tangent bundle of $X - \sum$: since X is Kaehler, the holomorphic forms are closed, hence the foliation is integrable.

We are going to show in particular that the leaves of the foliation are closed.

To this purpose, let $\mathbb{C}(\mathscr{F})$ be the field of meromorphic functions on X which are constant on the leaves of the foliation.

Furthemore, let Y' be a smooth birational model for $\mathbb{C}(\mathscr{F})$ and let $\pi : X \to Y'$ be the meromorphic map associated to the inclusion of fields $\mathbb{C}(\mathscr{F}) \subset \mathbb{C}(X)$.

We let $\mathscr{V}$ be the «good» open set where π is a holomorphic submersion, and where not all the k-forms of the subspace $\Lambda^k(U)$ of $H^0(\Omega_X^k)$ vanish.

We let now $\omega_1, \omega_2, \ldots, \omega_{k+1}$, be any basis of U, and let x be a point of $\mathscr{V}$ where $\omega_1 \wedge \omega_2 \wedge \ldots \wedge \omega_k \neq 0$: then, since the ω_i's are d-closed, there are local holomorphic coordinates $z_1, z_2, \ldots, z_n$ around x, and a function ψ such that

$$\omega_1 = dz_1, \omega_2 = dz_2, \ldots, \omega_k = dz_k, \omega_{k+1} = d\psi. \tag{2.22}$$

Moreover, $\omega_1 \wedge \omega_2 \wedge \ldots \wedge \omega_{k+1} = 0$ implies that $\psi = \psi(z_1, z_2, \ldots, z_k)$, i.e., ψ depends only upon the first k coordinates.

Another important remark is that, even if ψ is a local holomorphic function, its partial derivatives $\partial \psi / \partial z_j$ are global meromorphic functions, and thus in $\mathbb{C}(\mathscr{F})$.

In fact, $\omega_1 \wedge \omega_2 \wedge \ldots \wedge \omega_{j-1} \wedge \omega_{j+1} \wedge \ldots \wedge \omega_{k+1} = (-1)^{k-j}(\partial \psi / \partial z_j) \omega_1 \wedge \omega_2 \wedge \ldots \wedge \omega_k$. Similarly, for every meromorphic function $w \in \mathbb{C}(\mathscr{F}), (\partial w / \partial z_j) \in \mathbb{C}(\mathscr{F})$ (remember that $w = w(z_1, z_2, \ldots, z_k)$).

Let now $r = \dim(Y')$, and $w_1, w_2, \ldots, w_r$ be meromorphic functions $\in \mathbb{C}(\mathscr{F})$ which give local holomorphic coordinates at $\pi(x) \in Y'$. In particular, $w_1, w_2, \ldots, w_r$ are holomorphic and we may assume (up to replacing $\omega_1, \omega_2, \ldots, \omega_k$ by suitable linear combinations) that $w_1, w_2, \ldots, w_r, z_{r+1}, \ldots, z_n$ are local holomorphic coordinates at x. Since, though, we are dealing with local holomorphic functions constant on the leaves, it suffices to work in k variables, where we shall apply the following

LEMMA 2.23. *Let $\psi = \psi(z_1, z_2, \ldots, z_k)$, $w_i = w_i(z_1, z_2, \ldots, z_k)$, for $i = 1, \ldots, r$ be germs of holomorphic functions around the origin in $\mathbb{C}^k$, such that*

i) *$w_1, w_2, \ldots, w_r, z_{r+1}, \ldots, z_k$ are local holomorphic coordinates at the origin in $\mathbb{C}^k$.*

ii) *For each $i, j, (\partial \psi / \partial z_j)$ and also $(\partial w_i / \partial z_j)$ are functions of $(w_1, w_2, \ldots, w_r)$ only.*

Then, ψ can be written uniquely as $\varphi(w_1, w_2, \ldots, w_r) + \gamma$, where γ is a linear function of $z_{r+1}, \ldots, z_k$.

Proof. Denote by $y = (y_1, y_2, \ldots, y_r, y_{r+1}, \ldots, y_k)$ the new coordinate vector given by the functions $w_1, w_2, \ldots, w_r, z_{r+1}, \ldots, z_k$.

Irregular Surfaces

Then $(\partial \psi/\partial y_i) = \sum_{j=1,\ldots,k}(\partial \psi/\partial z_j)(\partial z_j/\partial y_i)$, but the matrix $(\partial z_j/\partial y_i)$ is the inverse of the matrix

$$(\partial y_i/\partial z_j) = \begin{pmatrix} \partial w_k/\partial z_j \\ 0 \quad I_{k-r} \end{pmatrix}$$

where I_{k-r} is the identity matrix of order $k - r$: therefore, $(\partial z_j/\partial y_i)$ and also, by ii), $(\partial \psi/\partial y_i)$ are functions of $w = (w_1, w_2, \ldots, w_r)$.

Set now $u = (z_{r+1}, \ldots, z_k)$, and write now ψ as a power series in the $y = (w, u)$ coordinate:

$$\psi(w, u) = \sum_{\alpha,\beta} \psi_{\alpha,\beta} w^\alpha u^\beta .$$

Since $(\partial \psi/\partial w_i) = (\partial \psi/\partial y_i)$ is just a function of w, we see that if the multiindex β is $\neq 0$, and if $\psi_{\alpha,\beta} \neq 0$, then necessarily $\alpha = 0$.

Hence we can write ψ uniquely as $\psi = \varphi(w) + \gamma(u)$; finally, since also $(\partial \psi/\partial u_i) = (\partial \psi/\partial y_i)$ is again a function of w alone, we see that $(\partial \psi/\partial u_i) = (\partial \gamma/\partial u_i)$ is a constant, whence γ is a linear function.

Q.E.D.

We can immediately apply the previous lemma, since it implies that, given any basis $\omega_1, \omega_2, \ldots, \omega_{k+1}$ of U, there are constants $c_1, c_2, \ldots, c_k$ such that $\omega_{k+1} - \sum_{j=1,\ldots,k} c_j \omega_j = \omega'_{k+1}$ is (at least locally) the pull back of a holomorphic differential on Y'. Repeating the same argument inductively for a new basis $\omega_1, \omega_2, \ldots, \omega_h, \omega'_{h+1}, \ldots, \omega'_{k+1}$ of U (letting ω_h play the role of ω_{k+1}), we find a basis of $U, \omega'_1, \omega'_2, \ldots, \omega'_{k+1}$ such that all the ω'_i's, hence all of U, pull back from Y'.

Hence $\dim Y' = k$, and the leaves of $\mathscr{F}$ are closed (and smooth) in the good open set $\mathscr{V}$, and thus also in $X - \sum$.

A standard argument (on the universal cover X' of X, the leaves are the inverse images of a holomorphic function $\psi : X' \to \mathbb{C}^{k+1}$, and we just showed that $\pi_1(X)$ operates properly discontinuously on the image $\psi(X'))$, shows that there is a holomorphic quotient $\pi : X \to Y$, with connected fibres equal to the closures of the leaves of $\mathscr{F}$, with Y a normal variety of dimension $= k$, and π a submersion on $X - \sum$. Moreover, since $\mathbb{C}(\mathscr{F})$ is algebraically closed inside $\mathbb{C}(X)$, $\mathbb{C}(\mathscr{F}) = \mathbb{C}(Y)$, and finally the holomorphic forms in U are pull-backs from Y.

Q.E.D.

The following is the generalization we were previously referring to

THEOREM 2.24. (Twisted GCdF) *Let X be a compact Kaehler manifold, and let $\omega_1, \omega_2, \ldots, \omega_{k+1}$ be respective sections of $H^0(\Omega^1_X \otimes \mathscr{L}_i)$, where $\mathscr{L}_i$, for $i = 1, k+1$, is an invertible sheaf in $\mathrm{Pic}^o(X)$. Assume that $\omega_1 \wedge \omega_2 \wedge \ldots \wedge \omega_{k+1} = 0$, while $\Lambda^k(\oplus_{i=1,\ldots,k+1} \mathbb{C}\omega_i)$ embeds into $H^0(\Omega^k_{X^*})$, where X^* is the universal cover of X. Then there do exist a holomorphic map $f : X \to Y$ to a k-dimensional normal variety, and respective sections $\eta_1, \eta_2, \ldots, \eta_{k+1}$ of $H^0(\Omega^1_Y \otimes \mathscr{L}'_i)$, where $\mathscr{L}'_i$, for $i = 1, \ldots, k+1$, is an invertible sheaf in $\mathrm{Pic}^o(Y)$, such that $f^*(\mathscr{L}'_i) \cong \mathscr{L}_i$, and $f^*(\eta_i) = \omega_i$.*

Proof. As in 2.23, we obtain a rank k saturated subsheaf $\mathscr{H}$ of Ω^1_X, and a foliation $\mathscr{F}$ defined by $\mathscr{H}$. We argue as in 2.23, replacing the field $\mathbb{C}(\mathscr{F})$ by the field of local meromorphic functions at x obtained by meromorphic sections of $\mathscr{L}$ which are constant on the leaves of the foliation.

We proceed with the same local argument mutatis mutandis, that is, we fix local flat trivializations of the $\mathscr{L}_i$'s, so that the ω_i's are represented locally by d-closed holomorphic 1-forms, and we can take linear combinations of them (with $\mathbb{C}$-coefficients). We choose $w_1, w_2, \ldots, w_r$ to be meromorphic functions $\in \mathbb{C}(\mathscr{F})$ such that they give a holomorphic map of highest possible rank at some point p in a neighbourhood U of x.

The only difference now is that the partial derivatives $\partial \psi / \partial z_j$, $\partial w / \partial z_j$ (in the chosen system of coordinates) are represented by sections σ of some flat bundle $\mathscr{L}$, and are constant on the leaves of $\mathscr{F}$.

But then $\mathrm{dlog}(\sigma)$ is a meromorphic 1-form belonging to the differentials of the field $\mathbb{C}(\mathscr{F})$, and thus σ can locally be written as a holomorphic function of $w_1, w_2, \ldots, w_r$. Then the argument is the same as in 2.23, yielding $r = k$.

Therefore the leaves are closed and the same argument applies: on the universal cover X' of X, the leaves are the inverse images of a holomorphic function $\psi : X' \to \mathbb{C}^{k+1}$, and since $\pi_1(X)$ operates properly discontinuously on the image $\psi(X')$, there is a holomorphic quotient $\pi : X \to Y$, with Y a normal variety of dimension $= k$, and with π having connected fibres equal to the closures of the leaves of $\mathscr{F}$ and being a submersion on $X - \sum$.

DEFINITION 2.25. *A holomorphic fibration $f : X \to Y$ to a k-dimensional normal variety of Albanese general type, as in theorem 2.23 (GCdF), shall be called an Albanese general k-fibration, or a higher irrational pencil.*

A compact Kaehler manifold X admitting no higher irrational pencil shall be said to be **primitive**.

The property of X being primitive is just a topological property of X and actually is a property of the cohomology algebra $H^*(X, \mathbb{C})$ endowed with complex conjugation.

The results we are now presenting show that the classification of irregular manifolds should be attacked via the study of primitive manifolds and via the study of the fibrations over them.

We have to introduce some terminology, in particular we let (cf. 2.11).
$\Lambda^{\mathrm{hol}}(X)$ be the exterior subalgebra generated by $H^0(\Omega_X^1)$.

DEFINITION 2.26. *Let A be a graded exterior algebra, and U be a vector suspace of A^1. Then U is said to be a k-wedge subspace if $\Lambda^k U$, the natural image of $\Lambda^k(U)$ into A^k, is non zero, whereas the dimension of U is at least $k + 1$ and $\Lambda^{k+1} U = 0$.*

U is said to be a strict k-wedge subspace if moreover the natural map of $\Lambda^k(U)$ into A^k is an isomorphism onto its image $\Lambda^k U$.

U is said to be an honest k-wedge subspace if it contains a strict k-wedge subspace.

REMARK 2.27. *Let U be a k-wedge subspace.*
Then one can prove:

i) there exists an integer $k' \leq k$, and a subspace $U' \subset U$ such that U' is a strict k'-wedge subspace

ii) if U is strict, there do not exist an integer $h < k$ and a subspace U'' of U which is a h-wedge

iii) if U is $(k + 1)$-dimensional, it contains a unique strict wedge subspace U' as in i).

Irregular Surfaces 73

More generally, a honest k-wedge subspace U is said to be a *primitive* k-wedge if the only wedge subspaces it contains are k-wedges.

REMARK 2.28. G.C.d.F. can thus be rephrased as: *there is a bijection between* { *maximal honest k-wedge subspaces U of $H^{1,0}(X)$*} *and* { *Albanese general type k-fibrations $f: X \to Y$*}.

A corollary of G.C.d.F. is that either there is a strict k-wedge vector subspace of $H^0(\Omega_X^1) = H^{1,0}(X)$, for some $k < n = \dim X$, or, letting $h = \min\{q, n\}$, every h-dimensional subspace U of $H^0(\Omega_X^1) = H^{1,0}(X)$ has $\Lambda^h U \neq 0$.

In particular, **if $q > n$, and X is primitive**, since $n(q - n)$ is the projective dimension of the Grassmann variety, **the subspace** $(\Lambda^n H^0(\Omega_X^1))$ **of** $H^0(\Omega_X^n)$ **has dimension at least** $n(q - n) + 1$.

We give now a few indications of the main ideas leading to theorems 2.32-2.34: by the Kaehler condition, the type decomposition of forms passes to cohomology, and therefore the Hodge decompostion, together with complex conjugation, relates properties of the exterior algebra $H^*(X, \mathbb{C})$ to properties of the («holomorphic») subalgebra $\Lambda^{\text{hol}}(X)$ generated by $H^0(\Omega_X^1)$.

Thus the existence of wedge subspaces in $H^1(X, \mathbb{C})$ implies the existence of wedge subspaces in $H^0(\Omega_X^1)$, even if we have to remark that there are some technical difficulties due to the fact that (as one can already see in the case where X is a curve) the natural homomorphism of $\Lambda^{\text{hol}}(X) \otimes \Lambda^{\text{hol}}(X)^*$ (* denoting complex conjugation), into $H^*(X, \mathbb{C})$ has a very huge kernel, hence recovering strict and maximal honest k-wedge subspaces of $H^0(\Omega_X^1)$ by inspecting wedge subspaces of $H^1(X, \mathbb{C})$, and using complex conjugation in $H^1(X, \mathbb{C}) = H^1(X, \mathbb{R}) \otimes_\mathbb{R} \mathbb{C}$, is not such an easy matter.

Essentially the main point is to translate properties of a subspace U of $H^0(\Omega_X^1)$ into properties of $U \oplus U^*$; the attempt we give (2.30 and foll.), through the definition of a «good» subspace, is a bit complicated by the fact that if V is good, and U is such that $U \oplus U^* = V \oplus V^*$, then it is not necessarily true that also U is good, as it is shown by the following

EXAMPLE 2.29. Let X be the hyperelliptic Riemann surface of equation $w^2 = P(z)$, where P is a polynomial with real roots, hence such that $|P(z)| = |P(z^*)|$.

We let $\varphi_1 = w^{-1} dz, \varphi_2 = zw^{-1} dz$, so that $\varphi_1 \wedge \varphi_2^* + \varphi_1^* \wedge \varphi_2$ is zero in cohomology, its integral on X being given by twice the integral over the Riemann sphere of the form $-2 i \text{Im}(z)|P(z)|^{-1} dz \wedge dz^*$, which vanishes being antisymmetrical for the involution exchanging z with z^*.

Here $k = 1, V$ is spanned by φ_1, φ_2, while U is spanned by $\varphi_1 + \varphi_1^*, \varphi_2 + \varphi_2^*$.

DEFINITION 2.30. *A k-wedge subspace of $H^1(X, \mathbb{C})$ is said to be a good k-wedge if $V \cap V^* = 0$ (V^* denoting as usual the complex conjugate of V), and if moreover $\Lambda^{2k} W \oplus W^* \neq 0$ for each k-dimensional subspace W of V*.

DEFINITION 2.31. *A $(2k$-wedge) self conjugate subspace F of $H^1(X, \mathbb{C})$ is said to be a real $2k$-wedge if there exists a k-wedge V such that $V \oplus V^* = F$*.

F is said to be a saturated good real $2k$-wedge if it is a maximal real $2k$-wedge, and one can find V as above which is good.

After setting up this terminology, we have a few theorems, for the proof of which we defer to [Cat4].

THEOREM 2.32. *Let k be the smallest integer for which there exists a k-wedge subspace V of $H^1(X, \mathbb{C})$. Then V uniquely determines a honest primitive k-wedge subspace of $H^0(\Omega^1_X) = H^{1,0}(X)$.*

THEOREM 2.33. *Let V be a k-wedge subspace of $H^1(X, \mathbb{C})$. Then there are subspaces U, W of $H^0(\Omega^1_X) = H^{1,0}(X)$ such that V is contained in $U \oplus W^*$, and such that $U + W$ is a k-wedge subspace.*

In particular, every real $2k$-wedge is contained in a unique maximal real $2k$-wedge subspace F of $H^1(X, \mathbb{C})$ (cf. 2.31).

Such an F can be uniquely written in the form $Y \oplus Y^*$, with Y a maximal k-wedge subspace of $H^0(\Omega^1_X)$.

THEOREM 2.34. *Every saturated real $(2k$-wedge$)$ F of $H^1(X, \mathbb{C})$ uniquely corresponds to an Albanese general type fibration.*

Moreover, setting $F = Y \oplus Y^*$ as above, Y is a honest k-wedge if and only if F does not contain any saturated real $2h$-wedge subspace F' of codimension $2(k - h)$ in F.

Finally, Y is a primitive k-wedge if and only if F does not contain any h-wedge subspace, for all $h < k$.

QUESTION. Is it true that Y is honest if and only if F contains a good real $2k$-wedge (i.e. $F \supset V \oplus V^*$ with V a good k-wedge)?

3. Moduli of algebraic surfaces and irrational pencils

Noether introduced in 1887 ([Noe]) the concept of number M of moduli of an algebraic surface S, and postulated an equality which in modern terminology reads out as $M = 10(1 + p_a) - 2K^2$, where K is a canonical divisor on S, i.e. the divisor of a meromorphic 2-form.

It is a fundamental contribution of F. Enriques ([En], pp. 209-215) the clarification of the role of the above postulation formula as a lower bound for the number of moduli:

$$10(1 + p_a) - 2K^2 \leq M. \tag{3.1}$$

We do not claim that Enriques gave a proof of the previous inequality (in fact he did not claim it himself, firstly because of a working hypothesis, secondly because he was not completely sure about the case of irregular surfaces).

But, in the years 1956-57 on, a complete clarification came through the theory of deformations developed by Kodaira and Spencer (see [K-M], [Kod1]), and when Kuranishi ([Ku]) proved his famous theorem, then the inequality followed as a direct consequence.

Only much later, in 1982, we remarked ([Cat1]), as a consequence of Freedman's theorem on the topological classification of simply connected 4-manifolds ([Fr]), that, contrary to the case of curves, there can be no postulation formula for the number M of moduli of a surface. In fact we proved that, for each integer d, there are d pairwise homeomorphic surfaces $S_1, S_2, \ldots, S_d$, such that their respective numbers of moduli $M_1, M_2, \ldots, M_d$, are pairwise distinct.

There exists anyhow an effective upper bound for the number of moduli (cf. [Cat1])

$$10(1 + p_a) - 2K^2 \leq M \leq 10(1 + p_a) + 3K^2 + \text{constant}. \tag{3.2}$$

Irregular Surfaces

It was rather clear that, as soon as a reasonable assumption would be made upon a given surface S, then immediately a much better inequality could be derived.

In this context, Bombieri brought to my attention a paper by Castelnuovo published in 1949 ([Cas2]), claiming that, if S is an irregular surface not admitting any irrational pencil, then
$$M \leq p_g + 2q. \tag{3.3}$$
Unfortunately the proof was based upon some very fishy proof by Severi ([Sev1]) of the following (deeply false) fact: if the Albanese map has as image a surface, then α is a local embedding.

In fact, by taking ramified covers of Abelian surfaces I gave ([Cat1]) an infinite series of examples of surfaces with $q = 2$, without irrational pencils, and with $M \geq 4p_g$.

On the other hand, I showed that if one would indeed assume the Albanese map to be unramified in codimension 1, i.e., a local embedding outside a finite subset of S (and with no assumption whatsoever about the existence of irrational pencils), then one would have two inequalities
$$\begin{aligned} &i)\, K^2 \geq 6(1+p_a) \\ &ii)\, M \leq p_g + 3q - 3. \end{aligned} \tag{3.4}$$

I asked ([Cat2]) in fact whether the sharper inequality asserted by Castelnuovo could indeed be proved with the same assumptions, and a very nice proof was given in 1985 by I. Reider ([Rei]).

I'll sketch now my argument of proof, because in fact it can lead to a better estimate, and motivates many of the questions I'll discuss at greater length in the next chapter.

Sketch of proof of 3.4:
First of all, the number of moduli M is the dimension of the local moduli space, given in this case by the base of the Kuranishi family, whose tangent space is $H^1(\Theta_S)$. By Serre duality
$$h^1(\Theta_S) = h^1(\Omega_S^1 \otimes \Omega_S^2) = h^1(\Omega_S^1(K)).$$

By assumption one can see that there are three 1-forms η_1, η_2, η_3 such that $C = \mathrm{div}(\eta_1 \wedge \eta_2)$ is a reduced irreducible canonical curve.

η_1, η_2 yield moreover an exact sequence
$$0 \to (\mathcal{O}_S)^2 \to \Omega_S^1 \to \mathcal{F} \to 0, \quad \text{while } \eta_3 \text{ induces} \tag{3.5}$$

$$0 \to \mathcal{O}_C \to \mathcal{F} \to \Delta \to 0, \tag{3.6}$$

where Δ is supported on a finite set, in particular its second Chern class is non positive.

The first inequality is derived by extracting Chern classes, while the second is gotten after tensoring the previous sequences with $\mathcal{O}_S(K)$, and taking the long cohomology sequences. What emerges is that one would have indeed the inequality
$$M \leq p_g + 2q - 2, \tag{3.7}$$
provided one could show that
$$h^0(\mathcal{O}_C(K)) \leq p_g. \tag{3.8}$$

To bound $h^0(\mathcal{O}_C(K))$, I looked at the exact sequence

$$0 \to \mathcal{O}_S \to \mathcal{O}_S(K) \to \mathcal{O}_C(K) \to 0$$

and then obtained only

$$h^0(\mathcal{O}_C(K)) \leq p_g + q - 1.$$

On the other hand, since $H^0(\mathcal{O}_C(K))$ is the space of first order deformations of the curve C, it was clear that there was floating around some inequality concerning the dimension of the space of deformations of C in S.

In fact, looking at Enriques' book, you can see that he asks the question whether the dimension of the **paracanonical system**, which he defined as the algebraic system of curves algebraically equivalent to a canonical divisor, is at most p_g.

But let's leave this matter temporarily aside, and let us see the different ideas Reider used.

He uses two exact sequences instead, the first of which is the Koszul sequence associated to one of the above 1-forms η (the zero scheme Z of η being of dimension zero by assumption):

$$0 \to \mathcal{O}_S \to \Omega_S^1 \to \mathcal{I}_Z \Omega_S^2 \to 0 \tag{3.10}$$

$$0 \to \mathcal{O}_S(-C) \to \mathcal{I}_Z \to \mathcal{J} \to 0. \tag{3.11}$$

He uses then Grothendieck duality on the possibly singular curve C, and then the non degenerate bilinear map trick (if $V \times W \to U$ is a non degenerate and bilinear map of vector spaces over $\mathbb{C}$, then $\dim U \geq \dim V + \dim W - 1$).

The nice aspects of his proof are two: first of all his proof generalizes to higher dimension, secondly he can prove the following

THEOREM 3.12. (Reider) *Assume that the surface S admits a holomorphic 1-form η with 0-dimensional zero scheme Z, and that there exists an irreducible curve in $|\mathcal{I}_Z(2K)|$. Then $K^2 \geq 4(1 + p_a) + (q - 5)/3$.*

I find this result rather important, it is a step in the direction of the following inequality asserted by Severi, and that we shall call **Severi conjecture**:

Severi conjecture (3.13) $K^2 \geq 4(1 + p_a)$ if the Albanese image of S is a surface.

I should point out that Severi's inequality has also been established in another case, namely by Xiao Gang.

THEOREM 3.14. (Xiao)*Assume that the surface S admits an irrational pencil and that the image of the Albanese map of S is a surface. Then $K^2 \geq 4(1 + p_a)$.*

Xiao uses the assumption that the surface S admits an irrational pencil $f : S \to C$ by analyzing the Harder-Narasimhan destabilizing filtration for the direct image $f_*(\omega_{X/C})$, using also Fujita's result ([Fu]) asserting that every locally free quotient of $f_*(\omega_{X/C})$ has positive degree.

Hence, to prove Severi's inequality, there remain some nasty cases, where for example S has no irrational pencil, but e.g. all 1-forms vanish on some curve.

On the other hand, the results mentioned in the previous chapter concerning irrational pencils have recently been used (cf. [Cat4]) to obtain bounds for the number M of moduli of a surface S fibred over a curve C of genus $b \geq 2$, and with fibres of genus $g \geq 2$, (then S is of generaly type, cf. [B-P-V], chapter III).

Irregular Surfaces 77

We recall that c_2, the second Chern class of S, equals the topological Euler-Poincare' characteristic of S, and (written in classical notation as $I+4$, where I is the Zeuthen-Segre invariant of S) it is related to K^2 and p_a by Noether's formula

$$c_2 = 12(1+p_a) - K^2 . \qquad (3.15)$$

THEOREM 3.16. *Let S be a complex surface fibred by a holomorphic map $f : S \to C$, with genus $(C) = b \geq 2$, and with a general fibre F of f of genus $= g \geq 2$.*
Then we have the following upper bounds for the number M of moduli of S

i) if the fibres of f have nonconstant moduli, then

$$M \leq c_2 - (b-1)(4g-7) \leq c_2 - (b-1)(4(q-b)-3) .$$

ii) if the fibres of f have constant moduli, then $M \leq c_2 - (2b-3)(2g-4) + 5$.
iii) if f is a holomorphic bundle, but not a product, then

$$M \leq 3(b-1) + 2(g-1) + 4 \leq 3/4 c_2 - 3(b-2)(g-2) + 6 .$$

iv) if f is a product projection, then $M = 3(b-1) + 3(g-1) \leq 3/4 c_2 - 3(b-2)(g-2) + 3$.

Idea of Proof. Notice first of all that any surface S' homeomorphic to S carries, by the results of the previous chapter, a fibration over a curve of genus b, and if there is a diffeomorphism carrying one saturated subspace to the corresponding one, then also the genus g of the fibres is the same.

Moreover, if the fibration has nonconstant moduli, the same occurs for any small deformation of S.

Let now $\sum \subset C$ be the set of critical values of f, and σ its cardinality. If $x \in C$, we set $F_x = f^{-1}(x)$, and we denote by $e(F_x)$ the topological Euler-Poincare' characteristic of the fibre F_x, which equals $-2(g-1)$ unless $x \in \sum$, otherwise we have $e(F_x) = -2(g-1) + \delta(x)$. $\delta(x)$ is called the defect, and it is a strictly positive number.

Applying the classical Zuthen-Segre formula, asserting that

$$c_2 = 4(b-1)(g-1) + \sum_{x \in \sum} \delta(x) , \qquad (3.17)$$

we derive $\sigma \leq c_2 - 4(b-1)(g-1)$.

Case i): set $C^* = C - \sum$ and apply Arakelov's theorem ([Ara]), cf. also [B-P-V]): given C^*, there are only a finite number of fibrations $f' : S' \to C$, with genus of a general fibre F' of f' equal to g, and with critical values of f' contained in $\sum$.

Then the number M of moduli is at most the number of moduli for the pairs $(\sum \subset C)$, which is $\sigma + 3(b-1) \leq c_2 - 4(b-1)(g-1) + 3(b-1)$, whence the first assertion (since $g + b \geq q$, equality holding if and only if f has constant moduli, cf. ([Be3]).

Cases ii), iii): **S** is a quotient $F \times C''/G$, where we can assume that G is a subgroup of $Aut(F)$.

By Hurwitz's theorem $\text{Aut}(F)$ has order at most $84(g-1)$, in particular there are only a finite number of choices for G, once g is fixed. The irregularity q of S equals the dimension $h^0(\Omega^1_{F\times C''})^G$ of the space of G-invariant 1-forms on $F \times C''$: since $C''/G = C$, we obtain that $q = b + h^0(\Omega^1_F)^G$, and thus genus $(F/G) = (q - b)$.

Then one plays with Hurwitz's formula for the map of F onto F/G, and concludes the proof easily.

Q.E.D.

It is an interesting question whether the previous bounds can be improved upon without making too special assumptions.

REMARK 3.18. *The inequalities become worse when b, or g, are low, say $= 2$. For instance, the bound can be improved when $g = 2$, as it was suggested by Rick Miranda, with whom we carried out the following computation. Let S be a double cover of $C \times \mathbb{P}^1$, branched on a smooth irreducible curve B. Then, if γ is the genus of B, then by Hurwitz's formula, since B is a $6 - 1$ cover of $\mathbb{P}^1$, we have $2\gamma - 2 = \mu + 12$, where μ is the number of ramification points of the projection of B onto $\mathbb{P}^1$. On the other hand, since $f : S \to C$ has nonconstant moduli, the Zeuthen-Segre formula yields $\mu \leq c_2 - 4(q - 1)$, and moreover the genus b of C is q or $q - 1$.*

To bound the number of moduli M of S, it suffices, by the considerations made previously, to add to $3(q - 1)$ the dimension $h^0(\mathcal{O}_B(B))$ of the characteristic series of B. We know that the canonical divisor on $C \times \mathbb{P}^1$ is algebraically equivalent to $2(b - 1)F_1 - 2F_2$, where F_1, F_2 are the fibres of the two projections of $C \times \mathbb{P}^1$; we can then bound as follows:

$$\gamma + 12 \geq h^0(\omega_B(2F_2)) \geq (\text{by Clifford's inequality}) \geq h^0(\mathcal{O}_B(B)) + (q - 1)d,$$

where d is the degree of the projection of B onto C.

Hence $h^0(\mathcal{O}_B(B)) \leq \mu/2 + 19 - (q - 1)d \leq 1/2(c_2) - (2 + d)(q - 1) + 19$, and finally $M \leq 1/2(c_2) - (d - 1)(q - 1) + 19$.

4. The paracanonical system and generic vanishing theorems

Let's go back to the inequality (3.8) $h^0(\mathcal{O}_C(K)) \leq p_g$, which is related to the algebraic system $[K]$ of effective divisors algebraically equivalent to K (it of course contains the linear system $|K|$, which has dimension $p_g - 1$).

On the grounds of examples, Enriques ([En] page 356) asked whether the dimension of $[K]$ would be equal to p_g.

In fact some of these examples, e.g. the symmetric square of a curve, were worked out by Severi in Bologna after the suggestion of Enriques (at least according to Enriques).

But, nevertheless, Enriques himself pointed out a counterexample, namely of a surface X whose Albanese map yields an elliptic fibration over a curve C of genus q.

A little reflection shows that more generally, any irrational pencil with base curve C of genus $b > (q + 1)/2$ provides a counterexample, since it implies the existence of exorbitant linear systems on X (cf. [Sev], and [Ca1], [Ca2], [Be2] for a modern account).

In fact, $[K]$ is a fibration of projective spaces over $\text{Pic}^0(X) = A^*$, the dual Abelian variety of $A = \text{Alb}(X) : [K] = \bigcup_{t \in A^*} |K + t|$, where we identify a point $t \in A^*$ with a Cartier divisor such that the associated sheaf $\mathscr{P}_t$ represents the isomorphism class t.

By the Riemann-Roch theorem, if $t \neq 0$, since $H^0(X, \mathcal{P}_t) = 0$,

$$\dim |K + t| = p_g - q + h^1(X, \mathcal{P}_t). \tag{4.1}$$

Hence the dimension of $[K]$ is $\geq p_g$, and indeed equal if $h^1(X, \mathcal{P}_t) = 0$ for $t \neq 0$. But, if $f : X \to C$ is an irrational pencil, with C of genus $b \geq 2$, since the natural homomorphism $f^* : \mathrm{Pic}^\circ(C) \to \mathrm{Pic}^\circ(X)$ is injective, and for $\mathscr{L}$ of degree 0 on C, $h^1(C, \mathscr{L}) = b - 1$, whereas $h^1(X, f^*\mathscr{L}) \geq h^1(C, \mathscr{L})$, we get on X a b-dimensional family of linear systems in $[K]$, each of dimension at least $p_g - q + b - 1$, so that $\dim [K]$ exceeds p_g if $2b > q+1$.

Thus at the Ravello Conference in 1982, entitled «Open problems in Algebraic Geometry», (cf. [Cat2]) I posed the following question, whether if X is an irregular surface without irrational pencils, then the dimension of the irreducible component of $[K]$ containing $|K|$ would be $\leq p_g$, and indeed whether under the same assumption one would have $h^1(X, \mathcal{P}_t) = 0$ for $t \neq 0$.

Soon after, Beauville showed that one could not have this strong form of vanishing result for $H^1(X, \mathcal{P}_t)$, even if X has no irrational pencils. Taking a sufficiently general surface section of a $\mathbb{Z}/2$ quotient of the product of two Abelian varieties, $A \neq B$, with $\mathbb{Z}/2$ acting as $+1$ on the tangent space of A, and as -1 on the tangent space of B, he produced an example where there are no irrational pencils, yet for t a point of order 2, $h^1(X, \mathcal{P}_t)$ is very big compared to $q = \dim A$, and $\dim |K + t|$ exceeds p_g.

Beauville also pointed out that the conjecture should be modified then in the following way: if the Albanese image of X is a surface, then, for t generic, $h^1(X, \mathcal{P}_t) = 0$.

He also generalized the conjecture in higher dimension as follows: if X has Albanese image of dimension r, then, if t is generic, $h^i(X, \mathcal{P}_t) = 0$ for $i < r$.

Concerning the structure of $[K]$, a few more words are needed.

According to Grothendieck, to a coherent sheaf $\mathscr{F}$ on Y, one can associate a fibration of projective spaces as follows:

if $\mathscr{L}_1 \to \mathscr{L}_0 \to \mathscr{F} \to 0$ is an exact sequence where the $\mathscr{L}_i$'s are locally free, to $\mathscr{F}$ we associate the schematic kernel of $\mathrm{Proj}(\mathscr{L}_0) \to \mathrm{Proj}(\mathscr{L}_1)$, denoted by $\mathrm{Proj}(\mathscr{F})$.

In this situation, let $\mathscr{P}$ be a Poincare' invertible sheaf on $X \times A^*$ (thus the restriction of $\mathscr{P}$ to $X \times \{t\}$ is the above $\mathscr{P}_t$); if $\pi_X : X \times A^* \to X$ is the projection onto the first factor, $\pi : X \times A^* \to A^*$ is the projection onto the second factor, then $\mathscr{H} = \pi_X^*([K]) \otimes \mathscr{P}$ is defined to be the **paracanonical sheaf** on $X \times A^*$, and we take as $\mathscr{F}$ on A^* the coherent sheaf $\mathscr{R}^2\pi_*\mathscr{P}$. This sheaf enjoyes the base change property, and since the dual vector space of $H^2(X, \mathcal{P}_t)$ is just $H^0(X, \mathcal{O}_X(K) \otimes \mathcal{P}_t)$, we see that $[K]$ is the fibration of projective spaces $\mathrm{Proj}(\mathscr{R}^2\pi_*\mathscr{P})$. In particular there is an open set U in A^* where $\mathscr{R}^2\pi_*\mathscr{P}$ is locally free and of minimal rank: the closure of the restriction of $[K]$ to U is now defined to be the **paracanonical system** $\{K\}$ of X.

Note that by relative duality for the map π, the torsion free sheaf $\pi_*\mathscr{H}$ is the dual of $\mathscr{R}^2\pi_*\mathscr{P}$, so another possibility could also be the one of defining $\{K\}$ as Proj of the reflexive sheaf dual of $\pi_*\mathscr{H}$, which is naturally also a subspace of $[K]$: but it is not at all clear that this second definition is the same as the previous one.

Not long thereafter, in 86-87, a beautiful and rather complete answer to these questions was given by M. Green and R. Lazarsfeld ([G-L1]).

These authors consider (for $i < n$) subvarieties

$$S^i(X) \subset \mathrm{Pic}^\circ(X), S^i(X) = \{\mathscr{L} \in \mathrm{Pic}^\circ(X) | H^i(X, \mathscr{L}) \neq 0\}, \tag{4.2}$$

and prove the following remarkable result

THEOREM 4.3. (Green-Lazarsfeld's Generic vanishing theorem) *If X is an irregular Kaehler manifold, α its Albanese map, then*

i) $\mathrm{codim}(S^i(X)) \geq \dim(\alpha(X)) - i$

ii) $H^i(X, \Omega_X^j \otimes \mathscr{L}) = 0$ *for* $i + j < \dim(\alpha(X))$ *and* $\mathscr{L}$ *generic*

iii) *If X does not possess any irrational pencil of genus ≥ 2, then 0 is an isolated point in $S^1(X)$, and in particular* $\dim\{K\} = p_g$.

Idea of Proof. First of all, by the base change theorem, there is a complex of vector bundles on $\mathrm{Pic}^\circ(X)$

$$0 \to E^0 \to E^1 \to \ldots \to E^{n-1} \to E^n \to 0 \qquad (*)$$

such that for $t \in \mathrm{Pic}^\circ(X)$, the cohomology groups $H^j(X, \mathscr{P}_t)$ are given as the cohomology of the complex of vector spaces obtained by tensoring $(*)$ with the residue field of the local ring of the point $t \in \mathrm{Pic}^\circ(X)$.

The new technique these authors introduce is the Derivative complex, to compute first order deformations of cohomology groups.

To explain the concept of the derivative complex, pick a tangent vector v at t, and consider the restriction of the complex $(*)$ to the double point determined by t and v.

We then have a complex of free sheaves over $\mathrm{Spec}\,(\mathbb{C}[x]/(x^2))$, and writing as $d + xD$, resp. $d' + xD'$ two successive differentials, the condition of having a complex reads out as $d'd = 0, D'd + d'D = 0$.

We can interpret the above formulae by saying that the operators $D, D', \ldots$ induce a complex on the cohomology of the complex of vector spaces whose differentials are given by the $d, d', \ldots$; the cohomology of this new complex shall be denoted by $H^i(D_v)$.

The two basic results are then the following ones:

a) defining $S_m^i(X) = \{\mathscr{L} \in \mathrm{Pic}^\circ(X) | h^i(X, \mathscr{L}) \geq m\}$, then, if $t \in S_m^i(X)$ and v is a tangent vector at t, then v is in the tangent cone at $S_m^i(X)$ if and only if $\dim H^i(D_v) \geq m$.

b) $D_v : H^{i-1}(X, \mathscr{L}) \to H^i(X, \mathscr{L})$ is given by cup product with $v \in h^1(X, \mathscr{O}_X)$.

It is rather clear how to use a), since at any point the dimension of a subvariety is at most the dimension of its tangent cone. The use of b) is slightly more delicate, and we shall show for instance how to prove, through it, part iii) of the theorem.

In fact, interpreting $\mathscr{L}$ as $= \mathbf{L} \otimes \mathscr{O}_X$, where $\mathbf{L}$ is a locally constant sheaf determined by a cocycle in $H^1(X, U(1))$, we have that, by Hodge theory with twisted coefficients, $H^i(X, \mathscr{L})$ is canonically isomorphic to the complex conjugate of $(H^0(\Omega_X^i \otimes \mathbf{L}^*))$.

Thus, by taking complex conjugates, the complex given by $D_v : H^{i-1}(X, \mathscr{L}) \to H^i(X, \mathscr{L})$ is replaced by the complex, $H^0(\Omega_X^{i-1} \otimes \mathbf{L}^*) \to (H^0(\Omega_X^i \otimes \mathbf{L}^*))$ given by exterior product with η, where $\eta \in H^0(\Omega_X^1)$ is such that $v = \eta^*$.

In the case where $t = 0, \mathscr{L} = \mathscr{O}_X$, and the exactness of D_v is equivalent to the exactness of exterior multiplication by η:

$$0 \to H^0(\mathscr{O}_X) \to H^0(\Omega_X^1) \to H^0(\Omega_X^2),$$

which is guaranteed by the classical Castelnuovo-De Franchis theorem if X has no irrational pencils. Thus under this assumption 0 is an isolated point of $S^1(X)$.

Q.E.D.

Further progress on the subject was made by Beauville in '87 ([Be2]), who found that, as Hodge theory extends to unitary flat bundles, also the Castelnuovo-De Franchis theorem admits a «twisted» generalization.

(4.4.) Beauville's extension of the CdF theorem

Let X be a compact Kaehler manifold, and let $\omega \in H^0(\Omega_X^1), \alpha \in H^0(\Omega_X^1 \otimes \mathbf{L})$ be such that $\omega \wedge \alpha = 0$: then there exists a holomorphic map $f: X \to C$ to a curve such that ω, α pull back from C.

Beauville's argument differs from the classical proof just replacing, in the case of α, the operator d by the flat connection on $\mathbf{L}$, and the meromorphic function φ by a meromorphic section φ of $\mathscr{L}$ such that $\alpha = \omega \otimes \varphi$. One notices that $\mathrm{dlog}\,\varphi$ and ω differ by a meromorphic function f, and then the proof runs exactly as in the classical case.

We have seen (2.24, cf. [Cat4]) that also this twisted version generalizes to higher rank exterior products.

In any case, what is really important is the consequence of this generalization.

In fact now, by this result, also the following sequence is exact, $0 \to (H^0(\mathscr{O}_X \otimes \mathbf{L}^*)) \to (H^0(\Omega_X^1 \otimes \mathbf{L}^*)) \to (H^0(\Omega_X^2 \otimes \mathbf{L}^*))$, provided there is no irrational pencil under which η pulls back. Beauville can thus give a complete description of $S^1(X)$.

THEOREM 4.5. (Beauville) *If X is an irregular Kaehler manifold*

i) $S^1(X) = \cup_{f:X \to C} f^*(\mathrm{Pic}^\circ(C)) \cup \Delta$, *where Δ is a finite set*

ii) *$[K]$ has as irreducible components the paracanonical system $\{K\}$, the components lying over $f^*(\mathrm{Pic}^\circ(C))$, with genus (C) at least $q/2+1$, and, possibly, some linear systems of the form $|K+t|$.*

iii) *If q is even, and there are no irrational pencils of genus $\geq q/2$, then $|K|$ is an irreducible component of $[K]$.*

We defer to [Be2] for the proof.
Part iii) deserves, though, some further comment.
Let's indeed go back to the exact sequence

$$0 \to \mathscr{O}_S \to \mathscr{O}_S(K) \to \mathscr{O}_C(K) \to 0 : \qquad (4.6)$$

its long cohomology sequence identifies $H^0(\mathscr{O}_C(K))/H^0(\mathscr{O}_S(K))$ as the kernel of β: $H^1(\mathscr{O}_S) \to H^1(\mathscr{O}_S(K))$.

But the second space is, by Serre duality, dual to $H^1(\mathscr{O}_S)$, and some computation shows that, if $C = \mathrm{div}(s), s \in H^0(\mathscr{O}_S(K)), w, v \in H^1(\mathscr{O}_S)$, then $\beta(v)$ evaluated on w equals simply $s(v \cup x)$, once we have identified the cohomology group $H^2(\mathscr{O}_S(K))$ with $\mathbb{C}$.

Since $H^1(\mathscr{O}_S)$ can be identified (cf. 2.2) with $(H^0(\Omega_S^1))^*$, we get thus a linear map of $H^0(\mathscr{O}_S(K))$ into the space of alternating forms on $H^0(\Omega_S^1))^*$, induced by esterior product and the perfect pairing of $(H^0(\Omega_S^2))^*$ with $H^0(\mathscr{O}_S(K))$.

a) By dimension counting, under the assumption of Beauville's theorem, there is a section s in $H^0(\mathscr{O}_S(K))$ whose associated alternating form does not have a two dimensional kernel. Therefore, if q is even, this form is non degenerate, β is invertible, and $H^0(\mathscr{O}_C(K))$ has dimension $= p_g - 1$: whence $\{K\}$ is smooth at C and coincides with $|K|$.

b) Since the subspace B, consisting of the alternating forms with kernel of dimension at least 2, has codimension $= 1$ or 3 (accordingly as q is even of odd) in the space of all alternating forms, and q at least 2 implies $p_g \geq 2q - 1$, we have the following interesting corollary, which was mentioned to us by Lazarsfeld in 1987:

COROLLARY 4.7. (Lazarsfeld) *If X is an irregular surface without irrational pencils of genus at least $q/2$, then there exists always an obstructed canonical curve (i.e., such that it gives a singular point of the Hilbert scheme of curves algebraically equivalent to K).*

Proof. Any canonical curve C belongs to a component of $\{K\}$ of dimension at most p_g, under our assumption. But if the corresponding linear map β has a 2-dimensional kernel, the dimension of the Zariski tangent space, $h^0(\mathcal{O}_C(K))$, is at least $p_g + 1$. On the other hand, by the previous remarks, there is always a curve $C \in |K|$ with dim Ker$\beta \geq 2$.

Q.E.D.

The previous result sheds some light on the classical theory of characteristic series of curves, and in particular it shows that the existence of obstructed curves should not be regarded as a property peculiar only to pathological algebraic surfaces.

Later on, see [Be4], Beauville derived from the previous result the above mentioned theorem 2.18 (cf. [Siu], [Be4])

THEOREM. *Let X be a compact Kaehler manifold: then there is a non constant holomorphic map $f : X \to C$, to some curve of genus $g \geq 2$, if and only if there is a surjective homomorphism $\pi_1(X) \to \pi_1(C')$, where C' is some curve of genus $g' \leq g$.*

Proof. (Beauville) Firstly, we can assume, going to the Stein factorization of f, and hence replacing C by a curve C'' of higher genus, that the fibres of f are connected.

Under this assumption, since f has the local lifting property, we have a surjection $\pi_1(X) \to \pi_1(C'')$. We conclude the proof in one direction observing that there are surjective homomorphisms $\pi_1(C'') \to \pi_1(C')$.

On the other hand, if such a homomorphism $\pi_1(X) \to \pi_1(C'')$ exists, take n to be a prime number and consider all the surjective homomorphisms $\beta : \pi_1(C'') \to \mathbb{Z}/n$. The important remark is that by composition we get at least $n^{2g'} - 1$ surjective homomorphisms $\gamma : \pi_1(X) \to \mathbb{Z}/n$, each of which gives a cyclic unramified covering $\pi : X' \to X$.

The number of such distinct coverings is $(n^{2g'} - 1)/(n - 1)$.

For each of those coverings, we have a natural splitting

$$\pi_*(\mathcal{O}_{X'}) = \mathcal{O}_X \oplus (\oplus_{i=1,\ldots,n-1} \mathscr{L}^i) , \qquad (4.8)$$

where $\mathscr{L}^i$, for $i = 1, \ldots, n-1$, is an element of n-torsion in Pic$^0(X)$.

Clearly the kernel of γ, i.e. $\pi_1(X')$, maps onto ker(β), which is isomorphic to the fundamental group of a curve of genus $n(g - 1) + 1$.

If n is large enough, then the irregularity of X' is strictly larger than the irregularity of X, hence for some $i = 1, \ldots, n-1$, $H^0(X, \mathscr{L}^i) \neq 0$.

We have thus produced at least $(n^{2g'} - 1)/(n-1)$ points in $S^1(X)$ which are elements of n-torsion in Pic$^0(X)$.

Since for a k dimensional subtorus T there are at most $(n^{2k} - 1)$ n-torsion points in T, we conclude, by theorem 4.5., that there is a non constant holomorphic map $f : X \to C$, to a curve of some genus $g \geq g'$.

Q.E.D.

Recently Green and Lazarsfeld have considerably improved Beauville's theorem 4.5. ([G-L2]) proving that the subvarieties $S^i(X)$ are translates of complex subtori.

THEOREM 4.9. (Green-Lazarsfeld) *If X is an irregular compact Kaehler manifold and Z is a positive dimensional irreducible component of $S^i(X)$, then*
  1) Z *is a translate of a complex subtorus of* $\mathrm{Pic}^0(X)$
  2) *there is a holomorphic map* $f : X \to Y$ *to a k-dimensional normal variety of maximal Albanese dimension* $a = k$, *with* $k \leq i$, *and such that Z is a translate of $f^*(\mathrm{Pic}^0(Y))$.*

Idea of Proof. Let $\mathscr{L}$ be a non isolated point in $S^i_m(X)$, and let $v \in H^1(X, \mathscr{O}_X)$ be a tangent vector in the normal cone of $S^i_m(X)$. We have seen in the course of the proof of theorem 4.3. that the cohomology group $H^i(X, \mathscr{L})$ continues to live over the infinitesimal neighbourhood determined by v if and only if cup product with v induces the zero map $H^i(X, \mathscr{L}) \to H^{i+1}(X, \mathscr{L})$. In this second paper essentially the authors observe that under the above condition the whole translate at $\mathscr{L}$ of the 1-parameter subgroup generated by v is contained in $S^i_m(X)$.

The reason for this is as follows: $\mathscr{L}$ is the holomorphic bundle associated to a local system $\mathbf{L}$, and varying $\mathscr{L}$ in the direction determined by v (cf. 2.4.) amounts to the following:

i) write $v = \omega^*$, with $\omega \in H^0(\Omega^1_X)$, take an open coordinate covering U_α such that, on U_α, there are holomorphic functions φ_α with $\omega = d\varphi_\alpha$.

ii) Let $g_{\alpha\beta} \in H^1(X, U(1))$ be a cocycle which determines $\mathbf{L}$, then moving the cocycle along the translate of the 1-parameter subgroup determined by v amounts to considering the constant cocycle $g_{\alpha\beta}\exp(t2\pi i f_{\alpha\beta})$, with $f_{\alpha\beta} \in H^1(X, \mathbb{R})$ (hence the cocycle is unitary iff $t \in \mathbb{R}$).

iii) $f_{\alpha\beta}$ is related to ω as follows: one can write $\omega = \partial(\varphi_\alpha + \zeta^*_\alpha)$, for suitable holomorphic functions ζ_α, and if we set $F_\alpha = (\varphi_\alpha + \zeta^*_\alpha)$, then

$$F_\alpha - F_\beta = (\varphi_\alpha + \zeta^*_\alpha) - (\varphi_\beta + \zeta^*_\beta) = f_{\alpha\beta} \qquad (4.10)$$

is a real constant.

Now, given $\psi \in H^i(X, \mathscr{L}) = (H^0(\Omega^i_X \otimes \mathbf{L}^*))^*$, there is a natural way to propagate ψ^* as a differentiable section of $(\Omega^i_X) \otimes \mathscr{L}_t^{-1}$ along the 1-parameter subgroup.

For this, one simply replaces ψ_α, on U_α, by $\psi_\alpha(t) = \psi_\alpha \exp(t2\pi i F_\alpha)$, since then, by (4.10), $\psi_\alpha(t) = \psi_\beta(t) g_{\alpha\beta} \exp(t2\pi i f_{\alpha\beta})$.

But then one has to verify whether $\psi^*_\alpha(t)$ is still holomorphic, i.e. whether $\partial(\psi_\alpha(t)) = 0$. But $\partial(\psi_\alpha(t)) = \partial(\psi_\alpha \exp(t2\pi i F_\alpha)) = \partial(\psi_\alpha) \cdot \exp(t2\pi i F_\alpha) + t(2\pi i .\psi_\alpha(t) \wedge \partial F_\alpha) = 0 + t(2\pi i .\psi_\alpha(t) \wedge \partial F_\alpha)$.
One can then conclude:

$$\psi^*_\alpha(t) \text{ is holomorphic for all } t \text{ iff } \psi_\alpha \wedge \omega = 0. \qquad (4.11)$$

Thus, if Z is a positive dimensional irreducible component of $S^i(X)$, then at any point $\mathscr{L}$ of Z, for every direction in v in the tangent cone of Z, one sees that Z contains the translate at $\mathscr{L}$ of the 1-parameter subgroup associated to v, hence Z is the translate of a complex subtorus T (note that it is enough to give the proof for a real 1-parameter subgroup).

Thus part 1) is proved.

To prove part 2), consider the inclusion $T \subset \mathrm{Pic}^0(X)$: dually we have an epimorphism $A = \mathrm{Alb}(X) \to T^\vee$, which we can compose with the Albanese map α, obtaining the desired $f : X \to Y \to T^\vee$ as the Stein factorization.

It remains to prove that the dimension of Y is at most i.

Indeed, using our previous notation, every tangent vector v to Z is a conjugate of a holomorphic 1-form ω such that $\psi_\alpha \wedge \omega = 0$; since ψ_α is locally a holomorphic i-form, it follows by an easy argument in linear algebra that if $\omega_1, \omega_2, \ldots, \omega_{i+1}$ are as above, then $\omega_1 \wedge \omega_2 \wedge \ldots \wedge \omega_{i+1} = 0$.

Q.E.D.

5. Classification of irregular manifolds, and char p analogues

DEFINITION 5.1. *An irregular Kaehler manifold X is said to be* **simple** *if $H^i(X, \mathscr{L}) = 0$, for $i < n = \dim X$, and $\mathscr{L} \in \mathrm{Pic}^\circ(X) - \{0\}$.*

We remark that if X is simple, then it is primitive (cf. 2.25), and if X is primitive, by the previous theorem the subvarieties $S^i(X)$ are 0-dimensional, that is, $S^i(X)$ is a finite set.

I have in more than one occasion (cf. [Cat4]), and in particular during the Conference in Bologna, raised the following

CONJECTURE 5.2. *Assume X is primitive: then $S^i(X)$ consist only of torsion points.*

The above conjecture was also, as far as I understand, formulated by Beauville: the evidence for conjecturing it was not great, but recently Beauville seems to have found a proof in the case of $S^1(X)$, and under the assumption that the commutator subgroup of $\pi_1(X)$ is finitely generated. His idea is very nicely based on the fact that the only complex numbers which remain unitary under all automorphism of $\mathbb{C}$ are the roots of unity.

QUESTION 5.3. *Assume X is primitive: then does there exist an unramified Abelian covering X' of X which is simple?*

If the answer would be positive, the study of irregular manifolds would be reduced, via fibrations and unramified coverings, to the study of simple manifolds (notice that both notions, by what we have seen, are homotopy invariants).

The simple manifolds enjoy the important property that the direct image sheaves

$$(\pi)_*(\pi_X^*([K])^{\otimes m} \otimes \mathscr{P}) \text{ on } A^* \text{ (with } \pi: X \times A^* \to A^*) \tag{5.4}$$

are locally free of rank equal to $\chi([mK])$, enjoy the base change property (with the exception of the case $m = 0$ at the point $t = 0$).

Thus the natural models of simple manifolds should live in the projectivization of certain vector bundles over Abelian varieties.

This approach has been used successfully for the classification of irregular surfaces with $q = 1$ (see [C-C1], [C-C2]).

Finally, an interesting object of study for the classification and the geometry of irregular manifolds, could be the paracanonical algebra of X:

$$\oplus_{t \in A^*}(\oplus_{m \in \mathbb{N}} H^\circ(X, [mK] \otimes \mathscr{P}_t)) . \tag{5.5}$$

The last is not a graded algebra in the usual sense, but should be viewed as an algebra of finitely supported functions defined on A^* and with values in a system of coefficients, endowed with a convolution product induced by the natural bilinear pairings

$$H^\circ(X, [mK] \otimes \mathscr{P}_t) \times H^\circ(X, [m'K] \otimes \mathscr{P}_{t'})) \to H^\circ(X, [(m+m')K] \otimes \mathscr{P}_{t+t'}) .$$

On the other hand, to pursue this approach further, a better knowledge about vector bundles on Abelian varieties seems desirable.

But let's now leave this topic aside and let's pass to varieties in char $= p$.

In order to treat the case of an algebraic surface defined over an algebraically closed field of positive characteristic $= p$, we cannot use topological considerations any longer, and one possible direction is to understand the deformation theoretic analogues of the arguments we employed ([Il 1-2], [Fl], [Ho 1-3], [Pa1], [Ran 2] are then general references).

We defer to [Cat4] for discussion and proof of the following results, concerning the space Def $(f: X \to Y)$ of deformations of a holomorphic map $f: X \to Y$, and which are infinitesimal analogues of the above results concerning the topological nature of Albanese general type fibrations.

THEOREM 5.6. *Let X, Y be compact complex manifolds, or complete smooth varieties of general type (of arbitrary characteristic), and let $f: X \to Y$ be a surjective morphism with connected fibres.*

Then Def $(f: X \to Y)$ *maps onto* Def (X) *provided* $H^0(\mathcal{R}^1 f_* \mathcal{O}_X \otimes \Theta_Y) = 0$.

The above hypothesis holds, in particular, if char $= 0$, Y is a curve of genus ≥ 2, and X is projective, but it is not completely clear as to which degree of generality this result can be extended: in a recent letter, Kollar points out that the hypothesis holds in char $= 0$ for Y with ample canonical bundle, or Y a surface of general type.

On the other hand $H^0(\mathcal{R}^1 f_* \mathcal{O}_X \otimes \Theta_Y)$ does not need to vanish in positive characteristic, even when Y is a curve of genus ≥ 2 (cf. [Cat4]).

PROBLEM 5.7. Are there algebraic surface in positive characteristic with an irrational pencil of genus ≥ 2 which is not stable by deformation?

Turning to the problem of bounding the number of moduli in positive characteristic, in [Cat4] some result is gotten under some supplementary hypothesis, and the following result is proved

THEOREM 5.8. *Let X be a complete smooth surface of general type (of arbitrary characteristic), and let $f: X \to C$ be a fibration, with C a curve of genus b, and where f has only a finite number of critical points, and a general fibre of genus g.*

Then we have the following inequalities for the tangent dimension M' of the space of moduli of $X(M' = h^1(\Theta_X))$ if the fibres of f have nonconstant moduli:

$$M' \leq c_2 - (b-1)(4g-7) \leq c_2 - (b-1)(4(q-b)-3).$$

It would be interesting to see whether also this result can extend to a more general situation.

Acknowledgements. It is a pleasure to acknowledge my indebtness to Ciro Ciliberto and Arnaud Beauville.

Ciro Ciliberto, since long now, stimulated my interest towards reading the classic Italian algebraic geometers, and has always been a precious reference point for me. Letters, articles and conversations with Arnaud Beauville have often been very inspiring for me.

Finally I want to thank Salvatore Coen for organizing the extremely enjoyable Conference, and Harry D'Souza for improving my English.

This research has been partly supported by M.U.R.S.T. and G.N.S.A.G.A. of C.N.R.

REFERENCES

[Ara] S.J. Arakelov, «Families of algebraic curves with fixed degeneracies», *Math. USSR Izv.* **5** (1971), 1277-1302.

[At] M.F. Atiyah, «Vector bundles over an elliptic curve», *Proc. London Math. Soc.* (3), **7**, (1957), 414-452.

[Be1] A. Beauville, «Surfaces algebriques complexes», Asterisque **54** Soc. Math. France, Paris (1978).

[Be2] A. Beauville, «Annullation du H^1 et systemes paracanoniques sur les surfaces», *Crelle Jour.* **388**, (1988), 149-157.

[Be3] A. Beauville, «L'inegalité $p_g \geq 2q - 4$ pour les surfaces de type general», *Bull. Soc. Math. France* **110** (1982), 344-346.

[Be4] A. Beauville, Appendix to [Ca4].

[Bog] F. Bogomolov, «Holomorphic tensors and vector bundles on projective varieties», *Math. USSR Izv.*, **13** (1979), 499-555.

[Bo] E. Bombieri, «Canonical models of surfaces of general type», *Publ. Math. I.H.E.S.*, **42** (1973), 171-219.

[B-M3] E. Bombieri, D. Mumford, «Enriques' classification of surfaces in char. p: III», *Inv. Math.* **35** (1976), 197-232.

[B-P-V] W. Barth, C. Peters, A. Van de Ven «Compact complex surfaces», Ergebnisse der Math., Springer Verlag, (1984).

[Cas1] G. Castelnuovo, «Sulle superficie aventi il genere aritmetico negativo», *Rend. Circ. Mat. Palermo*, **20** (1905), 55-60.

[Cas2] G. Castelnuovo, «Sul numero dei moduli di una superficie irregolare», I, II, *Rend. Acc. Lincei*, **7** (1949), 3-7, 8-11.

[Cat1] F. Catanese, «On the moduli spaces of surfaces of general type», *J. Diff. Geom.*, **19** (1984), 483-515.

[Cat2] F. Catanese, «Moduli of surfaces of general type», in "Algebraic Geometry Open Problems", *Proc. Ravello* 1982, Springer L.N.M. **997** (1983), 90-112.

[Cat3] F. Catanese, «Moduli of algebraic surfaces», *"Theory of Moduli"*, Proc. C.I.M.E. 1985, Springer L.N.M. 1337 (1988), 1-83.

[Cat4] F. Catanese, «Moduli and classification or irregular Kaehler manifolds (and algebraic varieties) with Albanese general type fibrations», Preprint, 1989.

[C-C1] F. Catanese, C. Ciliberto, «Surfaces with $p_g = q = 1$ », to appear in "Problems on surfaces and their classification", *Proc. Cortona* 1988, Symp. Math., INDAM, Academic Press.

[C-C2] F. Catanese, C. Ciliberto, «Symmetric products of elliptic curves and surfaces with $p_g = q = 1$ », to appear in the Proc. of the Conf. *"Projective Varieties"*, Trieste 1989, ed. by Springer L.N.M.

[D-F] M. De Franchis, «Sulle superficie algebriche le quali contengono un fascio irrazionale di curve», *Rend. Circ. Mat. Palermo*, **20** (1905), 49-54.

[D-F2] M. De Franchis, «Sulle varietà algebriche di n dimensioni trasformabili razionalmente in varietà a $p < n$ dimensioni, aventi il genere p-dimensionale maggiore di p», *Atti Acc. Gioenia di Scienze Nat.* in Catania, Ser. 5 vol. 3.

[Des] M. Deschamps, «Reduction semi-stable», in *"Seminaire sur les pinceaux de courbes de genre au moins deux, Asterique"*, **86**, Soc. Math. France, Paris, (1981), 1-34.

[En] F. Enriques, *«Le superficie algebriche»*, Zanichelli, Bologna, (1949).

[Fl] H. Flenner, «Ueber Deformationen Holomorpher Abbildungen», *"Osnabruecker Schriften zur Mathematik* (1979), I-VII and 1-142.

[Fos] R. Fossum, «Formes differentielles non fermées», in *"Seminaire sur les pinceaux de courbes de genre au moins deux" Asterisque*, **86**, Soc. Math. France, Paris (1981), 90-96.

[Fr] M.H. Freedman, «The topology of four dimensional manifolds», *Jour Diff. Geom.* **17** (1982), 357-453.

[Fu] T. Fujita, «On Kaehler fibres spaces over curves», *J. Math. Soc. Japan*, **30** (1978), 779-794.

[Ha] R. Hartshorne, «Algebraic Geometry», G.T.M. 52, Springer (1977).

[Ho1] E. Horikawa, «On deformation of holomorphic maps I», *J. Math. Soc. Japan* **25** (1973), 372-396.

[Ho2] E. Horikawa, «On deformation of holomorphic maps II», *J. Math. Soc. Japan* **26** (1974), 647-667.

[Ho3] E. Horikawa, «On deformation of holomorphic maps III», *Math. Ann.* **222** (1974), 275-282.

[G-L1] M. Green - R. Lazarsfeld, «Deformation theory, generic vanishing theorems and some conjectures of Enriques, Catanese and Beauville», *Inv. Math.* **90** (1987), 389-407.

[G-L2] M. Green - R. Lazarsfeld, «Higher obstructions to deforming cohomology groups of line bundles», preprint.

[Igu] J.I. Igusa, «On some problems in abstract algebraic geometry», *Proc. Nat. Acad. Sci. USA*, **41** (1955), 964.

[Il 1-2] L. Illusie, «Complexe cotangent et deformations I, II», Springer L.N.M. **239** (1971), **283** (1972).

[Il 3] L. Illusie, «Complexe de De Rham-Witt et Cohomologie Cristalline», *Ann. E.N.S.* (4) **12**, (1979), 501-661.

[Kaw] Y. Kawamata, «Characterization of Abelian varieties», *Comp. Math.* **43** (1981), 253-276.

[Kod1] K. Kodaira, «Collected Works», vols. I-III, Iwanami Shoten and Princeton Univ. Press (1975).

[Ku] M. Kuranishi, «On the locally complete families of complex analytic structures», *Ann. Math.* (2) **75** (1962), 536-577.

[K-M] K. Kodaira J. Morrow, «Complex Manifolds», Holt, Rinehart and Winston, New York (1971).

[Ko1] J. Kollar, «Higher direct images of dualizing sheaves», *Ann. of Math.* (2) **123**, (1986), 11-42.

[Kod2] K. Kodaira, «On the structure of compact complex analytic surfaces I-IV», *Amer. J. Math., I:* **86** (1964), 751-798, II: **88** (1966), 682-721, III: **90** (1968), 55-83, IV: **90** (1968), 1048-1066.

[Lg] S. Lang, «Abelian Varieties», Interscience-Wiley, New York (1959).

[La] W. Lang, «Quasi-elliptic surfaces in Characteristic 3», *Ann. E.N.S.* (4) **12**, (1979), 473-500.

[Me] R. Menegaux, «Un theoreme d'annulation en caracteristique positive», in *"Seminaire sur les pinceaux de courbes de genre au moins deux", Asterisque*, **86**, Soc. Math. France, Paris (1981), 35-43.

[Mi] J.S. Milne, «Etale cohomology», Princeton Mathematical Series 33 (1980), Princeton Univ. Press.

[Moi] B. Moishezon, «On n-dimensional compact varieties with n algebraically independent meromorphic functions», *Izv. Ak. Nauk. SSSR Ser. Mat.*, **30** (1966), 133-144, 345-386, 621-656, translated in Am. Math. Soc. Transl. (2) **63**, (1967), 51-177.

[Mu1] D. Mumford, «*Lectures on curves on an algebraic surface*», Annals of Mathematics Studies **59**, Princeton Univ. Press (1966).

[Mu2] D. Mumford, «Pathologies of modular surfaces», *Amer. J. Math* **83** (1961), 339-342.

[Noe] M. Noether, «Anzahl der Moduln einer Klasse algebraischer Flaechen», *Sb. Kgl. Preuss. Akad. Wiss. Math. - Nat. Kl.*, Berlin (1888), 123-127.

[Ny] N. Nygaard, «Closedness of regular 1-forms on algebraic surfaces», *Ann. E.N.S.*, (4) **12** (1979), 33-45.

[Pa1] V.P. Palamodov, «Deformations of complex spaces», *Uspecki Mat. Nauk.* 31.3 (1976), 129-194.

[Poi] H. Poincaré, «Sur les courbes tracées sur les surfaces algebriques», *Ann. Ecole Normale Sup.*, **27**, (1910).

[Ran 1] Z. Ran, «On subvarieties of Abelian varieties», *Inv. Math.* **62** (1981), 459-479.

[Ran 2] Z. Ran, «Deformations of maps», in "Algebraic Curves and Projective Geometry", *Proc. Trento 1988*, Springer L.N.M. **1389** (1989), 246-253.

[Rei] I. Reider, «Bounds for the number of moduli for irregular varieties of general type», *Manuscr. Math.*, **60** (1988), 221-233.

[Ser] J.P. Serre, «Sur la topologie des varietés algebriques en caracteristique p», *Symposium internacional de topologia algebraica Univ. Nac. Auton. de Mexico and UNESCO*, Mexico City (1958), 24-53.

[Sev1] F. Severi, «La serie canonica e la teoria delle serie principali di gruppi di punti sopra una superficie algebrica», *Comm. Math. Helv.*, **4** (1932), 268-326.

[Sev2] F. Severi, «Geometria dei sistemi algebrici sopra una superficie e sopra una varietà algebrica», vol. II and III, Cremonese, Roma, (1958).

[Sh] I. R. Shafarevich, «Algebraic surfaces», Steklov Institute Publ., Moskva, (1965).

[Siu] Y.T. Siu «Strong rigidity for Kaehler manifolds and the construction of bounded holomorphic functions», in "*Discrete groups and Analysis*", R. Howe ed., Birkhauser (1987), 124-151.

[Sz] L. Szpiro, «Proprietés numerique du faisceau dualisant relatif», in "*Seminaire sur les pinceaux de courbes de genre au moins deux*", Asterisque, **86**, Soc. Math. France, Paris (1981), 44-78.

[Ueno] K. Ueno, «*Classification theory of algebraic varieties and compact complex spaces*», Springer L.N.M. **439**, (1975).

A Few Comments on some Aspects of the Mathematical Work of F. Enriques

CIRO CILIBERTO Universita' di Roma II, Via O. Raimondo 00173, Roma (Italy)

INTRODUCTION

The present paper is a slightly expanded version of the talk I gave at the conference "Geometria e Variabile Complessa a Bologna", held in Bologna in the February 1989, in the occasion of the celebrations for the 900th year from the foundation of that University. I wanted to take this chance to pay a tribute to one of the greatest mathematicians who taught in that University: Federigo Enriques. I did it with the greatest pleasure: I have been in fact so lucky to be initiated soon, since I was an undergraduate student, to the richness and deepness of inspiration of the italian school of algebraic geometry, and in particular to Enriques' work. I had therefore the possibility of realizing how the roots of modern algebraic geometry deeply plunge in our tradition. The scope of my talk, and of this paper, has been to communicate part of these feelings. In order to do so, I decided to concentrate on a particular, but relevant, aspect of Enriques' investigations, namely the study of canonical and pluricanonical images of curves and surfaces (other parts of Enriques' work have been discussed in the talks by F. Catanese and I. Dolgachev). My idea has been to make an hystorical reconstruction of the contribution of Enriques and his school on this subject, but also to mention some of the main more recent developements. Enriques' papers are still an incredible source of problems: now and then I refer to some of them. In particular, at the end of the paper I concentrate, and make some comments, on one problem, often explicitely posed by Enriques: the study and classification of irregular surfaces of general type (which have been also the object of Catanese's talk).

1   PLURICANONICAL MAPS

Let V be a smooth irreducible projective variety of dimension $n \geq 1$ over the complex field. On V one can consider the <u>canonical linear system</u> $|K|$, or, more generally, for every integer $i \geq 1$, the <u>i-th pluricanonical system</u> $|iK|$. The integer $P_i = \dim |iK| + 1$ is called the <u>i-th plurigenus</u> of V. P_1 is usually denoted p_g and called the <u>geometric genus</u> of V.

If $P_i \geq 2$, associated with the linear system $|iK|$ there is a rational map, called the <u>i-th pluricanonical map</u>

$$\varphi_i : V \to \mathbb{P}^{P_i - 1}$$

which is a morphism if $|iK|$ is base points free. Let Σ_i be the image of φ_i, called the i-th pluricanonical image of V. If φ_i is birational onto its image, then the intrinsic birational geometry of V reflects into the projective geometry of Σ_i. And projective geometry, as was observed starting with M. Noether and his school (around 1870), is a powerful tool for studying algebraic varieties.

Moreover it has been one of the main discoveries of Castelnuovo and Enriques, around the end of last century, that plurigenera are birational invariants on which strongly depends the classification of varieties.

2 THE CASE OF CURVES

From the point of view of complex analysis the study of curves amounts to studying compact complex varieties of dimension 1. This was done for a great deal by Riemann [R] aroud 1860. Among many other things, Riemann discovered that curves are classified by a discrete invariant, the genus g of the underlying topological compact orientable surface, and continuous invariants, the so called moduli, parametrizing the complex structures which one can put on such a topological surface. Riemann also proved that g equals the number of linearly independent holomorphic 1-forms on the curve.

Brill, Clebsch and Noether translated Riemann's analytic theory in geometric form. This work started around 1870. In particular, in the famous paper [BN] of 1873, Brill and Noether gave the first algebro-geometric exposition of the theory of linear series on a curve, introducing the canonical series and proving in this context the theorem of Riemann-Roch. As a consequence one can prove some basic facts about the canonical and pluricanonical images of curves, which I collect in the following:

Proposition. Let C be a smooth curve of genus g. Then $g=p_g$ and moreover:

(i) all pluricanonical series of C are empty, namely $P_i=0$ for all $i \in \mathbb{N}$, if and only if C has genus g=0. This happens if and only if C is rational, i.e. isomorphic to $\mathbb{P}^1$;

(ii) all pluricanonical series are trivial (i.e. K is the zero divisor on C) if and only if C has genus g=1. This happens if and only if C is elliptic, i.e. a complex torus of dimension 1;

(iii) the consideration of canonical and pluricanonical models makes sense only for $g \geq 2$. In this case one has:

(a) the following formula for the plurigenera holds:

$$P_i = (2i-1)(g-1), \quad i \geq 2$$

(b) $|iK|$ is base points free for every $i \geq 1$, or, equivalently, all pluricanonical maps φ_i are morphisms;

(c) φ_1 is not birational if and only if C is hyperelliptic, i.e. a double covering of $\mathbb{P}^1$. In this case φ_1 realizes C as a double covering of its image Σ_1, which is a rational curve of degree g-1 in $\mathbb{P}^{g-1}$ (in particular any curve of genus 2 is hyperelliptic);

(d) if φ_1 is birational, it is an isomorphism onto Σ_1, which is a smooth curve of degree $2(g-1)$ in $\mathbb{P}^{g-1}$, called a canonical curve of genus g. Viceversa, any smooth curve of degree $2(g-1)$ in $\mathbb{P}^{g-1}$ is the image of φ_1 for a suitable curve C;

(e) φ_i is an isomorphism of C onto Σ_i for any $i \geq 2$, as soon as $g \geq 3$;

(f) if g=2, then φ_2 is still a double covering of its image, but φ_i is an isomorphism of C onto Σ_i if $i \geq 3$.

This proposition shows that, with the only exception of the hyperelliptic curves, the canonical image of a curve is the right object to look at, inasmuch as it is isomorphic to the abstract curve, and, as mentioned above, its projective geometry reflects the intrinsic theory of curves: I mention here Sernesi's proof of the unirationality of the moduli space of curves of genus 12 [S], and M. Green's results (culminated with a famous fascinating conjecture) on the resolution of the ideal of a canonical curve [G]. But Petri's ideas can also be applied to higher dimensional varieties, as already pointed out by Arbarello and Sernesi. This has been effectively used in the study of canonical images of regular surfaces, as I will explain later.

3   THE CASE OF SURFACES: THE CANONICAL MAP

In 1896 Castelnuovo [C_2] proved his celebrated theorem to the effect that a surface S is rational if and only if it is regular and $P_2=0$. In the same year Enriques [E_2] stated the possibility of finding minimal models of surfaces by contracting exceptional curves (but he gave the proof only for regular surfaces with $p_g>0$). After these results the skeleton of the classification of surfaces started to become clear. Castelnuovo and Enriques realized that minimal surfaces can be divided in four classes:

(A) surfaces for which $P_i=0$ for all integers i: this class contains only the plane and ruled surfaces;

(B) surfaces for which there is some i such that $P_i \neq 0$, but the corresponding i-th canonical system is trivial (thus $P_i=1$); the classification of the surfaces of this class into Enriques surfaces ($p_g=q=0$, 2K trivial), abelian surfaces ($p_g=1$, q=2, K trivial), what we call today K3 surfaces ($p_g=1$, q=0, K trivial) and bielliptic surfaces ($p_g=0$, q=1, second Betti number equal to 2) was achieved in the first years of this century by Enriques;

(C) surfaces for which there are integers i such that $P_i>1$, but the corresponding pluricanonical image Σ_i is always a curve: these surfaces were recognised by Enriques to possess a (rational or irrational) pencil of elliptic curves: today we call these surfaces elliptic surfaces;

(D) surfaces for which there is some integer i such that the i-th pluricanonical image Σ_i is a surface: these surfaces we call today surfaces of general type.

The above classification, explained at pg. 456 of Enriques' book [E_3], we know today as the classification of surfaces according to their Kodaira dimension (which is respectively $-\infty, 0, 1, 2$ for the surfaces of the four classes).

The investigations about the canonical and pluricanonical maps of surfaces of class (D) were started by Castelnuovo in a famous paper [C_3] of 1891 (before the classification was ready!) and then continued by Enriques. One of the main points here was to see if, similarly to the case of curves, some pluricanonical model is isomorphic, or at least birational, to the surface itself. A beautiful account on the work of Enriques and his school on this subject is contained in chapt. VIII of his book [E_3]. This work, developed along a period of about half a century and consisted in first producing a huge amount of examples, then deducing from them what the general behaviour was: the typical "eurystic" method loved by Enriques.

Enriques' main discovery was that the canonical map is by no means sufficient for studying surfaces of general type. A phenomenon which already occurs in dimension 1, but curves having a "pathological" canonical map, the hyperelliptic curves, are easily classified. In the case of surfaces the situation is more various and complicated. Here is a list of the possible phenomena which may occur.

a) Surfaces with low p_g.

The dimension of the canonical system can be too low to give a birational image: this happens if $p_g \leq 3$. The actual existence of surfaces with $p_g=1,2,3$ was soon recognised by Enriques: there are double planes or quintic surfaces with isolated triple points in $\mathbb{P}^3$ whose

properties of the abstract curve. In the hyperelliptic case the same consideration can be done for the bicanonical or tricanonical image.

It should be clear now why the interest of the old geometers was soon attracted by the study of properties of canonical curves. This was initiated by M. Noether (1871) who proved the following basic result:

Theorem (M. Noether, [N]).- Any canonical curve of genus $g \geq 3$ in $\mathbb{P}^{g-1}$ is contained in $[g(g+1)/2]-3(g-1)$ linearly independent quadrics.

Noether's theorem started a series of investigations about the homogeneous ideal of polynomials vanishing on a canonical curve, or, more generally of a curve in a projective space. In this circle of ideas an important contribution by Enriques has to be mentioned:

Theorem (Enriques,[E_1]).-A canonical curve C of genus $g \geq 4$ in $\mathbb{P}^{g-1}$ is the intersection of the quadrics containing it, unless one of the following two facts happens:
 (a) C is <u>trigonal</u> (i.e. a triple covering of $\mathbb{P}^1$);
 (b) C is isomorphic to a smooth plane quintic (in which case g=6). In the last two cases the intersection of the quadrics containing C is a rational surface F of degree g-2 in $\mathbb{P}^{g-1}$, and C is cut out on F by the cubic hypersurfaces containing C.

Enriques' proof relies on a lemma of projective geometry due to Castelnuovo [C_1], which characterizes the sets of points spanning a projective space but imposing few conditions on the quadrics containing them. Castelnuovo used the lemma in his classification of curves of maximal genus in a projective space, but it proved to be crucial in many instances concerning the geometry of projective curves (informations and results can be found in [CI_1] and [H]). A small gap in Enriques' original argument was later fixed by Babbage [B_1].

The most relevant contribution to the study of canonical curves is perhaps due to K. Petri, a student of M. Noether, who in 1922 generalized Enriques' theorem, proving the following:

Theorem (Petri, [P_1]).- The ideal of a canonical curve of genus $g \geq 4$ in $\mathbb{P}^{g-1}$ is generated by quadrics unless C is trigonal or isomorphic to a smooth plane quintic, in which cases it is generated by quadrics and cubics.

Petri's theorem was forgotten for a long time. A reason for this, I believe, is that perhaps most algebraic geometers at that time, and certainly the Italians, were not interested in ideal theoretic statements. In other words Enriques' theorem was probably considered sufficient for geometric purposes. This shows something which is perhaps interesting to observe. The italian school of algebraic geometry was strongly influenced in its beginnings, around 1880, by the german masters. In the following fourty years it developed acquiring a very specific, more and more "geometric" and "intuitive" character. By contrast the Germans still remained interested in subtler algebraic analyses, already present in Brill, Clebsch and Noether, a motivation for this probably being the influence of a great school of algebraists (Dedekind, Kronecker, Hensel, etc.) which had no equal in Italy. Anyhow in the end the two school diverged in such a way that they no more communicated with each other, although they were working on the same subjects. Petri's work gives an example of this situation, but there are also others, like Hensel's work on singularities (see an illuminating passage at pg. 652 of [EC]).

After the paper [P_1] was rediscovered, in the early 70's, another paper by Petri [P_2] was also brought again into light. It contains other beautiful ideas for finding suitable determinantal equations of a curve in a projective space, by using both the projective geometry of the curve and the homological algebra of certain modules over the ring of polynomials related to the curve itself. Arbarello and Sernesi explained Petri's point of view in [AS], refining his analysis and giving various applications, among which an extension, under suitable hypotheses, to higher dimensional varieties of Petri's theorem on the ideal of a canonical curve. Since then Petri's ideas have played a crucial role in many questions of the

minimal desingularizations have such invariants. It required more time to discover surfaces of general type with $p_g=0$. This was done by Campedelli [CA] and Godeaux [GO] essentially at the same time, in the early 30's.

It is perhaps worth mentioning that until Campedelli and Godeaux found these surfaces, their existence was considered rather controversial. If one thinks that M. Noether conjectured in 1875 that $p_g=0$ should characterize rational or ruled surfaces (see [E$_3$], pg. 138), one understands that the developements of the theory reserved rather big surprises!

One more word about surfaces of general type with $p_g=0$. Somebody could expect that, since they have low genus, they should be the simplest examples of surfaces of general type and therefore easy to classify. It is a rather embarassing fact however that the classification of these surfaces is still missing (an account on what is known on this subject is contained in the book [BPV] by Barth, Peters and Van de Ven; see also the discussion of §§ 16 and 20 of [E$_3$], which shows the difficulty of handling these surfaces).

b) <u>Base points of the canonical system</u>.

The canonical system of surfaces of general type may present, even if $p_g \geq 4$, "pathologies" which do not occur for curves. For instance $|K|$ may have base points, or even fixed components, in which case the canonical map φ_1 is not a morphism.

Enriques found examples of this type with $|K|$ having isolated base points: they are discussed in [E$_3$], chapt. VIII. All of them can be reduced to the case of a minimal surface of general type containing a curve E with genus $g \geq 1$ and $E^2=-1$. This is the standard example of a surface with $|K|$ not base point free: using the adjunction formula in fact one sees that $|K|$ must have a base point on E. There are infinitely many families of surfaces of this type: for instance desingularizations of double planes with a branch curve having a so-called [2n+1,2n+1] singularity (see [E$_3$], chapt. 3, § 8). A different example of surfaces with $|K|$ having isolated base points, is due to Bartalesi and Catanese [BC].

Finitely many examples of surfaces with $|K|$ having fixed components are also contained in Enriques' book [E$_2$], pgs. 156-157. An infinite number of such examples has been found only much later in [CI$_3$].

c) <u>Non birationality of the canonical map</u>.

Finally it may happen that φ_1 is not birational onto Σ_1, and more precisely:

(i) either $|K|$ may be composed with a pencil, so that the image Σ_1 of φ_1 is a curve;

(ii) or Σ_1 is a surface, but still φ_1 is not birational.

Surfaces of type (i) were not found until 1948, when Pompilj found infinitely many families of surfaces with $|K|$ composed with a rational pencil [PO$_1$]. However Enriques suspected the possibility that such examples should occur, as the note at pg. 158 of [E$_3$] witnesses. Families of surfaces with $|K|$ composed with an irrational pencil have been exhibited later by Beauville [BE] and Xiao [X].

As for surfaces of type (ii), the first examples of this type, and infinitely many families indeed, in which φ_1 is generically two to one onto its image were imediately found. In fact a result of 1875 by M. Noether says that for a surface of general type one has $K^2 \geq 2p_g-4$. Castelnuovo [C$_3$], on the other hand, observed that φ_1 birational yields $K^2 \geq 3p_g-7$. Thus, for $3p_g-6 \geq K^2 \geq 2p_g-4$, one certainly has φ_1 non birational, and Castelnuovo remarked that Σ_1 has to be a ruled surface and φ_1 has degree 2 in these cases. These surfaces were studied by Enriques in 1896 (see [E$_3$], pg 296).

Various examples where the degree of φ_1 is 3 or more were exhibited later, among others, by B. Segre [SE], Pompilj [PO$_2$] and Franchetta [F$_1$]. Another interesting example, with φ_1 of degree two onto Σ_1, was found by Enriques [E$_3$], pg. 300: the surface is regular, has $p_g=4$, $K^2=10$, the canonical map is two to one onto its image which is a quintic surface of $\mathbb{P}^3$, with 20 ordinary double points, which are isolated branch points of the double covering.

The remarkable fact, which <u>was never</u> noticed by Enriques, is that all the examples known at his time, fell into two categories:

(a) surfaces such that φ_1 is not birational, but Σ_1 has $p_g=0$;

(b) surfaces such that φ_1 is not birational, but Σ_1 is of general type, and in fact is a canonical model of some surface.

Surfaces with $3p_g-6 \geq K^2 \geq 2p_g-4$ are in the first category, and Σ_1 is indeed a scroll for them. For the second category Enriques only found <u>one single</u> example, the aforementioned double quintic.

The fact that the image Σ_1 of a canonical map of degree two can either have $p_g=0$ or be a canonical surface, had been proved by Babbage [B$_2$] in 1934. Anyhow the second possibility was excluded in [B$_2$] with an erroneous argument. Enriques, who produced the first counterexample to Babbage's statement, was probably not aware of Babbage's paper, since he never quotes it.

The situation about surfaces with Σ_1 a surface but φ_1 not birational has been clarified only recently by A. Beauville. In his paper [BE] of 1979, which contains the main results we know today about the canonical map of surfaces of general type, he proved that the two possibilities (a) and (b) are the only ones which can occur (whatever the degree of φ_1 is); moreover in case (b) the degree φ_1 onto its image is at most 3 if $p_a \geq 13$. Furthermore he noticed that <u>all</u> surfaces with $p_g=0$ can appear as images of a non birational, two to one, canonical map for some surface of general type. Finally he found again and clarified Enriques' counterexample to Babbage's assertion. It is remarkable that in the same year 1979 the same example was discovered also by Catanese [CT$_1$] and Van der Geer and Zagier [VZ].

Enriques' double quintic remained for some more years the only example of type (b). Other four examples where found by myself in 1980, but I never published them. These surfaces are regular with the invariants

$$p_g=5, K^2=16, 18$$
$$p_g=6, K^2=24$$
$$p_g=7, K^2=32$$

They are easy to describe, being natural extensions of Enriques' double quintic. This quintic is in fact defined by an equation $F_5=0$ where F_5 is a symmetric determinant of order 5 of homogeneous linear forms in the variables $x_0,...,x_3$. In general I define a <u>symmetroid</u> of degree d in $\mathbb{P}^r$ to be a hypersurface of degree d defined by an homogeneous equation $F_d(x_0,...,x_n)=0$, where F_d is a symmetric determinant of order d of homogeneous linear forms. A symmetroid of degree $d>1$ is singular, and a "general" such hypersurface has a double locus of codimenion 2 and degree $(d^2-1)/6$.

If one takes two general symmetroids of degrees 2 and 4 in $\mathbb{P}^4$ and considers their intersection, one obtains a canonical surface with $p_g=5$ and $K^2=8$, with 24 ordinary double points. It is easy to see that these nodes are <u>even</u> in the sense of Catanese [CT$_1$], thus there exists a double covering of the canonical surface of degree 8 in $\mathbb{P}^3$, branched only at the 24 double points: this surface is indeed a surface with $p_g=5$, $K^2=16$ whose canonical map is exactly the above mentioned double covering.

Similarly the other surfaces of the list can be obtained by intersecting two general symmetroids of degree 3 in $\mathbb{P}^4$, three general symmetroids of degrees 2, 2 and 3 in $\mathbb{P}^5$, four general symmetroids of degree 2 in $\mathbb{P}^6$, and taking the double coverings of these surfaces branched at the nodes.

More recently Beauville eventually found infinitely many families of counterexamples to Babbage's statement, as explained in [CT$_2$]. Beauville's examples are irregular, with q=2. It is still an open problem to find infinitely many regular examples of surfaces for which (b) happens, and in general to see for which invariants p_g, q, K^2 the phenomenon can occur.

Another interesting question is to find, if possible, surfaces of type (b) such that φ_1 realizes them as a triple covering of Σ_1.

But what about surfaces for which the canonical map φ_1 is "the best as it can be"? Enriques made investigations in this direction too and reported on them in the first paragraphs of chapt. VIII of [E$_3$].

He starts with the lowest value of p_g for which one may expect the canonical map to be birational, namely p_g=4, so that Σ_1 sits in $\mathbb{P}^3$. Then he looks for surfaces Σ_1 of degree d=K^2 in $\mathbb{P}^3$ which may be the images of a "well behaved" <u>canonical morphism</u>. It is a well known fact that a surface of degree d≤4 in $\mathbb{P}^3$ is never of general type, thus one finds d=K^2≥5, according to Castelnuovo's result to the effect that K^2≥3p_g-7=5 if φ_1 is birational. In fact for d=5 one finds smooth quintic surfaces, which are canonically embedded in $\mathbb{P}^3$: this is the simplest example of a "nice" canonical surface.

If d≥6, then Σ_1 must be singular, because smooth surfaces of degree d>5 have p_g≥10. More precisely a general hyperplane section of Σ_1, a curve of degree d, has to have genus g=K^2+1=d+1<(d-1)(d-2)/2, by the adjunction formula. Therefore the singularities of Σ_1 cannot be isolated (in other words the general hyperplane section of Σ_1 cannot be smooth).

As a first approximation Enriques looks for canonical surfaces with <u>ordinary singularities</u>, which, roughly speaking, consist of an irreducible nodal curve with finitely many triple points which are triple for the surface too[(*)]. Enriques' main remark is that the double curve Γ, of degree

$$[(d-1)(d-2)/2]-g=[(d-1)(d-2)/2]-(d+1)=d(d-5)/2$$

has to be a <u>contact curve</u> between Σ_1 and a surface " of degree d-5, namely Γ counted twice is the complete intersection of Σ_1 and ". In fact surfaces of degree d-4 passing through Γ, the so called <u>adjoint surfaces</u>, have to cut out on Σ_1, away from Γ, curves whose pull back on S via φ_1 fill up the complete canonical system |K|. On the other hand the planes enjoy the same property. Therefore the adjoint surfaces have to be reducible in a variable plane and in a fixed surface " of degree d-5 passing through Γ, and intersecting Σ_1 along Γ only.

For instance in case K^2=6, Enriques finds surfaces of degree 6 with an irreducible smooth double plane quintic. After one moment of reflection one sees that the equation of such a surface can be written as

---

(*) Enriques considered these singularities as the "nicest" singularites a surface in $\mathbb{P}^3$ can have, something similar to the nodes for a plane curve. They appear, for instance, in a generic projection into $\mathbb{P}^3$ of a smooth surface in $\mathbb{P}^r$, r≥4, with the only exception of the Veronese surface, whose projection has a reducible double curve but not worse singularities (see the discussion in the introduction of [E $_3$]).

(+) $$\begin{vmatrix} F_1 & F_3 \\ F_3 & F_5 \end{vmatrix} = 0$$

where $F_1=0$ is the equation of the adjoint plane ", $F_3=F_1=0$ is the equation of the double cubic Γ, and F_5 is a suitable quintic polynomial vanishing on Γ. This surface turns out to be regular.

Enriques works out in a similar way the case $K^2=7$, and briefly discusses the cases $K^2=8,9,10$, all of which turn out to give rise to regular surfaces. For increasing values of K^2 the analysis becomes more and more difficult, especially for what concerns the existence problem, which was solved by Franchetta [F$_3$] using a nice, but rather complicated, degeneration argument.

As I will explain later these surfaces with $p_g=4$, $q=0$, $K^2=5,...,10$ have been a nice test, in more recent times, for the application of techniques of Petri's type for writing the equations of surfaces of general type (see § 5).

Moreover the first counterexample to Babbage's claim is related to these surfaces: Enriques' double quintics in fact are limits of surfaces with $p_g=4$, $q=0$, $K^2=10$, for which φ_1 is birational, and this is in fact the way Enriques found them.

Finally the consideration of these surfaces gave Enriques the opportunity of studying their degenerations. In doing this he discovered the first examples, I believe, of reducible moduli space of surfaces of general type with fixed invariants p_a and K^2, a phenomenon which is new with respect to the case of curves. His work on this subject inspired a lot of important later work by Horikawa [HO$_1$], [HO$_2$] and others who studied surfaces of general type with small invariants. The subject of moduli of surfaces has been recently investigated by Catanese [CT$_3$], [CT$_4$], [CT$_5$].

4 PLURICANONICAL MAPS OF SURFACES AND THE CONCEPT OF ARITHMETIC CONNECTEDNESS

The study of pluricanonical images of surfaces of general type was started by Enriques, as usual, with the consideration of examples. He first looked at those surfaces, with $p_g \leq 3$, whose canonical map is forced to be non birational. This research started early in the papers [E$_4$], [E$_5$] of 1897. A more thorough (although still not complete) analysis is contained in §§14-21 of chapt. VIII of [E$_2$]. He studied the bicanonical and the threecanonical images of these surfaces, and found that only in two cases, corresponding to the values

$$p_g=p_a=2, K^2=1$$
$$p_g=p_a=3, K^2=2$$

of the invariants, the tricanonical map is not birational. Then he proved the following theorem:

<u>Theorem</u> (Enriques, [E$_2$], chapt. VII, § 21).- The tricanonical map of a regular surface of general type with $p_g>0$ is birational, with the only two exceptions mentioned above.

This is an important theorem, because it says that, despite the many "pathologies" which appear in the consideration of the canonical map, φ_i is uniformly "well behaved" for a low value of i and for <u>all</u> (regular) surfaces of general type. This suggests a deep similarity with the case of curves (remember, φ_3 is birational for all curves of genus $g \geq 2$, but not φ_2!), which is insuspectable if one looks at what happens for φ_1.

It is perhaps worth pointing out that such an important result was proved only in 1945, thus very late with respect to all other main theorems in the theory of surfaces, found before 1910.

Enriques' original proof uses projective geometry, and is certainly considered old fashoned by the modern experts who know it. The way this kind of results are nowadays proved is using suitable vanishing theorems for arithmetically connected curves. This technique too however is a discovery of Enriques and his school, although it was not yet completely developed at the time the book [E_3] was written, and therefore it was not possible for Enriques to make a systematic use it.

The concept of arithmetically connected curve on a surface occurred to Franchetta [F_3] in his researches, which go back to 1940, on the reducible exceptional curves of the first kind on a surface. A smooth irreducible curve E on a surface S is an <u>exceptional curve of the first kind</u>, i.e. it can be contracted to a smooth point, if and only if is rational with self-intersection $E^2 = -1$: this is a fundamental result by Castelnuovo. Of course there are also reducible exceptional curves which one gets by blowing up series of infinitely near points. Such reducible curves E possess in general multiple components, as was shown by Barber and Zariski [BZ] in 1935, but still have arithmetic genus $p_a(E)=0$ and self intersection $E^2=-1$. The purpose of Franchetta was to characterize these reducible exceptional curves, like Castelnuovo did with the irreducible ones, only by means of their arithmetical invariants $p_a(E)$ and E^2. Franchetta observed that a necessary condition for E to be exceptional is to be <u>arithmetically connected</u>: i.e. for any way of writing E=A+B, with A, B effective non zero, then A\B>0. He succeeded in proving in [F_4] that this is also a sufficient condition, together with $p_a(E)=0$ and $E^2=-1$, for E to be exceptional of the first kind on a non ruled surfaces.

Franchetta (and presumably Enriques too) immediately understood the importance of the concept of arithmetical connectedness, as stated in the introduction of [F_4]. In particular Enriques realized that arithmetical connectedness could be very useful in the so-called problem of regularity of the adjoint and in the study of pluricanonical systems. And in fact he wanted to report on this in some paragraphs of [E_2] (chapt. IV, § 17-20), although the work he and Franchetta were making on the subject was still in progress. I will explain in particular the contents of § 17 where Enriques works on a <u>regular</u> surface S.

(a) First Enriques remarks that if C is a reduced, connected curve on S, then the adjoint system |K+C| of C is <u>regular</u>, which translates in the laguage of sheaves by $h^1(S, \mathcal{O}_S(K+C)) = =h^1(S, \mathcal{O}_S(-C))=0$. If one uses cohomology this is trivial, but Enriques gave a direct proof of it.

(b) Secondly Enriques proves that the previous result is equivalent to the fact that the only regular functions on C are the constants, which in sheaf theoretic language translates in $h^0(S, \mathcal{O}_S(C))=1$. So Enriques says: in order to prove that on a regular surface arithmetical connectedness of C implies regularity of |K+C|, one should prove that C arithmetically connected yields $h^0(S, \mathcal{O}_S(C))=1$. He is not able to prove this, so he <u>conjectures</u> this to be true, and deduces form the conjecture some consequences (at the same time explicitely pointing out that these depend from the validity of the conjecture itself: the eurystic method again!).

(c) As a consequence of the conjecture Enriques computes the plurigenera. He proves in fact that on a minimal regular surface of general type the canonical and pluricanonical divisors are arithmetically connected. Whence he deduces that <u>if the conjecture holds</u> then all the pluricanonical systems |iK|, i≥2, are regular, and therefore

$$P_i = p_a + 1 + [i(i-1)/2]K^2$$

(d) Finally Enriques shows how to use connectedness in order to have finer informations about pluricanonical systems: as an example he proves that if |K| has no multiple fixed components on a regular surface, then |2K| has no fixed components.

The approach to the study of pluricanonical systems outlined above was realized one year after the publication of [E_2] (1949-50) by Franchetta in the series of notes [F_5], [F_6], [F_7], [F_8], in which he was able to work on any surface of general type, regular or not.

(a) Franchetta's first main result is the proof of Enriques' conjecture: C arithmetically connected yields $h^0(S,\mathcal{O}_S(C))=1$. This is the basic fact about arithmetical connectedness, and is contained in [F_6], § 3.

(b) As already mentioned this result yields the regularity of $|K+C|$ if C is arithmetically connected on a regular surface. Franchetta extends in [F_5] this result to irregular surfaces, by proving the following powerful vanishing theorem for $h^1(S,\mathcal{O}_S(-C))$:

Theorem (Franchetta, [F_5]).- If C is arithmetically connected and one of its components moves in a system which is not an irrational pencil, then $|K+C|$ is regular.

Refinements of this theorem were later found by Ramanujan [R_1], [R_2] and others (see for instance the article [MI_1]).

(c) Then Franchetta proves in [F_7] that pluricanonical divisors are arithmetically connected on irregular surfaces too (in order to do this he applies a nice criterion of connectedness he has in [F_6]), and agains deduces the formula for the plurigenera.

(d) Finally he uses the theory of connectedness in order to study the birationality of φ_i, $i \geq 2$. He proves in [F_8] the following extension of Enriques' theorem on φ_3:

Theorem (Franchetta, [F_8]).- Let S be a minimal surface of general type with $p_a > 0$. Then:

(i) $|3K|$ has no fixed components and is not composed of a pencil; moreover if φ_3 is not birational, then φ_3 is of degree 2 and Σ_1 is a scroll;

(ii) φ_i is a birational morphism onto its image if $i \geq 4$.

After Franchetta's contribution, the subject slept for almost twenty years, until in the 60's Zariski [Z], Mumford [M_1], Shafarevich [SH] and Kodaira [K] brought again the attention on the pluricanonical maps for surfaces. But only after Ramanujan in the early 70's rediscovered Franchetta's vanishing theorem, the ground was actually ready for a systematic exploration. This was done by Bombieri in his epochal paper [BO]. After Bombieri, relevant contribution have been given by Francia [FR] and Reider [RE] (see [CT_2], § 1). In particular Reider's paper contains a new idea, which, togheter with connectedness, seems to be very effective for the study of linear systems of curves on a surface: the use of the concept stability of vector bundles, which was first introduced in the theory of surfaces by Bogomolov [BO] in the attempt of proving the bound $K^2 \leq 9(p_a+1)$, actually proved by Miyaoka [MI_2] and Yau [Y].

In conclusion I refer to the expository papers [CI_2], [CT_2], and [PE] for references and more detailed informations about the developements of the subject from Bombieri's paper [BO] until now.

5 A FEW WORDS ON IRREGULAR SURFACES

I will conclude by spending some time on a problem which is frequently mentioned in Enriques' book [E_2]: the classification of irregular surfaces.

General properties of irregular surfaces are treated in chap. IX of [E$_2$]. The main result stated (but never proved!) is that on an irregular surface the <u>geometric irregularity</u> (i.e. the maximum dimension of a continuous system of non equivalent curves on the surface) equals the <u>arithmetical irregularity</u> $q=p_g-p_a$. This is the so-called <u>fundamental theorem</u> of irregular surfaces, first proved by Picard with trascendental methods. As pointed out by Catanese in his talk, the Italian school never succeeded in giving an algebro-geometric proof of it. This was found only later, in the 60's, (in characteristic 0, the theorem is false in positive characteristic!) by Grothendieck and Mumford [MU$_2$].

As for the classification of irregular surfaces, in particular of those of general type, one would like to know, for instance, what kind of restrictions the irregularity imposes on the other invariants K^2 and p_a, and also if the irregularity influences the behaviour of the canonical or pluricanonical mappings.

These problems are clearly asked in [E$_2$], pg. 458-59, and in particular in the case q=1 at pg. 397 and 450. The consideration Enriques makes at pg. 458 are particularly interesting. He says: we tried to systematically construct canonical surfaces of general type, for example with $p_g=4$ (see § 3). We found surfaces of degree $d=K^2=5,...,10$, but <u>all these surfaces</u> turn out to be regular. So apparently to be q>0 forces the surface either to have K^2 large with respect to p_g, or to have a "rather degenerate" canonical mapping. He points out this question to the attention of the reader. In a footnote then he reports on a private communication received by Castelnuovo, who claimed to have a proof of the following result, closely related with the above question: if a surface of general type S has irregularity q≥3 and |K| is not composed with the curves of a pencil, then $K^2 \geq 6p_a+3$. It should be mentioned that Severi [SV] too claimed a similar result: if S has no irrational pencil, then $K^2 \geq 4(p_a+1)$. But unfortunately Severi's argument was wrong and Castelnuovo probably used it (see [CT$_5$]).

What we know today on the subject shows that from a qualitative point of view Enriques, Castelnovo and Severi were right: K^2 tends to increase for irregular surfaces. For example Debarre [D] proved in 1982 a result stated by Jongmans [J] in 1947, which improves a theorem of Castelnuovo quoted above: if φ_1 is birational then $K^2 \geq 3p_g+q-7$. Debarre's paper also contains an appendix by Beauville, in which he proves that for a surface of general type one has $p_g \geq 2q-4$: a result claimed, but apparently never proved by the classics.

Moreover Catanese has in [CT$_5$] a result which goes in the direction of (and in fact was inspired by) Castelnuovo's inequality $K^2 \geq 6p_a+3$: if q≥3 and there are two holomorphic one forms ω_1 and ω_2 on the surface such that $\omega_1 \wedge \omega_2$ vanishes on an irreducible, reduced curve, then $K^2 \geq 6(p_a+1)$. Other results on this subject are due to Xiao and Horikawa-Reid (see [CT$_1$] for references). However it is still an open problem to prove (or disprove!) Severi's inequality $K^2 \geq 4(p_a+1)$.

Without spending more time on these generalities, which have been touched in Catanese's talk too, I will present now two possible way of attacking the construction and classification of irregular surfaces. This is still work in progress, which I hope will give some fruit in the near future.

a) <u>The approach via one-dimensional systems</u>.

A first set of ideas I want to present has its roots in what Enriques says in [E$_2$], pg. 366.

Let S be a surface, and let $\mathcal{C}=\{C_t\}_{t \in \Gamma}$ be a flat 1-dimensional system of curves on S, parametrized by a smooth irreducible curve Γ. To give $\mathcal{C}$ is equivalent to give the incidence correspondence

$$\Sigma=\{(x,t) \in S \times \Gamma: x \in C_t\}$$

with its projections p: $\Sigma \to S$ and $\pi: \Sigma \to \Gamma$. I will assume $\mathcal{C}$ has some good properties, like:

(i) the general curve of $\mathcal{C}$ is irreducible, non singular;
(ii) $\mathcal{C}$ has no base points (i.e. the morphism p is finite);
(iii) Σ is smooth.

One can then consider some numerical invariants attached to $\mathcal{C}$, like the <u>degree</u> $n=C_t^2$ of $\mathcal{C}$, the <u>genus</u> $g=[(K+C_t)\backslash C_t/2]+1$ of the curves of $\mathcal{C}$, the number ∂ of singular curves of $\mathcal{C}$ (which has to be properly counted) and the <u>index</u> ν of $\mathcal{C}$, namely the degree of the covering p: $\Sigma \to S$, or, in other words, the number of curves of $\mathcal{C}$ passing through the general point of S (or through any point of S, provided one takes into account multiplicities). Notice that $\nu=1$ if and only if $\mathcal{C}$ is a (rational or irrational) pencil.

All these invariants are related by a nice formula, first stated in 1906 by Torelli (who gave an incomplete proof of it) and recently proved in [CG]. The formula estimates the number δ and involves the Euler-Poincarè characteristic e(S) of S:

(*) $$\delta \leq \nu[n+4(g-1)+e(S)]$$

Torelli's theorem also says that the equality holds in (*) if and only if all the curves in $\mathcal{C}$ lie in one and the same linear system. The difference between the r.h.s. of (*) and δ is geometrically interpreted in [CG].

Irregular surfaces have therefore the pecularity, with respect to the regular ones, of possessing systems of curves $\mathcal{C}$ for which in (*) a strict inequality appears. To be more specific, it seems reasonable to expect that the more δ is far from Torelli's bound, the more $\mathcal{C}$ is general on S and "better reflects the irregularity of the surface". For example, the greatest lower bound $\beta(S)$ of the ratio $\delta/\nu[n+4(g-1)+e(S)]$, as $\mathcal{C}$ varies among all possible 1-dimensional systems, is an invariant of the surface S, which probably has to do with the classification, but in which way? In particular: how does it vary in a flat family? A related question is to study minimal surfaces for which $\beta(S)=0$, and in particular surfaces on which there are systems $\mathcal{C}$ with $\delta=0$ but $\nu>1$. The projective plane is the only regular surface enjoing this property; the abelian surfaces are irregular surfaces ejoing the property too. Is there any other?

Once $\mathcal{C}$ has been given, one may deduce from it some maps defined on the surfaces S. For instance, by associating to each point x of S the divisor $t_1+...+t_\nu$ of degree ν on Γ such that the curves of $\mathcal{C}$ passing through x are exactly $C_{t_1},...,C_{t_\nu}$, one gets a morphism

$$\varphi: S \to \Gamma^{(\nu)}$$

of S into the ν-th symmetric product $\Gamma^{(\nu)}$ of Γ. By composing with the Abel-Jacobi map $\Gamma^{(\nu)} \to J(\Gamma) = Pic^{(\nu)}(\Gamma)$ of Γ into its Jacobian $J(\Gamma)$, one finds a new map

$$\psi: S \to J(\Gamma)$$

which factors through the Albanese map of S

$$\begin{array}{ccc} S & \to & J(\Gamma) \\ & \searrow & \nearrow \\ & Alb(S) & \end{array}$$

Therefore some characters of the map ψ are independent of $\mathcal{C}$. For example, if S is regular, whatever $\mathcal{C}$ is, ψ is constant. If $\mathcal{C}$ is irregular, but the image of the Albanese map is a curve, the same happens for ψ, whatever $\mathcal{C}$ is. Essentially, the map ψ is a geometrical counterpart of the Albanese map.

There is a related construction one can do. Notice that $\mathcal{C}$ can be considered as a curve Γ described in a component $\mathcal{H}$ of the Hilbert scheme of curves on S. Consider the component $\mathcal{H}^{(v)}$ of the Hilbert scheme, which contains all v-ples of curves of $\mathcal{H}$. The datum of $\mathcal{C}$ determines a map $\gamma: S \to \mathcal{H}^{(v)}$, associating to $x \in S$ the point of $\mathcal{H}^{(v)}$ corresponding to $C_{t_1}+\ldots+C_{t_v}$, where $C_{t_1},\ldots,C_{t_v}$ are the curves of $\mathcal{C}$ containing x. Clearly γ factors through φ.

There is also a Abel-Jacobi map $\alpha: \mathcal{H} \to \text{Pic}^0(S)$, which is defined, up to a choice of a base curve $C_0 \in \mathcal{H}$, by sending a point $C \in \mathcal{H}$ to the class of $C-C_0$ in $\text{Pic}^0(S)$. Accordingly there is the Abel-Jacobi map $\alpha: \mathcal{H}^{(v)} \to \text{Pic}^0(S)$ and, by composing γ with α, one gets a map $\vartheta: S \to \text{Pic}^0(S)$, depending on $\mathcal{C}$, whose geometrical meaning is transparent: to each point $x \in S$ one associates the point of $\text{Pic}^0(S)$ corresponding to the class of $C_{t_1}+\ldots+C_{t_v}-vC_0$. Since there is a factorization

$$S \to \text{Pic}^0(S)$$
$$\searrow \quad \nearrow$$
$$\text{Alb}(S)$$

again some of the features of the map $\varnothing$ do not depend on $\mathcal{C}$.

But although ϑ can be rather degenerate (for instance its image could be a curve, what certainly happens if the image of the Albanese map of S is a curve), γ can be, by contrast, rather well behaved: examples of this sort will be mentioned later.

Anyhow if v is large enough, which can be managed by taking Γ "general enough" in $\mathcal{H}$, then $\alpha: \mathcal{H}^{(v)} \to \text{Pic}^0(S)$ is surjective, and the fibres are projective spaces (i.e. linear systems) all of the same dimension. In other words there is a suitable vector bundle $\mathcal{E}$ on $\text{Pic}^0(S)$ such that $\mathcal{H}^{(v)}=\text{Proj}(\mathcal{E})$. The upshot is that the datum of a system $\mathcal{C}$ enjoing good properties like (i), (ii) and (iii) above, gives rise to a morphism $\gamma: S \to \text{Proj}(\mathcal{E})$, where $\mathcal{E}$ is a suitable vector bundle on $\text{Pic}^0(S)$. It would be interesting to understand how γ depends on $\mathcal{C}$ and to see if, for "general" $\mathcal{C}$, γ is an isomorphism onto its image, or at least birational: if this were the case, then the natural places (or, may be, the "most natural places") to look at to find irregular surfaces would be the projective bundles on abelian varieties.

I will briefly describe now an example which illustrates, I hope, the strategy of constructing and classifying irregular surfaces as subvarieties of projective bundles on abelian varieties. Precisely I want to discuss surfaces with $p_g=q=1$ (and accordingly $2 \leq K^2 \leq 9$): this is a situation in which, as I will explain, a system $\mathcal{C}$ is naturally given, so that the resulting maps too appear to be "natural". I have been studying these surfaces in collaboration with F. Catanese (see [CTC]).

Example: surfaces of general type with $p_g = q = 1$.

Let S be a minimal surface with $p_g = q = 1$. Here $E = Alb(S) = Pic^0(S)$ is an elliptic curve. Consider the Albanese map $f: S \to E$. It determines a pencil of curves on S, which I call the Albanese pencil. I will denote by p the genus of the curves of the Albanese pencil.

On S one can consider the so called <u>paracanonical</u> system $\mathcal{K}$ which is the irreducible component of the Hilbert scheme of curves on S containing the unique canonical curve K of S. If $t \in E = Pic^0(S)$, consider $K_t = \mathcal{O}_S(K \otimes t)$. Since $h^0(S,K) = p_g = 1$, for a general $t \in E$ one has $h^0(S,K_t) = 1$, namely there is only one curve $K_t \in |\mathcal{O}_S(K \otimes t)|$. Therefore $\mathcal{K} = \{K_t\}_{t \in E}$ is 1-dimensional, described by the closure in the Hilbert scheme of the system of curves $\{K_t\}_{t \in E'}$, where E' is the open subset of E consisting of points t such that $h^0(S,K_t) = 1$. If $\Lambda = E - E'$ is the set of points $t \in E$ such that $h^0(S,K_t) > 1$, Λ can be given an appropriate scheme structure, and λ will denote the degree of such scheme.

Now, if ν is the index of the paracanonical system, we would like to consider the map φ: $S \to E^{(\nu)}$ described in general before. But in order to have it we need $\mathcal{K}$ to behave nicely, and first of all to be base points free. If this does not happen however, one only has a rational map $\varphi: S \to E^{(\nu)}$, which I call the <u>paracanonical map</u>. It gives us a way of sending S in $E^{(\nu)}$ which is a projective bundle over $E = Pic^0(S)$. If one understands the geometry of the map $\varphi: S \to E^{(\nu)}$, there is some hope one can classify these surfaces.

One of the main questions here is to find what relations there are between the invariants K^2, p, ν and λ. The simplest case $K^2 = 2$ was studied by Bombieri and Catanese [CT$_6$], who proved, by means of a careful study of $\mathcal{K}$, that in this case one has $\nu = K^2 = p = 2$, $\lambda = 0$, $\mathcal{K}$ is base points free and therefore φ is a morphism, which realizes S as a double covering of $E^{(2)}$ whose branch curve can be well described.

Catanese and I tried to figure out what the situation is for larger K^2 by working out examples and making efforts in order to generalize the approach of [CT$_6$] for the study of $\mathcal{K}$. We can prove the basic equality $p - \lambda = \nu$. We also prove that for $k^2 = p = 3$ the map φ is a morphism and that, in any event, $p \leq 3$ and $\lambda = 0$. Thus if $K^2 = 3$ the only possibilities are $\nu = p = 2$ and $\nu = p = 3$. Both cases are actually possible. For instance one has a surface with $\nu = K^2 = p = 3$, $\lambda = 0$ if one takes a general enough surface in $E^{(3)}$ which is homologous to a divisor of the type 4D-F, where F is a fibre of $E^{(3)} \to E$ and D is $E^{(2)}$ naturally embedded in $E^{(3)}$.

As a byproduct of our analysis of $\mathcal{K}$, we can prove that $|2K|$ is base point free for surfaces with $p_g = q = 1$, $K^2 = 3,4$. The attempt of proving this originally drew our attention on these surfaces: in fact as a consequence of results by Bombieri, Francia and Reider, one knows that $|2K|$ is base point free for any surface of general type with $p_g \geq 1$, with the possible exceptions $p_g = q = 1$, $K^2 = 3,4$. A possibility which, as we showed, does not occur.

b) <u>Equations of irregular surfaces in a projective space.</u>

The second approach to the study of irregular surfaces has more to do with the commutative algebra and the projective geometry of irregular surfaces, and is related to Petri's ideas mentioned in § 1. On this subject I benefited of some very useful discussions with E. Sernesi.

Let X be an irreducible surface in $\mathbb{P}^3$ and assume, for sake of simplicity, that X has ordinary singularities, so that its normalization S is non singular. Let H be a divisor on S which is the pull back of a plane section of X. If $\mathcal{O}$ is a line bundle on S, one can consider the following object:

$$\gamma(\mathfrak{X}) = \oplus_{n\in \mathbb{Z}} H^0(S,\mathfrak{X})(nH))$$

which can be considered in a natural way as a graded torsion module over A, the polynomial ring in four variables, i.e. the coordinate ring of $\mathbb{P}^3$. It can be proved (Catanese did it first in [CT$_1$]) that the projective dimension of $\gamma(\mathfrak{X})$ is 1 if and only if $h^1(S,\mathfrak{X}(nH))=0$ for all $n\in \mathbb{Z}$.

In this case $\gamma(\mathfrak{X})$ is a Cohen-Macauly A-module, a minimal resolution of which is of the type

$$0 \to M_1 \to M_0 \to \gamma(\mathfrak{X}) \to 0$$

The only relevant map $M_1 \to M_0$ is described by a square matrix **A** with homogeneous entries in A, and $|A|=0$ turns out to be the equation of X in $\mathbb{P}^3$.

In [CI$_4$] this approach was applied to the study of Enriques' examples of canonical surfaces with $p_g=4$, $q=0$, $K^2=5,\ldots,10$ in $\mathbb{P}^3$ mentioned in § 3, and extensions were made to the study of surfaces in higher dimensional projective spaces in [CI$_5$]. If one takes $\mathfrak{X}=\mathcal{O}_S(K)$, then the Cohen-Macaulay conditions $h^1(S,\mathcal{O}_S(nK))=0$, for all $n\in \mathbb{Z}$, on $\gamma(K)=\gamma(\mathcal{O}_S(K))$ are equivalent to the single condition $q=0$. Moreover a result by Catanese, consisting in a smart application of Serre duality, says that bases of the free modules M_0 and M_1 can be taken is such a way that the matrix **A** is symmetric. The equation $|A|=0$ for Enriques' canonical surfaces becomes then very simple and reflects the geometry of the surfaces, which I briefly explained in § 2. In particular for $K^2=6$ one finds again the equation (+) of § 2.

These ideas were later clarified by Catanese [CT$_7$], who related the structure of the matrix **A** with the ring structure of $\gamma(K)$, the so-called canonical ring of S. Parenthetically I mention that informations about the generators of this ring as a $\mathbb{C}$-algebra can be found in [CI$_3$].

In order to apply the above procedure one needs $\gamma(\mathfrak{X})$ to be Cohen-Macaulay, a fact that never happens, for instance, for irregular canonical surfaces. But even if $\gamma(\mathfrak{X})$ is not Cohen-Macaulay one can use its resolution to get information about the equation of X.

In fact it is not difficult to prove that the projective dimension of $\gamma(\mathfrak{X})$ is at most 2, so that it is exactly two if, as I will suppose from now on, $\gamma(\mathfrak{X})$ is not Cohen-Macaulay. Therefore one has a minimal free resolution of the type

(o) $$0 \to M_2 \to M_1 \to M_0 \to \gamma(\mathfrak{X}) \to 0$$

I will denote by r_i the rank of M_i and by A_i the matrix appearing in the map $M_i \to M_{i-1}$, $i=1,2$. In force of a general result by Mac Rae [MR], the determinantal equation $rk(A_1)<r_1$ defines, at least set-theoretically, X in $\mathbb{P}^3$, so again we have equations for X. The problem is to determine the sizes of A_1 and its entries.

We can split the resolution (o) in two exact pieces

$$0 \to K \to M_0 \to \gamma(\mathfrak{X}) \to 0$$
$$0 \to M_2 \to M_1 \to K \to 0$$

By dualizing and twisting (I refer to [FAC] for generalities, as well as for the notation I use), one has:

$$0 \to M_0^*(-4) \to K^*(-4) \to \mathrm{Ext}^1(\gamma(\mathfrak{X}),A(-4)) \to 0$$
$$0 \to K^*(-4) \to M_1^*(-4) \to M_2^*(-4) \to \mathrm{Ext}^1(K,A(-4)) \to 0$$

Moreover Serre duality gives

$$\mathrm{Ext}^1(K,A(-4)) = \mathrm{Ext}^1(\gamma(\mathfrak{X}),A(-4)) =$$
$$= \oplus_{m \in \mathbb{Z}} H^1(S,\mathfrak{X}(-mH))^* = \oplus_{m \in \mathbb{Z}} H^1(S,\mathfrak{X}^*(K+mH))$$

$$\mathrm{Ext}^1(\gamma(\mathfrak{X}),A(-4)) = \oplus_{m \in \mathbb{Z}} H^1(S,\mathfrak{X}(-mH))^* =$$
$$= \oplus_{m \in \mathbb{Z}} H^0(S,\mathfrak{X}^*(K+mH)) = \gamma(\mathfrak{X}^*(K))$$

If we put

$$\gamma^1(\mathfrak{X}) = \oplus_{m \in \mathbb{Z}} H^1(S,\mathfrak{X}(mH))$$

this is again a graded module, of finite lenght, over A. What we found is that

$$0 \to K^*(-4) \to M_1^*(-4) \to M_2^*(-4) \to \gamma^1(\mathfrak{X}^*(K)) \to 0$$

is "piece" of a (not necessarily minimal!) resolution of $\gamma^1(\mathfrak{X}*(K))$. Anyhow the minimality of the resolution (+) yields that r_2 equals the minimum number of generators of $\gamma^1(\mathfrak{X}^*(K))$. Since $r_1 = r_0 + r_2$, if we know generators, and their degrees, for $\gamma(\mathfrak{X})$ and $\gamma^1(\mathfrak{X}^*(K))$, we may hope to deduce how M_1 is done, whence deducing the structure of the matrix A_1 and therefore getting equations for X.

I made this computation for the simplest examples of irregular surfaces in $\mathbb{P}^3$ I know, namely projections of elliptic scrolls of degree r+1 in $\mathbb{P}^r$. I resolved $\gamma(K)$ and the result I got is that $r_2 = 1$, M_2 having a single generator in degree 4, M_0 has then r+1 generators in degree 2 and M_1 has one generator in degree 4 and r+1 generators in degree 3. The matrices A_1 and A_2 are respectively of the form

$$\begin{Vmatrix} [2] & [1] \end{Vmatrix} \quad r+1$$
$$1 \quad\ \ r+1$$

$$\begin{Vmatrix} [0] \\ [1] \end{Vmatrix} \quad 1$$
$$\ \ r+1$$
$$1$$

where each block [i] stays for a block of forms of degree i (forms of degree 0 being 0 by minimality) whose size I indicated on the right side or underneath the block.

The square block [1] appearing in the matrix A_1 has determinant 0, because $A_1 \backslash A_2 = 0$. All other minors of order r+1 of A_1 have degree r+2, one more than the degree of X: in fact each of them is the product of the equation of X times the corresponding linear polynomial appearing in A_2.

I hope it will be possible to work out similar computations for other (more interesting!) surfaces, like irregular canonical surfaces in $\mathbb{P}^3$ and generic projection in $\mathbb{P}^3$ of abelian surfaces: an interesting case is probably the projection in $\mathbb{P}^3$ of the smooth abelian surface of degree 10 in $\mathbb{P}^4$ which is related to the Horrocks-Mumford bundle.

References

[AS] E. Arbarello and E. Sernesi, Petri's approach to the study of the ideal associated to a special divisor, *Inventiones Math.*, 49 (1978), 99-119.

[B_1] D.W. Babbage, A note on the quadrics through a canonical curve, *J. London Math. Soc.*, 14 (1939), 310-315.

[B_2] D. W. Babbage, Multiple canonical surfaces, *Proc. Cambridge Phil. Soc.*, 30 (1934).

[BC] A. Bartalesi and F. Catanese, Surfaces with $K^2=6$, $p_g=4$, and with torsion, *Rend. Semin. Mat. Univ. e Polit. Torino*, 1985.

[BE] A. Beauville, L'application canonique pour les surfaces de type général, *Inventiones Math.*, 55 (1979), 121-140.

[BG] F. Bogomolov, Holomorphic tensors and vector bundles on projective varieties, *Math. USSR Izv.*, 13 (1979), 499-555.

[BN] A. Brill and M. Noether, Ueber die algebraischen Functionen und ihre Anwendung in der geometrie, *Math. Ann.*, 7 (1873), 269-310.

[BO] E. Bombieri, Canonical models of surfaces of general type, *Publ. Math. Inst. Hautes Etud. Sci..*, 42 (1973), 171-219.

[BPV] W. Barth, C. Peters, A. Van de Ven, *Compact complex surfaces*, Ergebnisse der Math., Springer Verlag, 1984.

[BZ] E. Barber and O. Zariski, Reducible exceptional curves of the first kind, *Amer. J. of Math.*, (1935), 119-141.

[C_1] G. Castelnuovo, Sui multipli di una serie lineare di gruppi di punti appartenente ad una curva algebrica, *Rend. Circ. Mat. Palermo*, 7, (1893).

[C_2] G. Castelnuovo, Sulle superficie di genere zero, *Soc. It. delle Scienze (detta dei XL)*, 1896, see Memorie scelte, pg. 307.

[C_3] G. Castelnuovo, Osservazioni intorno alla geometria sopra una superficie, *Rend. Ist. Lombardo*, 24 (1891).

[CA] L. Campedelli, Sopra alcuni piani doppi notevoli con curve di diramazione del decimo ordine, *Atti Accad. Nazionale Lincei*, 15 (1932), 536-542.

[CI_1] C. Ciliberto, Hilbert functions of finite sets of points and the genus of a curve in a projective space, *Springer L. N. in M.* 1266 (1987), 24-73.

[CI_2] C. Ciliberto, Superficie algebriche complesse: idee e metodi della classificazione, *Atti del Convegno di Geometria Algebrica*, Nervi, Tecnoprint, Bologna, 1984.

[CI_3] C. Ciliberto, Sul grado dei generatori dell'anello canonico di una superficie di tipo generale, *Rend. Semin. Mat. Univ. e Polit. Torino*, 41 (1983), 83-112.

[CI$_4$] C. Ciliberto, Canonical surfaces with $p_g=p_a=4$, $K^2=5,...,10$, *Duke Math. J.*, 48 (1981), 121-157.

[CI$_5$] C. Ciliberto, Canonical surfaces with $p_g=p_a=5$, $K^2=10$, *Annali S.N.S. Pisa*, 9 (1982), 287-336.

[CG] C. Ciliberto and F. Ghione, Serie algebriche di divisori su una curva e su una superficie algebrica, *Annali di Mat. Pura e Appl.*, 136 (1984), 329-353.

[CT$_1$] F. Catanese, Babbage's conjecture, contact of surfaces, symmetric determinantal varieties and applications, *Inventiones Math.*, 63 (1981), 433-465.

[CT$_2$] F. Catanese, Canonical ring and "special" surfaces of general type, *Proc. Symposia in pure Math.* 46 (1987), 175-194.

[CT$_3$] F. Catanese, On the moduli space of surfaces of general type, *J. Differential Geom.*, 19 (1984), 483-515.

[CT$_4$] F. Catanese, Connected components of moduli spaces, *J. Differential Geom.*, 24 (1986), 395-399.

[CT$_5$] F. Catanese, Moduli of surfaces of general type, *Springer L. N. in M.*, 997 (1983), 90-112.

[CT$_6$] F. Catanese, On a class of surfaces of general type, C.I.M.E. 1977, Algebraic Surfaces, Liguori Napoli, 1981, 269-284.

[CT$_7$] F. Catanese, Commutative algebra methods and equations of regular surfaces, Algebraic Geometry Bucharest 1982, *Springer L. N. in M*., 1056 (1984), 68-111.

[CTC] F. Catanese and C. Ciliberto, Surfaces with $p_g=q=1$, pre-print.

[D] O. Debarre, Inégalités numeriques pour les surfaces de type général, *Bull. Soc. Math. de France*, 110 (1982), 319-346.

[E$_1$] F.Enriques, Sulle curve canoniche di genere p dello spazio a p-1 dimensioni, *Rend, Accad. Sci. Ist. di Bologna*, 23 (1919), 80-81.

[E$_2$] F. Enriques, Introduzione alla geometria sopra le superficie algebriche, *Soc. It. delle Scienze (detta dei XL)*, 10 (1896), 1-81.

[E$_3$] F. Enriques, *Le superficie algebriche*, Zanichelli, Bologna, 1949.

[E$_4$] F. Enriques, Le superficie algebriche di genere lineare $p^{(1)}=2$, *Rend. Accad. Naz. Lincei*, 1987.

[E$_5$] F. Enriques, Le superficie algebriche di genere lineare $p^{(1)}=3$, *Rend. Accad, Naz, Lincei*, 1987.

[EC] F. Enriques and O. Chisini, *Lezioni sulla teoria geometrica delle equazioni e delle funzioni algebriche, Vol. II*, Zanichelli, Bologna, 1918.

[F_1] A. Franchetta, Sulle superficie le cui curve canoniche posseggono una g^1_3, B.U.M.I., 1942.

[F_2] A. Franchetta, Su alcuni esempi di superficie canoniche, *Semin. Mat. di Roma*, 1939.

[F_3] A. Franchetta, Sulle curve eccezionali riducibili di prima specie, *B.U.M.I.*, 4 (1940), 332-341.

[F_4] A. Franchetta, Sulla caratterizzazione delle curve eccezionali riducibili di prima specie, *B.U.M.I.*, 3 (1940-41), 372-375.

[F_5] A. Franchetta, Sul sistema aggiunto ad una curva riducibile, *Rend. Accad, Lincei*, 6 (1949), 685-687.

[F_6] A. Franchetta, Sulle curve riducibili appartenenti ad una superficie algebrica, *Rend. Mat. e Appl.*, 8 (1949), 378-398.

[F_7] A. Franchetta, Sui sistemi pluricanonici di una superficie algebrica, *Rend. Mat. e Appl.*, 8 (1949), 423-440.

[F_8] A. Franchetta, Sui modelli pluricanonici delle superficie algebriche, *Rend. Mat. e Appl.*, 9 (1950), 293-308.

[FAC] J. P. Serre, Faisceaux algébriques cohérents, *Ann. Math.*, 61 (1955), 197-278.

[FR] P. Francia, The bicanonical map for surfaces of general type, to appear.

[G] M. Green, Koszul cohomology and the geometry of projective varieties, *Duke Math. J.*, 49 (1982), 1087-1113.

[GO] L. Godeaux, Sur une surface algébrique de genre zero et de bigenre deux, *Atti Accad. Naz. Lincei*, 14 (1931), 479-481.

[H] J. Harris, *Curves in projective space*, Le presses de l'Universite' de Montrèal, 1982.

[HO_1] E. Horikawa, On deformations of quintic surfaces, *Inventiones Math.*, 31 (1975), 43-85.

[HO_2] E. Horikawa, Algebraic surfaces of general type with small c_1^2, I, *Ann. Math.*, 104 (1976), 375-387; II, *Inventiones Math.*, 37 (1976), 121-155; III, *Inventiones Math.*, 47 (1978), 209-248; IV, *Inventiones Math.*, 50 (1979), 103-128.

[J] F. Jongmans, Contributions a la théorie des variétés algèbriques, *Mem. Soc. Roy. de Sci. de Liège*, 7 (1947), 1-102.

[K] K. Kodaira, Pluricanonical systems on algebraic surfaces of general type, *J. Math. Soc. Japan*, 20 (1968), 170-192.

[M_1] D. Mumford, The canonical ring of an algebraic surface, *Ann. Math.*, 76 (1962), 612-615.

[M_2] D. Mumford, Lectures on curves on an algebraic surface, *Ann. Math. Studies* 59, Princeton Univ. Press, Princeton (1966).

[MI$_1$] I. Miyaoka, On the Mumford-Ramanujan theorem on a surface, in *Journées de géométrie algébrique d'Angers* (1979), Sijthoff and Noordoff, 1980.

[MI$_2$] I. Miyaoka, On the Chern numbers of surfaces of general type, *Inventiones Math.*, 42 (1977), 225-237.

[MR] R. E. Mac Rae, On an application of the Fitting invariants, *J. Algebra*, 2 (1965), 153-169.

[N] M. Noether, Ueber die algebraische Funktionen, *Gottingen Nachr.*, 7 (1871).

[P$_1$] K. Petri, Ueber die invariante Darstellung der algebraischen Funktionen einer Veranderlische, *Math. Ann.*, 88 (1922), 242-289.

[P$_2$] K. Petri, Ueber Spezialkurven I, *Math. Ann.*, 93 (1924), 182-209.

[PE] U. Persson, An introduction to the geography of surfaces of general type, *Proc. Symposia in Pure Math.*, 46 (1987), 195-218.

[PO$_1$] G. Pompilj, Alcuni esempi di superficie algebriche a sistema canonico puro degenere, *Rend. Acc. Lincei*, 4 (1948).

[PO$_2$] G. Pompilj, Sulle superficie algebriche le cui curve canoniche posseggono una g^1_3, *Rend. Ist. Lombardo*, 1941.

[R] B. Riemann, Theorie der Abelschen Funktionen, *J. Reine und Angew. Math.*, 54 (1859).

[RE] I. Reider, Vector bundles of rank 2 and linear systems on algebraic surfaces, *Ann. Math.*, 127 (1988), 309-316.

[RA$_1$] C. P. Ramanujan, Remarks on the Kodaira vanishing theorem, *J. Ind. Math. Soc.*, 36 (1972), 41-51.

[RA$_2$] C. P. Ramanujan, Supplement to the article "Remarks on the Kodaira vanishing theorem", *J. Ind. Math. Soc.*, 38 (1974), 121-124.

[S] E. Sernesi, L'unirazionalita' della varieta' dei moduli delle curve di genere dodici, *Ann. Sc. Norm. Super. Pisa*, 8 (1981), 405-439.

[SE] B. Segre, Sulle superficie algebriche aventi il sistema canonico composto con una involuzione, *Rend. Accad. Lincei*, 16 (1932).

[SH] I. P. Shafarevich, *Algebraic surfaces*, Moskva, 1965.

[SV] F. Severi, La serie canonica e la teoria delle serie principali dei gruppi di punti sopra una superficie algebrica, *Comm. Math. Helv.*, 4 (1932), 268-326.

[X] G. Xiao, L'irrégularité des surfaces de type général dont système canonique est composé d'un pinceu, *Compositio Math.*, 56 (1985), 251-257.

[Y] S. T. Yau, On the Ricci-curvature of a complex Kaeler manifold and the complex Monge-Ampere equation, *Comment. Pure Appl. Math.*, 31 (1978), 339-411.

[VZ] G. Van der Geer and D. Zagier, The Hilbert modular group for the field $\mathbb{Z}(13^{1/2})$, *Inventiones Math.*, 42 (1978), 93-134.

[Z] O. Zariski, The theorem of Riemann-Roch for high multiples of an effective divisor on an algebraic surface, *Ann. Math.*, 76 (1962), 560-612.

Geometry and Complex Variables in the Work of Beppo Levi

SALVATORE COEN, Dipartimento di Matematica Università di Bologna, Piazza di Porta San Donato 5, 40126, Bologna, Italy

Sommario. Scopo di questo contributo è fornire un'analisi della attività scientifica di Beppo Levi (Torino 1875, Rosario (Argentina) 1961) nell'ambito della teoria delle funzioni di variabile complessa ed in quello della geometria.

La produzione di Beppo Levi nel campo della geometria fu decisamente più ampia di quella nell'ambito della variabile complessa ed è certamente anche meglio conosciuta.

Tratteremo, pertanto, i due argomenti separatamente e secondo schemi diversi.

Nella prima parte di questo lavoro, dopo avere fornito nel §2 un elenco della produzione del Levi nel campo della teoria delle funzioni di variabile complessa, recensiremo tutti i singoli lavori di ricerca elencati. Dedicheremo, poi, gli ultimi due paragrafi di questa parte ad una esposizione particolarmente dettagliata di alcuni semplici risultati del Levi nel campo della teoria delle funzioni analitiche di più variabili, pubblicati quando egli era professore a Bologna.

Nella seconda parte forniremo nel §7 un elenco di lavori di Beppo Levi di argomento geometrico, poi illustreremo la produzione del Levi nell'ambito della geometria algebrica, soffermandoci sulla sua genesi e sulla considerazione in cui essa fu tenuta quando il Levi era ancora in vita. Nel §9 daremo una visione panoramica dell'ulteriore produzione di Beppo Levi in campo geometrico, con particolare riguardo alle ricerche sui fondamenti della geometria.

La bibliografia finale si riferirà ad entrambe le parti di questo lavoro.

The purpose of this paper is to present an analysis of the work of Beppo Levi (Turin (Italy) 1875, Rosario (Argentina) 1961)[1], on the complex function theory and in geometry.

---

[1] Beppo Levi was born on May 14th, 1875 in Turin, Italy and died August 28th, 1961 in Rosario, Argentina.

Beppo Levi's work in geometry was definitely more extensive than that on the complex function theory and is certainly generally better known. Because of this, we shall treat the two arguments separately following different outlines.

In the first part of this paper, after presenting in §2 a listing of Levi's studies on the complex function theory, we shall review each research work listed. The last two paragraphs of this section will then be dedicated to a detailed presentation of some simple results arrived at by Beppo Levi on the theory of several complex variables as published when he was a professor in Bologna.

In the second part, in §7, a listing of Beppo Levi's works in geometry is included. In §8 Beppo Levi's works in algebraic geometry are examined, pointing out their genesis and the acceptance which these received when Beppo Levi was still living. In §9, an overall view of other Beppo Levi's works in geometry is presented with particular emphasis on his research in foundations of geometry.

The bibliography provided refers to both parts of this paper.

Part one: The Work of Beppo Levi on Complex Function Theory

1. Introduction   112
2. List of Beppo Levi's Papers on Complex Function Theory   113
3. Reviews of Beppo Levi Research Papers on Complex Function Theory   114
4. Beppo Levi and the Theory of Analytic Functions of Several Variables:
   Construction of Vanishing Subsets   118
5. Beppo Levi and the Theory of Analytic Functions of Several Variables:
   a Geometrical Result   121

Part two: The Work of Beppo Levi in Geometry

6. Introduction   123
7. List of Beppo Levi's Papers on Geometry   124
8. Research in Algebraic Geometry   126
9. Other Research in Geometry   134
References   138

PART ONE: THE WORK OF BEPPO LEVI ON COMPLEX FUNCTION THEORY

1. Introduction

In 1928, when Beppo Levi became Professor of Function Theory (Teoria delle Funzioni) at the University of Bologna, he was fifty-three years old and had only been working on the complex function theory for a few years. Before this, he has been a professor of geometry (Geometria Proiettiva e Descrittiva) at the University of Cagliari (1906-1910) and a professor of «Analisi Algebrica» and of «Matematica Speciale» at the University of Parma (1910-1928). Of course, he had already used complex function theory methods in his papers, but on the basis of his known scientific work, we think that he started to do research in this field only around 1925.

In his Parma period, he wrote two papers on the subject, as far as we know. The first one is [1]. It is a non-research contribution to the «Annuario Scientifico ed Industriale», written to bring readers up-to-date on the then latest results on the theory of almost periodic functions and of ordinary trigonometric series. The paper contains a noteworthy account of the work of

Harold A. Bohr (1887-1951), in the period 1923-1925. Paper [2] was also written in the Parma period.

When in Bologna, Levi wrote four papers on our subject: [3], [4], [5], [6].

When in Argentina, in 1941, Beppo Levi gave an account of the analytic function theory in his monograph [8]. In this monograph the theory of systems of analytic equations is amply developed and finds several applications. For the most part, the book is concerned with the theory of analytic ordinary and partial differential equations; this book is perhaps a not sufficiently known treatise that rose from lectures given at the University of Genova in 1913-1914 by Eugenio Elia Levi, and from other lectures given by Beppo Levi himself on several occasions in Bologna (1931-1932, 1933-1934) and in Rosario (1941). Papers [9], ..., [13] written in the Rosario period may be considered more or less in the field of one variable's complex function theory.

During his long scientific period (more than 60 years), Beppo Levi wrote more than 150 papers, some of which are long and important monographs, at least 180 scientific reviews and several mathematics books. Beppo Levi made important contributions to many different branches of mathematics which include his papers on the resolution of singularities of algebraic surfaces, his contribution on the integration theory, his contribution on the Dirichlet problem, his ideas on the axiom of choice (see Levi's biography [T], written by A. Terracini).

In comparison, with such an important mathematical work, Levi's contribution specifically dedicated to complex function theory seems to be limited. Indeed, in the 13-items list presented in §2, we include papers containing interesting results in the complex function theory even if the main subject of these papers is different. Our choice of papers to review was somewhat personal. Please note, in particular, that the paper [C-L] cannot be considered, in the proper sense of the word, specifically as a paper on complex analysis. In spite of this, however, we are going to present a brief review because it gives evidence of Beppo Levi's interest in the theory of trigonometric functions.

Though it seems that Beppo Levi was not mainly concerned with the complex function theory, we note that he completely mastered difficult classical techniques and some of his proofs can be considered, in our opinion, really clever.

Unfortunately, we only know of one contribution to the theory of analytic functions of several variables; it was written when he was in Bologna (papers [5], [6] which are strongly related). As far as we know, he never followed the ideas of his brother Eugenio Elia in this field, and we do not know of any Beppo Levi contribution to the Levi problem. We are going to expose results of [5] and [6] in detail in the next chapters 4, 5 as our contribution to the history of the development of the theory of several complex variables: an historical subject that, in our opinion, has not been investigated enough up to now.

2. List of Beppo Levi's Papers on Complex Function Theory

[1] Beppo Levi, *Teoria delle Funzioni*, Annuario Scientifico ed Industriale, Anno **LXII** (1925), vol. II, 291-301.

[2] Beppo Levi, Sulla relazione $\eta_1 \omega_2 - \eta_2 \omega_1 = \pi i/2$, *Boll. Un. Mat. Ital.*, 6 (1927), 137-141.

[3] B. Levi, Sulla natura analitica di una classe di funzioni definite da serie di potenze convergenti sulla circonferenza di convergenza, *Boll. Un. Mat. Ital.*, 12 (1933), 1-5.

[4] B. Levi e T. Viola, Intorno ad un ragionamento fondamentale nella teoria delle famiglie normali di funzioni, *Boll. Un. Mat. Ital.*, 12 (1933), 197-203.

[5] Beppo Levi, Sul teorema di identità per le funzioni analitiche di più variabili, *Boll. Un. Mat. Ital.*, **13** (1934), 1-5 and 104.

[6] Beppo Levi, Sur les ensembles de points qui ne peuvent être ensembles de zéros d'une fonction analytique de plusieurs variables, *C.R. Acad. Sci. Paris*, **198** (Séance du 14 Mai 1934), 1735-1736.

[7] Beppo Levi, Osservazioni riguardo alla precedente nota di C. Biggeri, *Boll. Un. Mat. Ital.*, **15** (1936), 214-215.

[8] Beppo Levi, *Sistemas de equaciones analiticas en términos finitos, diferenciales y en derivadas parciales*, Monografias publicadas per la Facultad de Ciencias Matematicas..., **1** (1944), Rosario (R. Argentina), 1-216.

[9] B. Levi, Los polinomios de aproximación de sen x y cos x, *Math. Notæ*, **4** (1944), 156-163.

[10] Beppo Levi, Una ejercitación sobre integrales elipticas, *Math. Notæ*, **6** (1946), 167-190.

[11] Antonio Petracca y B. Levi, Estudio de una función polidroma (a proposito del ejercicio 171), *Math Notæ*, **10-11** (1951), 124-138.

[12] Antonio Petracca y Beppo Levi, Complemento a la Nota: Estudio de una función polidroma, *Math. Notæ*, **12-13** (1952), 48-49

[C-L] M. Cotlar y B. Levi, Ejercitaciones sobre la función coseno, *Math. Notæ*, **5** (1945), 1923-214.

3. Reviews of Beppo Levi's Research Papers on Complex Function Theory

We are going to omit the reviews of [1] and [8] because, as we have already mentioned, these are non-research papers.

[2]. Consider the Sigma-function of Weierstrass $\sigma(z)$ with half-periods ω_1, ω_2 and assume that Im $\omega_1/\omega_2 > 0$; $\sigma(z)$ is a pseudo-periodic function. The half-periods of $\sigma(z)$ of the second kind of are numbers η_1, η_2 so that

$$\sigma(z + 2\omega_j) = -e^{2\eta_j(z+\omega_j)}\sigma(z)$$

for $j = 1, 2$. Then η_1, η_2 must be known to get information about the behavior of σ when an element $2m\omega_1 + 2n\omega_2$ of the parallelogram of periods is added to z. Thus, it is useful to know the relationship between half-periods and half-periods of the second kind. It is easy to prove that

$$\eta_1\omega_2 - \eta_2\omega_1 = h\pi i/2$$

where $h \in \mathbb{Z}$ and h is odd. One has $h = 1$. Weierstrass proved this last equality by long computation. Another very short proof consists in integrating on the boundary of the parallelogram of periods of the Zeta-function of Weierstrass $\zeta(z)$. But, Beppo Levi remarks in [2] that this method breaks «*la purezza del metodo algebrico*» of Weierstrass. Then, he succeeds in giving a new proof that avoids integration methods, that saves the «algebraic purity» of Weierstrass' methods and that is considerably easier than Weierstrass' one. Indeed, we think that this proof is so simple that it can be used in an elementary exposition of properties of the Sigma-function $\sigma(z)$ of Weierstrass.

[3]. In the present paper the Author considers the sum $f(z; k, h)$ of the series

$$\sum_{n=1}^{\infty} \frac{z^k}{n^k + h}$$

where z belongs to the complex unitary disc, $k \in \mathbb{Z}, k > 1$ and $h \in \mathbb{C}$, proving that the functions $z \to f(z; k, h)$ cannot be expressed as rational functions of z or of elementary trascendent functions of z. First of all, he remarks that the functions $f(z; k, h)$ are the solutions of a known class of linear differential equations, then he studies the case $h = 0$ and, by recurrence on k, he gives an expression of the functions $f(z; k, 0)$ suitable for the study of the circulations of such functions around the point $z = 1$. This investigation proves that for each $k > 1$, the Riemann surface of $f(z; k, 0)$ cannot be the Riemann surface of a rational function of z or of an elementary trascendent function of z. Finally, he is able to reduce the general case $h \in \mathbb{C}$ to the case $h = 0$ and to prove that the functions $f(z; k, h)$ cannot be expressed as rational functions of z or of elementary trascendent functions of z. The paper gives a positive answer to a question proposed in the special case $h = 1, k = 2$ on the same *Bollettino Un. Mat. Ital.*, **11** (1932), p. 303.

[4]. This work may be considered more as a methodological paper than a work on the complex function theory. In the books: (A) «*Leçons sur les familles normales de fonctions analytiques et leurs applications*» Ed. Gauthier-Villars, 1927 and (B) «*Leçons sur les fonctions univalentes ou multivalentes*», Ed. Gauthier-Villars, 1933, P. Montel used several times an argument («ragionamento fondamentale») in which somebody perceived the use of the axiom of choice (on a countable family of sets). Montel used this argument in proving interesting results about normal families of holomorphic functions. Beppo Levi and Tullio Viola do not accept the axiom of choice in its whole generality, noting that each time one has to study the proof «nel senso di una sua possibile giustificazione». This is what they do in this paper according to previous Levi's works on the notion of deductive domain («dominio deduttivo»). Here, we cannot give an account of the delicate Levi notion of deductive domain; for this, see B. Levi *Sui procedimenti transfiniti* (Auszug aus einem Briefe an Herrn Hilbert), Math. Ann., **90** (1923), 164-173, B. Levi *La noción de «dominio deductivo» como elemento de orientación en las cuestiones de fundamentos de las teorìas matemàticas* Publ. Ins. Mat. Rosario, **2** (1940), 179-208 and [M] pp. 243-6. Levi and Viola think that the theory of normal families of functions concerns mainly the deductive domain of sequences of functions and that this is the deductive domain in which previous arguments must be set. In any event, they conclude that this «ragionamento fondamentale» may be accepted. Nevertheless, a proposition which is proved by such an argument, once it is also proven by independent tools (as happens when another deductive domain is used) must be considered a new proposition. They suggest also a slight modification in the definition of normality.

As an example, consider the family $\mathfrak{F}$ of all the analytic and univalent functions on the unitary disc D, which have an expansion of the form $z + a_2 z^2 + a_3 z^3 + \ldots$. Montel proved that $\mathfrak{F}$ is a normal family, then he used his above-quoted argument to show that a disc of $\mathbb{C}$ exists which is covered by $f(D)$ for each f in the family $\mathfrak{F}$ (see (A), p. 35 and (B), p. 28-31). But this is a classic result that was proven by other mathematicians by means of different methods (as computation of coefficients of particular series). In this last case, the deductive domain of the set of real numbers is used. Then Levi and Viola say that these results (the first of Montel and the others) must be considered as different results.

[5], [6]. These papers take rise from a talk that B. Levi had at the Institute of Mathematics of Bologna with some of his colleagues (perhaps with L. Fantappiè and T. Viola). The papers deal with the question of finding a suitable identity theorem for analytic functions of several variables that would be a good generalization of the theorem in the one dimensional case.

When Levi wrote these papers, other mathematicians had already worked on this subject, as Beppo Levi himself recalled. To present Levi's work in its historical background, please refer to the following works:

G. Bouligand, *Fonctions harmoniques. Principes de Picard et de Dirichlet*, Memorial de Sciences Mathématiques, fascicule XI, Gauthiers-Villars et Cie, 1926 (see p. 20).

C. Carathèodory, *Ein dem Vitalischen anologer Satz für analytische Funktionen von mehreren Veränderlichen*, J. Reine Angew. Math., **165** (1931), 180-183.

P. Montel, *Leçons sur les familles normales des fonctions analytiques et leurs applications* (recueilles et rédigées par J. Barbotte), Gauthiers-Villars et Cie, 1927 (see p. 246).

F. Severi, *Risultati, vedute e problemi nella teoria delle funzioni di due variabili complesse*, Rend. Mat., **VII**, 2 (1932), 1-58 (see p. 47).

Observe that Bouligand's work was interested in harmonic functions and that it seems that T. Viola drew inspiration from Levi's work for his paper.

T. Viola, *Sur le théorème d'identité pour les fonctions holomorphes de plusieurs variables* C.R. Academie de France, 198, séance du 19 février 1934, 705-707.

In [5] the Author succeeds in giving some different simple constructions of subsets E of $\mathbb{C}^n$ with the following properties: a) E is a countable subset of $\mathbb{C}^n$, b) E has only one accumulation point (and this is O), c) if $S(X_1,\ldots,X_n) \in \mathbb{C}\{X_1,\ldots,X_n\}$ and if $S(x_1,\ldots,x_n) = 0$ in every point $(x_1,\ldots,x_n)$ of E in which it is defined, then one has $S = 0$. Though we will give later in §4 a detailed account of some results of [5], we are going here to recall one of its constructions because of the clever geometrical ideas that support it. The proof is a straightforward recurrence proof on the dimension $n \geq 1$. Let H be the hyperplane $x_n = 1$ of $\mathbb{C}^n$. By the inductive assumption, a countable subset $W =: \{w^{(j)} | j \geq 1\}$ of H exists that fulfills properties a), b), c) and whose only accumulation point is the point $(0,\ldots,0,1) \in H$. For every $j \geq 1$, let r_j be the complex line of $\mathbb{C}^n$ containing 0 and $w^{(j)}$. For every $j \geq 1$ let E_j be a countable subset of R_j whose only accumulation point is O. Assume that the diameter of E_j is less than $1/j$. Then $E := \bigcup_{j \geq 1} E_j$ fulfills all the required properties a), b), c). The paper [6] gives new, more geometrical results of this kind and will be reviewed in the following chapter 5.

[7]. This short remark was published on the Bollettino dell'Unione Matematica Italiana just after Biggeri's paper «*Sulle singolarità delle funzioni analitiche*». Biggeri had furnished sufficient conditions for a complex power series with convergence radius $R > 0$ had in the point R a singularity. This was a generalization of some Vivanti and Dienes' results. First of all, Beppo Levi in his article places Biggeri's result in a correct bibliographic frame. Thus, he recalls some recent results of Pringsheim, Mandelbrojt and Hadamard and remarks that the main part of Biggeri's proof was stated in an article published on the Rendiconti dei Lincei in 1923 by Milos Kössler. But Kössler gave an outline of the proof, using a different technique than that used by Biggeri. Then, Beppo Levi indicates how it is possible to simplify Biggeri's proof and generalize the results somewhat. It may be noteworthy that also Oskar Perron gave a generalization of Biggeri's result in a paper of the same journal (*Osservazioni riguardo un teorema di C. Biggeri*, XVI, 82-84).

[9]. Here the Author recalls previous papers of L. A. Santaló (1942), and J.V. Uspenski (1944) appearing in the Math. Notæ about the question of finding simple tools to determine the power series expansion (centered in 0) of sen x, cos x. He also recalls some historical research on the mathematical work of I. Newton. His aim is to determine these series only by means of methods used in Newton's age. He succeeds, by means of elementary algebraic tools, using only four functional inequalities (that can be easily proven by geometrical tools), the De Moivre's equality (i.e. the expression of sen nx, cos nx as polynomial functions in sen x, cos x) and the property that for each $0 \leq \xi \leq 1/2$, one has $\sum_{j=1,\ldots,n} \xi^i < 2$ for every

$n \in \mathbb{N}$. The four functional inequalities may be collected in the following formula

$$1 \geq \left\{ \begin{array}{c} \frac{|\operatorname{sen} x|}{x} \\ |\cos x| \end{array} \right\} > 1 - \frac{x^2}{2} \qquad x > 0$$

We may remark that from this work one can deduce a characterization of $\operatorname{sen} x, \cos x$ by means of simple functional properties.

[10]. This paper is a nice exercise. Let C be an elliptic cylinder and let $\mathcal{E}$ be an ellipsoid with center on the axis of the cylinder. A normal section of C and the axes of $\mathcal{E}$ are known. One has to compute the volume of $C \cap \mathcal{E}$. This simple question (proposed on Math. Notæ, IV) gives rise to this work. First of all, the Author remarks that we are not losing generality by assuming E be the unitary ball, then he determines what integrals are to be computed. One has to compute some elliptic integrals of the first, the second and the third kind (according to Legendre). It is in the computation of these integrals that some difficulties may arise. It is difficult to find computation tables suitable for the problem and the use of Theta-functions of Jacobi, computed by Jacobi himself brings «una profundización teorica no indiferente», inconsistent with such an elementary problem. Thus, Beppo Levi, by means of a rather difficult calculation, succeeds in determining an explicit expression which allows him to find the numerical solution of the problem only using the computation tables of Legendre. This is the exercise of the title of this paper. The whole work is permeated by a strong sensitivity to the numerical aspects of the problem.

[11], [12]. Exercise 171 in Mat., Notæ, X, p. 49 asked to find the convergence domain of the following fractionary power series

$$z + z^{1+1/2} + z^{1+1/2+1/3} + z^{1+1/2+1/3+1/4} + \ldots$$

Thus, this question required to make a study of a special multiple-valued function. This suggests the investigation, for $\rho > 0$ and for $\varphi \in \mathbb{R}$ of the following series

$$S(\rho, \varphi, \{k_n\}) := \sum \rho^{H_n} e^{i(\varphi + 2\pi k_n) H_n}$$

where $H_n = 1 + 1/2 + \ldots + 1/n$ and $\{k_n\}_{n \geq 1}$ is a sequence of integers. In [11], first of all, B. Levi and A. Petracca prove that for every sequence of integers $\{k_n\}_{n \geq 1}$ and for any $\varphi \in \mathbb{R}$, the series $S(\rho, \varphi, \{k_n\})$ converges for $\rho < 1/e$ and that it diverges for $\rho > 1$. Thus, the main interest consists in testing the case ρ belongs to the interval $[1/e, 1]$. In this case the series absolutely diverges, but some remarkable facts appear. For every $\rho \in [1/e, 1]$ and for every $\varphi \in \mathbb{R}$, infinite sequences of integers $\{k_n\}_{n \geq 1}$ may be chosen so that the partial sums of $S(\rho, \varphi, \{k_n\})$ have a cluster point in an arbitrary complex number (for $\rho \neq 1$ we can ask that the cluster point is unique). Moreover, fix $\rho \in [1/e, 1], \varphi \in \mathbb{R}$ and a connected, compact subspace C of $\mathbb{C}$, then, at least a sequence $\{k_n\}_{n \geq 1}$ exists so that the derivate set of the set of partial sums of $S(\rho, \varphi, \{k_n\})$ [i.e. the subspace of cluster points of the set of partial sums] contains C. In [11] several other results are proven; e.g. a notiion of radius of convergence of the series $S(\rho, \varphi, \{k_n\})$ is given and deeply studied. In [12] continua in metric spaces M are proven to be the derivate sets of some particular countable sets (such a criterion was already be used in [11]).

[C-L]. In «M. Cotlar y B. Levi, *Consideraciones sobre una proposición de W. H. Young*, Math. Notæ, 4 (1944), 145-155», the Authors remarked that a proposition of W.H. Young stated in the paper «*Open sets and the theory of content*», Proc. London Math. Soc., II, 2 (1905), p. 25 is false. It may be noteworthy that E.W. Hobson had already reported the result of Young in the first volume of his treaty «*The Theory of Functions of a Real Variable and the Theory of Fourier's Series*», Cambridge Univ. Press., at p. 183. Precisely, Hobson gave the same proof of Young, but with a weaker and correct statement of Young's result.

The above-mentioned remark gave Cotlar and Levi a good opportunity to prove interesting results about trigonometric functions. One of their counter-examples is the following one. Consider the sequence $S = \{\cos nx | x \in [0, 2\pi]\}_{n \in \mathbb{N}}$. If Young's proposition was correct, then we could select a subsequence of S, uniformly convergent on a subset of $[0, 2\pi]$ of Lebesgue measure 2π. But no subsequence of S can uniformly converge on a subset of $[0, 2\pi]$ with finite Lebesgue-measure $\mu > 0$. The proof is achieved only by classic tools.

This last proposition was improved by R.P. Boas Jr. in a letter to B. Levi (Math. Notæ, V, p. 91) in the sense of omitting in the previous statement the word «uniformly». Indeed, he showed that the proposition is an easy consequence of a Steinhaus lemma (cfr. Zygmund, *Trigonometrical series*, Monografje Matematyczne, Tom V, 267-270).

First of all, in the paper [C-L], Cotlar and Levi remark that it is possible to avoid Steinhaus Lemma in the proof of Boas. Then, they give several results about the functions $A(n, \rho, x) := \rho(\cos nx + \alpha)$, where $x \in E$, E is a subset of $\mathbb{R}$ with strictly positive Lebesgue measure and where $n, \rho, \alpha \in \mathbb{R}, n, \rho > 0$.

As an example, consider the following proposition they prove. Let $E \subseteq \mathbb{R}$ be a subset of $[0, 2\pi]$ with strictly positive Lebesgue measure. Let $\{n_j\}_{j \geq 1}$ be a divergent increasing sequence of positive real numbers. Fix $\gamma \in \mathbb{R}$ with $0 < \gamma < 2$. Let $\lambda := \arccos \gamma/2, 0 < \lambda < \pi$. Then, there exist $\delta > 0$, δ depending only by E and γ, such that if n', n'' are items of the given sequence with $n' > \delta, n'' > \frac{2\pi n'}{\lambda}$, one has

$$\cos n'x - \cos n''x \geq \gamma$$

for suitable points of E. Moreover, in this paper the Authors give a new proof of the above quoted Steinhaus Lemma, and they introduce a new notion which is strictly related with the Lebesgue density. Finally, they state applications of their work to the Probability Theory.

4. Beppo Levi and the Theory of Analytic Functions of Several Variables: Construction of Vanishing Subsets

A. Introduction to Chapters 4,5.

Only in recent times have mathematicians started to study the contribution of Italians to the development of the theory of analytic functions of several variables. In this connection, let us recall Fichera's paper [F] and other papers of the present Meeting. Then, it is mainly because of historical reasons that we are going to furnish here and in the next chapter a detailed account of some of Beppo Levi's results in this field. As already mentioned, this paper was written when Beppo Levi was a professor at the University of Bologna.

These results are really very simple ones concerning the local theory of several complex variables. It is worth remarking that he only used quite elementary methods and a very simple statement of the Vorbereitungssatz.

Of course, we are going to use a slightly different language from Beppo Levi's and our proofs will be much more detailed than the original ones. In particular, observe that the notion of «vanishing subset of $\mathbb{C}^n$» was not stated explicitly by Beppo Levi. We shall give here only some of the results of [5], [6], but by means of their proofs and methods all the results of [5], [6] can be easily deduced. We shall try to stay as close as possible to Levi's methods and ideas. The reader will find easy generalizations of some of these results, as Beppo Levi himself realized. We decided not to give any of these generalizations to stay closer to Levi's work.

Denote by O the point $(0,\ldots,0) \in \mathbb{C}^n$. If $\varepsilon > 0$ and $x \in \mathbb{C}^n$, then $P(x,\varepsilon)$ will denote the polycylinder

$$P(x,\varepsilon) := \{z \in \mathbb{C}^n \mid |z_j - x_j| < \varepsilon,\ j = 1,\ldots,n\}.$$

Set $\mathbb{N}^* := \{n \in \mathbb{N} \mid n > 0\}$.

4.1. DEFINITION. *Let $E \subseteq \mathbb{C}^n, n \geq 1$; E si called to be a vanishing subset of $\mathbb{C}^n$ (in O), when the following conditions are fulfilled*
 a) *O is an accumulation point of E;*
 b) *let $S(X_1,\ldots,X_n) \in \mathbb{C}\{X_1,\ldots,X_n\}$ be a power series convergent in the polycylinder $P(O,\varepsilon)$ in $\mathbb{C}^n, \varepsilon > 0$. Let $S(x_1,\ldots,x_n) = 0$ for every $(x_1,\ldots,x_n) \in P(O,\varepsilon) \cap E$, then $S(X_1,\ldots,X_n) = 0$.*

In §3, we already saw how B. Levi succeeded in constructing vanishing subsets E of $\mathbb{C}^n$ whose only accumulation point is O and which are of the form $E := \underset{j \geq 1}{\cup} E_j$ where each E_j is a *countable* subset of a complex line r_j. Here, in §5, we shall show how he constructed countable vanishing subsets whose only accumulation point is O and that are of the form $E := \underset{j \geq 1}{\cup} E_j$ where each E_j is a *finite* subset of a complex line R_j. In §6 we shall see a more geometrical result in $\mathbb{C}^2$.

B. A particular construction of vanishing subsets of $\mathbb{C}^n$.
The present chapter contains an enlarged version of the brief §5 of [5].

4.2. PROPOSITION. *Let $E \subseteq \{(1,0,x_3,\ldots,x_n) \in \mathbb{C}^n \mid x_3,\ldots,x_n \in \mathbb{C}\} \cong \mathbb{C}^{n-2},\ n \geq 2$, be a vanishing subset of $\mathbb{C}^{n-2}$ in the origin O' of $\mathbb{C}^{n-2}$; $O' \cong (1,0,\ldots,0) \in \mathbb{C}^n$. Let $S(X_1,\ldots,X_n) \in \mathbb{C}\{X_1,\ldots,X_n\}, S \neq 0$, be a power series convergent in the polycylinder $P(O,\varepsilon),\varepsilon > 0$, of $\mathbb{C}^n$. Then*

$$S(X_1,\ldots,X_n) = X_2^j F(X_1,\ldots,X_n)$$

where $j \in \mathbb{N}$ and $F(X_1,\ldots,X_n) \in \mathbb{C}\{X_1,\ldots,X_n\}$ is a power series such that
 a) *F does not factorize in factors of the form $X_2^s, s > 0$.*
 b) *a point $(1,0,a_3,\ldots,a_n) \in E$ exists such that the analytic function $x \to F(x,0,a_3,\ldots,a_n)$ does not identically vanish on any neighborhood of O in $\mathbb{C}$.*

4.3. LEMMA. *Let E, S, ε be as in the previous Proposition. Assume $x^\circ \in \mathbb{C}, x^\circ \neq 0, |x^\circ| < \varepsilon$. Assume $S(x^\circ,0,a_3,\ldots,a_n) = 0$ when $|a_j| < \varepsilon, j = 3,\ldots,n$ and $(1,0,a_3,\ldots,a_n) \in E$.
Then, $S(x^\circ,0,y_3,\ldots,y_n) = 0$ for every $y_3,\ldots,y_n$ when $|y_j| < \varepsilon$.*

Proof of 4.3. Consider the power series $T(Y_3,\ldots,Y_n) \in \mathbb{C}\{Y_3,\ldots,Y_n\}$, defined by

$$T(Y_3,\ldots,Y_n) = S(x^\circ,0,Y_3,\ldots,Y_n).$$

This series is convergent in the polycylinder, $P := \{(y_3, \ldots, y_n) \in \mathbb{C}^{n-2} \,||y_j| < \varepsilon\}$. The set $A := E \cap \{(1, 0, y_3, \ldots, y_n) \in \mathbb{C}^n | (y_3, \ldots, y_n) \in P\}$ is non-empty. The assumption implies that for each $(1, 0, a_3, \ldots, a_n) \in A$ one has $T(a_3, \ldots, a_n) = S(x^\circ, 0, a_3, \ldots, a_n) = 0$. Recall that E is a vanishing subset of $\mathbb{C}^{n-2} \cong \{(1, 0, x_3, \ldots, x_n) \in \mathbb{C}^n | x_3, \ldots, x_n \in \mathbb{C}\}$ in O'; thus, $T = 0$. Then, $T(y_3, \ldots, y_n) = S(x^\circ, 0, y_3, \ldots, y_n) = 0$ for every y_j with $|y_j| < \varepsilon, j = 3, \ldots, n$. ∎

Proof of 4.2. Write
$$S(X_1, \ldots, X_n) = \sum b_{j_1 \cdots j_n} X_1^{j_1} \cdots X_n^{j_n}$$

There are points $(a_3, \ldots, a_n) \in \mathbb{C}^{n-2}$ such that $(1, 0, a_3, \ldots, a_n) \in E$ and $|a_j| < \varepsilon$. It is suitable to distinguish two different cases.

First case. Assume that for someone of the above considered points $(1, 0, a_3, \ldots, a_n)$ one has $S(x^\circ, 0, a_3, \ldots, a_n) \neq 0$ for a suitable $x^\circ \in \mathbb{C}, x^\circ \neq 0, |x^\circ| < \varepsilon$. Then, there are $j_1, j_3, \ldots, j_n \in \mathbb{N}$ with $b_{j_1 0 j_3 \cdots j_n} \neq 0$. Set $j = 0, F = S$.

Second case. Assume that for each one of the above considered points $(1, 0, a_3, \ldots, a_n)$ and for each $x, |x| < \varepsilon$, one has $S(x, 0, a_3, \ldots, a_n) = 0$. From 4.3 it follows $S(x, 0, y_3, \ldots, y_n) = 0$, when $|y_j| < \varepsilon, j = 3, \ldots, n$. Then, one has $S(x_1, 0, x_3, \ldots, x_n) = 0$ for $|x_j| < \varepsilon, j = 1, 3, \ldots n$ and $b_{j_1 \cdots j_n} = 0$ when $j_2 = 0$.

Consider the set $B := \{j_2 \in \mathbb{N} | j_1, j_3, \ldots, j_n \in \mathbb{N} \text{ exist so that } b_{j_1 j_2 \cdots j_n} \neq 0\}$. One has $B \neq \emptyset$. Set $j := \min B$. One has $j > 0$. Then

$$S(X_1, \ldots, X_n) = X_2^j F(X_1, \ldots, X_n)$$

and F, j fulfill condition a).

Now, we have to prove that also condition b) holds. Assume that one has $F(x, 0, a_3, \ldots a_n) = 0$ for each $x \neq 0, |x| < \varepsilon$ and for every $(a_3, \ldots, a_n) \in \mathbb{C}^{n-2}$ so that $(1, 0, a_3, \ldots, a_n) \in E$ and $|a_j| < \varepsilon$. By the same argument we already used, we deduce that a power series $T(X_1, \ldots, X_n) \in \mathbb{C}\{X_1, \ldots, X_n\}$ exists so that

$$F(X_1, \ldots, X_n) = X_2^i T(X_1, \ldots, X_n)$$

$i > 0$. This is a contradiction. It follows that a point $(a_3, \ldots, a_n) \in \mathbb{C}^{n-2}, |a_j| < \varepsilon$ exists so that $(1, 0, a_3, \ldots, a_n) \in E$ and $F(x^\circ, 0, a_3, \ldots, a_n) \neq 0$ for some $x^\circ \neq 0, |x^\circ| < \varepsilon$. Then, condition b) holds, too. ∎

4.4. PROPOSITION. (Beppo Levi, [5], §5). *Let* $W \subseteq \{(1, 0, x_3, \ldots, x_n) \in \mathbb{C}^n | x_3, \ldots, x_n \in \mathbb{C}\} \cong \mathbb{C}^{n-2}, n \geq 2$ *be a countable vanishing subset of* $\mathbb{C}^{n-2}$. *Assume that* $W := \{w^i | i \geq 1\}$ *has the only accumulation point* $O' = (1, 0, 0 \ldots, 0) \in \mathbb{C}^n$.

For every $i \geq 1$, *let* ρ^i *be the complex line containing the points* O *and* w^i *and let* $\{\rho_k^i\}_{k \geq 1}$ *a sequence of complex lines of* $\mathbb{C}^n$, *containing* O *and converging to* ρ^i; *suppose that no* ρ_k^i *is contained in the hyperplane* $z_2 = 0$.

Let the ordinary sequence $\{\tau_j\}_{j \geq 1}$ *be a rearrangement of the double sequence* $\{\rho_k^i\}_{i,k \geq 1}$. *For every* $j \geq 1$, *let* $E_j \subseteq \tau_j$ *be a finite subset of* τ_j *constituted by* N_j *points. Assume*
   a) $\lim_{j \to \infty} N_j = \infty$
   b) $\lim_{j \to \infty} \max\{|z| \, | z \in E_j\} = 0$.

Work of Beppo Levi

Then, the set $E := \bigcup_{j \geq 1} E_j$ is a countable vanishing subset of $\mathbb{C}^n$ in O and O is the only accumulation point of E.

Proof. Remark that E is a countable set of $\mathbb{C}^n$ and that O is the only accumulation point of E.

Let $S(X_1, \ldots, X_n) \in \mathbb{C}\{X_1, \ldots, X_n\}$. Assume that $S \neq 0$ converges in the polycylinder $P(O, \varepsilon)$ in $\mathbb{C}^n, \varepsilon > 0$, and that S vanishes on $P(O, \varepsilon) \cap E$. One has $S(0) = 0$
Consider the series

$$H(X_1, \ldots, X_n) := S(X_1, X_1 X_2, \ldots, X_1 X_n).$$

Observe that $H(X_1, \ldots, X_n) \in \mathbb{C}\{X_1, \ldots, X_n\}, H \neq 0, H(0, \ldots, 0) = 0$. Indeed, a point $(x_1^\circ, \ldots, x_n^\circ) \in \mathbb{C}^n, x_1^\circ \neq 0$, exists with $H(x_1^\circ, \frac{1}{x_1^\circ} x_2^\circ, \ldots, \frac{1}{x_1^\circ} x_n^\circ) = S(x_1^\circ, \ldots, x_n^\circ) \neq 0$.
Then, by 4.2, a power series $F(X_1, \ldots, X_n) \in \mathbb{C}\{X_1, \ldots, X_n\}$ and a natural number j exist such that

$$H(X_1, \ldots, X_n) = X_2^j F(X_1, \ldots, X_n);$$

moreover, a point $(1, 0, a_3, \ldots, a_n) \in W$ exists such that the function $x \to F(x, 0, a_3, \ldots, a_n)$ does not vanish identically.

Observe that $F(0, 0, a_3, \ldots, a_n) = 0$. Indeed, from $F(0, 0, a_3, \ldots, a_n) \neq 0$ one deduces $F(0, x_2, a_3, \ldots, a_n) \neq 0$ if $x_2 \neq 0, |x_2|$ small. This is a contradiction because of the equalities $0 = H(0, x_2, a_3, \ldots, a_n) = x_2^j F(0, x_2, a_3, \ldots, a_n)$.

Thus, F is X_1-regular of order $N \in \mathbb{N}^*$ at $(0, 0, a_3, a_n)$ and in a suitable polycylinder $P := P((0, 0, a_3, \ldots, a_n), \delta)$ of $\mathbb{C}^n, 0 < \delta < \min(1, \varepsilon)$, one has by the Weierstrass Vorbereitungssatz.

$$F(x_1, \ldots, x_n) = (x_1^N + h_{N-1}(x_2, \ldots, x_n) x_1^{N-1} + \ldots + h_0(x_2, \ldots, x_n)) G(x_1, \ldots, x_n)$$

where G is analytic and without zeros on P and the functions $h_0, \ldots, h_{N-1}$ are analytic in the polycylinder $P((0, a_3, \ldots, a_n), \delta)$ of $\mathbb{C}^{n-1}, h_j(0, a_3, \ldots, a_n) = 0$ if $j = 1, \ldots, N-1$.

Now, a complex line τ_j exists of equation $z = tl, l = (1, l_2, \ldots, l_n) \in \mathbb{C}^n, l_2 \neq 0, |l_2| < \delta, |l_j - a_j| < \delta$ for $j = 3, \ldots, n$ such that $E_j \cap P$ contains at least $N + 1$ points, different from 0. Let $t_1 l, \ldots, t_{N+1} l$ be such points with $|t_j| < \delta, t_i \neq t_j$ for $1 \leq i, j \leq N + 1$.

One has for every $j = 1, \ldots, N + 1$

$$S(t_j l) = H(t_j, l_2, \ldots, l_n) = l_2^j F(t_j, l_2 \ldots, l_n) = l_2^j (t_j^N + h_{N-1}(l_2, \ldots, l_n) t_j^{N-1} + \ldots + h_0(l_2, \ldots, l_n)) G(t_j, l_2, \ldots, l_n) = 0.$$

This is a contradiction, because of the inequalities $l_2 \neq 0, G(t_j, l_2, \ldots, l_n) \neq 0$. ∎

By means of preceding results, a new construction of simple vanishing subsets of $\mathbb{C}^n, n \geq 2$, can be reached by induction from the $(n-2)$-dimensional case to the n-dimensional case. Observe that previous methods are especially expressive in the case $n = 2$, where the set W of 4.4 becomes $W = \{(0, 1)\}$. This case will be examined again in the next chapter where a stronger result will be obtained.

5. Beppo Levi and the Theory of Analytic Functions of Several Variables: a Geometrical Result

At the beginning of the short paper [6], B. Levi states a Proposition whose proof «réussit très facile - he writes - comme application du théorème préparatoire de Weierstrass et sera donnée ailleurs».

It is easy to deduce from this proposition all other results of [6], but, as far as we know, Levi never gave its proof. Thus, we are going to prove this result (Th. 5.2) here, by very simple tools, the same ones Beppo Levi used in [5].

5.1. PROPOSITION. *Let $S(X_1, X_2) \in \mathbb{C}\{X_1, X_2\}, S \neq 0, S(0,0) = 0$. Let r be the complex line of $\mathbb{C}^2, ax_1 + bx_2 = 0, a, b \in \mathbb{C}, b \neq 0$.*

Then $N \in \mathbb{N}^*$ and $\delta, \varepsilon \in \mathbb{R}$ exist with $\delta > 0, \varepsilon > 0$ so that if $L \neq r$ is the complex line $a'x_1 + b'x_2 = 0$ with $|a - a'| < \varepsilon, |b - b'| < \varepsilon$, then S has not more than N zeros in the 1-dimensional domain $P(O, \delta) \cap L$.

Moreover, either S identically vanishes in $P(O, \delta) \cap r$ or S vanishes at the only point O of $P(O, \delta) \cap r$.

Proof. Consider the $\mathbb{C}$-linear isomorphism $\sigma : \mathbb{C}^2 \to \mathbb{C}^2, \sigma(x_1, x_2) = (x_1, ax_1 + bx_2)$.
Let us introduce the new auxiliary power series T, H, F as it follows:
Set $T := S \circ \sigma^{-1}$,
$T(Y_1, Y_2) = S(Y_1, \frac{-aY_1 + Y_2}{b})$.
Observe that $T(Y_1, Y_2) \in \mathbb{C}\{Y_1, Y_2\}, T \neq 0, T(0,0) = 0$. Now, set
$H(Y_1, Y_2) := T(Y_1, Y_1 Y_2)$.
Let $(y_1', y_2') \in \mathbb{C}^2, y_1' \neq 0$ be so that $T(y_1', y_2') \neq 0$; then, $H(y_1', y_2'/y_1') = T(y_1', y_2') \neq 0$. Thus, $H(Y_1, Y_2) \in \mathbb{C}\{Y_1, Y_2\}, H \neq 0, H(0,0) = 0$.

(1). Definition of j. Assume

$$H(Y_1, Y_2) = \sum a_{j_1 j_2} Y_1^{j_1} Y_2^{j_2}$$

and set $B := \{j_2 \in \mathbb{N} | j_1 \in \mathbb{N} \text{ exists so that } a_{j_1 j_2} \neq 0\}$. Let $j := \min B$. Remark that $j > 0$ if and only if $H(Y_1, 0) = 0$. In any case, a power series, $F(Y_1, Y_2) \in \mathbb{C}\{Y_1, Y_2\}$ exists so that

$$H(Y_1, Y_2) = Y_2^j F(Y_1, Y_2) ;$$

F does not factor in elements $Y_2^s, s > 0$ and $x \to F(x, 0)$ does not vanish identically in any neighborhood of 0 in $\mathbb{C}$.

Now, observe that $F(0,0) = 0$. Indeed, consider two cases: $j = 0, j > 0$. If $j = 0$, one has $H = F, F(0,0) = 0$. Let $j > 0$, observe that for every y_2 one has $H(0, y_2) = T(0,0) = S(0,0) = 0$. Now, if $F(0,0) \neq 0$, then $F(0, y_2) \neq 0$ for $|y_2|$ small enough; thus, $H(0, y_2) = y_2^j F(0, y_2) \neq 0$: a contradiction.

(2). We are going to introduce the numbers δ, N. Remark that F is Y_1-regular at $(0,0)$. Thus, by Weierstrass Vorbereitungssatz, $\delta > 0$ and $N \in \mathbb{N}^*$ exist so that in the polycylinder $P(O, \delta)$ of $\mathbb{C}^2$ one has

$$F(y_1, y_2) = (y_1^N + h_{N-1}(y_2) y_1^{N-1} + \ldots + h_0(y_2)) G(y_1, y_2)$$

where G si an analytic function without zeros on $P(O, \delta)$ and $h_0, \ldots, h_{N-1}$ are analytic functions in a neighborhood of 0 in $\mathbb{C}, h_v(0) = 0$, for $v = 1, \ldots, N - 1$. It is possible, also, to choose $\delta > 0$ so that S converges on $P(O, \delta)$.

(3). Definition of γ_L and of $\varepsilon = \varepsilon(\rho)$. Let ρ be a positive number. It is an easy computation to show that $\varepsilon = \varepsilon(\rho), |b| > \varepsilon > 0$, exists such that for every complex line

$L : a'x_1 + b'x_2 = 0$ with $|a - a'| < \varepsilon, |b - b'| < \varepsilon$ the corresponding complex line $\sigma(L)$ has equation $y_1 = t, y_2 = \gamma_L t$ with $\gamma_L = (ab' - a'b)/b', |\gamma_L| < \rho$.

(4). Now, we want to prove that the numbers N, δ defined in (1) and $\varepsilon = \varepsilon(\delta)$ defined in (3), are fulfilling the conditions of our statment.

Let L be a complex line, $a'x_1 + b'x_2 = 0$ with $|a - a'| < \varepsilon, |b - b'| < \varepsilon$.

Assume that at least $N + 1$ distinct points $(x_1^1, x_2^1), \ldots, (x_1^{N+1}, x_2^{N+1}) \in P(0, \delta) \cap L$ exist with $S(x_1^1, x_2^1) = 0, \ldots, S(x_1^{N+1}, x_2^{N+1}) = 0$. One has $L \neq \{x_1 = 0\}$; thus $x_1^1, \ldots, x_1^{N+1}$ are $N + 1$ distinct numbers.

Observe that $(x_1^i, \gamma_L) \in P(0, \delta) \subset \mathbb{C}^2$ for every $i = 1, \ldots, N + 1$. Then from (2) it follows for each $i = 1, \ldots, N + 1$.
$S(x_1^i, x_2^i) = T \circ \sigma(x_1^i, x_2^i) = T \circ \sigma(x_1^i, -a'x_1^i/b') = T(x_1^i, \gamma_L x_1^i) = H(x_1^i, \gamma_L) = \gamma_L^j F(x_1^i, \gamma_L) = \gamma_L^j((x_1^i)^N + h_{N-1}(\gamma_L) \cdot (x_1^i)^{N-1} + \ldots + h_0(\gamma_L))G(x_1^i, \gamma_L)$.

Consider the case $L \neq r$, i.e. $\gamma_L \neq 0$. It follows that $x_1^1, \ldots, x_1^{N+1}$ are $N + 1$ distinct roots of the polynomial $t^N + h_{N-1}(\gamma_L)t^{N-1} + \ldots + h_0(\gamma_L)$. This is a contradiction.

Now, consider the case $L = r$, i.e. $\gamma_L = 0$. If $j > 0$, then S identically vanishes on $P(0, \delta) \cap r$. If $j = 0$, one has $S(x_1, x_2) = x_1^N G(x_1, 0)(x_1^i, \gamma_L)$ for every $(x_1, x_2) \in P(0, \delta) \cap r$. Then, S vanishes at the only point $O = (0, 0) \in P(0, \delta) \cap r$ (with multiplicity N). ∎

5.2. THEOREM. (Beppo Levi, [6], p. 1735). *Let* $S(X_1, X_2) \in \mathbb{C}\{X_1, X_2\}, S \neq 0$, $S(0, 0) = 0$. *A natural number* $N \in \mathbb{N}^*$ *and a real one* $\delta > 0$ *exist so that for every complex line* $r : a X_1 + bX_2 = 0$ *on which* S *does not identically vanish,* S *has not more than* N *distinct zeros in* $P(O, \delta) \cap r$.

Proof. Let $(a, b) \in S^3 := \{(a, b) \in \mathbb{C}^2 | |a|^2 + |b|^2 = 1\}$. By previous Proposition there are three numbers $N = N(a, b) \in \mathbb{N}^*, \delta = \delta(a, b) > 0, \varepsilon = \varepsilon(\delta) > 0$ so that for every complex line $L, a'X_1 + b'X_2 = 0$, with $(a', b') \in P((a, b), \varepsilon) \cap S^3$, the domain $P(O, \delta) \cap L$ does not contain more than N zeros of S, except for the case that S vanishes on the line $aX_1 + bX_2 = 0$ and that $(a', b') = (\rho a, \rho b), \rho \in \mathbb{C}, |\rho| = 1$.

Then, points $(a_1, b_1), \ldots, (a_s, b_s) \in S^3$ exist so that the open family $\{P((a_j, b_j), \varepsilon_j) \cap S^3 | j = 1, \ldots, s\}$ covers S^3. Let $\delta_i = \delta(a_i, b_i), N_i = N(a_i, b_i)$ for $i = 1, \ldots, s$ and let $\delta = \min(\delta_1, \ldots, \delta_s), N = \max(N_1, \ldots, N_s)$.

Each complex line containing $O \in \mathbb{C}^2$ has equation of the form $ax_1 + bx_2 = 0$ with $(a, b) \in S^3$. Thus, S cannot have more than N zeros in $P(O, \delta) \cap r$ except for the case that S identically vanishes on r. ∎

This Theorem gave B. Levi some ideas about the discrepancy between properties of algebraic and analytic curves.

PART TWO: THE WORK OF BEPPO LEVI IN GEOMETRY

6. Introduction

Beppo Levi's work in geometry was greatly influenced by the ideas and scientific format of Corrado Segre while his work in logic was undoubtedly influenced by the scientific approach of his other great teacher, Giuseppe Peano. Beppo Levi got his degree in Turin, July 1896, having in fact the very Corrado Segre as his advisor. At that time, Corrado Segre was only thirty-six years old and was at the top of his scientific career. As we all know, Corrado Segre was teacher and inspiration for other well-known mathematicians like G. Castelnuovo, F. Enriques, G. Fano, F. Severi, A. Terracini, E. Togliatti and was one of the founders of the Italian school of geometry. Beppo Levi's work in geometry can certainly be included in this thinking. We shall, in fact, see that Levi is particularly faithful to this special view of geometry throughout his long scientific activity.

Beppo Levi's work in algebraic geometry, or at least that of his early years, was well-known to the mathematicians of his generation and was later, in its essential points, clearly summarized by Oscar Zariski in [Z]. We refer those of you who are interested in a detailed description of Beppo Levi's methods in algebraic geometry to [Z] or to the brief study [1902-1] written by Levi himself. It is our intention in this paper to list Levi's contributions in geometry indicating their origin and the acceptance which these received on the whole in the scientific world of the period, «end of 1800/1960», placing Levi's works within the framework of the research of his times.

7. List of Beppo Levi's Papers on Geometry

[1897-1] Beppo Levi *Sulla riduzione delle singolarità puntuali delle superficie algebriche dello spazio ordinario per trasformazioni quadratiche*, Ann. Mat. Pura Appl., (2) **XXVI**, 1897, 219-253.

[1897-2] Beppo Levi *Risoluzione delle singolarità puntuali delle superficie algebriche*, Atti R. Accad. Sci. Torino, **XXXIII**, 1897, 66-86.

[1898-1] Beppo Levi *Sulla varietà delle corde di una curva algebrica*, Mem. R. Accad. Sci. Torino Cl. Sci. Mat. Fis. Natur., (2), **XLVIII**, 1898, 83-142.

[1898-2] Beppo Levi, *Sulla trasformazione di una curva algebrica in un'altra priva di punti multipli*, Atti R. Accad. Lincei Rend. Cl. Sci. Fis. Mat. Nat. (5), **VII**, 1898, 111-113.

[1899-1] Beppo Levi, *Intorno alla composizione dei punti generici delle linee singolari delle superficie algebriche*, Ann. Mat. Pura Appl., (3), **2**, 1899, 127-138.

[1899-2] Beppo Levi, *Dell'intersezione di due varietà contenute in una varietà semplicemente infinita di spazi*, Atti R. Accad. Sci. Torino, **XXXIV**, 1898-99, 745-760.

[1899-3] Beppo Levi, *Sulla trasformazione dell'intorno di un punto per una corrispondenza birazionale tra due spazi*, Atti R. Accad. Sci. Torino, **XXXV**, 1899, 20-33.

[1902-1] Beppo Levi, *Sur la résolution des points singuliers des surfaces algébriques*, C. R. Accad. Sci. Paris, **134**, 1902, 222-225.

[1902-2] Beppo Levi, *Sur la théorie des fonctions algébriques de deux variables*, C. R. Accad. Sci. Paris, **134**, 1902, 642-644.

[1903] Beppo Levi, *Teoria geometrica delle proporzioni tra segmenti, indipendente dal postulato di Archimede*, Supplemento al Periodico di matematica, **6**, 1903, 114-117.

[1904-1] Beppo Levi, *Sull'uguaglianza diretta ed inversa delle figure*, Periodico di matematica, (III), **1**, 1904, 207-214.

[1904-2] Beppo Levi, *Fondamenti della metrica proiettiva*, Mem. R. Accad. Sci. Torino Cl. Sci. Mat. Fis. Natur., (2), **LIV**, 1904, 281-354.

[1904-3] Beppo Levi, *Sulle superficie del 4° ordine con 13 punti doppi (ricerche sintetiche e grafiche)*, ed. Bona, Torino, 1904, pp. 22.

[1905-1] Beppo Levi, *Punti doppo uniplanari delle superficie algebriche*, Atti R. Accad. Sci. Torino, **XL**, 1905, 139-167.

[1905-2] Beppo Levi, *Sur la géométrie et la trigonometrie sphériques*, Enseign. Math. **7**, 3, 1905, 193-206.

[1907-1] Beppo Levi, *Dalla pittura alla cartografia. Prolusione al corso di geometria proiettiva e descrittiva*, Anno 1906-1907, Cagliari, G. Dessì, pp. 28.

[1907-2] Beppo Levi, *Geometrie proiettive di congruenze e geometrie proiettive finite*, Trans. Amer. Math. Soc. **8**, 3, 1907, 354-365.

[1908] Beppo Levi, *Il teorema di Desargues, il teorema di Pappo e l'esistenza di una reciprocità o di una polarità*, Ann. Mat. Pura Appl., (3), **15**, 1908, 171-186.

[1910] Beppo Levi, *Sui postulati della metrica generale proiettiva*, Jahresber. Deutsch. Math. - Verein, **XIX**, Heft 11/12, 1910, 300-306.

[1910-1] Beppo Levi, Review to the volume «Friedrich Schur, *Grundlagen der Geometrie*, Teubner, 1909, pp. VII+192», Bollettino di Bibliografia e Storia delle Scienze Matematiche, **XII**, 2, 1910, 36-43.

[1924] Beppo Levi, *Calcolo differenziale assoluto*, Annuario Scientifico ed Industriale, **LXI**, 1924, II, Ed. Fratelli Treves, Bologna 1926, 215-244.

[1927] Beppo Levi, *Sur la résolution des points singuliers d'une variété algébrique à un nombre quelconque de dimensions*, In memoriam N. I. Lobatschevskii-Collection des mémoires présentés par les savantas de divers pays à la Société Physico-Mathématique de Kazan à l'occasion de la célébration de la découverte de la Géométrie Non-Euclidienne par N. I. Lobatcheffsky (12/24 Février 1826), **II**, 1927, 191-196.

[1931-1] Beppo Levi, *Intorno ad una interessante osservazione dell'Ing. Riccardo Cantoni*, Periodico di matematiche, (4), **11**, 1931, 301-302.

[1931-2] Beppo Levi, *Un problema di geometria*, Boll. Un. Mat. Ital., **X**, 1, 1931, 319-322.

[1932] Beppo Levi, *Postilla*, Boll. Un. Mat. Ital., **XI**, 1, 1932, 37-38 (remark: this is a postilla to a previous paper by A. Chiellini).

[1934] Beppo Levi, *Intorno ai reticoli spaziali*, Periodico di matematiche, (4), **14**, 1934, 235-239.

[1941] Beppo Levi, *Los poligonos planos y el teorema de Jordan*, Math. Notæ, **1**, 1941, 9-26.

[1942-1] Beppo Levi, *Definición y condiciones de existencia de la tangente y del circulo osculador en un punto de una curva*, Math. Notæ, **2**, 1942, 11-34.

[1942-2] Beppo Levi, *Apostilla*, Math. Notæ, **2**, 1942, 184-187 (remark: this is a postilla to a previous paper by Luis A. Santaló).

[1944] Beppo Levi, *El principio de corrispondencia de Chasles-Cremona y el orden de la reglada de las trisecantes de una curva*, Math. Notæ, **4**, 1944, 129-136.

[1946-1] Beppo Levi, *Propriedades del cuadrángolo base de un haz de cónicas*, Math. Notæ, **6**, 1946, 112-115.

[1946-2] B. Levi, C. De Maria, L. Santaló, *Estúdios numerativos sobre las variedades de contacto de las superficies en un espacio de n dimensiones*, Publicaciones del Instituto de Matematica della Facultad de Ciencias Matematicas etc. de la Universidad Nacional del Litoral, **8**, 1946, 3-72.

[1947] Beppo Levi, *Leyendo a Euclides*, Editoria Rosario, 1947, pp. 225 (the fist part is an essay on *La geometria y el pensamiento socratico*, 13-78; S.C.).

[1950] Beppo Levi, *Sobre una propriedad limite de la esfera en n dimensiones*, Math. Notæ, **10**, 1950, 36-40.

[1957] Beppo Levi, *Puntos y variedadas singulares sobre variedades algebricas y analiticas*, Math. Notæ, **6**, 1957, 1-62; 73-129.

8. Research in Algebraic Geometry

In 1897, the Corrado Segre paper [S] was published. About forty years later, when it was possible to better evaluate this research, Zariski wrote in the first chapter of [Z], devoted to the Theory and Reduction of Singularities: «*The foundations of a geometric theory of singularities of algebraic surfaces have been laid down by C. Segre in his important paper* (3) ([S], this is author's identification). *Using the method of successive quadratic transformations of the 1st kind he arrives at a theoretically satisfactory definition of infinitely near multiple points on a surface...*»

It is in [S] that for the first time we find a reference to Beppo Levi. In fact, Corrado Segre concludes [S] with these words[2]: «*Nel finire la revisione delle bozze, sento il dovere di ringraziare il Dr. Beppo Levi, mio discepolo, per l'aiuto intelligente che mi ha prestato in questa revisione. In pari tempo avvertirò che questo giovane nella sua dissertazione di laurea dedicata alle singolarità superiori delle curve algebriche sghembe (iperspaziali) ha studiato più profondamente quei caratteri relativi ai rami di dette curve dei quali ho fatto cenno qui nel n. 6; e che egli pure mi ha comunicato per la proposizione contenuta al n. 3 di questo lavoro, una dimostrazione geometrica, la quale si basa sul fatto che per gli h punti s-pli successivi di F ivi nominati si può sempre far passare una superficie che abbia in O un punto semplice, ed applica alla curva intersezione di questa superficie colla F una disuguaglianza che lega l'ordine alle multiplicità delle curve*».

---

[2] «While revising the final draft, I deem it my duty to thank my student Dr. Beppo Levi for his most intelligent help in this revision. At the same time I wish to point out that this young man in his thesis dissertation on the singularities of skew algebraic (hyperspatial) curves studied in depth those characters related to the branches of said curves which I have mentioned here in n. 6; and that he has given me for the proposition contained in n. 3 of this work, a geometrical proof based on the fact that through the h points s-fold successive of F there named, one can make a surface pass which has in O a simple point, and apply to the intersecting curve of this surface with F an inequality which brings together the order to the multiplicities of the curves».

Beppo Levi's work mentioned by Segre, first appeared in [1897-1] and was completed by [1899-1]. In the meantime, in 1898, the work [1898-1] which represented the development of his thesis dissertation of 1896 had been published.

The first part of [1897-1] was concerned with a rather detailed critical analysis of G. Kobb's work *Sur la Théorie des fonctions algébriques de deux variables*, Journ. de Math. pures et appl. (4), **8**, 1892, 385-412. In this work Beppo Levi found many inaccuracies. Continuing then with his own research, Levi was able to geometrically demonstrate some results of [S] related to the finiteness of the composition of a single point that Segre had reached by using Kobb's above-cited work. In this way, Levi was able to give a definitive answer to certain criticism which Pasquale del Pezzo in his work *Osservazioni su una memoria del Prof. Corrado Segre e risposta ad alcuni suoi appunti*, Atti Acc. Pontaniana, 27 (1897), 1-13 had made to Corrado Segre (we are referring to results contained in n. 3 of Segre's work in which Segre had already written that Beppo Levi had given him a new geometric proof). The main result of [1897-1], [1899-1] is the following (see [1897-1], p. 227)[3]: *Se $\mathcal{F}$ è una superficie algebrica e A un suo punto s^{plo}, e se, applicata ad A una trasformazione quadratica, si ottiene come sua trasformata una curva a_1 di una superficie Φ_1 trasformata di $\mathcal{F}$, sulla quale esistano punti s^{pli} per Φ_1; e se, detto A_1 uno di questi punti, si opera su esso come precedentemente su A, ottenendone una linea a_2 di una superficie Φ_2 sui cui possono esistere punti s^{pli} per Φ_2 uno dei quali sia A_2, e così si prosegue: la successione delle superficie $\Phi_1, \Phi_2, \ldots$ e dei punti $A_1, A_2, \ldots$ è finita, cioè esiste nella successione un'ultima superficie Φ_r su cui esiste una linea a_r trasformata di A_{r-1} su cui non esistono punti s^{pli} per Φ_r; purchè i punti $A_1, A_2, \ldots$ siano scelti in modo che, a cominciare da uno di essi in poi, non accada mai che uno di questi punti stia sulla linea trasformata di una linea s^{pla} passante pel punto precedente; in particolare da uno di essi in poi non stiano su una linea s^{pla} della superficie.*

But the author adds[4]: *può accadere che il numero dei punti s^{pli} $A_1, A_2, \ldots$ cresca oltre ogni limite quando si scelgano questi punti in modo che due o più (necessariamente consecutivi) appartengono a linee s^{ple} corrispondentisi (trasformate l'una dell'altra) delle superficie successive, pur essendo finito il numero dei punti che stanno su linee trasformate di una stessa linea s^{pla} ».*

Let us keep in mind that in Beppo Levi's terminology[5]: «*si applica una trasformazione*

---

[3] «If $\mathcal{F}$ is an algebraic surface and A one of its s-fold points and if, once a quadratic transformation is applied to A, we obtain as its transform a curve a_1 of a surface Φ_1, transform of $\mathcal{F}$ on which s-fold points for Φ_1 exist; and if, once one of the points is named A_1, we act on this as we did before on A, obtaining a curve a_2 of a surface Φ_2 on which s-fold points may exist for Φ_2 one of which is A_2 and continue as above, [then] the sequence of the surfaces $\Phi_1, \Phi_2 \ldots$ and of the points $A_1, A_2 \ldots$ is finite; i.e. a last surface Φ_r exists on which a curve a_r transform of A_{r-1} exists on which no s-fold points for Φ_r exist; this under the condition that the points $A_1, A_2 \ldots$ are chosen so that, starting from one of these it never happens that one of the points lies on a curve transform of a s-folde curve passing through the previous point, and, in particular, [provided that] starting from one of these, they do not lie on an s-fold curve of the surface».

[4] «It may happen that the number of the s-fold points $A_1, A_2, \ldots$ increase beyond any bound when these points are chosen so that two or more (necessarily consecutive) belong to corresponding (one is the transform of the other) s-fold curves of successive surfaces, even if the number of points lying on transformed curves of the same s-fold curve is finite».

[5] «A quadratic transformation is applied to a singular point when the surface or the curve to which the

quadratica ad un punto singolare quando si trasforma la superficie cui il punto appartiene, per mezzo di una trasformazione quadratica avente il punto come punto fondamentale isolato».

The natural conclusion of [1897-1] should have been the classic resolution theorem for algebraic surfaces. And this is what Beppo Levi tried to do in his works [1897-2], [1899-3]. At the time the reduction theorem was expressed as follows: «Let $\mathcal{F}$ be an algebraic surface in S_3. Let n be a natural number ≥ 3. Then, $\mathcal{F}$ can be birationally transformed into a surface $\mathcal{F}_n$ in S_n, with ordinary singularities only. More specifically, if $n = 3$, we can obtain that the only singularities of $\mathcal{F}_3$ are a biplanar double curve with a finite number of ordinary cuspidal points and a finite number of ordinary triple points for the curve which are triple and triplanar for the surface; if $n \geq 5$ that $\mathcal{F}_5$ is free from singularities».

Before discussing the acceptance of the above-mentioned works, it should be noted that Beppo Levi had in the meantime also developed, as was to be expected, some results about the reduction of singularities for algebraic curves. He returned many times to this topic, but his most interesting paper was probably [1898-2]. Here, he proved that each algebraic curve in the ordinary three dimensional space can be transformed by birational transformations *of the space* into another algebraic curve of the ordinary three dimensional space without singularities. This theorem is easily extendible to curves in n-dimensional spaces.

Much later, Beppo Levi returned to the subject of reduction of singularities in [1927]. Here, he presented the essential points for an extension of his results in the case of algebraic varieties of any dimension. For this, he proposed the use of methods already employed in [1897-2] and an inductive proof.

Let us now consider the opinions of contemporary mathematicians on these Beppo Levi papers.

To do this, we must remember that the final ten years of the last century were already important for results on the reduction of singularities of algebraic curves. Generally, we attribute to Max Nœther in his work *Über die algebraischen Functionen einer und zweier Variabeln*, Note 2, Nachr. Akad. Wiss. Göttingen Math. - Phys. Kl., 1871, **9**, 267-278, the correct positioning of the problem of reducting the singularities of algebraic curves and surfaces. Of course, some ideas and methods used can also be found in earlier works or communications by other authors (L. Kronecker, F. Klein, Hamburger). For an historical presentation of the problem of reduction of singularities of plane algebraic curves one can profitably consult the work of G.A. Bliss *The Reduction of Singularities of planar Curves by birational Transformations*, Bull. Am. Math. Soc. *XXIX*, April 1923, 161-183; see also F. Severi, *Trattato di Geometria Algebrica*, Vol. 1, Parte I, ed. Zanichelli, Bologna, 1926, pp. 332-335.

By the end of the nineteenth century, the first works concerned with the reduction of algebraic surface theorem begin to appear; let us mention among these, those of Del Pezzo, beginning with his *Estensione di un teorema di Noether*, Rend. Circ. Mat. Palermo, **2**, 1888, 139-144.

Thus, it seemed that 1897 was to be the decisive year for the proof of the theorem in question. In fact, it was at this time we find published the important memoir [S] of Corrado Segre, the first volume of the treatise [Pi-Si] of Picard-Simart with a demonstration of the theorem (see pp. 73-82) as well as the publications [1897-1] and [1897-2].

But perhaps the problem was more complicated than one might think. Segre's work, while containing original and important results and ideas, did not arrive at a conclusive solution of this

---

point belongs is transformed by means of a quadratic transformation having it as fundamental isolated point».

problem. The proof contained in the Picard-Simart work was not entirely acceptable. Picard, in fact, stated as much to the editor of the Bulletin des Sciences Mathématiques, Tome **XXII**, 1898 (see p. 27) and he, at the same time, announced the publication of [1897-2]. The sharp controversy between Del Pezzo and Corrado Segre, which reached its height in 1897, in some way indirectly proved the difficulty of the problem as well as the unsuitability of Del Pezzo's methods[6]. Another indication of the difficulty of the subject can be indirectly deduced from the paper [1902-2], where Beppo Levi, with his usual keen sense of criticism, found a serious gap in K. Hensel's proof of an important theorem of the same topic; see *Über eine neue Theorie der algebraischen Functionen zweier Variablen*, Acta Math., **XXIII**, 1900, 339-416 and Jahresber. Deutsch. Math. - Verein., **8**, 1900, 221-231.

As regards the memoirs [1897-1] and [1899-1], it seems that these were immediately accepted by the mathematicians of the day except for two small observations (one of which is a question of priority), moved by Max Noether to [1897-1] and accepted by Levi in [1899-1]. Oscar Zariski in [Z] summarizes the central problem which Beppo Levi solved in these works as follows: «*Under what conditions, if any, is a succession of infinitely near points, all of the same multiplicity s , necessarily finite?*» For Zariski «*the question is treated in Del Pezzo..., Kobb..., and especially B. Levi [...] whose proof is entirely rigorous and who arrives at the final correct result*». More recently in [D], J. Dieudonné writes «*Il faut noter que si l'on omet, après chaque éclatement, de «normaliser» la surface obtenue, il se peut qu'on n'aboutisse jamais à des points simples, comme l'avait remarqué B. Levi*», after having presented the Zariski method or reduction of singularities.

The work [1897-2] was certainly also received with much interest. We have already seen Picard's interest. In 1900, in the second volume of [Pi-Si], on p. 523, Picard and Simart with reference to their demonstration in their first volume write: «*La démonstration du théorème général relatif à la reduction des singularités d'une surface algébrique est incomplète. Aux Mémoires cités pag. 74, ajoutons les travaux de M. Beppo Levi (Annali di Matematica pura ed applicata, 2 Serie, t.* **XXVI***, 1897, et Comptes Rendus, t.* **CXXXIV***, 1902, p. 222 et 642) qui résolvent complètement la question*». Oddly enough, there is no mention of Levi's memoir [1897-2]. There was also some correspondence between Levi and Osgood who a few years later sent Levi C.W.M. Black's thesis, *The parametric representation of the neighborhood of a singular point of an analytic surface*, Proc. Am. Acad. Arts and Sciences Vol. **XXXVII**, 11, 1902, 279-330 (here Balck found new errors in the above-mentioned Kobb's paper).

But the difficulty of the subject, together with certain real obscurities, gave rise to mistrust among the mathematicans of the period on this particular Levi work. In [1902-1] Beppo Levi himself speaks of these «défiances». Here he laments the fact that H. Poincaré in his paper *Sur la connexion des surfaces algébriques*, C. R. Acad. Sci. Paris, **CXXXIII**, 2 sem. 1901, 969-973, wanting to use the reduction theorem, had cited Picard's proof rather than his, in spite of the fact that Picard had admitted this could not be applied in all cases. It is interesting to note that Levi's communication was presented by Picard himself to C. R. Certainly, even among Italian mathematicians, the difficulty encountered in interpreting Levi's papers gave rise to doubts as to validity of the proof[7].

---

[6] On the controversy between these two famous Italian mathematicians one can refer to the article by Paola Gario, *Resolution of Singularities of Surfaces by Del Pezzo. A Mathematical Controversy with Corrado Segre*, Archive for History of Exact Sciences **40**, 1989, 247-278. In this interesting work, the role which Beppo Levi played in this controversy is also mentioned.

[7] As regards this, we would like to recall an amusing episode remembered, among others, by Sandro Faedo in his article, *La rinascita della Scuola Matematica pisana dopo la seconda guerra mondiale*, Dipartimento

Nevertheless, the attitude of the mathematicans towards Levi's work changed to one of general acceptance.

Francesco Severi in *Trasformazione birazionale di una superficie algebrica qualunque, in una priva di punti multipli*, Rend. Accad. Lincei, **XXIII** (2° Sem.) (1914), 527-539 writes: «*a Beppo Levi è dovuta la completa dimostrazione della possibilità di tale trasformazione*». On the other hand, Federigo Enriques is a little less explicit in [E], Libro Terzo. Actually, Enriques was more interested in the resolution of the singularities of any surface by birational transformations of the ordinary three dimensional space (Cremona Transformations). This is what Oscar Chisini did in *La risoluzione delle singolarità di una superficie mediante trasformazioni birazionali dello spazio*, Mem. Acc. Sc. Bologna, Serie VII, **8**, (1921), 1-42. In this work, however, even Chisini showed his acceptance of the Beppo Levi proof. Levi's methods were also accepted by Hilda P. Hudson in the treatise «*Cremona Transformations in Plane and Space*», Cambridge Press, 1927.

But the attitude of the mathematicians towards [1897-2] was destined to change once again. In fact, in [Z], Zariski, after a detailed study of Beppo Levi's memoir, writes that «*Levi

---

di Matematica Università di Pisa, February 1990. It seems that a committee of geometry professors had gotten together for the purpose of assigning a university professorship. The committee was made up of, among others, F. Enriques, L. Bianchi, and E. Bertini. Beppo Levi was one of candidates.

As Faedo, when referring to the memoir [1897-2] writes, L. Bianchi and E. Bertini had been able to understand only the first few pages of B. Levi's work. While L. Bianchi and E. Bertini were trying in vain to understand Levi's written work, Enriques gave the candidate some particularly important example of singularities to solve. B. Levi explained to Enriques how he would have treated these cases and succeeded in finding the required birational transformations. After which, Enriques tried in vain to convince the other committee members of the validity of B. Levi's theorem, concluding with the words: «*I'll cut my head off if this theorem isn't valid*». To which E. Bertini, immediately replied: «*Not even this would prove it*». We think that the committee was the one charged with the decision of assigning the Projective and Descriptive Geometry position at the University of Cagliari which was then won by Beppo Levi himself. The committee consisted of E. Bertini, F. Enriques, C. Segre, A. Del Re and M. Pieri. Bianchi was not a member of this group. As regards the works on the reduction of singularities, the Committee reports in September of 1906: «... *trattano delle singolarità puntuali delle superficie algebriche e della loro risoluzione. L'autore con fine spirito critico osserva che, sebbene i rami uscenti da un punto singolare e non appartenenti alla superficie abbiano con questa un numero finito d'interruzioni successive, non si può assegnare un limite superiore a questo numero per i vari rami possibili* [...]. *Tale circostanza rende a priori dubbio che i consueti processi di riduzione permettano in tutti i casi di risolvere le singolarità; ma un esame approfondito conduce l'autore a dimostrare, che ogni superficie algebrica può trasformarsi birazionalmente in un'altra dotata di singolarità ordinarie* [...]. *Resta soltanto il desiderio che questo teorema fondamentale venga stabilito in forma più lucida e semplice*». [«... they deal with singularities of algebraic surfaces and their resolution. The author with his fine sense of criticism observes that even if the outgoing branches from a singular point and not belonging to the surface have with the surface a finite number of successive interruptions, one cannot assign an upper bound to this number for the various branches [...]. This circumstance gives rise to an a priori doubt that the usual reduction methods permit in all cases that the singularities be solved; but a better examination of the problem leads the author to prove that each algebraic surface can transform birationally into another with ordinary singularities. [...]. We are left only with the wish that this fundamental theorem be set up in a more lucid and simple form»].

It is this author's opinion that Levi, having been involved in his youth in very intensive mathematics activity, which was almost frenetic as regards very difficult problems, over the years and armed with extensive experience overcame a large part of his difficulty in expression and obscurities becoming, in fact, a fascinating teacher and excellent promoter of mathematics at the most varied levels.

first simplifies the singularities of the given surface f by transformations alternately quadratic of the first kind and birational, and obtains a surface f' whose only singularities are multiple curves, free from singularities and carrying no point of multiplicity higher than the multiplicity of the generic point of the curve. [...] To this new surface f' Levi applies transformations alternately monoidal and birational. [...] In regard to the «accidental» singularities introduced by the quadratic and the monoidal transformations and in regard to the manner in which they should be eliminated by birational transformations, Levi's proof is not sufficiently explicit»
Actually, Zariski did not spare any of the various theorem proofs written in his time, finding errors or at least insufficiencies even in the works of F. Severi, O. Chisini, and G. Albanese. He attributed to R. Walker's article, *Reduction of the Singularities of an Algebraic Surface*, Ann. of Math. 1935, **36**, 336-365 the first sure proof of the theorem. «*Walker's proof* – he wrote – *stands the most critical examination and settles the validity of the theorem beyond any doubt*».

When [Z] appeared on the scene, Beppo Levi was not in a condition to reply to Zariski. In fact, it seems that with the exception of the work [1927], Levi had not published anything on algebraic geometry for about thirty years. He had many diverse mathematical interests and during the period of his stay in Bologna (1928-1938) worked on other problems. At the Istituto Matematico of the University of Bologna, he had a very heavy teaching schedule which certainly did not leave him much time for research. Also the political situation in Italy was going from bad to worse until 1938, at which time he was released from his teaching post in accordance with the «racial laws». In 1939, Levi was already in Rosario (Argentina) where he was busy founding and fostering the Instituto de Matematicas of the Universidad del Litoral. In Argentina, he also founded and directed two mathematical journals the «*Publicaciones del Instituto de Matematica*» della Facultad de Ciencias Matematicas etc. de la Universidad Nacional del Litoral and the *Mathematicæ Notæ*. In addition, he directed the series *Monografias publicadas por la Facultad de Ciencias...* of the Universidad Nacional del Litoral. It is only in the work [1942] that we find Beppo Levi's interest returning to problems in algebraic geometry.

It is indeed surprising that in his large memoir [1957], divided into two parts, he once more takes up the problem of the reduction of singularities both in algebraic and analytic varieties. Let us remember that in 1957 Beppo Levi was already more than eighty years old and that this memoir was published sixty years after his first work on the subject. In the last pages of the new memoir [1957] he writes a dedication in Italian to his far away colleagues and friends where he frankly admits the «*difficile intelligenza*» (difficulty to understand) of the last part of [1897-2]. He also affirms that the last part of [1957] is certainly much clearer[8] «*e, nondimeno tale che forse, per raggiungere una completa comprensione, è tuttavia necessario che il lettore si adatti a svolgere qualche sviluppo ulteriore*». At the end of his work, Beppo Levi believed that the considerations he developed were at least sufficient to prove the resolution of singularities in the case of algebraic hypersurfaces. In his concluding pages of the memoir he, however, adds that the difficulty presented up to this point in proving a precise result absolutely general «*es con toda probabilidad insuperable por tender al infinito el numero de las ramificaciones que el problema presenta al crecer dimensiones y grados*». On this point it does not appear accidental that Levi in his title for [1957] made absolutely no reference to results of the resolution of singularities. He was always most precise in choosing the titles of his works and the use in [1897-1], [1897-2] of two different terms *riduzione* and *risoluzione* would not appear to be unintentional. Unfortunately, we must mention that memoir [1957] was not very well received

---

[8] «However, for better understanding, it may be necessary that the reader be prepared to carry out further development».

by the mathematicians of his days.

In the meantime, various other proofs of he resolution of singularities theorem had been published. Among these we find *The Reduction of the Singularities of an algebraic surface*, Ann. of Math., **40** (1939), 639-689 and in the same periodical *Simplified proof for the resolution of the singularities of a surface*, **43**, (1942), 583-593 by Zariski himself. But, most important, the problem had assumed, as we all know, a different and more general approach with greatly changed methodology.

Beppo Levi, although certainly aware of all this, nevertheless went ahead with the classic geometrical methods he had previously used. We find no mention in [1957] of Zariski's work. Recent L. Derwidué results are mentioned (we believe the reference is to the work *Le problème de la rèduction des singularités d'une variété algébrique*, Math. Ann., **123**, 1951, 302-330, followed by the brief *Réctification*, Math. Ann., **124**, 1951-52, 316) as well as B. Segre results[9] but only to point out their «little or no», connection with this final work.

At this point, one might ask what could be the thinking behind this special loyalty to methods which, for the problem under study, had already proven hardly efficient. On this point, it is interesting to read what F. Severi wrote in [Se] on the results and methods of the so-called Italian algebraic geometry. Severi takes the opportunity with this problem to present some considerations on the usefulness of «abstract» algebra. He, in fact writes: «*L'algèbre abstraite jusqu'ici a donné surtout des contributions de grande valeur à la systémization des fondaments et (ce qui est encore plus important) à l'extension des propriétes géometriques aux champs abstraits et à la meilleure application de notre géométrie à la théorie des nombres. Elle a enfin donné dernièrement la réponse à un problème essentiel et classique: la démonstration de l'existance d'un modèle projectif sans points multiples d'une variété algébrique quelconque (Zariski, Deuring)*». As we can see, the problem had in the meantime assumed a different image and, what is most surprising, he makes no mention of the solution of the classic problem; neither of Beppo Levi's nor his own contribution of the subject [10].

The answer to the question on the use of his classic methods by Levi can be a difficult one and may even involve psychological motives. Perhaps it was his wish to leave to future generations a proof *ab omne nævo vindicata*, demostrating worth of methods and ideas of sixty years earlier. Thus, indicating that the obscurities of earlier works were not basic errors and that one could go ahead considerably with the problem's solution even using his methods.

Certainly, the problem of adapting a language different from his own at his advanced age could have influenced his choice. Perhaps his being almost isolated in Rosario could also have its effect. In addition, his heavy workload in Rosario and his role as director of Mathematicæ Notæ could have influenced his approach.

We would also like to point out that Levi did not accept the unconditional use of Zermelo's

---

[9] Probably in reference to *Sullo scioglimento delle singolarità delle varietà algebriche*, Ann. di Mat., (4), **33**, 1952, 5-48. This work of Beniamino Segre was geometric in nature and fitted perfectly in with Italian School traditions. B. Segre, who worked only in the complex field was able to prove the reduction of singularities theorem for any algebraic surface. He also offered ideas on the possible generalization of his methods in cases of algebraic varieties of any dimension. See the contribution *On B. Segre and the Theory of Polar Varieties* of B. Teissier published on these proceedings; see also the long review of O. Zariski about the Derwidué's work on M.R. **13**, 1952, 67-70.

[10] Please note that in the first volume of *Serie, sistemi di equivalenza...*, ed. Cremonese, 1942, F. Severi once more attributes to Beppo Levi's memoir the solution of the problem of the reduction of singularities of an algebraic surface.

axiom, as we already mentioned in §3. As far as we know, Levi in his scientific work was always adamant in this, but of course, this position necessarily distanced Levi from current developments in algebraic geometry.

Actually, Levi who in his youth had always shown great interest in the more innovative and stimulating aspects of mathematics with personal important and profound results, now in his senior years assumed a most conservative position. He was convinced that when faced with mathematical developments with which he did not feel comfortable, it was best to refer back to great mathematicians of the past. We know that he lead his students in this direction, and he himself showed great familiarity with the works of I. Newton, K. Weierstrass, M. Noether from his quotes in the memoir [1957]... almost as if he wanted to dialogue with these great scientists of the past. Too much specialization was contrary to his personal view of mathematics, and he continued faithful throughout his life to a global view of mathematics, working on subjects which we today consider diverse and distant. Perhaps, even his book on Euclid [1947], with its stimulating observations both on mathematics and philosophy could be viewed as a consequence of this thinking.

At this point, another question comes to mind. Certainly, Corrado Segre was well aware of the fact that the problem of the reduction of singularities was an extremely difficult one. Then, we ask, why did a great teacher, and he certainly was that, suggest to such a young and consequently inexperienced mathematician like Beppo Levi so difficult a problem? Let us recall that Levi started to work on this subject before he was twenty-two. We could say that the subject of his thesis dissertation, as proposed by C. Segre, did not have the depth and the vastness which Levi brought to it (see [1898-1]). However, it is certain that Segre worked closely with Levi when he developed this study [11]. Here, the answer is supplied by Corrado Segre himself in his article [S-1], which is dedicated to his students and also deals with the problem of selecting a good research topic for a young mathematican. At the beginning of the article [S-1], Segre states[12]: «*Ma la facilità è una cattiva consigliera; e spesso i lavori a cui essa guida i giovani, se possono servire come esercizi, come preparazione a ricerche originali, non meriterebbero però di vedere la luce. [...]. Meglio, molto meglio, che il giovane, anziche produrre rapidamente una lunga serie di scritti di tal natura, si affatichi per molto tempo nella risoluzione anche di un sol problema, purché questo sia importante: meglio un risultato atto a*

---

[11] Corrado Segre followed his students' dissertations with much care. Let us quote from A. Terracini's book *Ricordi di un Matematico, un sessantennio di vita universitaria*, ed. Cremonese, Roma, 1968, pp. 12-13: «*Una menzione a sé vogliono le tesi di laurea, che Corrado Segre assegnava per scritto con una lunga e particolareggiata esposizione dello stato in cui si trovava la questione che il laureato doveva trattare [...] Durante la preparazione delle tesi, Segre le esaminava abbastanza spesso, formulando sempre per scritto le sue critiche ed eventuali consigli per la continuazione.* [«Special mention should be made on dissertation subjects which Corrado Segre assigned with a long and detailed up-to-date written presentation on the topic the student was to do develop. [...]. During the thesis preparations, Segre used check them rather often presenting written criticism and suggestions for their better development»].

[12] «But easy tasks are not good teaching tools: and often the work towards while they lead the young while useful as exercises preparatory to original research, do not, even merit coming to light [...]. It is better, much better, that the youth instead of producing a long series of work of this nature, should strive for a long period in the resolution of only one problem as long as it is an important one. It is better to reach one result which will prove scientifically valuable over the years than to have produced a thousand results destined to an early demise».

rimanere nella scienza che mille destinati a morire appena nati!» Later on pag. 57, he adds[13]: *«La teoria delle trasformazioni birazionali dello spazio [...] aspetta ancora di essere applicata alla grave questione della riduzione delle singolarità superiori delle curve sghembe e delle superficie, non che allo studio dei sistemi lineari di superficie, ecc. [...]. In tutti questi campi si trovano delle questioni vitalissime alle quali deve rivolgersi il giovane molto più che alle esercitazioni dianzi citate»*. Evidently, five years after this writing, Corrado Segre must have thought he had found in the young Levi a student must suited for such work.

Up to now, we have been studying the work of Levi on resolution of singularities. He wrote other, now almost-forgotten, algebraic geometry works. Most of these, however, were correlated with the theory of singularities. Certainly, the rather technical and lengthy work [1905-1] is in this group. It contains a detailed study of uniplanar double points of any algebraic surface; here Beppo Levi follows a synthetic approach to the subject. In [1899-2] he proves a formula which relates, in very special cases, order and multiplicity of two algebraic varieties at the order of their intersection, generalizing some known results. His research of [1904-3] consists of a synthetic study of the property of fourth order surfaces with 13 double points; this study refers earlier Rohn and Kummer works; the research of these two authors was, however, of an analytic nature. It should be noted that the work is accompanied by four beautiful tables.

Other algebraic geometry works belong to the Argentine period. They are the brief work [1944] (in which he supplies a particularly simple proof of the formula which expresses the degree of the ruled surface of all trisecants of a skew cubic) and the large work [1946-2] in collaboration with others. This last work, as the authors recall in a brief summary in Italian *«consiste in determinare le caratteristiche numerative (dimensione, ordine, molteplicità di sottovarietà...) delle varietà generate dagli spazi lineari aventi determinati ordini di contatto colle curve tracciate sopra una superficie immersa in uno spazio di un numero qualunque di dimensioni, per un punto semplice ordinario di essa»*. The case where the contact order is $p \leq 6$ is studied in detail, thus coming up once again to the case $p \leq 3$ Del Pezzo and Segre results. This is concerned with an important topic which Levi had already studied in his 1897 and early 1900 work (for an historical picture of the problem together with some corrections to [1946-2] see C. Longo *Studio numerativo sopra le varietà di contatto...*, Ann. Sc. Norm. Sup. 1950, 223-230).

9. Other Research in Geometry

Beppo Levi's work in the field of algebraic geometry and his studies in logic had to necessarily bring him to the problems in the foundations of geometry.

Let us remember that the end of the last century was a productive period for studies in the foundations of geometry. In Enriques' obituary, published on Math. Notæ, **6**, 118-123, Beppo Levi wrote: *«En el ultimo decenio del siglo pasado había alcanzado extraordinario interés el anàlisis de los fundamentos de la geometria proyectiva»*. At this time, in fact, the theory was already developed to such a point that together with many research works, important treatises

---

[13] «The theory of birational transformation in space. [...] is still awaiting applications to the serious problems of the reduction of singularities of skew curves and of surfaces, as well as the study of linear systems of surfaces etc. [...]. In all these fields we find vital problems to which the young should turn rather than to the exercise already mentioned».

were already being published. First in importance, if not in order of time, we find the treatise [H] of David Hilbert, published in its first edition in 1899, and in a second enlarged edition in 1903. Before [H] there had been published the *Vorlesungen* of M. Pasch [Pa], in 1882. These will greatly influence Levi's work either directly or through the Peano research [14].

Even in Italy, research and treatises in the field of the principles of geometry were quite intensive. Already in 1880, the paper of R. De Paolis *Sui fondamenti della geometria proiettiva*, Mem. R. Accad. Lincei (3), **IX**, 1880-81, 489-503 had been published. We must also mention the various works of G. Peano (we find, for example, the volumes 2 and 3 of the collected works of Peano [Pe] and, in particular, the studies *I principi di geometria logicamente esposti*, Fratelli Bocca Editori Torino, 1889, pp. 40 and *Sui fondamenti della Geometria*, Rivista di matematica, **IV**, 1894, 51-94) and the treatise [V] of G. Veronese, published in 1891 (in his introduction to the first volume of the Corrado Segre collected works, ed. Cremonese 1957, F. Severi attributes to Veronese's book the discovery of non-Archimedian geometry).

Once again, we owe to Corrado Segre's influence and indisputable prestige, much of the work in developing the theory of principles of geometry in Italy. Gino Fano wrote at the beginning of the article *Sui postulati fondamentali della Geometria proiettiva in uno spazio lineare a un numero qualunque di dimensioni*, Giornale di Matematiche ad uso… (del Battaglini), **XXX**, 1892, 106-132,: «*il Chiar.^{mo} Prof. Corrado Segre in un corso di lezioni di Geometria sopra un ente algebrico dettate nello scorso anno accademico 1890-91* [...] *proponeva appunto ai suoi studenti (fra i quali ho l'onore di annoverarmi) la questione seguente:* «*Definire lo spazio S_r non già mediante coordinate, ma con una serie di proprietà dalle quali la rappresentazione per coordinate si possa dedurre come conseguenza*…»[15]». In 1898 with his *Lezioni di Geometria Proiettiva* (ed. Zanichelli, Bologna, a text preceded by two lithographed volumes of the same title, 1893-4, 1894-5), F. Enriques already had made available to his first year students in Bologna the results of some of his work in the field of foundations of geometry, originating

---

[14] In the article *Osservazioni del Direttore sull'articolo precedente*, Rivista di Matematica, Vol. **I**, 1891, 66-69, Giuseppe Peano wrote: «*Fra gli autori che hanno discusso i fondamenti della geometria eccelle, a nostro avviso, il Pasch* (Vorlesungen über neuere Geometrie, Leipzig 1882). *Limitandoci ai principi su cui si basa la geometria di posizione (cioè fatta astrazione dall'idea di moto), questi postulati furono da noi trasformati in simboli di logica nei* «*Principi di geometria*» *e si devono, fino a prova contraria, ritenere indipendenti*». [«Among the authors who have discussed the foundations of geometry, in our opinion, Pasch (*Vorlesungen über neuere Geometrie*, Leipzig 1882) is outstanding. Limiting ourselves to the principle on which the «geometria di posizione» is based (that is, disregarding the idea of motion), these axioms were transformed by us into symbols of logic in the «Principi di geometria» and must be considered, until proven otherwise, *independent*»]. It should be noted that the «precedente» article to which Peano refers (in his title) is the work of Corrado Segre [S-1].

[15] Segre himself in [S-1] pp. 60-61, expressed the same ideas writing: «*Non è ancora stato assegnato e discusso (ch'io sappia) un sistema di postulati indipendenti che serva a caratterizzare lo spazio lineare ad n dimensioni, sì che se ne possano dedurre la rappresentazione dei punti di questo con coordinate. Sarebbe conveniente che qualche giovane si occupasse di questa questione (che non sembra difficile)*». [«As far as I know, an independent system of axioms has not yet been defined and discussed which would serve to characterize the n-dimensional linear space so as to be able to arrive at the representation of the points of this space by coordinates. It would be opportune that some youth take up this problem (which does not appear difficult)»] Even the work of Amodeo *Quali possono essere i postulati fondamentali della geometria proiettiva di un S_r*, Atti R. Accad. Sci. Torino, **XXVI**, 1891, 741-770, was influenced by Corrado Segre's thinking.

from his paper *Sui fondamenti della geometria proiettiva*, Rendiconti Ist. Lombardo II, vol. 27, 1894, 550-567. Let us also remember the intensive activity of Mario Pieri's research in foundations (see the volume of collected works by Pieri edited by the Unione Matematica Italiana [Pi], where Beppo Levi's memorial at the time of Pieri's death is included as introduction)[16].

The greatest and most significant work of Beppo Levi in foundations of proiective geometry was certainly the memoir [1904-2] of seventy-four pages. Here he introduced a new system of axioms, valid for the geometry of Euclid, Lobachevsky and Riemann and also valid for other geometries which had been introduced in that period. He assumed as *idee primitive* (primary concepts) that of point and congruency of two pairs of points. On the basis of this system, he was able to have follow: «*la rappresentazione per coordinate dello spazio proiettivo e la geometria analitica, e di conseguenza tutta la geometria projettiva nelle sue parti essenziali, indipendentemente da ogni nozione circa la potenza dell'aggregato dei punti (...), e circa l'ordine degli elementi in una forma di prima specie*». It can be noted that Beppo Levi once again demonstrates here his desire to faithfully follow the program already proposed by Segre. He did this, however, while manifesting great originality and independence in his treatment. We wish to mention an observation in logic which he exposed at the beginning of the memoir. Here, indeed, he proved that it is always possible to substitute for a system of axioms ordinately independent an equivalent system absolutely independent. With his systems of 22 axioms, Beppo Levi proposed to continue and to improve the systems proposed by Mario Pieri in *Della Geometria elementare come sistema ipotetico deduttivo*, Mem. R. Accad. Sci. Torino Cl. Sci. Mat. Fis. Natur., (2), **XLIX**, 1898-99, 173-222 and by Friedrich Schur in *Über die Grundlagen der Geometrie*, Math. Ann. Bd. **55**, 1902, 265-292.

Actually, all of Levi's memoir [1904-2] is strongly influenced by the reading of [Pa] of Pasch. In fact, one of his objectives is to overcome a contradiction he thought he recognized in [Pa] when dealing in elliptic geometry.

In 1906, Oscar Veblen and W.H. Bussey published in Trans. A.M.S., 7, 1906, the paper *Finite Projective Geometries*, 241-259. Levi noted that the work did not contain any mention of his memoir [1904-2] which had covered almost the same geometries and even more generally. Therefore, he wrote to Transactions in order to remind them of the more noteworthy results of [1904-2]. Actually, he took the opportunity to extend some of those results, comparing them with the work of Veblen and Bussey. This writing (identified by us [1907-2]) is where Beppo Levi's command of algebraic techniques is most evident. Because of this, [1907-2] can be placed with works on the number theory which Levi was writing at the same period. The article [1907-2] is, as far as we know, the only mathematics work by Beppo Levi appearing in

---

[16] We can find much affinity in the scientific personalities of Pieri and Levi. Let us recall that Pieri, born in Lucca (Italy) on June 22, 1860, started his studies in mathematics at the University of Bologna, transferring later to Pisa, where he attended the Scuola Normale. The two mathematicians, separated by fifteen years in age, had therefore a different scientific education. Then, in spite of these differences, they find themselves colleagues Turin; and finally working together as professors at the University of Parma. In Turin, they had occasion to frequent the same exceptional mathematics environment and were both profoundly influnced by this. So, we find both working on algebraic geometry problems (among other things, Pieri studied the problem of resolution of singularities of curves) and on problems of foundations of geometry. Beppo Levi mentioned many times in his writings both on algebraic geometry and foundations of geometry Pieri's results. On the other hand, Pieri remembers Beppo Levi in the memoir *La geometria elementare istituita sulle nozioni di «punto» e «sfera»*, Memorie della Società Italiana delle Scienze (detta dei XL), (3), **15**, 1908, 345-450. Pieri died prematurely in 1913 when he was still a professor in Parma.

a U.S.A. journal. The work was written in Italian.

Veblen's reply didn't take long in coming and consisted in another article entitled *Collineations in a finite projective geometry*, published in the same issue of Trans. A.M.S., at pp. 366-368. Veblen referred briefly to the memoir [1904-2] and others on the subject and then continued using a result form Beppo Levi's article [1907-2], to further generalize his earlier work with Bussey.

Beppo Levi referred many times to the findings of [1904-2] to extend or explain them further. In [1910] he thus demonstrated how his own system of axioms could satisfactorily complete certain Klein observations. In [1908] starting from an oversight which he himself found in [1904-2], he showed the relationship between the existence of a reciprocity and the existence of a polarity in the geometries under consideration. In this work there is a brief and slight critical mention of Lebesgue, regarding the problem of the resolution of the functional system $\phi(x+y) = \phi(x) + \phi(y), \phi(x^2) = [\phi(x)]^2$.[17]

Levi's research in the field of foundations of geometry started prior to [1904-2] with the brief work [1903]. There, as Enriques writes in [E-1], «*à donné une forme élémentaire aux développements qui sont nécessaires à l'établissement du cas particulier du théorème de Pappus correspondant à la possibilité de l'interversion des termes moyens d'une proportion*».

It appers from the list we have presented here that Beppo Levi was the author of various other works in the field of geometry. We are not considering the brief works [1905-2] and [1950] which dealt respectively with problems of spherical geometry (independently of Euclid's axiom of paralleles) as well as geometric probability. We would, however, like to touch briefly another of Beppo Levi's works which some might define «minor». Levi always had a strong vocation for the promotion and teaching of mathematics and didn't disdain to face even elementary problems if this could prove useful in the diffusion of the mathematical method of thought. With this in mind, we want to bring up a simple incident which is rather indicative of that certain mathematics atmosphere which then permeated the Istituto di Matematica in Bologna. In Am. Math. Monthly, **XXXVIII**, May 1931, p. 288, J.G. Deutsch presented the following problem: «Given four points in a plane, no three of which are collinear, what is the necessary and sufficient condition that there exists a square containing the given points

---

[17] The remark concerns the paper of H. Lebesgue, *Sur les transformations ponctuelles transformant les plans en plans qu' on peut définir par des procédés analytiques*, Extrait d'une lettre addressée par Mr. H. Lebesgue à Mr. C. Segre, Atti Accad. Sci. Torino, **XLII**, 1907, 3-10.

The subject had actually already been studied by Levi in his paper *Sulla teoria delle funzioni e degli insiemi*, Atti R. Accad. Lincei Rend. Cl. Sci. Fis. Mat. Nat. (5), **IX**, 2° Sem. 1900, 72-79 and had been inspired by a geometric problem studied by Corrado Segre in the first part, p. 288 of the large memoir *Un nuovo campo di ricerche geometriche*, divided in four sequential parts: Nota I, Atti R. Accad. Sci. Torino, **XXV**, 1890, 276-301, Nota II *Delle antinvoluzioni e delle catene*, ibidem, 430-457, Nota III *Delle antipolarità e delle iperquadriche*, ibidem, 592-611, Nota IV *Sistemi lineari ed intersezioni d'iperconiche e d'iperquadriche*, Atti R. Accad. Sci. Torino, **XXVI**, 1891, 35-71. This extensive work by C. Segre was not included in the «Opere» edited by the Unione Matematica Italiana, ed. Cremonese, Roma, 1957-1963.

As regards this interesting subject and its relationship with the axiom of choice, see also [M].

We can better understand Levi's remark when we recall that in the same period there had arisen controversy between H. Lebesgue and B. Levi on other points (the problem having started with the first of B. Levi's three notes *Ricerche sulle funzioni derivate*, appearing in Atti R. Accad. Lincei Rend. Cl. Sci. Fis. Mat. Nat. (5), **15**, 1° sem., 1906, 433-438, 551-558, 674-684; Lebesgue replied on the same journal Atti R. Accad. Lincei Rend. Cl. Sci. Fis. Mat. Nat. (5), **15**, 2° sem. 1906 with the note «*Sur les fonctions dérivées*», pp. 3-8...).

on its perimeter such that no two points are in the same side?» The problem was immediately re-proposed in the Boll. Un. Mat. Ital., **X**, 1931, p. 245. And it was to this very problem that Beppo Levi, who was then intensely involved in the Unione Matematica Italiana and its Bulletin, dedicated two small references-precisely [1931-2] and [1932]. It should now be noted that the problem was also solved by Giuseppe Vitali, who at that time was a colleague of Beppo Levi at the University of Bologna, as reported by Tullio Viola in his moving remembrance of Vitali *Ricordo di Giuseppe Vitali a 50 anni dalla sua scomparsa*, in *La Storia delle matematiche in Italia*, a cura di O. Montaldo e L. Grugnetti, Università di Cagliari, Istituti di Matematica delle Facoltà di Scienze e Ingegneria, 1982, pp. 536-537[18]. We believe that the works [1904-1], [1907-1], [1924], [1931-1] and [1934] also belong to this production. In fact, [1907-1] is the text of his inaugural lecture at the University of Cagliari illustrating to his collegues and students the usefulness of descriptive geometry; it is also important to recall the importance which he gave to aerial photography for map-making problems. In [1926] he undertook to explain to a learned scientific audience, but not made up necessarily of mathematicians, the elements of absolute differential calculus from the point of view of his related applications which was, actually, not a simple task.

This role of promoter would increase after his transfer to Rosario. It seems, in fact, that he almost felt called to spread the mathematical way of thinking in an environment where an Institute of Mathematics was being created for the first time. Actually, the founding of the Institute of Mathematics and of the two above mentioned periodicals are undoubtedly correlated.

A basic role in fostering this point of view was to be carried out by the periodical *Mathematicæ Notæ*. It was its purpose to spread mathematical thinking particularly among the young. The magazine, at least up to the demise of the already mentioned *Publicaciones*, was not to contain particularly advanced mathematical research. It was essential, if it was to succeed in its objective, that it publishes articles easily understable even by the less learned mathematicians, but on the other hand it was to offer very serious and well-founded mathematical problems and to establish a sort of scientific dialog with its readers. The first article in the journal was a geometry/topology article [1941] where Levi explained in the first few lines of the work why he made this choice: «*Tratamos con este ensayo de ofrecer un ejemplo de un argumento que, por los medios matemàticos utilizados se presenta extremadamente simple, por lo cual también la preparación que requeire en el lector puede decirse nula (requiriéndose sin embargo un poco de refléxion), mientras al contrario, como se verà por algo de bibliografía que daremos al final, se enlaza a cuestiones fundamentales para el análisis moderno*». In the work he demonstrated a form of the Jordan theorem valid for poligonal curves and used this to transmit to his readers the difficulty and delicacy of the Jordan theorem on plane curves in all its generality. In the paper [1946-1] he studies the notion of pseudo-orthogonality between conics, already covered in the second volume of Baker, *Principles of Geometry*, Cambridge, 1923. More precisely, [1946-1] constitutes a very brief purely geometric presentation of a note previoulsy published earlier in the same issue of Math. Notæ, **6** (J. Mateo *Costruccion de conicas...*, 96-111). Even in the works [1942-1] and [1942-2] we can see the intent to develop rather delicate mathematics disquisitions starting from points and using methods that were quite elementary.

---

[18] For the sake of completeness, we remember that in the same journal, Am. Math. Montly, **XXXIX**, April 1932, p. 242-243, another solution was given by Roger A. Johnson.

REFERENCES

[D] Jean Dieudonné, *Cours de géométrie algébrique*, I *Aperçu historique sur le développement de la géometrie algébrique*, Presses Universitaires de France, 1974.

[E] Federigo Enriques, *Lezioni sulla teoria geometrica delle equazioni e delle funzioni algebriche* (pubblicato a cura del dott. Oscar Chisini), Ed. Zanichelli, Bologna, 1915.

[E-1] Federigo Enriques, Principes de la Géométrie, «*Encyclopédie des Sciences Mathématiques*», III, 1911, 1-147.

[F] G. Fichera, I contributi di Francesco Severi e di Guido Fubini alla teoria delle funzioni di più variabili complesse, Atti del Convegno Matematico in celebrazione del centenario della nascita di Guido Fubini e di Francesco Severi, Torino 8-10 ottobre 1979, *Atti Accad. Sci. Torino Cl. Sci. Fis. Mat. Natur.* **115** (1981), supplemento (1982), 23-44.

[M] Gregory H. Moore, *Zermelo's Axiom of Choice. Its Origins, Development, and Influence*, Studies in the History of Mathematics and Physical Sciences 8, Springer, New York (1982).

[H] David Hilbert *Grundlagen der Geometrie*, In *Festschrift zur feier der Enthüllung des Gauss-Weber-Denkmals in Göttingen*, Leipzig; B.G. Teubner, 1899, 1-92.

[Pa] Moritz Pasch, *Vorlesungen über neuere Geometrie*, Leipzig; B.G. Teubner, 1882.

[Pe] Giuseppe Peano, *Opere Scelte*, a cura dell'Unione Matematica Italiana, 3 Voll., Edizioni Cremonese, 1959.

[Pi] Mario Pieri, *Opere sui Fondamenti della Matematica*, a cura dell'Unione Matematica Italiana, Ed. Cremonese, 1980.

[Pi-Si] Emile Picard et Georges Simart, *Théorie des Fonctiones Algébriques de deux variables indépendantes*, Gauthier-Villars, Paris, Tome I, 1897, Tome II, 1900.

[S] Corrado Segre, Sulla scomposizione dei punti singolari delle superficie algebriche, *Ann. di Mat. Pura ed Appl.*, Serie II, **XXV**, 1897, 1-54.

[S-1] Corrado Segre, Su alcuni indirizzi nelle investigazioni geometriche. Ad uso dei miei studenti, *Rivista di matematica*, Vol. I, 1891, 42-66.

[Se] Francesco Severi, La géometrie algébrique italienne, sa rigeur, ses méthods, ses problèmes, in *Colloque de Géomètrie Algébrique*, CBRM, George Thone, Liège, 1950.

[T] A. Terracini, Commemorazione del Corrispondente Beppo Levi, *Atti Accad. Naz. Lincei Rend. Cl. Fis. Mat. Nat.*, **XXXIV** (1963), 590-606.

[V] Giuseppe Veronese *Fondamenti di geometria a più dimensioni e a più specie di unità rettilinee esposti in forma elementare*, Padova, Tipografia del Seminario, 1891 (translation into german by A. Schapp., *Grundzüge der Geometrie...*, Leipizig, B.G. Teubner, 1894).

[Z] Oscar Zariski, *Algebraic Surfaces*, Ergebnisse der Mathematik und ihre Grenzgebiete, Band III, Heft 5, Springer, 1935.

When Is a Conic Bundle Rational?

ALBERTO CONTE, Dipartimento di Matematica, Università di Torino
V. Carlo Alberto 10, 10123 Torino, Italy

In this paper I shall report on some new results and conjectures about the problem stated in the title, which is however far to be solved.

Let V be an algebraic variety of dimension r smooth and irreducible over $\mathbf{C}$. V is said to be:

$$\left.\begin{array}{c}\text{uniruled}\\\text{ruled}\\\text{unirational}\\\text{rational}\end{array}\right\} \text{ iff } \left\{\begin{array}{l}\exists\, \mathbf{P}^1 \times S \dashrightarrow V\\ \exists\, \mathbf{P}^1 \times S \stackrel{\sim}{\dashrightarrow} V\\ \exists\, \mathbf{P}^r \dashrightarrow V\\ \exists\, \mathbf{P}^r \stackrel{\sim}{\dashrightarrow} V\end{array}\right.$$

where $\dashrightarrow$ (respectively $\stackrel{\sim}{\dashrightarrow}$) denotes a rational dominant (birational) map and S is a smooth irreducible variety of dimension $r - 1$.

All varieties of the above types have the Kodaira dimension $\kappa(V) = -\infty$, i.e. for all $n > 0$:

$$P_n(V) = \dim_{\mathbf{C}} |nK_V| + 1 = 0 \ (\iff |nK_V| = \emptyset),$$

where K_V is the canonical divisor of V and $P_n(V)$ its n-th plurigenus.

If V is rational (or even only unirational), then also $q = h^1(\mathcal{O}_V) = 0$.

For curves and surfaces $q = 0$ and $\kappa(V) = -\infty$ are also sufficient conditions for rationality. More precisely:

C is a rational curve $\iff q = 0$;
S is a rational surface $\iff q = P_2 = 0$ (Castelnuovo's criterion of rationality).

Note that, for a curve C, $q = g$ is the genus, whilst for a surface S q is its irregularity.

If $r = 3$ and V is a threefold, no such criteria of rationality are known (for many years, up to 1971, it was not even known whether unirational $\neq$ rational (Lüroth's problem), which is in fact the case).

However, one knows that:

$$\kappa(V) = -\infty \iff V \text{ is uniruled} \qquad \text{(Mori-Miyaoka)}$$

and moreover V is birationally equivalent to a variety W with at most terminal singularities such that, either:

(i) W is a minimal $\mathbf{Q}$-Fano variety (i.e., $-K_W$ is ample and Pic $W \simeq \mathbf{Z}$),
or :

(ii) $\exists$ a morphism $\delta : W \to C$ onto a smooth curve C whose generic fiber is a Del Pezzo surface and Pic $W \simeq \delta^*(\text{Pic } C) \oplus \mathbf{Z}$,
or :

(iii) W is a minimal *conic bundle*, i.e. $\exists$ a morphism $\pi : W \to S$, with S a non-singular projective surface (up to birational equivalence) whose fibers are rational curves isomorphic to conics in $\mathbf{P}^2$.

One can therefore look for separate rationality criteria for each of the above classes of varieties.

For Fano varieties there is a fairly complete classification and it is known which of them is rational (up to minor details).

Nothing is known about the rationality of varieties of type (ii), which are called *Del Pezzo fiber spaces* (see, however, the recent article by D'Souza, [D]).

In the present paper we will discuss *rationality criteria for 3-dimensional conic bundles*.

Let $\pi : V \to S$ be a conic bundle with V and S smooth and irreducible (and projective). If V is rational, then S is obviously unirational, therefore rational, which we will assume from now on.

Let us recall that a conic bundle $\pi : V \to S$ is said to be *standard* if, for all irreducible curves $D \subset S, \pi^{-1}(D)$ is an irreducible surface. Two conic bundles $\pi : V \to S$ and $\pi' : V' \to S'$ are said to be *birationally equivalent* if there exists a commutative diagram:

$$\begin{array}{ccc} V & \stackrel{\simeq}{\longrightarrow} & V' \\ \pi \downarrow & & \downarrow \pi' \\ S & \stackrel{\simeq}{\longrightarrow} & S'. \end{array}$$

Every conic bundle is birationally equivalent to a standard one ([Sa]).

If $\pi : V \to S$ is standard, one can prove that there exists a *discriminant curve* $C \subset S$, which is a reduced normal crossings divisor with at most double points as singularities, such that:

$$\pi^{-1}(s) = \begin{cases} c & \text{if } s \in S - C \\ l_1 + l_2 & \text{if } s \in C - \text{Sing } C \\ 2l & \text{if } s \in \text{Sing } C, \end{cases}$$

where c is an irreducible conic and l_1, l_2, l are lines ($l_1 \neq l_2$).

If $\tilde{C}$ is the curve parametrizing the components of the reducible conics which are the fibers of π over C, one has a double covering $\tilde{\pi} : \tilde{C} \to C$ which will be defined by a theta-characteristic $K_C \otimes \eta$, where η is a point of order two in the jacobian $J(C)$. Moreover, if V is rational then C is connected ([A-M]).

The archetype of all conic bundles is the hypersurface $V(n) \subset \mathbf{P}^4$ of degree n with a line l_o of multiplicity $n - 2$. Here the discriminant curve is (in the general case) irreducible and smooth of degree $3n - 4$.

At the present moment, no rationality criterion for conic bundles is known. However, in 1987 Iskovskikh ([Is 2]) has made the following:

CONJECTURE I. - *Let $\pi : V \to S$ be a standard conic bundle with C connected. Then*:

$$V \text{ is rational} \iff (i) \text{ or } (ii) \text{ holds },$$

where:

(i) ∃ *a standard conic bundle* $\pi' : V' \to S'$ *birationally equivalent to* $\pi : V \to S$ *and a base point free pencil* $|L'|$ *of curves of genus* 0 *on* S' *such that* $(L' \cdot C') \leq 3$, *where* C' *is the discriminant curve of* S';

(ii) $\pi : V \to S$ *is birationally equivalent to a standard conic bundle* $\pi_o : V_o \to \mathbf{P}^2$ *with discriminant curve* $C_o \subset \mathbf{P}^2$ *of degree* 5 *and theta-characteristic* $K_C \otimes \eta_o$ *even*.

Note that either of the above conditions implies the rationality of V.

In fact:

(i) $\Rightarrow$ *V is rational*

Take L on S such that $(L \cdot C) \leq 3$. Let L_ξ be the generic fibre of L over $k(\xi)$, where ξ is the generic point of $\mathbf{P}^1$ over C. By Tsen's theorem $L_\xi \simeq \mathbf{P}^1_{k(\xi)}$. Let $F_\xi = \pi^{-1}(L_\xi)$. Then, $\pi_\xi : F_\xi \to \mathbf{P}^1_{k(\xi)}$ defines a standard conic bundle structure on F_ξ over the non-algebraically closed field $k(\xi)$. The degeneracy divisor of π_ξ is $C_\xi = C \cdot L_\xi$ and deg $C_\xi \leq 3$ by hypothesis. Moreover:

$$K^2_{F_\xi} = 8 - \deg C_\xi \leq 5,$$

so that F_ξ is rational by [Is 1]. Since F_ξ is the generic fiber of a pencil of surfaces on V with base $\mathbf{P}^1_{k(\xi)}$, it follows that V is rational.

(ii) $\Rightarrow$ *V is rational*

All conic bundles $\pi_o : V_o \to \mathbf{P}^2$ with C_o of degree 5 and even theta-characteristic are birationally equivalent, so that it is sufficient to prove the rationality of any of them. Since $\tilde{\pi}_o : \tilde{C}_o \to C_o$ corresponds to an even theta-characteristic, by [Sho] the Prym variety $\Pr(\tilde{C}_o, C_o)$ is isomorphic to the jacobian $J(Y)$ of a smooth curve Y of genus 5, non hyperelliptic nor trigonal, so that (the canonical image of) Y is the intersection of three quadrics $Q_1 \cap Q_2 \cap Q_3 \cap \mathbf{P}^4$ and the base locus of the net $\mathbf{P}^2 = |2H - Y|$ (H hyperplane in $\mathbf{P}^4$) spanned by them. If we fix $O \in Y$ and for all $s \in \mathbf{P}^2$ denote by Q_s the corresponding quadric in the net, let V_s be the projectivized tangent cone in O to Q_s. We get in this way a conic bundle V over $\mathbf{P}^2$ having the same theta-characteristic (i.e., the same double covering) as V_o and therefore birationally equivalent to V_o. It is therefore sufficient to prove that V is rational. To see this, let:

$$p : V \to \mathbf{P}^4$$

be the map sending the generators of cones in V to the corresponding lines in $\mathbf{P}^4$ passing through $O \in Y$. p is surjective since each line in $\mathbf{P}^4$ through O lies on at least one quadric of the net $|2H - Y|$. Furthermore, the general line l through O is contained in exactly one quadric of the net because, if $l \subset Q_1 \cap Q_2$ and Q_3 is any quadric of $|2H - Y|$ linearly independent from Q_1 and Q_2, then $l \cap Q_3 = \{O, y\}$ with $y \in Y$, so that l is tangent or a chord to Y, which is not the case if l is general. p is therefore birational. Compounding it with the projection from O to $\mathbf{P}^3$ one gets the required birational morphism $\beta : V \dashrightarrow \mathbf{P}^3$. V is therefore rational.

As to the necessity of conditions (i) and (ii), it was proved by Shokurov [Sho] in the particular cas in which $S = F_n$ or $\mathbf{P}^2$ is a minimal rational surface. He proves the following:

THEOREM. - *Let* $\pi : V \to S$ *be a standard conic bundle over a minimal rational surface* S. *Then*, V *is rational if and only if the intermediate jacobian* $J(V)$ *is isomorphic (as a principally polarized abelian variety) to a product* $J(C_1) \times \ldots \times J(C_m)$ *of jacobians of non-singular curves* $C_1, \ldots, C_m$.

The above result amounts essentially to the following criterion:

(*) V *is rational* $\Longleftrightarrow |2K_S + C| = \emptyset$ (+ theta even if C is a plane quintic).

In the particular case in which $S = \mathbf{P}^2$ this means:

V *is rational* $\iff$ $\deg C \leq 5$ (+ theta even if it is =5).

Note that the above criterion does not apply to the cubic hypersurface $V(3) \leq \mathbf{P}^4$, which has a discriminant curve of degree 5, but whose double covering is defined by an odd theta characteristic.

In the case in which $S = \mathbf{P}^2$ one sees easily that (*) is equivalent to Conjecture I, since in this case the pencil $| \, L \, |$ cannot be nothing else than a pencil of lines, so that $(L \cdot C) \leq 3$ if and only if $\deg C \leq 4$ (if it is = 4, the pencil has its center on C). Then, blow up the center of the pencil to get the conic bundle $\pi' : V' \to S'$.

Shokurov has conjectured that (*) holds for every conic bundle and every S.

Going back to Conjecture I, let us remark that in both cases there exists a birational map $\beta : V \overset{\sim}{\dashrightarrow} \mathbf{P}^3$ which takes fibers $V_s = \pi^{-1}(s), s \in S$, into conics in $\mathbf{P}^3$:

CASE (i). - β is induced by the birational maps on the fibers:

$$\beta_\xi : F_\xi \overset{\sim}{\dashrightarrow} \mathbf{P}^2_{k(\xi)}$$

defined by the linear system $| \, -f_\xi - K_{F_\xi} \, |$, where f_ξ is a fiber of the morphism $\pi_\xi : F_\xi \to \mathbf{P}^1_{k(\xi)}$. Since:

$$(-f_\xi \cdot K_{F_\xi}) = (V_s \cdot K_V) = 2,$$

the image of V_s is a conic in $\mathbf{P}^3$.

CASE (ii).- The projection $\mathbf{P}^4 \to \mathbf{P}^3$ from O induces linear maps on the fibers of $\pi : V \to \mathbf{P}^2$, which are conics by construction. Therefore, their images in $\mathbf{P}^3$ are conics, which one could see can be obtained as residual intersections of a net of cubics $| \, 3H - Y' \, |$ going through the curve Y' which is the projection of Y (and which has degree 7).

One is therefore led to the following conjecture, which Iskovskikh proves to be equivalent to the necessity $|\Rightarrow$ in Conjecture I, and which is therefore equivalent to it:

CONJECTURE II . - *For any rational standard conic bundle $\pi : V \to S$ there exists a birational map $\beta : V \overset{\sim}{\dashrightarrow} \mathbf{P}^3$ taking the fibers V_s of β to conics in $\mathbf{P}^3$.*

Another way of formulating Conjecture II is the following:

CONJECTURE II' . - *Every 2-dimensional family (congruence) of index 1 of rational curves in $\mathbf{P}^3$ can be transformed in a congruence of conics (or of lines through one point) by a birational transformation of $\mathbf{P}^3$.*

In this form it was "proved" by Kantor [K] in 1901. Its proof was accepted, even if it is very obscure and hard to follow, up to the end of the thirties, when doubts started to arise. In 1960 Millevoi [M] gave a counterexample to it. The question about this conjecture seemed therefore to have been settled for ever in the negative. However, in [Is 2] Iskovskikh reports that Gizatullin has found that Millevoi's counterexample is wrong, but no proof of this statement has, as far as I know, yet appeared. The question is therefore still open.

In any case, Millevoi's (counter) example is as follows:

Let $| \, F \, |$ be the net of quartic surfaces in $\mathbf{P}^3$ having a tacnode in a point $O \in \mathbf{P}^3$ with fixed tangent plane ω and going through a curve C of degree 8 and genus 6 with a double point in O, and through 4 lines passing through O and contained into ω. Let's consider the congruence of quartic rational curves obtained as residual intersections from $| \, F \, |$.

By a suitable birational map, $\mathbf{P}^3$ can be mapped onto a sextic hypersurface $V(6) \leq \mathbf{P}^4$ on which the quartics correspond to conics and the curve representing reducible conics has a non trigonal component of genus 10.

On the other hand, the classical classification of congruences of conics in $\mathbf{P}^3$ gives the following three possible types:

(i) *Congruences reducible to stars of lines*

This is not possible in our case, because it should exist a unisecant surface S^n meeting the F's in $4n - 1$ fixed points.

(ii) *Congruences of conics lying in the planes of a pencil and generating a pencil of conics in each plane*

This is impossible too, because the curve of reducible conics of such a congruence is trigonal, since in every plane of the pencil there are three reducible conics.

(iii) *Congruences of conics which are the residual intersections of the cubics of a net $|3H - Y'|$, where Y' is a curve of degree 7.*

Here the curve of reducible conics is non trigonal, but of genus 6 and not 10, so also this case is impossible.

Finally, let me end these few remarks by noting that it is not known whether all conic bundles (even with base $\mathbf{P}^2$) are unirational or not. It is conjectured that the answer to this question is negative, the most likely non unirational candidate being the $V(n) \subset \mathbf{P}^4$ with a line of multiplicity $n - 2$ for $n \geq 5$.

REFERENCES

[A-M]. M. Artin, D. Mumford, *Some elementary examples of unirational varieties which are not rational*, Proc. London Math. Soc. **25** (1972), 79–95.

[D]. H. D'Souza, *Threefolds whose hyperplane sections are elliptic surfaces*, Pacific J. Math. **134** (1988), 57–78.

[Is 1]. V.A. Iskovskikh, *Rational surfaces with a pencil of rational curves and with positive square of the canonical class*, Mat. Sb. **83** (1970); = Math. URSS-Sb. **12** (1970), 91–117.

[Is 2]. V.A. Iskovskikh, *On the rationality problem for conic bundles*, Duke Math. J. **54** (1987), 271–294.

[K]. S. Kantor, *Die Typen der linearen Complexe retionaler Curven in R_r*, Amer. J. of Math. **23** (1901), 1–28.

[M]. T. Millevoi, *Sulla classificazione Cremoniana delle congruenze di curve razionali dello spazio*, Rend. Sem. Mat. Univ. Padova **30** (1960), 194–214.

[Sa]. V.G. Sarkisov, *Birational automorphisms of conical fibrations*, Izv. Akad. Nauk SSSR **44** (1980), 918–945; = Math. USSR-Izv., **17** (1981).

[Sho]. V.V. Shokurov, *Prym varieties: theory and applications*, Izv. Akad. Nauk SSSR **47** (1983); = Math. USSR-Izv. **23** (1984), 83–147.

Quadratic Inequalities for Hilbert Transforms and Hankel Forms in the Spaces $\mathcal{L}^2$ ($\mathcal{F}$) and $\mathcal{L}^2$ ($\mathcal{B}$)

MISCHA COTLAR Central University of Venezuela, Facultad de Ciencias, U.C.V., Caracas
A.P. 20513 - Venezuela.

REMINISCENCES OF BEPPO LEVI

After forty years of intense mathematical activity in his native Italy, Beppo Levi had to emigrate to Argentina at the age of sixty four in 1939. He was appointed Director of the Mathematical Institute at Rosario where he started, with the help of L. Santaló, two new mathematical journals: Mathematicae Notae and Publications of the Institute. Twenty five volumes of these journals appeared during his life, and he himself contributed in these volumes with 50 papers and two monographs. The Argentinian mathematical Union recognized his exceptional contribution to the development of mathematical studies in that country in a special 350 pages long volume celebrating his 80th birthday. Soon after the Accademia dei Lincei conferred him the Antonio Feltrinelli prize for his mathematical work.

Personally I remember Beppo Levi with deep admiration and gratitude. The year of his arrival I just was beginning my mathematical work and he encouraged me in a most generous way. Thanks to his protection I could publish most of my work of the 1940-1950 period in the above-mentioned journals, and even if some of the papers were not to his taste or contained wrong proofs he didn't reject them but patiently suggested corrections and offered advice. At that time I was much attracted to some recent developments in Harmonic and Functional Analysis. Knowing this, Beppo Levi sent me once a long letter advising not to lose sight of classical writings of great masters, and in particular of some ideas of V. Volterra and L. Fantappié in Functional Analysis. Such suggestions, and the personal contact with Beppo Levi, certainly influenced my work at least in an indirect way. For instance, although his comments on Fantappié ideas didn't turn me toward the classical papers he had in mind, they lead me nevertheless to the Grothendieck-Koethe-Silva and other developments of the Fantappié indicatrices that are used in the paper that follows.

Quantum analogues of the notion of Hilbert transform and that of Hankel forms can be formulated by replacing the ordinary Fourier operator in $L^2(\mathbb{R}^2)$ by the Weyl or the Bargman-Weyl transforms which map isomorphically $L^2(\mathbb{R}^2)$ onto $\mathcal{L}^2(L^2(\mathbb{R}))$ or $\mathcal{L}^2(\mathcal{F})$ respectively, $\mathcal{F}$ the Fock space. Quadratic inequalities for such Hilbert operators or Hankel forms were studied in [6] through a general lifting theorem closely related to classical theorems of Nagy-Foias and Bergman-Shiffer, and through representations of the liftings by operators in the Fock space.

Here we summarize those results and point out another variant based on the Martineau-Aizenberg representation of functionals in spaces $A(D)$ of analytic functions, which applies also to $\mathcal{L}^2(B)$, for B the Bergman or other reproducing kernel spaces.

The M-A formula arose in the theory of analytic functionals of L. *Fantappié*, one of the famous mathematicians who taught at the University of Bologna, who published this theory in 1930 ([7]; for a detailed account see [15]). A basic idea of this theory is to express analytic functionals F on an analytic curve $y(t,\alpha)$ through the indicatrice $v(\alpha) = F[1/(t-\alpha)]$: $F(y(t,\alpha)) = 1/2\pi i \int v(\xi)y(\xi,\alpha)\,d\xi$. This theory gave rise to the duality theorems of Silva, Koethe and Grothendieck for $\mathbb{C}^1$-domains, and later to the generalized Cauchy-Fantappié-Leray formula and the M-A theorem (see [1], [10]). While these theorems deal with linear functionals in $A(D)$, the lifting theorem considered here deals with invariant sesquilinear forms in $\mathcal{L}^2(\mathcal{F})$, which under supplementary conditions reduce to the functionals in the M-A theorem.

A general theory of Hankel forms in Fock spaces was developed by Janson, Peetre and Rochberg [8]; see also [14]. Some other works closely related to this paper are [3], [5], [9], [11], [12].

1. LIFTING OF QUADRATIC INEQUALITIES AND THE THEOREMS OF NEHARI AND HELSON-SZEGO

Let V be the set of all trigonometric polynomials $f(t) = \sum a_n e^{int} = \sum \hat{f}(n) e^{int}$, $t \in T \sim [0, 2\pi)$, $W_1 = \{f \in V : \hat{f}(n) = 0 \text{ for } n < 0\}$, $W_2 = \{f \in V : \hat{f}(n) = 0 \text{ for } n \geq 0\}$, and $\tau : V \to V$ the shift operator $f(t) \to e^{it} f(t)$. A sesquilinear form $B : V \times V \to \mathbb{C}$ is said Toeplitz if $B(\tau f, \tau g) = B(f, g)$, and positive if $B(f, f) \geq 0$, $\forall f$. The following properties are well known.

1.1. If B is Toeplitz then $B(f, g) = L(f\bar{g})$, for some linear functional $L : V \to \mathbb{C}$.

1.2. If B is a positive Toeplitz form, then $B(f, g) = \int f\bar{g}\,d\mu$ for some positive finite measure μ in T (Herglotz-Bochner).

1.3. If B is a Toeplitz form continuous in the L^∞-topology then $B(f, g) = \int f\bar{g}\,d\mu$ for some complex measure μ.

The above subspaces W_1 and W_2 satisfy

$$\tau W_1 \subset W_1, \quad \tau^{-1} W_2 \subset W_2, \tag{1}$$

and give rise to the following notions. A form $B_0 : W_1 \times W_2 \to \mathbb{C}$ is called Hankel if it is the restriction to $W_1 \times W_2$ of a Toeplitz form $B : V \times V \to \mathbb{C}$, or equivalently if $B_0(\tau f, g) = B_0(f, \tau^{-1} g)$ for $(f, g) \in W_1 \times W_2$. On the other hand, every $f \in V$ has a unique representation $f = f_1 + f_2, f_1 \in W_1, f_2 \in W_2$, and the Hilbert operator $S : V \to V$ is defined by $S(f) = S(f_1 + f_2) = -if_1 + if_2, \forall (f_1, f_2) \in W_1 \times W_2$. An important question, first solved by Helson and Szego, is to characterize the measure $\mu \geq 0$ such that

$$\int |Sf|^2 d\mu \leq c \int |f|^2 d\mu, \quad \forall f \in V. \tag{2}$$

Setting $B(f, g) = \int f\bar{g} \, d\mu$, and $f = f_1 + f_2$, $Sf = -i(f_1 - f_2)$, (2) can be rewritten as $(c - 1)B(f_1, f_1) + (c + 1)B(f_1, f_2) + (c + 1)B(f_2, f_1) + (c - 1)B(f_2, f_2) \geq 0$. Replacing f_1 by $\lambda_1 f_1$ and f_2 by $\lambda_2 f_2$, we get a positive quadratic form in λ_1, λ_2, and (2) is equivalent to the following inequality expressing the positivity of the determinant of that form:

$$|B(f_1, f_2)| \leq (1 - \varepsilon) B(f_1, f_1)^{1/2} B(f_2, f_2)^{1/2}, \quad \forall (f_1, f_2) \in W_1 \times W_2, \tag{2a}$$

where $0 < 1 - \varepsilon < 1$, $\varepsilon = 1 - (c - 1)^2/(c + 1)^2$. This question is a special case of the following one: Given two measures $\mu_1 \geq 0$, $\mu_2 \geq 0$ and a Hankel form $B_0 : W_1 \times W_2 \to \mathbb{C}$, when is B_0 bounded with respect to the norms of $L^2(\mu_2)$ and $L^2(\mu_1)$, and more precisely when is

$$|B_0(f, g)| \leq (\int |f|^2 \, d\mu_1)^{\frac{1}{2}} (\int |g|^2 \, d\mu_2)^{\frac{1}{2}}, \quad \forall (f, g) \in W_1 \times W_2. \tag{3}$$

In fact setting, $B_j(f, g) = \int f\bar{g} \, d\mu_j$, $j = 1, 2$, (3) becomes

$$|B_0(f, g)| \leq B_1(f, f)^{1/2} B_2(g, g)^{1/2}, \quad \forall (f, g) \in W_1 \times W_2, \tag{3a}$$

which reduces to (2a) when $B_0 = B$ and $B_1 = B_2 = (1 - \varepsilon)B$.

In the special case where $\mu_1 = \mu_2 =$ the Lebesgue measure, problem (3) was solved by Nehari.

If $B_1 \geq 0$, $B_2 \geq 0$ and B_0 satisfies (3a) then we write $B_0 \leq (B_1, B_2)$ in $W_1 \times W_2$ and say that the Hankel form $B_0|_{W_1 \times W_2}$ is bounded with respect to the quadratic seminorms $B_1(f, f)^{1/2}$ and $B_2(f, f)^{1/2}$. In particular (2a) means that $B \leq (1 - \varepsilon)(B, B)$. Similarly is defined the notion $B_0 \leq (B_1, B_2)$ in $V \times V$.

Observe that, unless $B = 0$, (2a) cannot hold for all $(f_1, f_2) \in V \times V$. However if (2a) holds in $W_1 \times W_2$ then there is a Toeplitz form B' satisfying $B' \leq (1 - \varepsilon)(B, B)$ in $V \times V$ and coinciding with B in $W_1 \times W_2$. More generally:

1.4. (Lifting theorem, [6]). Let V be an abstract vector space, $\tau : V \to V$ a linear isomorphis, and W_1, W_2 two subspaces of V satisfying (1). If $B_j : V \times V \to \mathbb{C}, j = 0, 1, 2$ are three τ-invariant forms, $B_j(\tau f, \tau g) = B_j(f, g)$, satisfying $B_0 \leq (B_1, B_2)$ in $W_1 \times W_2$, then there exists an invariant form $B' : V \times V \to \mathbb{C}$ such that $B' = B$ in $W_1 \times W_2$ and $B' \leq (B_1, B_2)$ in $V \times V$.

In the above trigonometric case the theorems 1.1, 1.2, 1.3 can be used to obtain the following more precise result.

1.5. Generalized Bochner theorem [6]. Let V be the trigonometric system as above and τ the ordinary shift. If the Toeplitz forms B_0, B_1, B_2 satisfy $B_0 \leq (B_1, B_2)$ in $W_1 \times W_2$, and B' is the lifting of B_0 as in 1.4, then there exist three measures μ', μ_1, μ_2 such that

$$\mu_1 \geq 0, \quad \mu_2 \geq 0, \quad |\mu'(E)| \leq \mu_1(E)^{1/2} \mu_2(E)^{1/2}, \quad \forall E \subset T, \tag{4}$$

and $B_j(f,g) = \int f\bar{g}\,d\mu_j$, $\forall f,g \in V$, $j = 1,2$,

$$B'(f,g) = \int f\bar{g}\,d\mu', \quad \forall f,g \in V; \quad B_0(f,g) = \int f\bar{g}\,d\mu', \quad \forall (f,g) \in W_1 \times W_2. \quad (4a)$$

Theorems 1.5 allows to deduce by a unified method the two above-mentioned theorems of Nehari and Helson-Szego, as follows. If $\mu_1 = \mu_2 = $ the Lebesgue Measure then (4) gives that $d\mu' = \sigma(t)\,dt$ with $\|\sigma\|_\infty \leq 1$, and we get the Nehari theorem that says that in this case $B_0(f,g) = \int f\bar{g}\sigma\,dt$, with $\|\sigma\|_\infty \leq 1$. Consider now the problem (2), where $B_1(f,g) = B_2(f,g) = (1-\varepsilon)B_0(f,g) = (1-\varepsilon)\int f\bar{g}\,d\mu$, $\mu \geq 0$, so that (4) gives $|\mu'(E)| \leq (1-\varepsilon)\mu(E)$, and since $\int f\bar{g}\,d\mu' = \int f\bar{g}\,d\mu$ in $W_1 \times W_2$, it follows from the F. and M. Riesz theorem that $d\mu' = d\mu + h\,dt$, $h \in H^1$. Thus we get that $d\mu = w(t)\,dt$ and $|w(t)+h(t)| \leq (1-\varepsilon)w(t)$. After some algebraic transformations one arrives at the Helson-Szego theorem: μ satisfies (2) iff $d\mu = w\,dt$ and $w = \exp(u + Hv)$ with $u,v \in L^\infty$ and $\|v\|_\infty < \pi/2$, where $Hf = Sf - \hat{f}(0)$.

REMARK 1. Theorem 1.4 can be deduced from a special case of the Nagy-Foias commutant theorem and conversely, [1], [6]b, the N-F theorem can be deduced from 1.4. However in special cases (as in 1.5) theorem 1.4 can be complemented by integral representations of the forms, going thus beyond the N-F theorem.

REMARK 2. Let X be an abstract set, X_1, X_2 two subsets of X, $\tau : X \to X$ an 1-1 transformation of X such that $\tau X_1 \subset X_1$, $\tau^{-1} X_2 \subset X_2$, and set $V = \{$the sequences $f : X \to \mathbb{C}$ of finite support$\}$, $W_1 = \{f \in V : \text{supp } f \subset X_1\}$, $W_2 = \{f \in V : \text{supp } f \subset X_2\}$. Every kernel $K : X \times X \to \mathbb{C}$ gives rise to a sesquilinear form $B_K : V \times V \to \mathbb{C}$ given by $B_K(f,g) = \sum_{x,y \in X} K(x,y)f(x)\overline{g(y)}$. K is a Toeplitz kernel if $K(\tau f, \tau g) = K(x,y)$, or equivalently if B_K is Toeplitz, and K is positive definite if $B_K \geq 0$, Theorem 1.4 becomes then: if K_0, K_1, K_2 are Toeplitz kernels such that K_1, K_2 are positive definite and $|B_{K_0}(f,g)| \leq B_{K_1}(f,f)^{1/2} B_{K_2}(g,g)^{1/2}$, $\forall (f,g) \in W_1 \times W_2$, then there is a Toeplitz kernel K' such that $B_{K'} = B_{K_0}$ in $W_1 \times W_2$ and $B_{K'} \leq (B_{K_1}, B_{K_2})$ in $V \times V$.

In the special case where τ is the identity, so that every kernel is Toeplitz, this corollary reduces to a theorem of Beatrous and Burbea [4]. Every positive definite kernel K gives rise to a Hilbert space H such that K is its reproducing kernel, i.e. the elements of H are functions defined in X, for every $y \in X$ is $K_y(.) = K(.,y) \in H$ and $f(x) = (f, K_y)$, $\forall f \in H$. Moreover if $B_{K_0} \leq (B_K, B_K)$ in $V \times V$, then $K_0(.,x)$ and $\overline{K_0(x,.)}$ are in H, $\forall x$, and $(Tf)(y) = (f, K_0(.,y))_H$ defines a bounded operator in H. In the special case where K is a sesquianalytic kernel the above corollary gives the Bergman-Shiffer theorem on analytic continuation of functions of two variables in terms of quadratic inequalities (see [4]).

2. GENERALIZED BOCHNER THEOREM IN $\mathcal{L}^2(L^2(\mathbb{R}))$ and $\mathcal{L}^2(\mathcal{F})$

The polynomials $f = \sum a_n e^{int} \in V$ in preceding section, are the (anti) Fourier transforms of the sequences $a_n = \hat{f}(n)$, $n \in \mathbb{Z}$, and $f \in W_1$ if $a = \hat{f}$ has support in $Z_1 = \{n \in \mathbb{Z} : n \geq 0\}$, and the shift operator τ is the Fourier transform of the translation operator $n \to n+1$. We shall replace now Z by $\mathbb{R}^2$ and the Fourier transform in $L^2(\mathbb{R}^2)$

Hilbert Transforms and Hankel Forms

by the Weyl transform, which gives the quantization of L^2, and then extend the results of section 1 as follows. A unitary representation of the symplectic plane $(\mathbb{R}^2, [\,,])$ in the Hilbert space H, is a function ψ which assigns to each $(x,y) \in \mathbb{R}^2$ a unitary operator $\psi(x,y)$ in H such that $\psi(x,y)\psi(x',y') = \exp(\pi i[(x,y),(x',y')])\psi(x+x',y+y')$, where $[(x,y),(x',y')] = xy' - x'y$ is the simplectic form in $\mathbb{R}^2$. All the irreducible unitary representations of $(\mathbb{R}^2, [\,,])$ are unitarly equivalent to the Schrodinger representation $[\phi, H_s]$, where $H_S = L^2(\mathbb{R})$ and $(\phi(x,y)f)(\xi) = \exp(2\pi i y\xi + \pi i x y)f(\xi + x)$, for all $f \in L^2(\mathbb{R})$. For every $f \in L^2(\mathbb{R}^2) \cap L^1(\mathbb{R}^2)$ the Weyl transform of f is the bounded operator $\phi(f)$ in $L^2(\mathbb{R})$ defined by $\phi(f) = \int f(x,y) \cdot \phi(-x,-y)\,dx\,dy$. The Plancherel theorem for the Weyl transform asserts that $\phi : f \to \phi(f)$ extends to an isometric map from $L^2(\mathbb{R})$ onto the space $\mathcal{L}^2(L^2(\mathbb{R}))$ of the Hilbert-Schmidt operators in $L^2(\mathbb{R})$, so that $\int f\bar{g}\,dx\,dy = \mathrm{tr}\phi(g)^*\phi(f)$ (cfr. [17]). We take now $V = \mathcal{L}^2(L^2(\mathbb{R}))$, $W_1 = \{A \in V : A = \phi(f), \mathrm{supp}\ \mathrm{of}\ f \subset \Delta \equiv \{(x,y) \in \mathbb{R}^2 : x \geq 0, y \geq 0\}\}$, $W_2 = \{\phi(f) : \mathrm{supp}\ f \subset \Delta' = \mathrm{the\ complement\ of}\ \Delta\}$, and say that the sesquilinear form $B : V \times V \to \mathbb{C}$ is Toeplitz if $B(\phi(x,y)A_1, \phi(x,y)A_2) = B(A_1, A_2), \forall A_1, A_2 \in V, (x,y) \in \mathbb{R}^2$. Let $S(L^2(\mathbb{R})) = \{\phi(f) : f \in S(\mathbb{R}^2) \subset L^2(\mathbb{R}^2)\}$, $S(\mathbb{R}^2)$ the Schwartz space of $\mathbb{R}^2$. Then from a theorem of Fokko du Cloux [16] and some unpublished results of N. Wallach, the following analogues of 1.1 and 1.2 can be deduced (cfr. [6]).

2.1. If $B : V \times V$ is a Toeplitz form whose restriction to $S(L^2(\mathbb{R}))$ is continuous in the Schwartz topology S, then $B(A_1, A_2) = L(A_2^* A_1)$ where $L : S(L^2(\mathbb{R})) \to \mathbb{C}$ is a continuous linear form.

2.2. If $B : V \times V$ is a Toeplitz form continuous in the $\mathcal{L}^2(L^2(\mathbb{R}))$ topology then $B(A_1, A_2) = \mathrm{tr}\,\Lambda\, A_2^* A_1$ for some bounded operator Λ in $L^2(\mathbb{R})$. Moreover, if $B \geq 0$ then $\Lambda \geq 0$, and if B is continuous in the in the $\mathcal{L}^\infty$-topology then Λ is a nuclear (trace class) operator.

The analogue of 1.5 has a somewhat modified statement. Set $W_2' = \{\phi(f) \in W_2 : \mathrm{supp}\ f \subset \{(x,y) \in \mathbb{R}^2 : y < 0\}\}$, $W_2'' = \{\phi(f) \in W_2 : \mathrm{supp}\ f \subset \{x < 0\}\}$, so that $W_2 = W_2' + W_2''$. Then:

2.3. (Generalized Bochner theorem in $\mathcal{L}^2$, [6]). Let B_0, B_1, B_2 be three Toeplitz forms such that B_1, and B_2 are continuous in the $\mathcal{L}^\infty$-topology and $B_0 \leq (B_1, B_2)$ in $W_1 \times W_2$. Then there exist four nuclear operators $\Lambda_1, \Lambda_2, \Lambda', \Lambda''$ such that $B_j(A_1, A_2) = \mathrm{tr}\,\Lambda_j A_2^* A_1$, for $j = 1, 2$ and $(A_1, A_2) \in V \times V$; $B_0(A_1, A_2) = \mathrm{tr}\,\Lambda' A_2^* A_1$ for $(A_1, A_2) \in W_1 \times W_2'$, $B_0(A_1, A_2) = \mathrm{tr}\,\Lambda'' A_2^* A_1$ for $(A_1, A_2) \in W_1 \times W_2''$, and such that the matrix operators

$$\begin{bmatrix} \Lambda_1 & \Lambda' \\ \Lambda'^* & \Lambda_2 \end{bmatrix} \quad \text{and} \quad \begin{bmatrix} \Lambda_1 & \Lambda'' \\ \Lambda''^* & \Lambda_2 \end{bmatrix}$$

are positive as operators in $L^2(\mathbb{R}) \times L^2(\mathbb{R})$.

The classical double Hilbert transform S in $L^2(\mathbb{R}^2)$ is defined as $S = S_x S_y$, i.e. $Sf(x,y) = S_x|S_y f(x,y)|$ where $S_x(S_y)$ is the 1-dimensional Hilbert transform in the variable $x(y)$. The corresponding operator in $V = \mathcal{L}^2(L^2(\mathbb{R}))$ is defined as follows.

Let $\Delta_{11} = \Delta$, $\Delta_{21} = \{(x,y) : (-x,y) \in \Delta\}$, $\Delta_{12} = \{(x,y) : (x,-y) \in \Delta\}$, $\Delta_{22} = \{(x,y) : (-x,-y) \in \Delta\}$, and $W_{ij} = \{\phi(f) : \mathrm{supp}\ f \subset \Delta_{ij}\}, i,j = 1,2$, so that each $A \in V$ has a unique decomposition $A = A_{11} + A_{12} + A_{21} + A_{22}$. We set then $S(A) = A_{11} - A_{21} - A_{12} + A_{22}$.

As in section 1 one sees that if Λ is a nuclear positive operator in $L^2(\mathbb{R})$ then it satisfies $\operatorname{tr}\Lambda(SA)^*SA \leq c\operatorname{tr}\Lambda A^*A, \forall A \in V$, iff $B \leq (1-\varepsilon)(B,B)$ in $W_1 \times W_2, 0 < 1-\varepsilon$, where $B(A_1, A_2) = \operatorname{tr}\Lambda A_2^* A_1$. Thus in the special case where $B_1 = B_2 = (1-\varepsilon)^{1/2} B_0$ theorem 2.3 allows to treat the problem of the boundedness of S with respect to the quadratic form $\operatorname{tr}\Lambda A^*A$, while for the special case where $B_1(A_1, A_2) = B_2(A_1, A_2) = \operatorname{tr} A_2^* A_1$, theorem 2.3 gives the quantum analogue of the Nehari theorem. (For details see [6]). Finally, let us remark that a variant of theorems 2.2 and 2.3 can be given where the forms B_j are represented by formulae of the Martineau-Aizenberg type, instead of traces with respect to nuclear operators. In fact, the Bargman transform β defined by $\beta f(z) = F(z) = 2^{1/4} \int f(x) e^{2\pi x z - \pi x^2 - (\pi/2) z^2} dx$, maps isometrically $L^2(\mathbb{R})$ onto the Fock space $\mathcal{F} = \{\text{analytic functions } F: \mathbb{C} \to \mathbb{C} : \|F\|^2 = \int |F(z)|^2 e^{-\pi|z|^2} dz < \infty\}$. Under the unitary isomorphism β the space $\mathcal{L}^2(L^2(\mathbb{R}))$ is transformed into $\mathcal{L}^2(\mathcal{F})$, and the operators $\phi(x,y)$ are transformed into the operators

$$\beta(x,y) F(z) = e^{-(\pi/2)(x^2+y^2) - \pi z(x-iy)} F(z + (x+iy)). \tag{5}$$

More precisely, each operator $A \in \mathcal{L}^2(L^2(\mathbb{R}))$ is given by a kernel $K \in L^2(\mathbb{R})$, $A\varphi(\xi) = \int\int K(\xi, \eta) \varphi(\eta) d\eta$, and the corresponding operator $\tilde{A}$ in $\mathcal{L}^2(\mathcal{F})$ is given by

$$\hat{A} F(z) = \int \tilde{K}(z, \bar{w}) F(w) e^{-\pi|w|^2} dw, \tag{6}$$

where $\tilde{K}$ is the 2-dimensional Bargman transform of K. Thus $\tilde{A}$ can be identified with the analytic function $\tilde{K}(z,w)$ in $\mathbb{C}^2$, or with a function in $A(D), D$ a ball in $\mathbb{C}^2$. To each Toeplitz form $\hat{B}: V \times V \to \mathbb{C}$, $V = \mathcal{L}^2(L^2(\mathbb{R}))$, corresponds a Toeplitz form B in $\mathcal{L}^2(\mathcal{F}) \times \mathcal{L}^2(\mathcal{F})$ or in $A(D) \times A(D)$. If B is continuous in the $A(D)$-topology then by 2.1 it follows that $\tilde{B}(\tilde{A}_1, \tilde{A}_2) = \tilde{L}(\tilde{A}_3)$, where $\tilde{A}_3 = (A_2^* A_1)^{\tilde{}}$ and $\tilde{L}$ is a continuous linear form in $A(D)$. By the Martineau-Aizenberg formula, mentioned in the introduction, $\tilde{L}$ is given by

$$\tilde{L}(F) = \int_{\partial D_m} F(z) \varphi(\sigma(\phi_m)) \omega(z, \operatorname{grad} \phi_m(z)).$$

In this formula φ is the Fantappié indicatrice $\tilde{L}_z[(1-\langle z, w\rangle)^{-2}], z \in D, w \in \tilde{D}, \tilde{D} = \{w : \langle w, z\rangle \neq 1, \forall z \in D\}$, $D_m = \{z : \phi_m(z) < 0\}$, $\sigma(\phi) = (\sigma_1(\phi), \sigma_2(\phi))$, $\sigma_j(\phi) = \phi'_{z_j} \langle \operatorname{grad} \phi, z\rangle^{-1}$, and ϕ_m depends only on D and $\tilde{L}$.

As already mentioned, if A_j is given by the kernel K_j, $j = 1, 2$, then A_3 is given by the kernel $K_3(\xi, \zeta) = \int K_1(\xi, \eta) \overline{K_2(\eta, \zeta)} d\eta$, and $\tilde{A}_3$ is obtained from A_3 through (6).

The formula (6) is a consequences of the fact that $\mathcal{F}$ is the Hilbert space associated with the sesquianalytic kernel $K(z,w) = e^{\pi z \bar{w}}$ (see Remark 2 in section 1). Thus the above considerations of this section apply also to the Bergman space $B = (A(D), \mu)$, where $\mu \gg$ the Lebesgue measure, which are also reproducing kernel spaces. The detailes will be given elswhere.

REMARK. The considerations of this section apply also to the Heisenberg group (see [6]).

REFERENCES

[1] L. Aizenberg and A. Iujakov, *Integral Representations and Residuals*, Novo Sibirsk (1979) (Russian).
[2] P. Alegria, On the Adamjan-Arov-Krein and Arocena parametrizations: a constructive and generalized version, *Acta Cient. Venezolana*, 39 (1988), 107-116.
[3] R. Arocena, Generalized Toeplitz kernels and dilations of intertwining operators, *Integral Eq. and Operator Theory*, 6 (1983), 759-778.
[4] F. Beatrous and J. Burbea, Positive-definiteness and its applications to interpolation problems for holomorphic functions, *Trans. Amer. Math. Soc.*, 284 (1984), 247-270.
[5] R. Bruzual, Local Semigroups of contractions and some applications to Fourier representation theorems, *Integral Eq. and Operator Theory*, 10 (1987), 780-801.
[6] M. Cotlar and C. Sadosky, a) *Two parameter Lifting Theorems and Double Hilbert Transforms in Commutative Settings, MSRI 03821-88*, Berkeley, C.A. b) A lifting theorem, *J. Func. Anal.*, 67 (1986), 345-359.
[7] L. Fantappié, *Funzionali Analitici*, Mem. Accad. d. L. Roma (1930), 453-683.
[8] S. Janson, J. Peetre and R. Rochenberg, Hankel forms on the Fock space, *Revista Mat. Iberoamericana, 3* (1987), 61-138.
[9] B. Fritzsche and B. Kirstein, *On Generalized Nehari Problem*, Karl-Marx-Universität, preprint.
[10] N. Kerzman, «Singular Integrals in Complex Analysis», *Proc. of Symposia Pure Math.* AMS. XXXV, Part 2, 3-41.
[11] N. Morán, Parametrization of Generalized Toeplitz Kernels, to appear, in *J. of Math. and App.*
[12] M. Domínguez, Weighted Inequalities of Hilbert Transform, to appear, *Studia Mat.*
[13] B.Sz. Nagy and C. Foias, *Analyse harmonique des opérateurs*, Paris 1967.
[14] R. Rochberg, Toeplitz and Hankel Operators, Wavetets, NWO sequences, ..., Operator Algebras and Applications, AMS SRI, Durham NH. 1988, 1-27.
[15] F. Pellegrino, *La Théorie des fonctionelles analytiques*, in the book by P. Levy, *Problèmes concrets d'Analyse*, F. Paris 1951, 357-447.
[16] F. du Cloux, Representations temperées des groupes de Lie nilpotents, *J. Funct. Anal.* (in press).
[17] G.B. Folland, *Harmonic Analysis in Phase Space*, to appear.

Some New Global Results in Twistor Geometry

PAOLO DE BARTOLOMEIS Istituto di Matematica applicata «G. Sansone», Via di S. Marta 3, Firenze

1. INTRODUCTION

A basic problem in complex geometry is the following: when a $2n$-dimensional oriented differentiable manifold M can be endowed with a complex structure, i.e. one can find an atlas $\mathcal{U} = (U_i, \phi_i)_{i \in I}$ in the differentiable structure of M, in such a way the changes of coordinates are given by holomorphic functions?

Of course, this question is essentially:

what is a complex manifold?

Consider the simplest case: $n = 1$, i.e. M is an oriented surface.

It is well known that every oriented surface admits complex structures: let us recall some of the steps of the proof:

a) first of all one introduces a Riemannian metric g on M and realizes that the existence of local holomorphic coordinates is equivalent to the existence of the so called *isothermal parameters* i.e. coordinates $x\,y$ such that

$$g = \lambda^2 (dx \otimes dx + dy \otimes dy)$$

b) the existence of isothermal parameters is, of course, a purely local question and it is reduced to the solution of a Beltrami's equation of the form

$$\frac{\partial z}{\partial \overline{w}} = \mu(w) \frac{\partial z}{\partial w}$$

c) finally, the complex structure depends only on the conformal class of g and, moreover, there is a one-to-one correspondence between conformal classes of Riemannian metrics on M and complex structures.

If M is a complex manifold, then its tangent bundle TM can be endowed with an almost complex structure J i.e. a section of $\text{End}(TM)$ such that $J^2 = -I$.

It is easy to realize this is a much weaker requirement: it is equivalent to ask for a collections of local frames which are $GL(n, \mathbb{C})$-related instead to be simply $GL^+(2n, \mathbb{R})$-related; in a more precise form, it is equivalent to ask that the principal $GL^+(2n, \mathbb{R})$-bundle of linear oriented frames on M, $P(M, GL^+(2n, \mathbb{R}))$, has a reduction to $GL(n, \mathbb{C})$, or, as it is well known, the bundle

$$Q(M) = P(M, GL^+(2n, \mathbb{R}))/GL(n, \mathbb{C})$$

admits global sections.

Again, it is a good idea to introduce a Riemannian metric g on M: we consider the $SO(2n)$-bundle of oriented orthonormal frames on M, $P(M, SO(2n))$; the bundle

$$Z(M) = P(M, SO(2n))/U(n)$$

is a bundle deformation retraction of $Q(M)$ and so $Q(M)$ admits global sections if and only if $Z(M)$ admits global sections.

It is clear that the global sections of $Z(M)$ correspond to those almost complex structures on M which are compatible with the metric g and induce the given orientation.

Note that, if $\dim_{\mathbb{R}} M = 2$, then $Z(M) = M$.

Therefore, to find an almost complex structure on M is equivalent to find a global section of $Z(M)$: this is a well understood algebraic-topological problem; for istance, if $n = 2$, there are topological obstructions to the existence of almost complex structures.

in particular, if M is a compact 4-manifold, we have the following:

THEOREM 1.1. (W.T. Wu [15]) *Given $h \in H^2(M, \mathbb{Z})$, there exists an almost complex structure J with first Chern class $c_1(T^{1,0} M) = h$ if and only if:*

a) $\quad h \equiv w_2(M) \pmod{2}$

b) $\quad h^2 = 3\tau(M) + 2\chi(M)$

where w_2 is the second Stiefel-Whitney class, τ is the signature and χ is the Euler class. Moreover, equivalence classes, via homotopy, of almost complex structures are in one-to-one correspondence with such h's.

It is easy to check that condition b) of Theorem 1.1 is not satisfied e.g. for $M = S^4$.

There are examples of 4-dimensional compact oriented manifolds that carry almost complex structures but no complex structures (cf. [1]): e.g.

$$S^1 \times S^3 \# S^1 \times S^3 \# P^2(\mathbb{C})$$

For $n > 2$ no such examples have been found; in particular, it is not known if S^6, which carries almost complex structures, has a complex structure.

Of course, there is another natural problem: when an almost complex structure J is integrable, i.e. it is induced by a complex structure?

This is completely solved by the following

Twistor Geometry

THEOREM 1.2. (Newlander-Nirenberg [11], Kohn [8]) *J is integrable iff and only if it has no torsion.*

The torsion of J is defined by the Nijenhuis tensor:

$$N(J)(X,Y) = \frac{1}{2}([JX,JY] - [X,Y] - J[JX,Y] - J[X,JY])$$

Remember finally that, in an almost complex manifold (M,J), we have:

$$d \wedge^{p,q} \subset \wedge^{p-1,q+2} \oplus \wedge^{p,q+1} \oplus \wedge^{p+1,q} \oplus \wedge^{p+2,q-1}$$

and, by projection, we can define:

$$\partial : \wedge^{p,q} \to \wedge^{p+1,q} \qquad \bar\partial : \wedge^{p,q} \to \wedge^{p,q+1};$$

it is easy to check that:

$$N(J) \equiv 0 \Leftrightarrow d = \partial + \bar\partial \Leftrightarrow \bar\partial^2 = 0$$

2. $Z(M)$ AND TWISTOR GEOMETRY

Let us recap: let (M, g) be a $2n$-dimensional oriented Riemannian manifold and let $P(M, SO(2n))$ be the principal $SO(2n)$-bundle of oriented orthonormal frames over M.

We consider $Z(M) = P(M, SO(2n))/U(n)$ with projection

$$q : P(M, SO(2n)) \to Z(M);$$

$Z(M)$ is called the *Twistor space* of M: it is a bundle over M with standard fibre $Z(n) = SO(2n)/U(n)$ and structure group $SO(2n)$.

Let $r : Z(M) \to M$ be the bundle projection; if $x \in M$, then:

$$Z_x = r^{-1}(x) = \{P \in SO(T_xM, g(x)) | P = -{}^tP \text{ giving the fixed orientation}\}$$

i.e. $P \in Z(M)$ represents a complex structure on $T_{r(P)}M$, compatible with the metric and the orientation.

We have just seen that (local) sections of $Z(M)$ correspond to (local) almost complex structures on M compatible with the metric and the orientation; actually, we can say much more: we can endow $Z(M)$ itself with an almost complex structure in such a way that *integrable* almost complex structures correspond to «holomorphic» sections.

Let us perform the standard construction: the Levi-Civita connection allows us to define, for every $P \in Z(M)$ a decomposition into horizontal and vertical components

$$T_PZ(M) = H_P \oplus F_P$$

where, if $x = r(P)$

$$F_P = T_PZ_x = \{X \in so(T_xM, g(x)) | PX = -XP\}$$

Therefore, we can define, on $Z(M)$, J in the following way: if $X \in T_PZ(M)$ and $X = X_h + X_v$, with $X_h \in H_P$, $X_v \in F_P$, we set

$$J[P](X) = r_*^{-1} \circ P \circ r_*(X_h) + P \circ X_v$$

where $r_* : H_P \to T_{r(P)}M$ is the isomorphism induced by the projection.

Now we have the following:

THEOREM 2.1. ([5], [12])
1) J depends only on the conformal class of g
2) J is integrable if and only if
   a) for $n = 2$: (M, g) is anti self-dual
   b) for $n > 2$: (M, g) is conformally flat
3) Let $U \subset M$ be an open set and let $\sigma : U \to Z(M)$ be a local section over U : then the following facts are equivalent:
   a) σ, as almost complex structure on U, is integrable
   b) $\sigma(U)$ is an almost complex (local) submanifold of $Z(M)$
   c) $\sigma(U)$ is a complex (local) submanifold of $Z(M)$

The previous theorem part 3) provides a nice proof of the following fact:

PROPOSITION 2.2. *There are no complex structures on S^6 compatible with the standard metric.*

In fact:
a) it is easy to check that $Z(S^{2n}) = Z(n+1) = SO(2n+2)/U(n+1)$ with the standard structure: in particular $Z(S^6) = Q_6$, the non singular hyperquadric in $P^7(\mathbb{C})$
b) if $\sigma : S^6 \to Z(S^6)$ were a section representing a complex structure on S^6 compatible with the standard metric, then $\sigma(S^6) \approx S^6$ would be a complex, and hence Kählerian, submanifold of Q_6 : absurde (cf. also [10]).

If $X, Y \in T_P Z(M)$ and $X = X_h + X_v$, $Y = Y_h + Y_v$ with $X_h, Y_h \in H_P$, $X_v, Y_v \in F_P$ we set
$$G[P](X, Y) = g[r(P)](r_*(X_h), r_*(Y_h)) + g[r(P)](X_v, Y_v)$$
Among others properties of G, we recall the folllowing:

PROPOSITION 2.3. [5] *G is a Kählerian if and only if*

$$\begin{array}{ll} a) & \text{for } n = 2 : \quad R_{|\Lambda_+^2} = I \\ b) & \text{for } n > 2 : \quad R_{|\Lambda^2} = I \end{array}$$

where $R : \Lambda^2(M) \to \Lambda^2(M)$ is the curvature operator and, for $n = 2$, $\Lambda^2 = \Lambda_+^2 \oplus \Lambda_-^2$ is the spectral decomposition with respect to the $*$ Hodge operator.

Theorem 2.1 is also an example of what is Twistor Geometry:

to establish a sort of dictionary between conformal properties of $(M, [g])$ and complex properties of $(Z(M), J)$.

Another word in this dictionary is the following:

THEOREM 2.4. ([3], [4]) *Let (M, g) be a compact oriented Riemannian manifold of dimension $2n > 4$; then the following facts are equivalent:*

1) $\pi_1(M) = 0$ and (M, g) is conformally flat
2) $Z(M)$ is Kählerian

Of course 1) is a complicate way to say (M, g) is conformally equivalent to the standard sphere $(S^{2n}, \text{std.})$ (cf. [9]).

Twistor Geometry

Actually, we can say more; let (M,g) be a m-dimensional $(m > 2)$, oriented, conformally flat Riemannian manifold: therefore, there exists a conformal immersion

$$\Phi : (\hat{M}, g) \rightarrow (S^m, \text{std.})$$

with $\hat{M} \rightarrow M$ covering such that

$$M = \hat{M} / \Gamma$$

for $\Gamma \subset O(1, m+1) / \pm I$, the Moebius group of S^m (cf. [13]); if $m = 2n$ at the twistor level we have;

$$Z(M) = Z(\hat{M}) / \hat{\Gamma}$$

where $\hat{\Gamma}$ is a subgroup of $\text{Aut}(Z(M)) = SO(2n+2, \mathbb{C}) / \pm I$.

By proposition 2.3, $Z(M)$ is Kählerian and so

> any integrable twistor space, for $n > 2$, arises as a quotient of a Kählerian twistor space by the free, properly discontinuous action of a group of holomorphic transformations;

Theorem 2.4 says that there are no non trivial Kählerian quotients; Theorem 2.4 provides also new examples of non Kähler manifolds; in fact, if M and N are conformally flat, then $M \# N$ is conformally flat and $Z(M \# N)$ is not Kählerian; as buildings blocks we can quote:

    a) $S^1 \times E$ with $R_E = \lambda I$
    b) $E \times F$ with $R_F = -\lambda I$
    c) $[P^2(\mathbb{R}) \# \ldots \# P^2(\mathbb{R})]^\approx$,

where $\approx$ denotes the orientation covering.

Three open problems

a) It is possible to find *open* non trivial Kählerian quotients of the form

$$Z(M) = Z(\hat{M}) / \hat{\Gamma} \quad ?$$

b) A fundamental invariant of a compact, conformally flat manifolds is the sign of scalar curvature; how to translate it in *twistorian*?

c) Which are the compact, conformally flat, orientable Riemannian manifolds (M,g) such that $Z(\hat{M})$ admits a *complete* Kähler metric?
Or, more in general, which are the domains $\Omega \subset \mathbb{R}^{2n}$ such that $Z(\Omega)$ admits a *complete* Kähler metric?
(e.g. one can easily prove that $Z(B^{2n})$ does not admit any complete Kähler metric)

3. GAUGE-INVARIANT CONSTRUCTIONS AND NON REALISTIC PHYSICS (?)

There is another circle of ideas one can imagine behind the Twistor Geometry: the so called generalized Kaluza-Klein theories, i.e., roughly speaking, how to encode the physics of the universe M into the geometry of a principal bundle $P(M,G) \to M$, or, more in general, of a bundle of the form

$$P(M,G)/H \to M;$$

as a very vague example, we can quote the following fact:

Let $P(M,G)$ be a principal bundle over the Riemannian manifold (M,g) and let ω be a connection 1-form on $P(M,G)$:

a *geodesic* γ on $P(M,G)$ with respect to the bundle metric

$$h = \pi^*(g) + \omega \otimes \omega$$

projects over a curve $\hat{\gamma}$ in M representing the
path of a particle moving under the influence of a force
described by the *curvature* of the connection.

This is a long distance program; here, let us perform a very small step: discuss a gauge invariant construction of almost complex structures on $Z(M)$.

First of all, we observe that any connection ω on $P(M,SO(2n))$ gives rise to an almost complex structure J_ω over $Z(M)$.

It is easy to check that this construction is not gauge invariant, in the sense that, if we take an element f of the group of automorphism of $P(M,SO(2n))$, $G(P(M,SO(2n)))$, then: $(Z(M), J_\omega)$ and $(Z(M), J_{f^*(\omega)})$ are not biholomorphically equivalent.

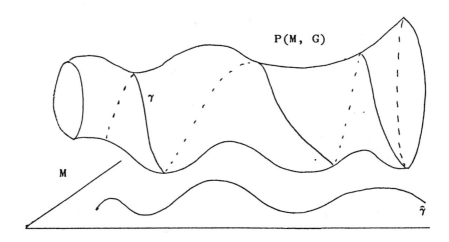

Twistor Geometry

Let us generalize our previous construction; we need some preliminaries:

a) If (M, g) is any m-dimensional oriented Riemannian manifold, it is well known that $P(M) = P(M, SO(m))$ is parallelizable i. e., as vector bundles

$$TP(M) \approx P(M) \times so(m+1)$$

where, of course,

$$so(m+1) \approx \mathbb{R}^m \oplus so(m) \quad \text{via} \quad \begin{bmatrix} K & \xi \\ -{}^t\xi & 0 \end{bmatrix} \mapsto (\xi, K)$$

Now, $SO(m) \subset SO(m+1)$ acts on the right both on $TP(M)$ and $P(M) \times so(m+1)$, the first being the action induced from $P(M)$, the second defined as: $R_a(u, X) = (ua, a^{-1}Xa)$; let $\chi : TP(M) \to P(M) \times so(m+1)$ be a bundle isomorphism such that:

i) χ is $SO(m)$-equivariant
ii) for every $X \in so(m)$, $\chi(X^*(u)) = (u, X)$ (where X^* is the *fundamental vector field associated to* X (cf. [7]))

χ is called a *regular parallelization*.
Let $\mathcal{G}$ be the set of all regular parallelizations.

b) Consider on $SO(m) \times so(m)$ the group structure given by the following multiplication

$$(a, \alpha) \cdot (b, \beta) = (ab, b^{-1}\alpha b + \beta)$$

therefore $SO(m) \times so(m)$ acts on the right on $P(M) \times so(m+1)$ as:

$$R_{(a,\alpha)}(u, X) = (ua, a^{-1}Xa + \alpha).$$

c) assume $m = 2n$: therefore $U(n) \times u(n)$ is a subgroup of $SO(2n) \times so(2n)$; consider:

$$E(M) = P(M) \times so(2n+1) / U(n) \times u(n)$$

with projection $\epsilon : P(M) \times so(2n+1) \to E(M)$. We have

$$E(M) = P(M) \times \mathbb{R}^{2n}/U(n) \oplus P(M) \times so(2n)/U(n) \times u(n) = E_h(M) \oplus E_v(M)$$

now:

$$E_h(M) = r^{-1}(TM)$$

$$E_v(M) = P(M) \times s(n) / U(n) = \{u[X, J_n]u^{-1} | u \in P(M), X \in so(2n)\}$$

where

$$s(n) = so(2n) / u(n) = \{X \in so(2n) | XJ_n = -J_n X\}$$

Therefore $E(M)$ has an intrisic structure of vector bundle over $Z(M)$ of rank $n^2 + n$; moreover:

$$E_h(M) = P(M) \times \mathbb{C}^n / U(n) \qquad E_v(M) = P(M) \times \Lambda^2 \mathbb{C}^n / U(n)$$

and so $E(M)$ is a complex vector bundle, therefore there exists a section J of $\text{End}(E(M))$ such that $J^2 = -I$; we have also:

$$J\left[u, \begin{bmatrix} K & \xi \\ -{}^t\xi & 0 \end{bmatrix}\right] = \left[u, \begin{bmatrix} \frac{1}{2}[J_n, K] & J_n\xi \\ -{}^t(J_n\xi) & 0 \end{bmatrix}\right]$$

d) A gauge-invariant construction of almost complex structures on $Z(M)$ is given in the following

PROPOSITION 3.1. *Let $\chi \in \mathcal{G}$; we have the diagram:*

$$\begin{array}{ccc} TP(M) & \xrightarrow{\chi} & P(M) \times so(2n+1) \\ \downarrow q_* & & \downarrow \epsilon \\ TZ(M) & \xrightarrow{\hat{\chi}} & E(M) \end{array}$$

therefore

1) $\hat{\chi}: TZ(M) \to E(M)$ *given as* $\hat{\chi}(q_*(X)) = \epsilon\chi(X)$ *is a well defined bundle isomorphism*
2) $J_\chi = \hat{\chi}^{-1} \circ J \circ \hat{\chi}$ *defines an almost complex structure on* $Z(M)$
3) *let* $f \in G(P(M))$ *and let* $\hat{f}: Z(M) \to Z(M)$ *be the induced map; define* $f^\#(\chi) = \chi \circ f_*^{-1}$; *then* $f^\#(\chi) \in \mathcal{G}$ *and* $J_{f^\#(\chi)} = \hat{f}_* \circ J_\chi \circ \hat{f}_*^{-1}$ *i.e.*

$$\hat{f}: (Z(M), J_\chi) \to (Z(M), J_{f^\#(\chi)})$$

is a biholomorphic equivalence and so the almost complex structure J_χ is gauge invariant.

Of course, elements in $\mathcal{G}$ have a lot to do with connections; in fact: if $\chi \in \mathcal{G}$ then

$$\chi = (\alpha, \omega)$$

with ω connection 1-form on $P(M)$ and α $\mathbb{R}^m$-valued tensorial 1-form on $P(M)$; moreover, if θ is the canonical $\mathbb{R}^m$-valued tensorial 1-form on $P(M)$ given by

$$\theta[u](X) = u^{-1}(\pi_*(X)),$$

for any connection 1-form ω on $P(M)$, we have:

$$\chi(\omega) = (\theta, \omega) \in \mathcal{G}$$

It is easy to check that:

$$J_{\chi(\omega)} = J_\omega$$

i.e. the construction through regular parallelizations generalizes the construction through connections.

Finally, it is clear that there are two others $G(P(M))$-actions on $\mathcal{G}$: for $f \in G(P(M))$ and $\chi = (\alpha, \omega) \in \mathcal{G}$, define:

$$f^0(\xi) = (\alpha, \omega \circ f_*^{-1}) \quad \text{and} \quad f_0(\chi) = (\alpha \circ f_*^{-1}, \omega)$$

evidently:

$$f_0 \circ f^0 = f^0 \circ f_0 = f^\#$$

and so:

$$\chi(\omega) = f_0^{-1} \circ f^\#(\chi(f^*(\omega)))$$

We like to think at $f^\#$ as a *supersimmetry* and at f_0 and f_0 as a *spontaneous simmetry breaking*.

At the end, we like to remember shortly a remarkable, unfortunately not very well known mathematician who can be considered as one of the Italian forerunners of relations between Geometry and Physics: Enea Bortolotti (Roma, Sept. 28, 1896, Firenze, June 22, 1942).

He was Associated Professor at University of Bologna from the a.y. 1923-24 to the a.y. 1928-29.

His main contributions can be found on differential geometric description of General Relativity and even from a quick look to his papers it clearly appears that he was one of the very few Italian mathematicians of his time really aquainted with the great mathematical revolution of the thirties.

It is easy to understand how his untimely death caused a serious delay on the development of our Geometry.

REFERENCES

[1] W. Barth C. Peters and A. Van de Ven, Compact Complex Surfaces, *Erg. d. Math.* 4, Springer, Berlin (1984).
[2] D. Bleecker, Gauge Theory and Variational Principles, *Global Analysis Pure and Applied*, Addison-Wesley, London (1981).
[3] F. Campana, preprint.
[4] P. de Bartolomeis, L. Migliorini and A. Nannicini, *Espaces de Twisteurs Kähleriens*, C.R.A.S. Paris, 307 s. I, 259-261 (1988).
[5] P. de Bartolomeis and A. Nannicini, *Handbook of Twistor Geometry*, to appear.
[6] N. Hitchin, Kählerian Twistor Spaces, *Proc. London Math. Soc.* (3) 43 133-150 (1981).
[7] S. Kobayashi and K. Nomizu, *Fondations of Differential Geometry*, Tracts in Math. 15, Interscience, New York (1963).
[8] J.J. Kohn, Harmonic Integrals on strongly pseudo-convex manifolds I, *Ann. of Math.* (2) 78 206-213 (1963).
[9] N. Kuiper, On Conformally flat spaces in the large, *Ann. of Math.* 50 916-924 (1949).
[10] C. Le Brun, Orthogonal Complex Structures on S^6, *Proc. AMS* 101 136-138 (1987).
[11] A. Newlander and L. Nirenberg, Complex Analytic Coordinates in Almost Complex Manifolds, *Ann. of Math.* (2) 65 391-404 (1957).
[12] S. Salamon, Harmonic and Holomorphic Maps, *Geometry Seminar «Luigi Bianchi»* II, 1984, *Lecture Notes in Math.* 1164, Springer, Berlin (1985).
[13] R. Schoen and S.T. Yau, Conformally flat Manifolds, Kleinian Groups and Scalar Curvature, *Inv. Math.* 92 47-71 (1988).
[14] M. Slupinski, Espaces de Twisteurs Kähleriens en dimension $4k$, $k > 1$, *J. London Math. Soc.* (2) 33 535-542 (1986).
[15] Wu Wen Tsun, *Sur les Espaces Fibrés et les Variétés Feuilletés*, Act. Sci. Ind. 1183, Hermann, Paris (1952).

Enriques Surfaces: Old and New

IGOR V. DOLGACHEV, Department of Mathematics, University of Michigan, Ann Arbor, Michigan, 48109, U S A

> "I ventotto anni trascorsi colà [a Bologna] furono forse i più lieti e fecondi della sua vita. In quella dotta città, dove lo spazio ristretto rende facili e frequenti i contatti tra i professori delle varie facoltà, egli trovò l'ambiente più favorevole per lo scambio delle idee e l'incremento della sua cultura…"
> G. Castelnuovo, "Commemorazione di F. Enriques", Atti Accad. Naz. Lincei, 1947.

Introduction.

In many aspects this talk repeats my recent talk at a conference in Cortona [Do3]. As is appropriate for the occasion, more emphasis is placed on the history of Enriques surfaces. It is easy to guess that Enriques surfaces were introduced first by Federigo Enriques. We begin with a brief story of how he was led to the discovery of such surfaces.

It is known that a rational algebraic curve X is characterized by the condition that its genus is equal to zero. This was first proven by A. Clebsch [CG]. Following Riemann the genus was defined as the maximal number of linearly independent differential 1-forms of the first kind (i.e. regular everywhere). At that time a standard model of a curve was a plane one, i.e. X was defined by an equation $F(x,y) = 0$, or, in homogeneous coordinates, by a homogeneous equation $F_n(x,y,z) = 0$, where F_n is an irreducible homogeneous polynomial of degree n. A differential form of the first kind can be written in the form $\Phi(x,y)dy/F'_x(x,y)$, where $\Phi(x,y) = 0$ is the affine equation of a curve of degree n-3 which passes through singular points (including those infinitely near) of multiplicity m > 1 with order m-1. Each such a curve was called an adjoint curve to X. Since passing through a point with order k imposes k(k-1)/2 conditions, the <u>expected</u> value of the genus is given by the formula:

$$g = (n-1)(n-2)/2 - \Sigma m_p(m_p-1)/2,$$

where the sum is taken over all singular points p∈ X including those infinitely near, and m_p denotes the corresponding multiplicity. It was proven by Max Noether [**No1**] that this is the right number, that is, there are no excessive adjoint curves. Also it was shown that this definition coincides with the topological definition of the genus given by B. Riemann. In [**Cl**] Clebsch generalizes this definition to the case of surfaces which were considered as the loci $F_n(x,y,z,w) = 0$ in the projective 3-dimensional space. The genus ("Flachengeschlecht") was defined as the maximal number of linearly independent double integrals of the first kind, or, geometrically, as the maximal number of linearly independent adjoints. Similarily to the case of curves, an adjoint surface is a surface of degree n-4 passing with order m-1 through any singular curve of X of multiplicity m, and passing with order r-2 through any singular point of multiplicity r of X. In [**No1**] Noether proved that the genus is a birational invariant. Again there is a formula as above (called a <u>postulation formula</u>) which gives the expected number of such adjoints. For example, if the singular locus of X consists of an irreducible curve of degree d and genus ρ with t double points and τ triple points (that can always be achieved by projecting a nonsingular model of a surface from $\mathbb{P}^5$), the formula looks like:

$$g = \binom{n-1}{3} - (n-4)\,d + 2t + \tau + \rho - 1.$$

However, it is not true anymore that this number is always equal to the number of linearly independent adjoints. For example, if X is a ruled surface, the number g is negative. For this reason, the first number was called the <u>numerical genus</u> and denoted by by p_n and the number of adjoints was called the <u>geometric genus</u> and denoted by p_g. The first example of a surface X with $p_n = p_g$, and $p_n \geq 0$ was constructed by Castelnuovo in 1891 [**Ca1**]. In modern terms, the difference $p_g - p_n$ is equal to q, the irregularity of the surface, and the numerical genus p_n is equal to p_a-1, where $p_a = \chi(X, \mathcal{O}_X)$ is the arithmetic genus of X. For every rational surface X, both p_n and p_g are equal to zero, hence a natural question arose as to whether this characterizes a rational surface. In Spring of 1894, Castelnuovo started his investigation of this question by using his idea of termination of adjoints. To prove the theorem he needed to verify that if the linear system of curves |2C| is contained in the linear system |2C'|, where C' is an adjoint curve, then |C| is contained in |C'|. At the same time, Enriques showed that the bicanonical linear system |2K| = |2C'-2C| is invariant with respect to birational transformations,

Enriques Surfaces

hence its dimension P, called the bigenus, is a birational invariant of a surface [En1]. Thus the problem led to the question of whether there exists a non-rational surface with $p_n = p_g = 0$, $P \neq 0$. Castelnuovo asked Enriques for help with this question, and, in July of the same year, Enriques suggested that he consider a surface of degree 6 passing doubly through the edges of the coordinate tetrahedron. The paper of Castelnuovo, containing his famous rationality criterion (X is rational if and only if $p_n = p_g = P = 0$) appeared in 1896 [Ca2]. In it, he presented the examples of Enriques. These surfaces can be given by homogeneous equations:

$$F_2(x_2x_3x_4, x_1x_3x_4, x_1x_2x_4, x_1x_2x_3) + x_1x_2x_3x_4 G_2(x_1, x_2, x_3, x_4) = 0,$$

where F_2 and G_2 are homogeneous polynomials of degree 2. It is easy to see that these equations represent a 10-dimensional family (up to a linear transformation of variables). For every $m \neq 1$, the m-canonical linear system $|mK|$ of this surface has dimension 0. Castelnuovo gave another example, this time his own, of a non-rational surface with $p_n = p_g = 0$. In his example $P_m = \dim |mK|$ is unbounded. It turned out later, in course of the classification of surfaces, that Enriques's example is essentially the only one for non-rational surfaces with the properties that $p_n = p_g = 0$ and dim $|mK|$ is bounded (in modern terms, of Kodaira dimension 0). The Castelnuovo example belongs to another class of surfaces, for which dim $|mK|$ is unbounded but grows at most linearly (of Kodaira dimension 1). These surfaces always contain an elliptic pencil. Their theory, founded by Enriques himself, was completed in the works of K. Kodaira and I. Shafarevich in the sixties of this century. It allows one to classify all such surfaces (cf. [Do1], [CD3]). Only much later, in 1932, L. Campedelli and L. Godeaux have found independently different examples of surfaces with $p_n = p_g = 0$ for which dim $|mK|$ grows quadratically (of Kodaira dimension 2). The classification of such surfaces is very far from being completed (cf. [Do1], [BPV]).

1. Models. Most of the classical results about Enriques surfaces, which from now on mean non-rational surfaces with $p_n = p_g = 0$ and of Kodaira dimension 0, were obtained by Enriques himself [En1],[En2], and later summarized in his book [En6]. In [En2] he starts with proving that every Enriques surface contains a pencil of elliptic curves $|F|$ without base points (in modern terminology, an elliptic fibration). In fact, by a very ingenious method, he proves that every curve on an Enriques surface S is linearly equivalent to a positive linear combination of elliptic or smooth rational curves. Every elliptic curve C either moves in a pencil, or taken doubly moves in such a pencil. There are exactly two double elliptic curves 2F and 2F' (which may degenerate) in every elliptic pencil. Then Enriques proves that for every elliptic pencil $|2F_1|$ one can find either i) an elliptic pencil $|2F_2|$ with $F_1 \cdot F_2 = 1$, or ii) a smooth rational curve

R with $F_1 \cdot R = 1$. In case i), the linear system $|2F_1+F_2|$ has two base points and maps S birationally onto a double plane $z^2 = F_8(x,y)$ with the branch curve of degree 8 composed of two lines and a sextic which has a node at the point of intersection of the two lines and has two tacnodes whose tangents are the two lines. In case ii) the linear system $|3F_1+R|$ does the same but the branch curve degenerates into the union of two lines and a sextic with 3 tacnodes; one of them is situated at the intersection point of the two lines, and another is infinitely near to it. We will refer to these double planes as an Enriques double plane (resp. an Enriques degenerate double plane). It is difficult to understand on what grounds Enriques asserted that a "general" Enriques surface has a representation as a non-degenerate double plane. The only explanation is that he was able to count the number of parameters for isomorphism classes of degenerate double planes (which is equal to 9) and compare it with the number of parameters of non-degenerate planes (which is equal to 10). Note that the formula of Noether [No2] for the number of moduli of an algebraic surface which existed at the time of writing [En2] could not be applied (one of its assumptions was the condition $p_g \geq 3$). A more general formula for the number of moduli was obtained by Enriques himself a little later [En5]. It agrees with counting of constants and gives the answer 10 for the moduli of Enriques surfaces. Next Enriques shows that, essentially, all Enriques surfaces arise from his earlier construction of a sextic surface passing doubly through the edges of the coordinate tetrahedron (an Enriques sextic). First he checks that the number of constants for the sextic construction is also 10. To show directly that a general Enriques surface is birationally isomorphic to an Enriques sextic, he, starting from the non-degenerate double plane construction, finds an elliptic curve F_1 such that $F_1 \cdot F_2 = F_1 \cdot F_3 = 1$ and shows that the linear system $|F_1+F_2+F_3|$ maps S birationally onto an Enriques sextic. The proof of the existence of such a curve is one of the most obscure points in Enriques's memoir. This argument has been reconstructed much later in the theses of Michael Artin at M.I.T [Ar] and Boris Averbukh in Moscow [AS], [Av]. It had been observed by Castelnuovo that the sextic construction may degenerate but also gives an Enriques surface. One of these degenerations corresponds to the case when the edges of the tetrahedron pass through one point. The sextic acquires a quadruple point and can be represented by an equation of the form:

$$F_2(l_2l_3l_4,l_1l_3l_4,l_1l_2l_4,l_1l_2l_3)+l_1l_2l_3l_4G_2(l_1,l_2,l_3,l_4) = 0,$$

where the l_i's are linearly dependent homogeneous linear forms in projective coordinates x_1, x_2, x_3 and x_4. This case occurs when the adjoint linear system $|F_1'+F_2+F_3|$ does not give a birational map. In the second degeneration two opposite edges become infinitely near. In this case the sextic acquires a triple line and a double line infinitely near to it. Its equation can be given in the form:

$$a(l_1l_2l_3)^2+bl_1l_2^2l_3^2l_4+cl_1^4l_2^2+dl_1^4l_3^2+l_1^2l_2l_3G_2(l_1,l_2,l_3,l_4) = 0,$$

where l_1, l_2, l_3 and l_4 are some linear forms in projective coordinates, a, b, c, d are constants. This degeneration occurs when one starts with a degenerate Enriques double plane. The corresponding linear system is $|2F_1+F_2+R|$, where F_1 and F_2 are elliptic curves, R is a smooth rational curve, and $F_1 \cdot F_2 = 1$, $F_1 \cdot R = 1$, $F_2 \cdot R = 0$. This result of Enriques was also reconstructed in the theses of Artin and Averbukh. Only recently, it was shown that the last case can always be avoided (see [**CD3**], Corollary 4.9.2).

In 1901 Gino Fano (born on the same day as Enriques) observed that the congruence of lines in $\mathbb{P}^3$ which are contained in a subpencil of quadrics in a fixed general web of quadrics is an Enriques surface [**Fa1**]. These congruences, called nowadays Reye congruences (cf. [**Co1**]), were introduced much earlier by Darboux [**Da**] and then were studied by Reye [**Re**]. Via its Plücker embedding this congruence is isomorphic to a surface of degree 10 in $\mathbb{P}^5$ lying on a quadric. This construction gives only a 9-parameter family of Enriques surfaces, and later Fano proved that a general Enriques surface can be embedded into $\mathbb{P}^5$ as a surface of degree 10 not necessarily lying on a quadric [**Fa2**]. This surface contains 20 plane cubic curves $F_{\pm i}$ with $F_i \cdot F_j = 1$ if $i+j \neq 0$ and $F_i \cdot F_{-j} = 0$, $i = 1,\ldots,10$. The linear system $|F_1+F_2+F_3|$ maps the surface birationally onto an Enriques sextic. Conversely, starting from an Enriques sextic surface, Fano finds a quintic elliptic curve C lying on it which together with two elliptic curves F_1 and F_2 coming from the edges of the tetrahedron form the linear system $|F_1+F_2+F_3|$ which maps the surface onto a surface of degree 10 in $\mathbb{P}^5$. Again this is true only generically. Only recently, it was proven that for every Enriques surface S one can find a birational morphism from S onto a surface of degree 10 in $\mathbb{P}^5$ with at most double rational points as singularities [**Co1**], [**CD3**]. In particular, every Enriques surface which does not contain smooth rational curves is isomorphic to a surface of degree 10 in $\mathbb{P}^5$.

The double plane construction of Enriques can be modified by considering a morphism of degree 2 onto some other rational surface. This can be obtained by applying some birational transformations to the plane. Thus one may consider any Enriques surface as a double cover of a quadric (the degenerate case corresponds to a singular quadric) (cf. [**Ho**]), or as a double cover of a 4-nodal Del Pezzo surface of degree 4 (in the degenerate case the Del Pezzo surface acquires 2 nodes and one double rational point of type A_3) (cf. [**BP**], [**Do2**]). There is a generalization of this construction to the case of arbitrary characteristic [**CD3**]. A systematic study of linear systems on Enriques surfaces was undertaken by F. Cossec in [**Co1**], [**Co2**].

In particular, all classic models of Enriques and Fano were reconsidered from the uniform approach by using the arithmetic of the quadratic form defined on the Picard group of a surface. It was shown that <u>any</u> Enriques surface admits a non-degenerate double plane construction. This was certainly unknown to Enriques.

2. Enriques surfaces and K3-surfaces. A non-singular quartic surface in $\mathbb{P}^3$ has all the genera p_n, p_g, $P_m = 1$ equal to 1. The class of surfaces with such invariants (later christened by André Weil as surfaces of type K3) has a long history. The ubiquitous Kummer surfaces belong to this class. Other examples of such surfaces can be obtained by taking a complete intersection of three quadrics in $\mathbb{P}^5$. In his joint work with F. Severi on hyperelliptic surfaces (surfaces which are covered by a complex torus) Enriques discovered an example of such a surface which admits a fixed-point-free involution with quotient isomorphic to an Enriques surface. In **[En4]** he proved that a general Enriques surface S can be obtained as a quotient of a K3-surface by a fixed-point-free involution. To show this he considers the double cover of $\mathbb{P}^3$ branched along the coordinate tetrahedron and shows that the induced cover of the sextic surface is a K3-surface. Nowadays this result is almost trivial (true if the ground field is of characteristic different from 2) and, by using standard arguments, follows from the fact that the canonical class K on an Enriques surface is a non-trivial 2-torsion element in the Picard group. In this way the study of Enriques surfaces over a field of characteristic different from 2 is reduced to the study of K3-surfaces with fixed-point-free involutions. This relationship plays very important role in the recent work on Enriques surfaces (see **[BPV]**). Many examples of Enriques surfaces obtained as quotients of K3-surfaces were given in works of L. Godeaux (cf. **[Go]**). Only recently it was shown by F. Cossec **[Co2]** and A. Verra **[Ve]** that every Enriques surface is birationally isomorphic to the quotient of the intersection of three quadrics in $\mathbb{P}^5$ by a fixed-point-free involution.

3. Automorphisms. Enriques was the first who observed that a general Enriques surface S admits infinitely many birational automorphisms. His argument goes as follows. Take the linear system $|F_1+F_2+F_3|$ which maps the surface onto a non-degenerate Enriques sextic. Then the pencils $|2F_2|$ and $|2F_3|$ cut out on every curve F from $|2F_1|$ two linear pencils of degree 4. Their difference defines on each F a divisor class ε of degree 0. For every point $x \in F$ the sum $x+\varepsilon$ is linearly equivalent to a unique point y on F, and the correspondence $x \to y$ defines a birational automorphism of S. He shows that ε is of infinite order, hence the obtained automorphism of S is of infinite order. Of course, this argument requires a justification,

Enriques Surfaces

certainly lacking in his paper. It is not clear why all the translation automorphisms of each member of a pencil are induced by an automorphism of the whole surface. The necessary technical tool for the needed justification is the notion of the Jacobian variety and its principal homogeneous spaces for an elliptic curve over a functional field. This was used in the later work of I. Shafarevich [AS]. In fact, Enriques applied this argument to a larger class of algebraic surfaces, namely, surfaces admitting an elliptic fibration. His study of such surfaces is, in my opinion, is one of the best proofs of Enriques' genius. Careful reading of his work on these surfaces reveals that, a long time before the fundamental works on elliptic surfaces by K. Kodaira and I. Shafarevich, he was aware of such concepts as principal homogeneous spaces, logarithmic transformation and Jacobian surfaces.

It is not clear whether Enriques and his contemporaries understood that the group of birational isomorphisms of a minimal non-ruled surface is equal to the group of biregular automorphisms. After all, the theory of minimal models was clarified only much later in the works of Oskar Zariski.

At the very end of [En 2] Enriques asks whether there exists an Enriques surface with only finitely many automorphisms. This question was answered in 1910 by Fano. In [Fa 2] he constructs a special web of quadrics in $\mathbb{P}^3$ whose Reye congruence is an Enriques surface with finitely many automorphisms. I have discovered this result of Fano only very recently, while visiting the University of Torino and looking through Fano's archive. Just prior to this I have published a paper [Do2] with an example of an Enriques surface with finitely many automorphisms, wrongly believing that this was the first example of such a kind. The example of Fano is different, his argument for the proof of finiteness needs justification, and his claim about the structure of the corresponding finite group is wrong. Very recently, V. Nikulin [Ni2] classified all Enriques surfaces with finitely many automorphisms from the point of view of the structure of the Picard group of its K3-cover. An explicit geometric classification (by means of equations) was given later by S. Kondō [Ko]. There are 7 classes of such surfaces.

The first explicit computation of the (infinite) group of automorphisms of a general Enriques surface was given by W. Barth and C. Peters in [BP]. Independently, this result, in a much more general context, was obtained by Nikulin [Ni1]. By acting on the Picard group the automorphism group is represented in the orthogonal group of a certain even unimodular lattice of signature (1,9). Its image is equal to its level 2 congruence subgroup factored by the subgroup generated by the transformation $x \to -x$. A purely geometric calculation of the automorphism group of a generic Reye congruence was given in [CD2] (though, again the result can be deduced from Nikulin's results based on the transcendental techniques of the period spaces of K3-surfaces).

4. Rational and elliptic curves on Enriques surfaces. Another of "belle questioni concernenti la superficie F_6" asked in **[En2]** was the following: What is the distribution of linear systems of a given order, especially isolated elliptic curves, on an Enriques sextic? Today we understand this question as a more general question about the Picard group of an Enriques surface. It was studied in detail in many recent works (**[Co2],[Co3],[CD1],[CD3]**). Let N_S be the Picard lattice of an Enriques surface S, i.e. the group of divisors modulo numerical equivalence equipped with the intersection form. It can be shown (rather easily, if the ground field is the field of complex numbers, and much less easily in the general case) that this lattice is an even unimodular lattice of signature (1,9), and as such is isomorphic to the lattice E_{10}, the direct sum of the standard hyperbolic plane U and the root lattice E_8 of a simple Lie algebra of type E_8 (taken with the opposite sign). Though the lattices of all Enriques surfaces S are isomorphic as abstract lattices, their semigroups of positive or ample divisor classes depend essentially on S. By Riemann-Roch, every divisor class D with $D^2 \geq 0$ is effective, however, if $D^2 < 0$, D is never effective unless the surface contains smooth rational curves R with $R^2 = -2$ (in which case the surface is said to be <u>nodal</u>). For instance, every surface represented by a degenerate Enriques sextic or a degenerate double plane is nodal. The converse was proven only recently by Cossec (**[Co3],[CD3]**). Applying reflection transformations $x \to x+(x \cdot e)e$ of N_S, where e is the class of a smooth rational curve (a nodal curve), allows one to transform every divisor D to a divisor D' which is numerically effective (i.e. $D' \cdot C \geq 0$ for every curve C'). If D is such a divisor, $D^2 = 0$, and D is not divisible by an integer, then D is an isolated curve of arithmetic genus 1. For such a curve $|2D|$ is an elliptic pencil (if the characteristic is different from 2). If D is numerically effective (nef) and $D^2 > 0$ then the property of the map f_D given by the linear system $|D|$ depends very much on the number $\Phi(D)$ which is equal to the minimum of the intersection numbers $D \cdot F$, where F is any elliptic curve. Thus $\Phi(D) = 1$ if and only if f_D has base points, $\Phi(D) = 2$ if and only if f_D is a birational map onto a non-normal surface, or $D^2 = 4, 6$ or 8, and f_D is a 4 to 1 map onto $\mathbb{P}^2$ if $D^2 = 4$, and f_D is a 2 to 1 map onto a rational surface if $D^2 = 6$ or 8. Finally, if $\Phi(D) \geq 3$, the map f_D is a birational map onto a surface with only double rational points as singularities. In particular, if F is unnodal, f_D is an isomorphism onto its image. The study of the arithmetic of the lattice E_{10} shows that $\Phi(D) \geq 3$ could happen only if $D^2 \geq 10$. The Fano model of a surface of degree 10 in $\mathbb{P}^5$ corresponds to such a divisor with

$D^2 = 10$. The sextic model corresponds to a divisor D with $D^2 = 6$ and $\Phi(D) = 2$, and the double plane model corresponds to D with $D^2 = 4$ and $\Phi(D) = 1$.

Returning to the original question of Enriques, we see that it asks (the part concerning isolated elliptic curves) about the description of all elliptic curves F whose classes are primitive vectors in N_S and $F \cdot D = n$ for a fixed class of a nef divisor D with $\Phi(D) = 2$, $D^2 = 6$, and a fixed positive integer n (degree of F). Let $E(D)_n$ be the set of such F's. Each such a set is finite. If S is general, then the set of n for which $E(D)_n \neq \emptyset$ is infinite. We do not know how to compute the function $f(n) = \#(E(D))_n$. We know that $f(2) = 6$ (the elements of $E(D)_2$ are mapped to the six edges of the tetrahedron), $f(3) = 98$, and $f(4) = 756$. The complete answer to this question can be obtained by further study of the arithmetic of the lattice E_{10}.

Another interesting question is the distribution of nodal curves on a nodal Enriques surface. This question was never considered in the classic literature and has been studied only recently. We refer to [Do3] for a survey of some results and problems concerning this question. Assuming the automorphism group of a surface is known, one can compute the number of its orbits in the set of linear systems of given genus. For example, the number of the orbits in the set of isolated elliptic curves on a general Enriques surface is equal to 1054, the number of linear systems |D| with $D^2 = 6$ and $\Phi(D) = 2$ on a general Enriques surface (i.e. the number of different representations as an Enriques sextic) is equal to 10,792,910 (see [BP]). Similar computations can be made for a general nodal surface ([CD2]). We could be proud to show these kinds of results to Enriques.

5. Moduli. Nothing was known in the old days about the moduli space of Enriques surfaces except that the number of moduli is equal to 10. We refer to a survey of the modern development in [Do3].

6. Enriques surfaces over fields of positive characteristic. All classic work on surfaces silently assumed that the ground field was the field of complex numbers. However, transcendental methods were never popular among Italian algebraic geometers. Thus many of the proofs could be extended almost word by word to any characteristic. In the seventies the Enriques classification of surfaces was extended to the case of arbitrary characteristic in the works of E. Bombieri and D. Mumford [M], [BM1], [BM2]. The case of Enriques surfaces was given special attention. It turned out in their work that only the case of characteristic 2 presents special difficulties. Since then much work was done on the study of this case. We refer for complete references to [CD3,CD4].

References.

[Art]　　　M. Artin, *On Enriques surfaces*, Harvard thesis. 1960.

[AS]　　　*Algebraic surfaces*　(ed. by I. Shafarevich), Proc. Steklov Math. Inst., v.75, 1964. Engl.transl.: AMS, Providence.R.I. 1967].

[Av]　　　B. Averbukh, Kummer and Enriques surfaces of special type. *Izv. Akad. Nauk SSSR*, Ser. Mat. 29 (1965), 1095-1118.

[BP]　　　W. Barth, C. Peters, Automorphisms of Enriques surfaces, *Inv. Math.* 73 (1983), 383-411.

[BPV]　　W. Barth, C. Peters, A. van de Ven, *Complex algebraic surfaces*, Springer-Verlag. 1984.

[BM1]　　E. Bombieri, D. Mumford, Enriques classification in char. p, II, in "Complex Analysis and Algebraic Geometry", *Iwanami-Shoten*, Tokyo. 1977, 23-42.

[BM2]　　E. Bombieri, D. Mumford, Enriques classification in char. p, III, *Invent. Math.* 35 (1976), 197-232.

[Ca1]　　　G. Castelnuovo, Osservazione intorno alla geometria sopra una superficie, *Rendiconti del R. Istituto Lombardo*, ser. 2, vol. 24 (1891).

[Ca2]　　　G. Castelnuovo, Sulle superficie di genere zero, *Mem. delle Soc. Ital. delle Scienze*, ser. III, 10 (1895).

[CG]　　　A. Clebsch, P. Gordan, *Theorie der abelschen Functionen*, Leipzig, 1866.

[Cl]　　　A. Clebsch, Sur les surfaces algébriques, *C. R. Acad. Sci. Paris*, 67 (1868).

[Co1]　　　F. Cossec, On the Picard group of Enriques surfaces, *Math. Ann.* 271 (1985),577-600.

[Co2]　　　F. Cossec, Projective models of Enriques surfaces, *Math. Ann.* 265 (1983), 283-334.

[Co3]　　　F. Cossec, Reye congruences, *Trans. Am. Math. Soc.* 280 (1983),737-751.

[CD1]　　F. Cossec, I. Dolgachev, Rational curves on Enriques surfaces, *Math. Ann.* 272 (1985), 369-384.

[CD2]　　F. Cossec, I. Dolgachev, On automorphisms of nodal Enriques surfaces, *Bull. Amer. Math. Soc.* 12 (1985), 247-249.

[CD3]　　F. Cossec, I. Dolgachev, *Enriques surfaces* I, Birkhauser Boston. 1989.

[CD4]　　F. Cossec, I. Dolgachev, *Enriques surfaces* II (in preparation).

[Da] G. Darboux, *Bull. Soc. Math. de France*, 1 (1870), p.438.

[Do1] I. Dolgachev, Algebraic surfaces with $p_g = q = 0$, in "Algebraic surfaces", *Proc.* CIME *Summer School in Cortona*, Liguori, Napoli, 1981, pp. 97-216.

[Do2] I. Dolgachev, Automorphisms of Enriques surfaces, *Invent. Math.* 76 (1984), 63-177.

[Do3] I. Dolgachev, Enriques surfaces: what is left, in *"Proc. Conference on algebraic surfaces"*. Cortona, 1988.

[En1] F. Enriques, Introduzione alla geometria sopra le superficie algebriche, *Mem.Soc. Ital. delle scienze*, ser. 3a, 10 (1896), 1-81.

[En2] F. Enriques, Sopra le superficie algebriche di bigenere uno, *Mem Soc. Ital. delle Scienze*, Ser. 3a, 14 (1906), 39-366.

[En3] F. Enriques, Sulle superficie algebriche che ammettono una serie discontinua di transformazioni birazionali, *Rend. Accad. Lincei*, s. V, t. XV (1906), 665-669.

[En4] F. Enriques, Un' osservazione relativa alle superficie di bigenere uno, *Rend. Acad. Scienze Inst. Bologna*, 12 (1908), 40-45.

[En5] F. Enriques, Sui moduli delle superficie algebriche, *Rendiconti Accad. Lincei,* ser. V, vol. XVII (1908).

[En6] F. Enriques, *Le superficie algebriche*, Zanichelli, Bologna. 1949.

[Fa1] G. Fano, Nuovo ricerche sulle congruenze di retta del 3 ordine, *Mem. Acad.Sci. Torino*, 50 (1901), 1-79.

[Fa2] G. Fano, Superficie algebriche di genere zero e bigenere uno e loro casi particulari, *Rend. Circ. Mat. Palermo*, 29 (1910), 98-118.

[Fa3] G. Fano, Superficie regolari di genere zero e bigenere uno, *Revista de Matematica y fisica teoretico. Univ. de Tucuman* Argentina, v. 4 (1944), 69-79.

[Go] L. Godeaux, *Les surfaces algébriques non rationnelles*, Paris. Hermann. 1934.

[Ho] E. Horikawa, On the periods of Enriques surfaces, I. *Math. Ann.* 234 (1978), 73-108; II, ibid. 235 (1978), 217-246.

[Ko] S. Kondō, Enriques surfaces with finite automorphism group, *Japan J. Math.* 12 (1986), 192-282.

[Mu] D. Mumford, Enriques' classification of surfaces in char. p, I, in "Global analysis", *Princeton. Univ. Press*, Princeton. 1969, pp. 325-339.

[Ni1]   V. Nikulin, On the quotient groups of the automorphism groups of hyperbolic forms modulo subgroups generated by 2-reflections, in "Current Problems of Mathematics", t. 18, *VINITI, Moscow,* 1981, pp. 3-114 [Engl. Transl.:*J. Soviet Math.* 22 (1983), 1401-1476].

[Ni2]   V. Nikulin, On a description of the automorphism groups of an Enriques surfaces, *Dokl. Akad. Nauk SSSR,* 277 (1984), 1324-1327 [Engl. Transl. *Soviet Math. Doklady* 30 (1984), 282-285].

[No1]   M. Noether, Zur Theorie des eindeutigen Entsprechens algebraischer Gebilde, *Math. Ann.* 2 (1869), 8 (1874).

[No2]   M. Noether, Anzahl der Moduln einer Klasse algebraischer Flächen, *Sitzungsberichte Akademie zu Berlin,* 1888.

[Re]   T. Reye, *Geometrie der Lage,* 3 ed., vol. 3, Leipzig, 1892.

[Ve]   A. Verra, The étale double covering of an Enriques' surface, *Rend. Sem. Mat. Univ. Polit. Torino,* 41 (1983), 131-166.

Some Connections Between Equations and Geometric Properties of Curves in $\mathbb{P}^3$

PHILIPPE ELLIA and MONICA IDÀ

Ph. ELLIA : Dépt. Math., Université de Nice, Parc Valrose, F-06034 Nice Cedex.
M. IDA' : Dip. Sci. Mat., Università di Trieste, Piazzale Europa, I-34100 Trieste.

Introduction.

In [E] Enriques started the study of equations defining a canonical curve of genus g in $\mathbb{P}^{g-1}$. Soon after this was taken up by Babbage [B] and Petri [P] . Today, thanks to the work of M.Green and others (see 5 for references) this subject is still alive; in particular Green's conjecture on canonical curves is a very stimulating problem.

The work on canonical curves shows a connection between the geometry of $C \subseteq \mathbb{P}^{g-1}$ and the minimal free resolution (m.f.r. for short) of C (that is to say, of its graded ideal).

Naturally we are led to ask which informations the m.f.r. of a curve in $\mathbb{P}^n$ encodes.

Moreover one could try to use m.f.r. for the existence and classification problems (see for ex. [PS], [gE], [I 1,2,3], [EH]).

In this paper we will review various connections between the m.f.r. and other geometrical properties of curves in $\mathbb{P}^3$.

Both the authors are indebted with A.Hirschowitz for the talks on this subject.

Notations.

We work over an algebraically closed field of characteristic zero. We shall use freely the following notations: $\mathcal{O}(k) := \mathcal{O}_{\mathbb{P}^3}(k)$, $n.\mathcal{O}(k) := \mathcal{O}(k)^{\oplus n}$, and $R := K[x_0,...,x_3]$.

A curve $X \subset \mathbb{P}^3$ is a closed subscheme of dimension one, equidimensional and locally Cohen-Macaulay.

This paper is the text of two lectures given by the authors on February 24, 1989, at the Università degli Studi di Bologna during the meeting :"Geometry and complex variables in Bologna" (in the occasion of the IX centenary of its founding).

A curve $X \subset \mathbb{P}^3$ is of <u>maximal rank</u> if $h^0(I_X(m)).h^1(I_X(m)) = 0$ for any m.

We denote by I(X) the saturated homogeneous ideal of X, i.e. $I(X) := \oplus H^0(I_X(m))$. We recall that I(X) has, as an R - graded module, a minimal free resolution:

(*) $0 \to \oplus^{r_3} R(-n_{3i}) \xrightarrow{f_2} \oplus^{r_2} R(-n_{2i}) \xrightarrow{f_1} \oplus^{r_1} R(-n_{1i}) \xrightarrow{f_0} I(X) \to 0$

where the f_i are given by matrices with entries in $m := (x_0,...,x_3)$. This resolution is essentially unique. Sheafifying we get an exact sequence:

(*)' $0 \to \oplus^{r_3} \mathcal{O}(-n_{3i}) \xrightarrow{f_2} \oplus^{r_2} \mathcal{O}(-n_{2i}) \xrightarrow{f_1} \oplus^{r_1} \mathcal{O}(-n_{1i}) \xrightarrow{f_0} I_X \to 0$

which we still call the minimal free resolution of I_X (m.f.r. of I_X).

Moreover $\oplus^{r_3} R(-n_{3i}) = 0$ iff X is aritmethically Cohen-Macaulay (a.C.M.).

Definition. The sequence $(n_{ji})_{1 \le j \le 3, 1 \le i \le r_j}$, is called the numerical mode of X. We set $n_j^+ := \max \{n_{ji}, 1 \le i \le r_j\}$, $n_j^- := \min \{n_{ji}, 1 \le i \le r_j\}$.

We recall briefly how to obtain such a m.f.r. in a constructive way. First of all define $s(X) := \min \{k \mid h^0(I_X(k)) \ne 0\}$. Now for $m \ge s$ consider the natural map:

$\sigma_X(m): H^0(I_X(m)) \otimes V \to H^0(I_X(m+1))$: $P \otimes H \to PH$.

A minimal system of generators of I(X) is obtained in the following way: in degree s one take a basis of $H^0(I_X(s))$, in degree s+1 one take a basis for a supplementar of $\text{Im}(\sigma_X(s))$ in $H^0(I_X(s+1))$, and so on; for $k > s$ one take a basis for a supplementar of $\text{Im}(\sigma_X(k-1))$ in $H^0(I_X(k))$. The process will stop for k big enough. For example this can be seen as follows: Let $0 \to \Omega(1) \to \mathcal{O}_{\mathbb{P}^3} \otimes V \to \mathcal{O}_{\mathbb{P}^3}(1) \to 0$ be the Euler sequence. Twisting by $I_X(m)$ and taking cohomology we get:

$$0 \to H^0(\Omega(m+1) \otimes I_X) \to H^0(I_X(m)) \otimes V \xrightarrow{\sigma_X(m)} H^0(I_X(m+1)) \to H^1(\Omega(m+1) \otimes I_X) \to$$

Since for $m >> 0$ $h^1(\Omega(m+1) \otimes I_X) = 0$ by Serre's theorem, $\sigma_X(m)$ is surjective and the process ends (i.e. I(X) is finitely generated).

Anyway it turns out that $\dim \text{coker } \sigma_X(m) =$ number of generators in degree m+1 in any minimal system of generators of I(X).

Hence the knowledge of $\dim \text{coker } \sigma_X(m)$ determines (up to automorphisms) the first piece of the m.f. r.: $\oplus^{r_1} \mathcal{O}(-n_{1i}) \xrightarrow{f_0} I_X \to 0$, $f_0 = (F_{11},...,F_{1r_1})$.

Now let $E := \ker(f_0)$. Since X has codimension two and is locally Cohen-Macaulay, E is locally free. By the way, observe that $H^1_*(E) = 0$, hence E is a very particular vector bundle (a so called "syzygies bundle"). The R-graded module $H^0_*(E) := \oplus H^\circ(E(m))$ is the module of relations among the generators F_{1j} of I(X). We can repeat the previous process and consider the maps: $\sigma_E(m): H^0(E(m)) \otimes V \to H^0(E(m+1))$ to get a minimal system of generators of

$H^0_*(E): \oplus^{r_2} \mathcal{O}(-n_{2i}) \xrightarrow{n} E \to 0$.

Curves in $\mathbf{P}^3$

Let F:=ker(n). Since E is locally free, F too is locally free. By definition of n, $H^1_*(F) = 0$. Moreover since $H^1_*(E) = 0$ we also have $H^2_*(F) = 0$. It follows (Horrocks criterion) that $F \cong \oplus^{r_3} \mathcal{O}(-n_{3i})$, and we have the whole m.f.r.:

$$0 \to \oplus^{r_3} \mathcal{O}(-n_{3i}) \xrightarrow{f_2} \oplus^{r_2} \mathcal{O}(-n_{2i}) \xrightarrow{f_1} \oplus^{r_1} \mathcal{O}(-n_{1i}) \xrightarrow{f_0} I_X \to 0$$

Also observe that X is a.C.M. iff E splits into a direct sum of line bunbles. This allows the following remarks:

1) if E is a non free syzygies bundle then $\text{rk}(E) \geq 3$ (because for a rank two bundle $H^1_*(E) = 0$ $\Leftrightarrow H^2_*(E) = 0$)

2) a curve $X \subset \mathbf{P}^3$ is a quasi-complete intersection (q.c.i.) if there exist three surfaces, F_i, $1 \leq i \leq 3$, yielding a surjection: $\oplus^3 \mathcal{O}(-n_i) \xrightarrow{f_0} I_X \to 0$, $f_1 = (F_1, F_2, F_3)$; if moreover these surfaces generate I(X) then X is an almost complete intersection (a.c.i.). A result of Fiorentini - Lascu [FL] states that a curve in $\mathbf{P}^n$ which is a q.c.i. is an a.c.i. iff it is a.C.M..

If n=3 this can be seen as follows: by assumption we have $\oplus^3 \mathcal{O}(-n_i) \xrightarrow{f_0} I_X \to 0$, $f_1 = (F_1, F_2, F_3)$. Let $E = \ker(f_1)$; E is locally free of rank two. If X is an a.c.i. the F_i's generate I(X). This implies $H^1_*(E) = 0$, hence E splits and X is a.C.M.. If X is a.C.M. then $H^1_*(I_X) = 0$ implies $H^2_*(E) = 0$, hence E splits and $H^1_*(E) = 0$ which shows that the F_i's generate I(X).

The minimal free resolution and its connections with other properties of the curve.

1 Cohomology and minimal free resolution.

The numerical mode (n_{ij}) of X determines $\{h^0(I_X(k))\}_{k \in \mathbf{Z}}$, in fact:

$$h^0(I_X(k)) = \sum_{j=1}^{3} (-1)^{j+1} . h^0(\oplus_{i=1}^{r_j} \mathcal{O}(-n_{ji}+k)).$$ It follows that the numerical mode determines the Hilbert function and, a fortiori, the Hilbert polynomial of X. In particular it determines the degree and the arithmetic genus of X. On the other hand the numerical mode doesn't determine the whole cohomology of X, only the difference $h^1(I_X(k)) - h^2(I_X(k))$ can be computed from the n_{ji}'s. The complete cohomology of I_X can be deduced from that of the syzygies bundle E and from the numerical mode. Indeed cutting the resolution into two pieces:

$$0 \to E \to \oplus \mathcal{O}(-n_{1i}) \to I_X \to 0 \qquad (E)$$

$$0 \to \oplus \mathcal{O}(-n_{3i}) \xrightarrow{f_2} \oplus \mathcal{O}(-n_{2i}) \to E \to 0 \qquad (W)$$

we see that: $h^1(I_X(k)) = h^2(E(k))$, $k \in \mathbf{Z}$ while $h^2(I_X(k))$ is given by $h^3(E(k))$ and the n_{1i}'s. From the west side (W) we see that $h^\circ(E(k))$ is given by the n_{3i}'s and the n_{2i}'s while $h^i(E(k))$ $2 \leq i \leq 3$ depends on the rank of the map $H^3(f_2(k))$. This shows that not only the numerical data but also the morphisms (in particular $H^3(f_2(k))$ or their duals) of the m.f.r. are of great importance.

2 Indices and minimal free resolution.

As usual we define $s(X) := \min \{m \mid h^0(I_X(m)) \neq 0\}$, $c(X) := \max\{m \mid h^1(I_X(m)) \neq 0\}$, ($c(X) = -\infty$ if X is a.C.M.), $e(X) := \max \{m \mid h^2(I_X(m)) \neq 0\}$; these are respectively the index of postulation, completeness, speciality. If no confusion can arise, we drop the X. These indices are connected with the m.f.r. by the following well known lemma:

2.1 Lemma. Let $X \subset \mathbb{P}^3$ be a non a.C.M. curve of numerical mode (n_{ji}) $1 \le j \le 3$, $1 \le i \le r_j$ and indices s,e,c. Then:

(1) $r_3 + r_1 = r_2 + 1$, $\sum^{r_3} n_{3i} + \sum^{r_1} n_{1i} = \sum^{r_2} n_{2i}$
(2) $n_1^- = s$, $n_1^- < n_2^- < n_3^-$
(3) $n_2^+ > n_1^+$
(4) $n_3^+ = c + 4$, moreover set $r_3^+ := \#\{ n_{3i} = n_3^+\}$, then $h^1(I_X(c)) = r_3^+$; $e + 4 \le n_2^+$
(5) If $n_3^+ \le n_2^+$ then $\oplus_{n_{2i} \ge n_3^+} \mathcal{O}(-n_{2i})$ is a direct summand of the syzygies bundle, $e + 4 = n_2^+$ and $e \ge c$.

Proof. (1) Just compute ranks and first Chern classes.
(2) Let $(F_{11},...,F_{1r_1})$ be a minimal system of generators of I(X). Assume $\deg(F_{11}) \le ... \le \deg(F_{1r_1})$. By definition $n_1^- = \deg(F_{11}) = s$. If $\sum F_{1i} \cdot P_i = 0$ is a relation then $P_i = 0$ or $\deg(P_i) \ge 1$ (minimality). It follows that $n_1^- < n_2^-$. Similarly considering a minimal system of generators of the module of relations we get $n_2^- < n_3^-$.
(3) We have the trivial relation $(-F_{1r_1},0,...,0,F_{11})$. Hence among the minimal relations there is one of the form $(P_1,...,P_{r_1})$ with $P_{r_1} \neq 0$, hence with $\deg(P_{r_1}) \ge 1$. It follows that $\deg P_{r_1} + \deg F_{1r_1} > \deg F_{1r_1} = n_1^+$, which shows $n_2^+ > n_1^+$.
(4) The inequality $c + 4 \le n_3^+$ follows immediately $(h^1(I_X(k)) = h^2(E(k))$, see 1). By twisting the west side (i.e. the sequence (W)) by $n_3^+ - 4$ we get :

$$H^3(r_3^+ \cdot \mathcal{O}(-4)) \xrightarrow{f_3'} H^3(\oplus_{n_{2i} \ge n_3^+} \mathcal{O}(-n_{2i} + n_3^+ - 4)).$$

The matrix of f_3' has for coefficients polynomials of negative degree hence by minimality they are zero. It follows that the map $H^3(f_3')$ is zero, hence $h^1(I_X(n_3^+ - 4)) = h^2(E(n_3^+ - 4)) = r_3^+$. If $h^2(I_X(e)) \neq 0$ the "east side" ((E) of 1) shows $h^3(E(e)) \neq 0$ and looking at (W) we get $H^3(\oplus \mathcal{O}(-n_{2i} + e)) \neq 0$, i.e. $e + 4 \le n_2^+$.
(5) Assume $n_{2i} \ge n_3^+$ if $1 \le i \le t$ and $n_{2i} < n_3^+$ if $t < i \le r_2$. Dualizing (W) we get:

$$0 \to E^{\vee} \to \oplus_{1 \le i \le t} \mathcal{O}(-n_{2i}) \oplus \oplus_{t < i \le r_2} \mathcal{O}(-n_{2i}) \xrightarrow{(f,g)} \oplus \mathcal{O}(-n_{3i}) \to 0$$

By minimality $f \equiv 0$. It follows that $\oplus_{1 \le i \le t} \mathcal{O}(-n_{2i})$ is direct summand in $E^{\vee}$. Now twisting (E) by $n_2^+ - 4$ and taking cohomology we see that: $h^3(E(n_2^+ - 4)) \neq 0$, $H^3(\oplus \mathcal{O}(-n_{1i} + n_2^+ - 4)) = 0$ (because $n_1^+ < n_2^+$ by (3)). It follows that $h^2(I_X(n_2^+ - 4)) \neq 0$, hence $e \ge n_2^+ - 4 \ge n_3^+ - 4 = c$. Finally by (4) we conclude that $e = n_2^+ - 4$.

Sometimes this lemma can be used to determine the m.f.r. of certain curves. Here is an example:

2.2 Curves on a smooth quadric.

Let X be a curve of bidegree (a, b), $a \geq b + 2$, on a smooth quadric Q. It is well known that $c = a - 2$, $e = b - 2$ (use $0 \to \mathcal{O}(-2) \to I_X \to \mathcal{O}_Q(-a, -b) \to 0$ (+)). From Castelnuovo-Mumford lemma's ([M] p.99) it follows that I(X) is generated in degrees $\leq a$. The exact sequence (+) shows that, besides Q, there are no generators in degree $< a$ and that there are $a - b + 1$ new generators in degree a. Hence the m.f.r. starts like: $0 \to E \to (a - b + 1).\mathcal{O}(-a) \oplus \oplus \mathcal{O}(-2) \to I_X \to 0$. From lemma 2.1: $n_3^+ = a + 2 > n_2^+ > n_1^+ = a$. Hence $n_2^+ = a + 1$. It is also clear that $n_{2i} \geq a + 1$, for all i; thus $n_{2i} = a + 1$, for all i. Since $a + 2 = n_3^+ \geq n_3^- > n_2^- = = a + 1$ it follows that $n_{3i} = a + 2$, for all i. Finally $\#\{n_{3i} = a + 2\} = h^1(I_X(c)) = a - b + 1$. Therefore the m.f.r. looks like: $0 \to (a-b+1).\mathcal{O}(-a-2) \to x.\mathcal{O}(-a-1) \to (a - b + 1).\mathcal{O}(-a) \oplus \oplus \mathcal{O}(-2) \to I_X \to 0$. Counting ranks we get $x = 2(a-b)$ and the final form of the m.f.r.:
$$0 \to (a-b+1).\mathcal{O}(-a-2) \to 2(a-b).\mathcal{O}(-a-1) \to (a - b + 1).\mathcal{O}(-a) \oplus \mathcal{O}(-2) \to I_X \to 0.$$
Finally if $b \geq a - 1$ then X is a.C.M.; more precisely if $b = a$, X is the complete intersection of Q with a surface of degree a; if $b = a - 1$, X is linked to a line by such a complete intersection and the m.f.r. is:
$$0 \to 2.\mathcal{O}(-a-1) \to 2.\mathcal{O}(-a) \oplus \mathcal{O}(-2) \to I_X \to 0.$$

3 Normal bundle and minimal free resolution.

One can get informations on the Hilbert scheme from the m.f.r. via the normal bundle or looking at the deformations of the m.f.r. An example of this latter point of view is provided by Ellingsrud's theorem ([gE]) asserting that Hilb$\mathbb{P}^3$ is smooth at [X] if X is an a.C.M. curve (note that in general $h^1(N_X) \neq 0$ for an a.C.M. curve). As for the first point of view, the m.f.r. can be a useful tool to compute the cohomology of the (twisted) normal bundle. There are two ways for relating N_X to the m.f.r. The first one is to use the isomorphism: $Ext^{i+1}(I_X, I_X(n)) = H^i(N_X(n))$, the second one is to apply $Hom(-, \mathcal{O}_X)$ to the m.f.r. bearing in mind that $Hom(I_X, \mathcal{O}_X) = N_X$. Hence one obtains:
$$H^0(N_X) = \ker (H^0(L_0 \otimes \mathcal{O}_X) \xrightarrow{f_1^* \otimes 1_X} H^0(L_1 \otimes \mathcal{O}_X)).$$ Here are some examples:

3.1 Linear a.C.M. curves.

For any $s \geq 1$ there are (smooth) a.C.M. curves with m.f.r.:
$0 \to s.\mathcal{O}(-s-1) \to (s+1).\mathcal{O}(-s) \to I_X \to 0$ (*). For any such curve $h^1(N_X(-2)) = 0$ (this follows easily using $Ext^2(I_X, I_X(-2)) = H^1(N_X(-2))$), in particular N_X is semi-stable (see [pE]). We have $\deg(X) = s(s+1)/2$. A consequence of the vanishing of $h^1(N_X(-2))$ is (see [Pe]): given $s(s+1)$ general points of $\mathbb{P}^3$ (or on a smooth quadric surface) there exists an a.C.M. curve of degree $s(s+1)/2$ with m.f.r. like (*) through these points (for applications see for ex. [Pa]).

3.2 Double lines

Let H(2,-a) denote the irreducible component of Hilb$\mathbb{P}^3$ parametrizing double lines of arithmetic genus -a. Then H(2,-a) is generically smooth of dimension 2a+5 (a ≥ 2), see [pE2] Prop.V.2. The proof goes as follows: first observe that it is enough to show $h^0(N_X) = 2a+5$ for some double line X with $p_a(X) = -a$. Then compute the m.f.r. of I_X; this is not too hard since it is easy to see that $I(X) = (x^2, y^2, xy, xF(z, t)-yG(z, t))$; F, G homogeneous of degree a without common zeroes. One finds:

$$0 \to \mathcal{O}(-a-3) \to 2.\mathcal{O}(-a-2) \oplus 2.\mathcal{O}(-3) \to \mathcal{O}(-a-1) \oplus 3.\mathcal{O}(-2) \to I_X \to 0;$$

the first map is given by (xF-yG, x^2, xy, $-y^2$) and the second by:

$$f_1 = \begin{pmatrix} x & y & 0 & 0 \\ -F & 0 & y & 0 \\ G & -F & -x & -y \\ 0 & G & 0 & -x \end{pmatrix}$$

Then take $F = z^a$, $G = t^a$; apply $Hom(-, \mathcal{O}_X)$ and compute! It is a little bit tricky but it works.

3.3 Arithmetically Buchsbaum curves of maximal rank.

Recall that a space curve, X, is arithmetically Buchsbaum (a.B.) if its Rao module, $M(X) = \oplus H^1(I_X(k))$, is annihilated by the maximal ideal m; i.e. M(X) has a trivial S-graded module structure. Starting from a resolution of the base field K (Koszul complex) we get a resolution of M(X) which in turn gives informations on the m.f.r. of X:

3.3.1 Lemma. Let X be an a.B.(non a.C.M.) space curve. Assume $h^1(I_X(k_i)) = n_i$, 1≤i≤r; $h^1(I_X(n)) = 0$ if $n \notin \{k_1,...,k_r\}$. Then the m.f.r. of X has the form:

$$0 \to \oplus_{i=1}^r n_i.\mathcal{O}(-k_i-4) \to (\oplus 4n_i.\mathcal{O}(-k_i-3)) \oplus (\oplus \mathcal{O}(-n_{2i})) \to \oplus \mathcal{O}(-n_{1i}) \xrightarrow{f} I_X \to 0$$

with $\ker(f) \cong (\oplus_{i=1}^r n_i.T\mathbb{P}(-k_i-4)) \oplus (\oplus \mathcal{O}(-n_{2i}))$.

Proof. The resolution of K is given by Koszul:
$$0 \to R(-4) \to 4R(-3) \to 6R(-2) \to 4R(-1) \to R \to K \to 0.$$
Sheafifying this yields:
$$0 \to \mathcal{O}(-4) \xrightarrow{g} 4.\mathcal{O}(-3) \to 6.\mathcal{O}(-2) \to 4.\mathcal{O}(-1) \xrightarrow{h} \mathcal{O} \to 0,$$
with ker (h) ≅ Ω, and coker (g) ≅ T$\mathbb{P}$(-4)).
By direct summand and shift we deduce the m.f.r. of $H^1_*(I_X)$:
$$0 \to \oplus_{i=1}^r n_i R(-k_i-4) \to \oplus_{i=1}^r 4n_i R(-k_i-3) \to ... \to \oplus_{i=1}^r R(-k_i) \to H^1_*(I_X) \to 0.$$
Then we conclude with [R] th.2.5.

3.3.2 Remark. For results on m.f.r. of a.B. curves see [A1, 2], [GM1, 2].

3.3.3 Proposition. Let $X \subset \mathbb{P}^3$ be an a.B. curve of maximal rank. If e ≤ s-2, then $h^1(N_X) = 0$.

Curves in $\mathbf{P}^3$

Proof. We have $n_3^+ = c+4 \leq s+3$ by 2.1 and maximal rank. Moreover, since $e \leq s-2$ and $c \leq s-1$, by Castelnuovo-Mumford lemma we get: $s+1 \geq n_{1i} \geq s$. By 2.1 it follows that the m.f.r. looks like:
$$0 \to r.\mathcal{O}(-s-3) \oplus h.\mathcal{O}(-s-2) \to t.\mathcal{O}(-s-2) \oplus z.\mathcal{O}(-s-1) \to y.\mathcal{O}(-s-1) \oplus x.\mathcal{O}(-s) \to I_X \to 0.$$
From the previous lemma the syzygies bundle is $E = r.T\mathbf{P}(-s-3) \oplus h.T\mathbf{P}(-s-2) \oplus \delta.\mathcal{O}(-s-2) \oplus \delta'.\mathcal{O}(-s-1)$; hence we have:

(*) $0 \to E \to F \to I_X \to 0$, with $F := y\mathcal{O}(-s-1) \oplus x\mathcal{O}(-s)$.

Applying Hom($-,I_X$) we get: $\mathrm{Ext}^1(E,I_X) \to \mathrm{Ext}^2(I_X,I_X) \to \mathrm{Ext}^2(F,I_X)$.

Now $\mathrm{Ext}^2(F,I_X) = y.H^2(I_X(s+1)) \oplus x.H^2(I_X(s)) = 0$. Since $\mathrm{Ext}^1(E,I_X) = H^1(E^\vee \otimes I_X)$, to show $\mathrm{Ext}^1(E,I_X) = 0$ it is enough to show: (a) $h^1(I_X \otimes \Omega(s+3)) = 0$; (b) $h^1(I_X \otimes \Omega(s+2)) = 0$; (c) $h^1(I_X(s+2)) = 0 = h^1(I_X(s+1))$. Clearly (c) follows from the maximal rank assumption. To prove (a) twist (*) by $\Omega(s+3)$: $0 \to E \otimes \Omega(s+3) \to F \otimes \Omega(s+3) \to I_X \otimes \Omega(s+3) \to 0$. It is easy to see that $h^1(F \otimes \Omega(s+3)) = 0$. Then $h^2(E \otimes \Omega(s+3)) = 0$ amounts essentially to $h^2(T\mathbf{P} \otimes \Omega) = h^2(T\mathbf{P} \otimes \Omega(1)) = 0$. This can be seen for instance by twisting Euler's sequence by Ω and inspecting cohomology. The proof of (b) is similar.

Remark. For further results on unobstructed a.B. curves, see [MR].

4 Number of moduli and minimal free resolution.

Let H be an irreducible component of the Hilbert scheme H(d, g). Let $p: H \dashrightarrow M_g$ denote the functorial rational map. The number of moduli of H is $\dim(p(H))$. Now let [X] be a point of H and consider the natural exact sequence: $0 \to TX \to T\mathbf{P}|_X \to N_X \to 0$. Taking cohomology we have $H^0(N_X) \to H^1(TX)$ which is nothing else than dp at [X]. Hence computing the number of moduli of H via differential methods envolves (especially if $H^1(N_X) = 0$) the cohomology of $T\mathbf{P}|_X$. Sometimes this cohomology can be read off the m.f.r. For the curves of 3.1, see [Pa].

5 Geometry and minimal free resolution: secant curves, stratifications,...

If X has a k-secant line then I(X) can't be generated by forms of degree $\leq k-1$. This simple but basic fact establishes a direct link between the m.f.r. and the geometry of X. More generally one may ask to what extent the geometry of X (existence of special linear series, of high secant curves; problems of completeness...) is reflected in the m.f.r. For canonical curves in $\mathbf{P}^{g-1}$, M.Green made a striking conjecture in this direction. We refer to [G], [GL], [S], [V], [PR] for further details on this subject. In $\mathbf{P}^3$ the situation seems a little bit different but something can be said.

First of all one would like to know what to expect in the generic case. A first guess on this topic has been the following conjecture of Hirschowitz (1985; see [I1] introduction):

Conjecture: If d is sufficiently big with respect to g then there exists a smooth curve of degree d, arithmetic genus g which is of maximal rank and minimally generated.

We recall that X is minimally generated if, for any m, the map $\sigma_X(m): H^0(I_X(m)) \otimes V \to H^0(I_X(m+1))$ is of maximal rank.

In [HH] (resp. [H]) maximal rank is proved for the generic union of d lines (resp. the generic rational curve of degree d).
In [I1, 4] minimal generation is proved for the generic union of d lines, $d \neq 4$, and for the generic rational curve of degree d, $d \neq 5$, $d \leq 100$. Notice that sometimes the whole m.f.r. can be deduced from these results. Indeed we have:

5.1: Lemma: Assume X is a maximal rank curve of degree d, aritmetic genus g, with $e < c$.
(a) The m.f.r. looks like: $0 \to a.\mathcal{O}(-s-3) \oplus t.\mathcal{O}(-s-2) \to b.\mathcal{O}(-s-2) \oplus z.\mathcal{O}(-s-1) \to$
$\to y.\mathcal{O}(-s-1) \oplus h.\mathcal{O}(-s) \to I_X \to 0$ where $h = (s+3)(s+2)(s+1)/3 - ds + g - 1$, $a = d(s-1) - g + 1 - (s+2)(s+1)s/3$, $z = 4h + y - (s+4)(s+3)(s+2)/3 + d(s+1) - g + 1$, $a + t + y + h = b + z + 1$.
(b) If $\sigma_X(s)$ is injective the m.f.r. looks like: $0 \to a.\mathcal{O}(-s-3) \to b.\mathcal{O}(-s-2) \to y.\mathcal{O}(-s-1) \oplus$
$\oplus h.\mathcal{O}(-s) \to I_X \to 0$ where h and a are as above, $y = (s+4)(s+3)(s+2)/3 - d(s+1) + g - 1 - 4h$, $a + y + h = b + 1$.
(c) If $c = s-2$ the m.f.r. is linear i.e.: $0 \to a.\mathcal{O}(-s-2) \to b.\mathcal{O}(-s-1) \to h.\mathcal{O}(-s) \to I_X \to 0$, and $a = d(s-2) - g + 1 - (s+1)s(s-1)/3$, $h = (s+3)(s+2)(s+1)/3 - ds + g - 1$, $a + h = b + 1$.

Proof: (a) since X is of maximal rank $n_3^+ \leq s+3$ (2.1); since $e < c$, $s+1 \geq n_{1i} \geq s$ (Castelnuovo-Mumford lemma) and $n_3^+ > n_2^+ > n_1^+$; moreover, $n_3^- > n_2^- > n_1^- = s$ (2.1), hence the m.f.r. is as wanted. Now notice that $z = h^0(E(s+1))$, where E is the syzygies bundle of X; from the east side of the m.f.r., we have $z = 4h + y - h^0(I_X(s+1))$. Finally since $e < s-1$, $h = h^0(I_X(s))$, $a = h^1(I_X(s-1))$ (2.1), and z are computable using Riemann-Roch.
(b) By (a) we have the general form of the m.f.r.. The injectivity of $\sigma_X(s)$ means that there are no relations of degree one between the generators hence $n_{2i} = s+2$, all i. This implies $n_{3i} = s+3$, all i. Finally $y = h^0(I_X(s+1)) - 4.h^0(I_X(s))$ and $a = h^1(I_X(s-1))$ (2.1).
(c) It follows immediately from 2.1 that the resolution is linear. Also it is clear that $a = h^1(I_X(s-2))$, $h = h^0(I_X(s))$, and again since $e < s-2$, these numbers are computable using Riemann-Roch.

Now define S:={d / there exists a union, X, of d skew lines of maximal rank with $\sigma_X(s)$ injective}, S':={d / there exists a union, Y, of d skew lines of maximal rank with $h^1(I_Y(s-1)) = 0$}. By easy computations it is not hard to see that S, S' are infinite sets. By 5.1 if d belongs to S$\cup$S', the m.f.r. of d general skew lines is known.
Similarly put S":={(d, g) / $g \geq 0$, $d \geq (3g+12)/4$ and there exists a smooth maximal rank curve, X, of degree d, genus g with general moduli and with $h^1(I_X(s-1)) = 0$}. From the maximal rank theorem [BE1,2] and easy computations, S" is infinite. For (d, g) in S" the m.f.r. of the general curve with general moduli is known.
For other results and a more precise conjecture about m.f.r. in the generic case, see [EH].

Now let's turn to "absolute" results i.e. results valid for <u>any</u> curve. At least for what concerns the first step of the m.f.r., we would like to be able to do the following:
Fix an irreducible component, H, of H(d,g); since $h^0(I_X(k))$ and $h^0(I_X \otimes \Omega(k))$ are semicontinuous functions on H, we can stratify H with respect to both functions and then study the intersection of these strata. This means that in each subset of H with fixed postulation, we look at all the possibilities for the number and degrees of generators in a minimal system. In [I2] one can find a detailed analysis in the case H = H(6, -5), the Hilbert scheme of unions of six skew lines. Unfortunately general results are very scarse and it seems difficult to formulate reasonable conjectures.

Curves in $\mathbf{P}^3$

Let us see what is known in general. First of all we have Castelnuovo-Mumford lemma ([M], p.99) which establishes a link between postulation and generation. We recall that in particular this lemma gives that if X is a curve and n is a positive integer such that $e(X) < n-2$ and $h^1(I_X(n-1)) = 0$ then: (i) $c(X) < n-1$ (hence, 2.1, $n_3^+ \le n+2$) and: (ii) $\sigma_X(k)$ is surjective for $k \ge n$ (i.e. $n_1^+ \le n$).

In the case of integral curves there are vanishing theorems for the cohomology of X. First we have the well known Castelnuovo's theorem [C] saying that $h^1(I_X(d-2)) = h^2(I_X(d-3)) = 0$ for a smooth connected space curve of degree d.

This has been improved by Gruson-Lazarsfeld-Peskine [GLP] who proved that for an integral curve, X, of degree d, $h^1(I_X(d-3)) \ne 0$ iff X is a smooth rational curve with a (d-1)-secant line. In [GLP] there is also a result for disconnected curves. For example, for any union, Y, of d skew lines, $h^1(I_Y(d-1)) = 0$.

Following [GLP], D'Almeida [D] proved that $h^1(I_X(d-4)) \ne 0$ iff X has a (d-2)-secant line (with a few exceptions of low degree).

Finally Gruson-Peskine [GP] proved that if X is smooth, connected then $h^1(I_X(n)) = 0$ for $n \ge d-e-3$.

So if we consider for example U in H(d,0), the open subset containing smooth curves in the Hilbert scheme of rational curves in $\mathbf{P}^3$, and set $S(i,k) = \{[X] \in U / h^0(I_X(k)) \ge [h^0(\mathcal{O}(k)) - (dk + 1)] + i\}$, we find $S(i,d-2) = \emptyset$ for $i > 0$, $S(1,d-3) = \{[X] \in U / X \text{ has a (d-1)-secant}\}$ and, for d sufficiently high, $S(1,d-4) = \{[X] \in U / X \text{ has a (d-2)-secant line}\}$.

Moreover, for $[X] \in U$, $\sigma_X(d-1)$ is surjective. If $[X] \in S(1,d-3)$ then $\sigma_X(d-2)$ is not surjective (there is a (d-1) secant) but $h^0(I_X \otimes \Omega(d-1))$ and hence the number of generators of degree d-1 is not a priori clear (although easy to compute). Things get more difficult at the level below (how many generators of degree d-2 for a curve with a d-2 secant ?).

The connection between postulation, generation and secants seems to be still tighter. For example, let Y be a union of d lines with s distinct d-secants. Then $s \le 2$ or Y lies on a smooth quadric. If Y is sufficiently general and $s \le 2$ then $h^1(I_Y(d-2)) = s$ and dim(coker $\sigma_Y(d-1)$) = s ([I2]) (this is in fact the reason why four skew lines are not minimally generated; the degree being so low the generic curve is allowed to have multisecants which influence the extremal $\sigma(k)$ which is also the critical $\sigma(k)$).

Now observe that s is exactly the number of superfluous conditions imposed by the d-secants to: (a) the sheaf $\mathcal{O}(d-2)$, (b) the sheaf $\Omega(d)$. Moreover, as we have seen, this number is also the difference between (a) the postulation at level d-2 and (b) the number of generators at level d of Y, and of the generic union of d lines.

Now consider the union X of d skew lines on a quadric Q. Then: (a) $h^1(I_X(d-2)) = d-1$, (b) dim(coker $\sigma_X(d-1)$) = d+1 (see 2.2). These numbers are again respectively the number of superfluous conditions imposed by the complex of d-secants (i.e. one family of lines in the quadric) to (a) $\mathcal{O}(d-2)$, (b) $\Omega(d)$.

So the first idea which comes up to mind is that for a union, X, of d skew lines, $h^0(I_X \otimes \Omega(k))$ is given by the integer $\mu(k) = h^0(\Omega(k)) - h^0(\Omega(k)|_X)$ (which is only function of k), plus the number of superfluous conditions imposed to $\Omega(k)$ by the complex of multisecant lines. Unfortunately this is not true. For example, if X is the general union of 15 1-secants and one 2-secant to a smooth conic, Z, then $\mu(9) = 24$ but $h^0(I_X \otimes \Omega(9)) = 26$; and 2 is exactly the number of superfluous conditions imposed by the 17-secant conic Z to $\Omega(9)$ (see [I3]).

Hence a good conjecture should take into account all the secant objects (curves and surfaces of secant curves).

Bibliography:

[A1] Amasaki, M.:"On the structure of arithmetically Buchsbaum curves in $\mathbb{P}^3$", *Publ. Res. Inst. Math. Sci.* 20, 793-837 (1984).

[A2] Amasaki, M.:"Integral arithmetically Buchsbaum curves in $\mathbb{P}^3$", preprint.

[B] Babbage, D.W.:"A note on the quadrics through a canonical curve", *J. London Math. Soc.* 14, 310-315 (1939).

[BE1] Ballico, E.-Ellia, Ph.: "The maximal rank conjecture for non special curves in $\mathbb{P}^3$" *Invent. Math.* 79, 541-555 (1985).

[BE2] Ballico,E-Ellia,Ph:"Beyond the maximal rank conjecture for curves in $\mathbb{P}^3$", in "*Space curves,* Proceedings Rocca di Papa, 1985" Lect. Notes in Math. 1266, 1-23 (1987) Springer-Verlag.

[C] Castelnuovo, G.:"Sui multipli di una serie lineare di gruppi di punti appartenente ad una curva algebrica", *Rend. Circ. Mat. Palermo* 7, 89-110 (1893).

[D] D'Almeida, J.:"Courbes de l'espace projectif: séries linéaires incomplètes et multisécantes", *J. Reine Angew. Math.* 370, 30-51 (1986).

[EH] Ellia, Ph-Hirschowitz, A:"Voie ouest", in preparation.

[pE] Ellia, Ph:"Exemples de courbes de $\mathbb{P}^3$ à fibré normal semi-stable, stable", *Math. Ann.* 264, 389-396 (1983).

[pE2] Ellia, Ph.:"Sur les lacunes d'Halphen", in "*Algebraic curves and projective geometry* ",Proc. Trento 1988, Lect. Notes in Math. 1389, 43-65 (1989) Springer-Verlag.

[gE] Ellingsrud, G.:"Sur le schéma de Hilbert des variétés de codimension deux dans $\mathbb{P}^e$ à cone de Cohen-Macaulay", *Ann. E.N.S.*, 4e série t.8, fasc. 4, 423-431 (1975).

[E] Enriques, F.:"Sulle curve canoniche di genere p dello spazio a p-1 dimensioni", *Rend. dell'Acc. delle scienze dell'Istituto di Bologna*, vol XXIII, 80-82 (1919).

[FL] Fiorentini, M-Lascu, A.:"On the homogeneous ideal of a quasi-complete intersection in the projective space", *Ann. Univ. Ferrara* XXIX, 211-219 (1983).

[GM1] Geramita, A.V-Migliore, J.:"On the ideal of an arithmetically Buchsbaum curve", *J. Pure & Applied Algebra*, 54, 215-247 (1988).

[GM2] Geramita, A.V-Migliore, J.:"Generators for the ideal of an arithmetically Buchsbaum curve", to appear.

[G] Green, M.:"Koszul cohomology and the geometry of projective varieties", *J. Diff. Geom.* 19, 125-171 (1984).

[GL] Green, M-Lazarsfeld, R.:"A simple proof of Petri's theorem on canonical curves", in "*Geometry today*", Giornate di geometria, Roma 1984, Birkhauser Boston, 129-142 (1985).

[GLP] Gruson, L.-Lazarsfeld, R.-Peskine, Ch.: On a theorem of Castelnuovo and the equations defining space curves" *Invent. Math.* 72, 491-506 (1983).

[GP] Gruson, L-Peskine, Ch.:"Space curves: complete series and speciality" in "*Space curves*", Proc. Rocca di Papa 1985, Lect. Notes in Math. 1266, 108-123, Springer Verlag (1987).

[HH] Hartshorne, R-Hirschowitz, A.:"Droites en position générale dans l'espace projectif", in "*Algebraic Geometry* ", Proc. La Rabida 1981, Lect. Notes in Math. 961, 169-189, Springer Verlag (1982).

[H] Hirschowitz, A.:"Sur la postulation générique des courbes rationnelles", *Acta Math.* 146, 209-230 (1981).

[I1] Idà, M.:"On the homogeneous ideal of the generic union of lines in $\mathbb{P}^3$". Thése, Université de Nice, 1986. To appear on *J. Reine Angew. Math.*.

[I2] Idà, M.:"Generating six skew lines in $\mathbb{P}^3$", in "*Algebraic curves and projective geometry*",Proc. Trento 1988, Lect. Notes in Math. 1389, 112-127 (1989) Springer-Verlag.

[I3] Idà, M.:"Maximal rank and minimal generation", *Arch. Math.*, 52, 186-190 (1989).

[I4] Idà, M.: work in progress.

[MR] Miró-Roig, R.M.:"Unobstructed arithmetically Buchsbaum curves", in "*Algebraic curves and projective geometry* ",Proc. Trento 1988, Lect. Notes in Math. 1389, 235-241 (1989) Springer-Verlag.

[M] Mumford, D.:"*Lectures on curves over an algebraic surface*" Ann. of Math. Studies, 59, p.200, Princeton Univ. Press (1966).

[PR] Paranjape, K- Ramanan, S.:"On the canonical ring of a curve", preprint.

[Pa] Pareschi, G.:"Components of the Hilbert scheme of smooth space curves with the expected number of moduli", *Manuscripta Math.* 63, 1-16 (1989)

[Pe] Perrin, D.:"*Courbes passant par m points généraux de* $\mathbb{P}^3$", Mémoire de la Soc. Math. de France, nouv. série 28/29, 1-137 (1987).

[PS] Peskine, Ch-Szpiro, L.:"Liaison des variétés algébriques", *Invent. Math.*, 26, 271-302 (1973).

[P] Petri, K.:"Uber die invariante Darstellung algebraischer Funktionen einer Variablen", *Math. Ann.* 88, 243-289 (1923).

[R] Rao, A.P.:"Liaison among curves in $\mathbb{P}^3$ ", *Invent. Math.*, 50, 205-217 (1979).

[S] Schreyer, F.O.:"Syzygies of canonical curves and special linear series", *Math. Ann.*, 275, 105-137 (1986).

[S2] Schreyer, F.O.:"Green's conjecture for general p-gonal curves of large genus", in *"Algebraic curves and projective geometry* ",Proc. Trento 1988, Lect. Notes in Math. 1389, 254-260 (1989) Springer-Verlag.

[V] Voisin, C.:"Courbes tétragonales et cohomologie de Koszul", *J. reine u. angew. Math.* 387, 111-121 (1988).

Biholomorphic Invariants of Real Submanifolds of $\mathbb{C}^n$

F. GHERARDELLI Istituto Matematico «U. Dini», Università di Firenze, Florence, Italy

1. In a book published some years ago H. Clemens defined the real projective space as «the unifier» and the complex one as «the great unifier». Nobody was ever more persuaded by this than the italian geometers of the beginning of this century. Every geometric problem was framed in the projective space, also if the natural ambient was another one, e.g. $\mathbb{R}^n$ or $\mathbb{C}^n$.

In spite of this, it was a point of view which gave interesting results. A remarkable example is offered by the work of B. Segre in complex analysis. He was concerned with Poincaré problem: to find pseudo-conformal invariants of a real hypersurface of $\mathbb{C}^2$.

The compactification of $\mathbb{C}^2 \cong \mathbb{R}^4$ with the real projective space P^4, suggests to B. Segre the definition of some projective invariants, which, almost by a miracle, are in fact biholomorphic invariants.

This paper could be an appendix to Segre's work. Here too the main idea comes from projective geometry. We remark the analogy between the Levi form of a real submanifold of $\mathbb{C}^n$ and a linear system of quadrics. Moduli of the base locus of the system of quadrics, suitably read on the Levi form are the pseudo-conformal invariants we are concerned with.

2. Let $f = 0$ be a local equation of a real hypersurface M of $\mathbb{C}^n (n \geq 2)$. The discriminant, $L(f)$, of the Levi form of f, is invariant under biholomorphic maps whose jacobian is unimodular. $L(f)$ may be also defined as the restriction to $f = 0$ of the coefficient of the (n, n) form $\partial f \wedge \bar{\partial} f \wedge (\partial \bar{\partial} f)^{(n-1)}$ (or $f^{n+1} (\partial \bar{\partial} \log f)^{(n)}$).

If $z = (z^1, \ldots, z^n)$ are complex coordinates in $\mathbb{C}^n$, its explicit expression is

$$L_z(f) = \begin{vmatrix} 0 & f_1 & \cdots & f_n \\ f_{\bar{1}} & f_{1\bar{1}} & \cdots & f_{n\bar{1}} \\ \cdots & \cdots & \cdots & \cdots \\ f_{\bar{n}} & f_{1\bar{n}} & \cdots & f_{n\bar{n}} \end{vmatrix}_{f=0} \tag{1}$$

where

$$f_i = \frac{\partial f}{\partial z^i}, \quad f_{\bar{j}} = \frac{\partial f}{\partial \bar{z}^j}, \quad f_{i\bar{j}} = \frac{\partial^2 f}{\partial z^i \partial \bar{z}^j}.$$

By a biholomorphic change of coordinates $w = w(z)$ we have

$$L_z(f) = \left| J\left(\frac{w}{z}\right) \right|^2 L_w(f),$$

where $J\left(\dfrac{w}{z}\right)$ is the determinant of the jacobian matrix.

If $g = h f = 0$ is another (local) equation for $M (h = \bar{h} \neq 0)$, then

$$L_z(h f) = h^{n+1} L_z(f).$$

The previous definitions can be easily generalized to real submanifolds of $\mathbb{C}^n$ of real codim $m \geq 2$ (§§3, 4) in order to get pseudo-conformal, pointwise, invariants for these manifolds.

The discriminant of the Levi form is the complex analogue of the discriminant of the second fundamental form of a hypersurface of $\mathbb{R}^n$ of codim ≥ 2.

3. Let E be a n-dimensional complex vector space; $H : E \times E \to \mathbb{C}$ a hermitian form, $\psi : E \to \mathbb{C}^m$ a complex linear map of maximal rank and let $W = \ker \psi$. In a fixed basis $(e_1, \ldots, e_n)$ of E, H and ψ are given by

$$H(v, w) = \sum_{i,j} h_{ij} z^i \bar{w}^j$$

and

$$\psi(v) = \sum_i a_i^r z^i \qquad r = 1, \ldots, m$$

$(v = \sum z^i e_i, w = \sum w^i e_i)$.

In these conditions, we have the following

LEMMA 1. The discriminant of the hermitian form H, restricted to W is given (up to a factor $\neq 0$) by

$$\begin{vmatrix} 0 & \cdots & 0 & a_1^1 & \cdots & a_n^1 \\ & & & & & \\ 0 & \cdots & 0 & a_1^m & \cdots & a_n^m \\ \bar{a}_1^1 & & \bar{a}_1^m & & & \\ & & & & h_{ij} & \\ \bar{a}_n^1 & & \bar{a}_n^m & & & \end{vmatrix}$$

Let now M be a real submanifold of $\mathbb{C}^n$, defined by the equations

$$f^1 = 0, \ldots, f^m = 0.$$

Suppose M (smooth and) analytic and that in every of its points (or in the generic one) the m complex hyperplanes tangent to $f^1 = 0, \ldots, f^m = 0$ are independent. For short I shall write sometime F instead of ${}^t(f^1, \ldots, f^m)$.

Let us consider in $\mathbb{C}^n$ the (n,n) form

$$\Omega_{(m)} := \partial f^1 \wedge \overline{\partial} f^1 \wedge \ldots \wedge \partial f^m \wedge \overline{\partial} f^m \wedge \left(\sum_{i=1}^{m} \lambda_i \partial \overline{\partial} f^m\right)^{(n-m)}$$

where $\lambda_1, \ldots, \lambda_m$ are m real constants never zero simultaneously.

The restriction to M of the coefficient of $\Omega_{(m)}$ is given the coordinates $z, \overline{z}$ by

$$P_z(F, \Lambda) := \begin{vmatrix} 0 & \cdots & 0 & f_1^1 & \cdots & f_n^1 \\ 0 & \cdots & 0 & f_1^m & \cdots & f_n^m \\ f_{\overline{1}}^1 & & f_{\overline{1}}^m & & & \\ & & & \sum_{k=1}^{m} \lambda_k f_{ij}^k & & \\ f_{\overline{n}}^1 & & f_{\overline{n}}^m & & & \end{vmatrix}_{F=0}, \quad \Lambda := (\lambda_1, \ldots, \lambda_m)$$

$P_z(F, \Lambda)$, if not identically zero[1], is a homogeneous polynomial in $\lambda_1, \ldots, \lambda_m$ of degree $n - m$, which is the analogue of the discriminant of the Levi form of a hypersurface. In fact, $P_z(F, \Lambda)$ has also the following definition. If $x = (z_0^i)$ is a point of M, consider the hermitian form

$$\sum_{i=1}^{m}\sum_{i,j=1}^{m} \lambda_r f_{ij}^r (z^i - z_0^i)(\overline{z}^j - \overline{z}_0^i)$$

and its restriction to the complex tangent space to M in x:

$$\sum_{i=1}^{n} f_i^r(z^i - z_0^i) = 0 \qquad r = 1, \ldots, m;$$

by the Lemma, the discriminant of such a restricition is (up to a factor $\neq 0$) just the polinomial $P_z(F, \Lambda)$.

If $w = w(z)$ is a biholomorphic change of coordiantes it is easy to verify that

$$P_z(F, \Lambda) = \left|J\binom{w}{z}\right|^2 P_w(F, \Lambda). \tag{5}$$

Moreover, if

$$g^r = \sum_{s=1}^{m} a_s^r f^s = 0, \qquad r = 1, \ldots, m,$$

are new local equation for M (a_s^r real analytic functions such that $\det(a_s^r) \neq 0$)

$$P_z(AF, \Lambda) = |A|^2 P_z(F, \Lambda A) \tag{3}$$

where $A := (a_s^r)$ and $|A| = \det A$.

---

[1] Compare the remark at the end of this §.

From (2) and (3) follows the main result of this note:

THEOREM 1. *Every projective invariant (absolute invariant) of the polynomial P (or of the algebraic hypersurface $P = 0$ of P^{m-1}) is a pointwise pseudoconformal invariant of M.*

REMARK. Looking at the polynomial P as the discriminant of a Levi form, we get a sufficient condition for the nonvanishing of P: it is enough that one (at least) of the hypersurface $f^r = 0$ be strictly pseudoconvex.

On the other hand, as it does, $P \equiv 0$ if M is a complex manifold (in this case biholomorphic maps have no local invariant).

4. Relation (3) says that in every point of M the linear group $GL(m, \mathbb{R})$ operates on the λ. May be of interest to consider pseudo-conformal invariants of the intersection of two or more *given* submanifolds, V_r, of codim m_r ($\sum m_r = m$). In this case the group $GL(m, \mathbb{R})$ is replaced by a subgroup, direct product of suitable $GL(m_r, \mathbb{R})$.

Let us consider e.g., the intersection of a hypersurface $V_1 = [f^1 = 0]$ with a submanifold $V_{m-1} = \{f^2 = \ldots f^m = 0\}$ of codim $m - 1$. The equation of V_1 may be changed by a multiplicative factor: $g^1 = af^1$ ($a \neq 0$) and new local equation for V_{m-1} are of the type

$$g^q = \sum_{p=z}^{m} a_p^q f^p \qquad q = 2, \ldots, m, \qquad |a_p^q| \neq 0.$$

The matrix A defined in §3 is now

$$\begin{pmatrix} a & 0 \\ 0 & a_p^q \end{pmatrix}$$

and in every point of $M = V_1 \cap V_{m-1}$ the group acting on the λ is $GL(1) \times GL(m-1)$ [2] $P(F, \Lambda)$ has to be considered as a polynomial in λ_1 with coefficients polynomials in $\lambda_2, \ldots, \lambda_m$. On these polynomials acts the group $GL(m-1)$; their invariants and the simultaneous invariants between some of them are relative invariants for the group $GL(1) \times GL(m-1)$. It follows the

COROLLARY. *Given four of these relative invariants an appropriate product of their powers is an absolute invariant and a pointwise pseudoconformal invariant of $V_1 \cap V_{m-1}$.*

In some cases the conclusion of the corollary holds also if we know only two or three relative invariants for the group $GL(1) \times GL(m-1)$.

---

[2] More generally on $V_1 \cap V_{m-1}$ the equation of V_{m-1} are of the type

$$g^q = a_1^q f^1 + \sum_{p=z}^{m} a_p^Q f^p, \qquad q = 2, \ldots, m.$$

Now the matrix

$$A = \begin{pmatrix} a & 0 \\ a_1^q & a_p^q \end{pmatrix}, \qquad |A| \neq 0$$

belongs, for any $x \in M$, to the affine group $GL(1, m-1; \mathbb{R})$. The affine invariants of the polynomial $P(F, \Lambda)$ are pseudo-conformal invariants of $V_1 \cap V_{m-1}$.

EXAMPLES. a) Let $n = 4$, $m = 2$: in $\mathbb{C}^4$ are given 2 real hypersurfaces $V_1 = \{f = 0\}$ and $V_2 = \{g = 0\}$; P is a second degree polynomial

$$a_0 \lambda_1^2 + a_1 \lambda_1 \lambda_2 + a_2 \lambda_2^2$$

where

$$a_0 = \begin{vmatrix} 0 & 0 & f_i \\ 0 & 0 & g_i \\ f_{\bar{j}} & g_{\bar{j}} & f_{i\bar{j}} \end{vmatrix} \qquad a_2 = \begin{vmatrix} 0 & 0 & f_i \\ 0 & 0 & g_i \\ f_{\bar{j}} & g_{\bar{j}} & g_{i\bar{j}} \end{vmatrix}$$

$$a_1 = \sum_{r \neq j} (-1)^r \begin{vmatrix} 0 & 0 & f_i \\ 0 & 0 & g_i \\ 0 & 0 & f_{i\bar{r}} \\ f_{\bar{j}} & g_{\bar{j}} & g_{i\bar{j}} \end{vmatrix}$$

$$I = \frac{a_0 a_2}{a_1^2} \quad \left(\text{or} \quad \frac{4 a_0 a_2 - a_1^2}{a_1^2} \right)$$

is a pseudo-conformal invariant of $V \cap V_2$ and it depends only on the two hypersurfaces V_1 and V_2 and not on their equations.

We notice that I is not an invariant of the manifold $M = \{f = 0 = g\}$. An invariant of M is the sign of the discriminant $\Delta := 4 a_0 a_2 - a_1^2$ and so the points of M are divided in three classes according $\Delta \gtreqless 0$.

b) More generally, let $M = \{f = 0 = g\}$ be a real (non complex) submanifold of $\mathbb{C}^n$, any n. $P(F, \Lambda)$ is now a binary form of degree $n - 2$.

From the classical theory of invariants, we know that if $n-2 \geq 4$ it has relative invariants (entire rational functions of its coefficients), whose ratios with suitable powers are absolute invariants for the action of $GL(2, \mathbb{R})$.

In every point $x \in M$ we consider the graded algebra $\mathcal{A}_x$ of relative invariants of $P(F, \Lambda)(x)$. If $\mathcal{I}_x$ is the ideal of those which are zero in x, $\mathcal{A}_x / \mathcal{I}_x$ is the graded ring from which to derive pseudo-conformal invariants of M.

For ex., if $n = 6$ $P(f, \Lambda)$ is a forth degree polynomial

$$a_0 \lambda^4 + 4 a_1 \lambda^3 \mu + 6 a_2 \lambda^2 \mu^2 + 4 a_3 \lambda \mu^3 + a_4 \mu^4,$$

which, as it is well known, has two independent relative invariants

$$i = a_0 a_4 - 4 a_1 a_3 + 3 a_2^2$$

$$j = \begin{bmatrix} a_0 & a_1 & a_2 \\ a_1 & a_2 & a_3 \\ a_2 & a_3 & a_4 \end{bmatrix}.$$

If $j(x) \neq 0$ $i^3(x)/j^2(x)$ is a pseudo-conformal invariant.

c) If $m > 2$ explicit computation are generally impossible. Nevertheless one has at least another interesting example for $n = 6$, $m = 3$. M is the intersection of three real hypersurface $f^1 = f^2 = f^3 = 0$. $P(F, \Lambda)$ is a homogeneous polynomial of third degree in three variables. The classical modular invariant J is a pseudo-conformal invariant of M; it is real because real is the cubic $P(F, \Lambda) = 0$. Another invariant is the number of cycles (or circuits) (1 or 2) of which the cubic is composed.

5. Let M be a «generic», codim m, submanifold of the n-euclidean space $\mathbb{R}^n$ $\left(m < \binom{n-m+1}{2}\right)$, locally defined by $f^1 = 0, \ldots, f^m = 0$. In each of its point the normal space has dim m and the second fundamental form is a quadratic form depending linearly on m parameters $\lambda_1, \ldots, \lambda_m$. The discriminant of this quadratic form is a polynomial in the λ. Obvious variants of the previous remarks, provide the definition of orthogonal (or linear affine) invariants of M. Their interest is much lower than in the pseudoconformal case, because now, at least if M is compact, much more important global invariants are defined.

The second fundamental form and its discriminant may be defined also for submanifolds M of the real or of the complex projective space. If codim $M = m$ and M is generic in a sense that may be easily specified, again the discriminant is a homogeneous polynomial in m variables. The projective invariants of this polynomial could be useful in the study, for instance, of the projective rigidity of submanifolds of $\mathbb{C}P^n$ (in particular if $m = 2$).

Le Prime Ricerche di Pietro Mengoli:
La Somma delle Serie

ENRICO GIUSTI   Dipartimento di Matematica, Università di Firenze.[1]

1. LA CATTEDRA ALL'UNIVERSITÀ DI BOLOGNA.

Della vita di Mengoli ben poco si sa, e comunque non dovette essere ricca di novità e di avventure. Anche l'anno di nascita (1625 o 1626) è incerto, e si può dedurre solo dalle brevi biografie che ci sono pervenute, come ad esempio quella inserita da Giovanni Fantuzzi nelle sue *Notizie degli scrittori bolognesi*[2], di cui riportiamo per esteso le parti salienti.

> Suo padre fu Simone Mengoli, e la madre Lucia Uccelli, onesti, e civili Cittadini Bolognesi. Applicatosi Pietro da giovinetto alla filosofia, e fatto il corso della medesima con moltissimo profitto, venne in essa laureato l'anno 1650[3]. Indi passò allo studio delle Leggi, e ne ricevette la laurea in amendue le facoltà, Civile e Canonica, l'anno 1653[4]. Incamminato poi per la via ecclesiastica, ne prese tutti gli ordini, e si diede totalmente allo studio delle Matematiche alla scuola del padre Bonaventura Cavalieri, e dopo la morte di questo approfittò moltissimo nella geometria per mezzo dell'amicizia, e carteggio con Gio. Antonio Rocca[5] da Reggio [...] e richiesta dal Senato

---

[1] Viale Morgagni 67/a, 50134 Firenze. Lavoro eseguito nell'ambito del programma di ricerca *Storia delle Matematiche* del M. P. I. (40%).

[2] Bologna, Stamperia di S. Tommaso d'Aquino, 1788.

[3] Il 18 gennaio. Si veda Serafino Mazzetti, *Repertorio di tutti i professori antichi e moderni della celebre Università e del celebre Istituto delle Scienze di Bologna*, Bologna, Tipografia di S. Tommaso d'Aquino, 1847.

[4] Il 7 giugno. Vedi Mazzetti, op. cit.

[5] Giovanni Antonio Rocca (1607–1656) fu amico e corrispondente di Cavalieri, e di numerosi altri

una cattedra vacante di Matematica nel pubblico Studio l'ottenne. Indi nell'anno 1660, li 19 aprile, divenne parroco, e priore della chiesa di S. Maria Maddalena in via S. Donato, e fu aggregato alla sacra scuola de' Confortatori.

Fu sua continua applicazione la Geometria, e l'Algebra; ed a sollevarsi da queste serie applicazioni si rivolgeva allo studio della musica, che da giovinetto aveva appreso, non come un semplice pratico dilettante, ma penetrando in essa come un erudito matematico. [...] Finalmente morì d'anni 60 adì 7 giugno 1686, e fu seppellito nella sua chiesa priorale, ove gli furono fatte solennissime esequie, e recitata funebre orazione[6].

Sulla carriera accademica di Mengoli abbiamo maggiori notizie, ricavabili da documenti conservati per lo più negli Atti dell'Assunteria di Studio[7]. Alla morte di Cavalieri, avvenuta il 30 novembre 1647, era rimasta vacante la lettura pomeridiana di Matematica, mentre le letture mattutine erano tenute da Ovidio Montalbani e da Giovanni Ricci[8]. La possibilità di ricoprire la cattedra del maestro dovette sembrare non tanto remota, se già nel gennaio dell'anno successivo uno dei suoi allievi, il gesuato Stefano Angeli[9], presentò una domanda in questo senso, della quale si trova traccia negli Atti dell'Assunteria di Studio:

Adì 25 Gennaro 1648.

Memoriale del Padre Stefano Angeli Giesuato che supplica di una lettura di Matematica essendo alievo del Padre Bonaventura Cavalieri Defunto. Per esser forestiero si stima necessaria l'habilitatione per conseguire detta lettura. Ordina però, che si vedano le Constitutioni dello Studio, et quello è solito praticarsi in simile occasione per portare il tutto alla Congregazione.[10]

Non sembra che vi siano stati altri atti riguardanti la domanda di Angeli, che comunque non ebbe seguito. Forse reso esperto dal fallimento del suo condiscepolo,

---

scienziati contemporanei. Di lui non restano opere stampate, se si eccettua un lemma inserito da Cavalieri e da Torricelli nelle loro opere. La sua corrispondenza fu pubblicata nel 1785 sotto il titolo *Lettere d'uomini illustri del XVII secolo a G. A. R.*, Modena. Si veda A. Favaro, *Amici e corrispondenti di Galileo Galilei. XIX. Giannantonio Rocca*, ristampa anastatica a cura dell'Istituto e Museo di Storia della Scienza, Salimbeni, Firenze, 1983, vol II.

[6]Fantuzzi, *Notizie*, op. cit., vol. VI, pag. 9–11. La stessa data viene riportata dal Mazzetti. Al contrario, Pellegrino Orlandi, *Notizie degli scrittori bolognesi e dell'opere loro stampate e manoscritte*, Bologna, Pisarri, 1714, fissa la data di morte al 16 giugno, e lo dice morto a 61 anni.

[7]Tutte le carte relative allo Studio bolognese sono conservate all'Archivio di Stato di Bologna.

[8]Ovidio Montalbani (1601–1671 o 72) tenne la lettura "*ad Mathematicam*" dal 1633/34 al 1650/51, anno in cui passò alla cattedra di Filosofia Morale, che coprì fino al 1670/71. Il carmelitano Giovanni Ricci (1607–1664) fu allievo di Cavalieri e poi lettore di Matematica dal 1642/43 al 1664/65. Su entrambi si possono vedere le opere citate di Fantuzzi, Orlandi e Mazzetti. Per le loro pubblicazioni matematiche, si veda P. Riccardi, *Biblioteca Matematica Italiana*, rist. anast. Gorlich, Milano.

[9]Stefano Angeli, o degli Angeli (1623–1697), gesuato, fu professore di matematica nello Studio di Padova dal 1662 all'anno della sua morte. I suoi lavori geometrici, per l'elenco dei quali rimandiamo alla *Biblioteca* del Riccardi, si muovono tutti nell'ambito della teoria cavalieriana degli indivisibili. Su Angeli si potrà vedere il *Dizionario Biografico degli Italiani*, Istit. Encicl. Ital., Roma 1961, vol. 3.

[10]*Atti dell'Assunteria di Studio*, vol. 12, c. 4^V.

Mengoli evita di richiedere la cattedra di Matematica, accontentandosi invece di aspirare ad un incarico più modesto, per il quale la sua qualifica di cittadino bolognese gli avrebbe dato più concrete possibilità. In effetti, già sul finire del 1648, dunque prima di ottenere la laurea in filosofia, Mengoli richiede la lettura di Aritmetica:

> Ill^{mi} SS^{ri}
>
> Pietro Mengoli Divotissimo Oratore delle SS. VV. Ill^{me}, allevato sotto la disciplina del Padre Cavalliero nelle Matematiche, desiderando d'incaminarsi con li debiti mezzi a servire alla Patria, et alle SS. VV. Ill^{me} nelle medesime professioni o in altre conforme a' talenti che gli concederà la Divina Bontà, le supplica riverentemente a farlo descrivere fra li Aritmetici nel Rotolo dello Studio, offerendosi pronto di essercitarsi nel modo, che gli verrà imposto; Che della grazia ne restarà obbligatissimo alle SS. VV. Ill^{me}. Quas Deus &c.[11]

Il foglio porta in alto il numero $\frac{3}{4}$ e il nome, Pietro Mengoli, e in calce le annotazioni seguenti:

> 16 Ottobre 1648 letto in Senato in n° di 18. I Signori Assunti di Studio considerino l'instanza et le qualità del soggetto per portare i loro sentimenti in Senato. Adì 24 novembre 1648 lettosi a' S^{ri} Assunti dello Studio, congregati in n° di tre compreso l'Illmo S. Confaloniere. Si lega la relazione nel aggionto foglio esistente.

Troviamo inoltre sul retro la scritta "Memoriale con relazione. A V. S^{rie} Ill^{me} per il Dott. Pietro Mengoli", e nelle pagine interne la minuta della relazione, cancellata da un tratto di penna:

> Riferiscono alle SS. VV. Ill^{me} gli Assunti dello Studio che l'Oratore è cittadino di quasta Patria, et ben fondato nell'Aritmetica e fu alievo del defunto Padre Bonaventura Cavallieri, soggietto del valore noto alle SS. VV. Ill^{me}. Laonde essendo tal lettura stimata molto utile e necessaria e letta da pochi, però si come altre volte fu simil materia letta in Casa a' tempi adietro da Alessandro e Galeazzo Bonasoni[12], e dippoi da Giovanni Maria Cambij, al quale dall'anno 1553 sotto gli undici di Dicembre successe per partito delle SS. VV. Ill^{me} Scipione Dattari[13] et fu rotulato a tal

---

[11] Ibidem.

[12] Né Alessandro né Galeazzo Bonasoni risultano iscritti ai rotuli tra i lettori di aritmetica. Vi si trovano invece un Petronio Bonasoni (dal 1563/64 al 1591/92 e un Antonio Maria Bonasoni (dal 1593/94 al 1630/31). La famiglia Bonasoni diede non pochi lettori all'ateneo bolognese: un Antonio Bonasoni fu professore di logica, di filosofia e infine di medicina dal 1532/33 al 1556/57, mentre Giovanni e Giulio furono professori il primo di Diritto canonico dal 1497/98 al 1505/06 e il secondo di Istituzioni civili dal 1589/90 al 1590/91.

[13] Giovanni Maria Cambij fu lettore *ad Arithmeticam et Geometriam* dal 1509/10 al 1512/13, e poi dal 1518/19 al 1554/55; gli successe Scipione Datari (che troviamo indicato anche con i nomi Datarus, Dactilus, Dattilus, Datilis, Dactilius) che tenne la lettura dal 1555/56 al 1604/05.

lettura con annuo stipendio di L. 175. Così anche crederiano gli Assunti suddetti che le SS. VV. Ill^{me} potessero camminare coll'istessa regola con l'Oratore, soggetto meritevole et tale per quanto attestano persone di dottrina e fede, et le SS. VV. Ill^{me} devono pensare di restare honoratamente servite poiché sarà un maggiormente animarlo in servizio della Patria et nel faticarsi virtuosamente.

Si rimettono però gli Assunti suddetti al solita prudenza delle SS. VV. Ill^{me} et a quella si rassegnano.

Pochi giorni dopo, la relazione degli Assunti veniva approvata dal Senato, che assegnava a Mengoli la cattedra di Aritmetica. Peraltro questa non era che una delle letture fantasma riservate a cittadini bolognesi, e che a un onorario irrisorio (a titolo di esempio si consideri che il salario di Cavalieri nel 1635 era di 1500 lire, aumentate poi fino a 1900) unisce l'impegno minimo di insegnare gratuitamente a quattro fanciulli. In ogni caso si trattava di un incarico che non comportava pubbliche letture allo Studio, e si limitava, oltre all'insegnamento gratuito di cui sopra, alla possibilità di dare lezioni private in casa, come risulta charamente dalla scritta *"Domi"* che accompagnava nei Rotuli la lettura *"ad Arithmeticam"* e soprattutto dal fatto che nei *Quartironi* degli stipendi l'Aritmetico Pietro Mengoli non compare tra i lettori, ma tra i *Salariati diversi*[14]. È solo alla fine del 1650, ottenuta la laurea in filosofia, che Mengoli può richiedere la lettura di Matematica, stavolta una vera cattedra che comportava lezioni ufficiali anche se non un sostaniale aumento di stipendio. La richiesta di Mengoli è datata 29 ottobre 1650 e porta in calce il parere favorevole degli Assunti, che però nello stesso tempo richiedono che legga nelle matematiche applicate:

Adi 13 novembre 1650 in Congregazione di Studio di n° 4.

Riferiscono gli Assonti a V.S. Ill^{me} che l'oratore fu altre volte sotto il dì 9 Gen 1649 honorato dall'Illmo Reggimento del Carattere d'aritmetico dello studio per insegnare detta professione in Casa, e provisionato di L. [ ][15] annue per tal fatica. Si è poi dottorato in filosofia, e pochi giorni sono difese con molta sua lode su le scuole conclusioni miste fra la filosofia e matematica stampate, e dedicate a V.S. Ill^{me}. Finalmente le attestazioni, che si hanno dalli intendenti della professione lo predicano per molto sacente nelle matematiche, e ne permettono ogni miglior riuscita; però stimano gli Assunti che meriti la grazia da lui dimandata, et che parendo bene a V.S. Ill^{me} compiacernelo sia per esser servigio publico l'obligarlo a leggere la sua professione in ordine alla matematica, et in somma alla matematica in pratica dalla quale si cava poi l'utile per le occorrenze del publico. Così riferiscono gli Assonti, & rimettendosi &c.

Sotto la stessa data si legge negli Atti:

---
[14]Archivio di Stato di Bologna, *Quartironi*, Anno 1650.
[15]Nel documento la cifra è lasciata in bianco.

Adì 13 novembre 1650.

Al memoriale del dottor Mengoli relatione favorevole con applicarlo dichiaratamente alla materia in ordine alla Machinatoria, et servizio della pratica[16].

La delibera fu presa alcuni giorni dopo, come risulta dalla seguente scrittura in calce alla domanda:

Adì 29 novembre 1650 in Regimento di n° 23.

Commesso partito di andare la suddetta lettura al detto Mengoli secondo la relatione degli Assunti di studio.

Ottenuto per voti 21.

Di conseguenza, viene assegnata a Mengoli la lettura di Meccanica, che egli occupa nello stesso anno, o al più tardi all'inizio del successivo 1651, e che terrà ininterrottamente fino alla sua morte.

2. LA TEORIA MENGOLIANA DELLE SERIE.

La relazione degli Assunti parla di "conclusioni miste tra la filosofia e matematica", che Mengoli avrebbe sostenuto pubblicamente, e poi stampate dedicandole ai Senatori. Di tali conclusioni non si è trovata traccia, ma non è escluso, anche se in verità è poco probabile, che in realtà si tratti delle *Novae quadraturae arithmeticae,* un'opera stampata nel 1650 e dedicata per l'appunto ai senatori bolognesi[17].

A differenza di Stefano Angeli, suo condiscepolo alla scuola di Bonaventura Cavalieri, la cui ricerca si situa quasi del tutto sotto il segno della teoria cavalieriana, fin dal suo primo lavoro Mengoli mostra una spiccata indipendenza scientifica, distaccandosi completamente e definitivamente dalla teoria degli indivisibili creata dal matematico milanese. E se il titolo dell'opera prima mengoliana porta ancora tracce dei temi cari al suo maestro, con il riferimento alle quadrature, basta una rapida occhiata al contenuto del volume per verificare che ogni contatto colle ricerche archimedee è troncato, e che il proposito di Mengoli non è di calcolare, sia pure per via aritmetica, aree o volumi di figure geometriche, ma solo di sommare un certo numero di serie, e nel contempo di delineare un quadro teorico generale nel quale i suoi risultati possano essere dimostrati rigorosamente.

Più che alle ricerche geometriche, il libro di Mengoli si colloca dunque sul versante degli studi algebrici, o più precisamente di una certa algebra geometrizzata, che pur conservando la flessibilità delle notazioni sincopate, cercava nel metodo geometrico la risposta alle esigenze di rigore per molto tempo sacrificate alla potenza dell'algoritmo

---

[16] *Atti dell'Assunteria di Studio*, vol. 12, c. 44^v.

[17] *Novae Quadraturae Arithmeticae, seu de Additione Fractionum:* Petri Mengoli Art. & Phil. Doct. Illustrissimis, & Sapientissimis Civitatis Bononiae Senatoribus. Bononiae, ex Typographia Iacobi Montij, 1650.

algebrico. Così, pur trattando esclusivamente di serie numeriche, Mengoli userà continuamente di linguaggio e termini tratti dalla geometria: *magnitudo, planum,* ecc., e terminerà il volume con una serie di definizioni, nelle quali si possono trovare echi delle formulazioni di Viéte e Descartes:

> Definitiones
>
> Exposita rationalis, & datis quotlibet, si rationalis ad aliam, quae invenitur habeat proportionem compositam ex proportionibus eiusdem rationalis ad singulas datas; vocetur inventa, productus datarum.
>
> Et datae lineae, dicantur, latera producti.
>
> Exposita rationali, & datis duabus alijs magnitudinibus; si ut prima datarum ad rationalem, ita fiat secunda ad aliam, quae invenitur; vocetur inventa, fractio facta ex denominatione secundae per primam.
>
> Et ipsa secunda magnitudo, numerator fractionis.
>
> Prima vero, denominator.
>
> Et exposita rationalis, unitas appelletur[18]

e che permettono di trasformare gli enunciati da aritmetici in geometrici:

> Huiusmodi definitiones in Arithmetici voluminis calce appositas esse volui, ut faciliter quisquis possit Arithmetica theoremata in Geometricos usus convertere, & demonstrationes in quantitate discreta expositas, in quantitate continua mutatis nominum interpretationibus adhibere.[19]

Ma veniamo al contenuto delle *Novae Quadraturae*. Mengoli inaugura un metodo espositivo, che userà ripetutamente nel seguito, e che consiste nel far precedere la trattazione da una lunga prefazione, nella quale espone brevemente i risultati principali ai quali perverrà nel corso del trattato. Sono queste delle prefazioni che indulgono molto poco all'uso di presentare l'opera sotto la luce migliore e più accetta al lettore, e che pertanto il più delle volte nascondono più di quanto rivelino delle vere intenzioni dell'autore. Al contrario non di rado le prefazioni mengoliane, prive come sono di intenti apologetici da una parte, e di tecnicismi matematici dall'altra, possono servire a comprendere appieno il contenuto e la portata dell'opera. Nel nostro caso poi, la

---

[18] *Novae quadraturae*, pag. 129: "Definizioni. Fissata un'unità, e date altre grandezze in numero arbitrario, se l'unità ha ad un'altra grandezza da trovare la proporzione composta di quelle dell'unità alle singole grandezze date; la grandezza trovata si chiama: *prodotto* delle date. E le linee date si diranno *lati* del prodotto. Data l'unità e due altre grandezze, se come la prima delle grandezze date all'unità, così la seconda ad un'altra, quest'ultima si dirà: *frazione composta denominando la seconda con la prima*. La seconda grandezza, *numeratore* della frazione. E la prima, *denominatore.*" Conformemente all'ultima definizione di Mengoli, abbiamo tradotto *rationalis* con *unità*.

[19] Ivi, pag.130: "Ho voluto apporre queste definizione in calce a un volume aritmetico, in modo che chiunque possa facilmente convertire i teoremi aritmetici agli usi geometrici, e adattare alle grandezze continue, cambiando l'interpretazione delle parole, le dimostrazioni condotte per le grandezze discrete."

prefazione contiene più del testo, in quanto è solo in essa che troviamo la dimostrazione della divergenza della serie armonica.

Nella prefazione troviamo anche il legame, tenue ma presente, con i problemi di quadratura:

> Meditanti mihi persaepe Archimedis parabolae quadraturam, propterquam infinita triangula in continue quadrupla proportione existentia certos limites quantitatis non excedunt; occurrit universalis illa quadratura eiusdem argumenti occasione a Geometris demonstrata, qua magnitudines infinitae continuam quamlibet proportionem maioris inaequalitatis possidentes in praefinitas homogeneas quantitates colliguntur.[20]

La quadratura della parabola è solo un pretesto, e si passa subito alla considerazione di più generali serie geometriche, e di qui senza soluzione di continuità a quelle armoniche. Ma l'interesse di questo brano iniziale, più che nel collegamento, del tutto formale, tra le sue ricerche e i problemi archimedei, sta nell'uso di un termine a prima vista innocuo. Dice Mengoli che nelle serie geometriche, un numero infinito di grandezze si raccolgono in una grandezza finita e *omogenea*. A prima vista sembra trattarsi di un pleonasma, e che il termine possa essere eliminato senza alterare il senso della frase. Non dimentichiamo però che Mengoli è un allievo di Cavalieri, e che questi ha insegnato come infinite grandezze (come *tutte le linee* di una figura) possano costituire una grandezza finita, ma eterogenea: le linee formeranno una superficie, i piani un volume[21].

Letta in questa luce, la parola *omogenee* è molto di più di una precisazione superflua; al contrario essa segna la presa di distanza dell'allievo dall'insegnamento di Cavalieri, la rivendicazione di un proprio spazio, e ancor più di un proprio metodo scientifico. Si tratta di una presa di distanza definitiva: se non di rado Mengoli rivolgerà il proprio pensiero riverente al suo illustre maestro[22], ciò sarà solo per ricordarne la persona e i meriti, mai per riprenderne e svilupparne il metodo scientifico.

Sulle ragioni di questo distacco non possiamo fare che delle ipotesi: di certo non si tratta, come avverrà più tardi con Stefano Angeli, di un abbandono a posteriori, derivante dall'estenuarsi del metodo cavalieriano; un tale esito non si sarebbe verificato così presto né in maniera così perentoria. Piuttosto, c'è da pensare a un'opposizione di principio: le dottrine di Cavalieri vengono rifiutate non perché sterili, ché anzi nel momento in cui Mengoli scrive esse sono nel pieno della loro fioritura, ma perché poco

---

[20] *Novae quadraturae*, Praefatio: "Meditando sovente sulla quadratura della parabola di Archimede, in cui infiniti triangoli in proporzione continua quadrupla non superano limiti fissati di quantità; mi sovvenne quell'universale quadratura dimostrata dai geometri mediante lo stesso argomento, in cui infinite grandezze in qualsivoglia proporzione continua di maggiore disuguaglianza, si raccolgono in quantità finite omogenee".

[21] Per una esposizione più completa della teoria degli indivisibili di Cavalieri, si veda il mio *Bonaventura Cavalieri and the theory of indivisibles*, Bologna, Cremonese 1980, e K. Andersen, *Cavalieri's method of indivisibles*, Archive Hist. Ex. Sci. **31** (1985) 291–367.

[22] Ricordiamo che fu proprio Mengoli a raccogliere e pubblicare sotto il titolo *Le lagrime d'Urania* (Bologna, Ferroni, 1647), alcune poesie celebrative di vari autori, scritte in occasione della morte di Cavalieri.

fondate. È questo il senso della precisazione iniziale: le grandezze devono essere omogenee; di conseguenza la teoria degli indivisibili deve essere abbandonata per sostituirvi un metodo più sicuro.

Quale sia questo metodo, ancora non è chiaro, o almeno non emerge dalla lettura delle *Novae quadraturae*; quello che però emerge compiutamente è un tratto caratteristico della matematica mengoliana: la ricerca di formulazioni rigorose e sicure, anche a prezzo di un restringimento dell'orizzonte scientifico e della formazione di un linguaggio espositivo personale e non di rado oscuro e contorto.

Una volta liberatosi dalla tutela di Cavalieri, Mengoli può continuare la narrazione del cammino dalla quadratura della parabola alle serie armoniche. La convergenza delle serie geometriche lo spinge infatti a chiedersi se ogni successione di grandezze decrescenti all'infinito abbia somma finita.

> Admirabile sane Theorema: cuius contemplatione in eam quaestionem inductus sum, utrum magnitudines ea quacumque lege dispositae, ut aliqua possit assumi minor qualibet data, vel ut deficientes in infinitum evanescant, infinitae compositae omnem propositam quantitatem valeant superare.[23]

e in particolare se la serie armonica

$$\frac{1}{2} + \frac{1}{3} + \frac{1}{4} + \frac{1}{5} + \cdots$$

dia luogo a una grandezza finita o infinita. Vale la pena di leggere direttamente le parole di Mengoli, anche per familiarizzarsi con una terminologia che spesso saremo costretti, per brevità, a sostituire con equivalenti moderni.

> In huiusmodi causa experimentum Arithmeticas fractiones tentare agressus, eas ita disposui, ut singulas unitates singulis post unitatem numeris denominarem, in qua quidem dispositione sumi potest magnitudo minor qualibet assignata, & propterea ipsae magnitudines ad ordinis incrementum quantitate decrescentes in infinitum evanescunt.
>
> $$\frac{1}{2}\ \frac{1}{3}\ \frac{1}{4}\ \frac{1}{5}\ \frac{1}{6}\ \frac{1}{7}\ \frac{1}{8}\ \frac{1}{9}\ \frac{1}{10}\ \frac{1}{11}\ \frac{1}{12}\ \frac{1}{13}\ \frac{1}{14}$$
>
> Causam igitur in assumptae dispositionis terminis proponens quaerebam, utrum unitates denominatae singulis numeris post unitatem in infinitum dispositae, & aggregatae infinitam aliquam, vel finitam componerent extensionem. Pro finita extensione respondendum videbatur; quod numerorum, & fractionum contrariae sint potestates, numerorum quidem in multiplicatione, qua magnitudines versus infinitum progrediantur, fractionum vero in

---

[23] *Novae quadraturae*, Prefatio: "Ammirabile teorema invero, e dalla cui contemplazione sono stato portato alla questione, se delle grandezze disposte con una legge qualunque, in modo che se ne possa prendere una minore di una qualsiasi data, ovvero che diminuendo vadano ad annullarsi all'infinito, composte in numero infinito valgano a superare ogni data quantità."

divisione, qua res ad ipsa indivisibilia reducitur: aggregati autem numeri superant quamlibet propositam quantitatem; ergo a contrario sensu aggregatae fractiones non videntur posse quamlibet propositam magnitudinem excedere. Hoc sophisma toto fere mense fuit expectationis argumentum, quod pro hac parte Geometricam in causa ferre posse sententiam: at qui dum processum demonstrationis examino, iudicium in alterius partis favorem convertitur.[24]

Appare qui chiaramente uno dei tratti distintivi dell'elaborazione mengoliana: la negazione della validità di qualsiasi ragionamento analogico, ed in contrasto la ricerca costante di prove "geometriche". Si tratta, è evidente, di un luogo comune delle ricerche matematiche: quale scienziato potrebbe affermare il contrario senza per ciò stesso svalutare la sua opera? Ma quando dalle enunciazioni generali si passa alla prassi matematica, e cioè agli enunciati e alle dimostrazioni, non è raro, ed anzi potremo dire che sia la regola, il caso in cui l'argomentare analogico ha la meglio sul procedere geometrico, e le difficoltà del dimostrare vengono aggirate mediante l'ambiguità dei concetti e dei metodi. Non così avviene in Mengoli, che spinge la precisione dei concetti fino e talvolta oltre il limite della pedanteria, definendo ogni termine, anche i più usati e conosciuti, e mai procedendo al di là del terreno sicuro del dimostrato geometricamente. Naturalmente ciò non vuol dire che egli sfugga alle trappole della definizione implicita e dell'assunzione nascosta; e però tra gli scrittori contemporanei, Mengoli è quello che si avvicina di più all'ideale sempre sfuggente della dimostrazione totale, basata solo su ipotesi chiare e su proprietà esplicitamente enunciate. Quando ciò viene meno, è il segno che siamo giunti al limite estremo in cui entrano in gioco le proprietà fondamentali dei numeri reali, ed in particolare la completezza; un territorio questo inaccessibile ai geometri del seicento e oltre.

Ma torniamo alla dimostrazione della divergenza della serie armonica. Mengoli osserva che gli inversi degli interi formano appunto una proporzione armonica, e dunque per una nota proprietà di quest'ultima la somma di tre termini consecutivi è maggiore del triplo del termine intermedio. Ma allora

fractiones in proposita dispositione sumptae ternae a prima sunt maiores

$$\frac{1}{2}\frac{1}{3}\frac{1}{4} \quad \frac{1}{5}\frac{1}{6}\frac{1}{7} \quad \frac{1}{8}\frac{1}{9}\frac{1}{10} \quad \frac{1}{11}\frac{1}{12}\frac{1}{13} \quad \frac{1}{14}\frac{1}{15}\frac{1}{16}$$

---

[24] Ibidem: "Così, tentando a mo' di esperimento le frazioni aritmetiche, le disposi in modo da denominare le singole unità con i numeri successivi all'unità, nella quale disposizione si può trovare una quantità minore di una qualsiasi data, e pertanto tali grandezze al crescere dell'ordine decrescono in quantità e all'infinito si annullano. Istruita dunque la causa per i termini di questa disposizione, mi chiedevo se le unità denominate dai numeri successivi a partire da uno, disposte fino all'infinito, ed aggregate, componessero un'estensione finita o infinita. Sembrava doversi concludere per l'estensione finita; poiché i numeri e le frazioni sono operazioni contrarie, i numeri nella moltiplicazione, con la quale le grandezze vanno verso l'infinito; le frazioni nella divisione, con la quale l'oggetto si riduce agli stessi indivisibili. Ora i numeri aggregati superano qualsiasi quantità data; dunque per converso le frazioni riunite non sembrano poter eccedere ogni grandezza assegnata. Questo sofisma per quasi tutto un mese mi fece sperare che si potesse trovare una dimostrazione geometrica in questa direzione; ma l'esame del procedimento dimostrativo fece pendere il giudizio a favore dell'altra alternativa.". Nel seguito useremo spesso il termine "serie" per rendere quello che Mengoli chiama "dispositio terminorum", e parleremo di "somma" dove egli dice "extensio".

triplis medijs: & mediae sunt unitates denominatae numeris a ternario multiplicatis $\frac{1}{3}$, $\frac{1}{6}$, $\frac{1}{9}$, $\frac{1}{12}$; & earundem triplae sunt $1, \frac{1}{2}, \frac{1}{3}, \frac{1}{4}$, quae eodem, quo supra argumento ternae sunt maiores tripli medijs[25].

In altre parole, raggruppando i termini della serie a tre a tre si ottiene l'unità più la stessa serie di partenza.

> Ergo fractiones propositae dispositionis assumptae totidem semper secundum numerus proportionis continue subtriplae 3, 9, 27, 81, singulas unitates excedunt. Possunt autem sumi, pro quovis assignato numero, totidem in continua proportione subtripla numeri a ternario, iuxta quorum aggregatum sumptae fractiones dispositionis propositae ipsum assignatum numerum superabunt: ergo propositae fractiones in infinitum dispositae, & aggregatae infinitam extensionem valent implere.[26]

Un esempio servirà, se ce ne fosse ancora bisogno, a chiarire l'argomentazione di Mengoli:

> Sit exempli gratia numerus assignatus 4: & sumantur a ternario quatuor continue proportionales in subtripla 3, 9, 27, 81, quorum summa 120: igitur sumptae fractiones in multitudine numeri 120 superant assignatum numerum 4; nam tres primae superant triplum $\frac{1}{3}$, videlicet unitatem: novem deinceps superant triplum aggregati $\frac{1}{6}$, $\frac{1}{9}$, $\frac{1}{12}$, videlicet aggregatum $\frac{1}{2}$, $\frac{1}{3}$, $\frac{1}{4}$; sed huiusmodi aggregatum superat unitatem, ut ostendi; ergo novem deinceps superant unitatem: & propter eandem demonstrationem 27, & 81 subsequentes singulas unitates excedunt.[27]

Da questo risultato seguono due corollari. Il primo, che iniziando a sommare da un qualsiasi termine della serie armonica si ottiene sempre una somma infinita, dato i termini trascurati hanno somma finita. In secondo luogo, è infinita la somma degli inversi di una qualsiasi progressione aritmetica $Cn+B$. Infatti si ha[28]

---

[25] Ibidem: "le frazioni nella disposizione data, prese a tre a tre a partire dalla prima, sono maggiore di tre volte le medie. Ora le medie sono l'unità denominata dai numeri moltiplicati per tre: $\frac{1}{3}$, $\frac{1}{6}$, $\frac{1}{9}$, $\frac{1}{12}$; e le loro triple sono 1, $\frac{1}{2}$, $\frac{1}{3}$, $\frac{1}{4}$, le quali, per lo stesso ragionamento di sopra, prese a tre a tre sono maggiori del triplo delle medie".

[26] Ibidem: "Dunque le frazioni della serie data, prese via via secondo la progressione geometrica di ragione tre: 3, 9, 27, 81, ogni volta superano l'unità. Dato dunque un numero qualsiasi, si possono prendere tanti numeri secondo la progressione geometrica di ragione 3 a partire da 3, che l'aggregato di altrettante frazioni della disposizione proposta superino il numero dato. Di conseguenza le frazioni proposte, disposte in infinito ed aggregate, valgono a raggiungere una grandezza infinita."

[27] Ibidem: "Sia ad esempio 4 il numero dato, e si prendano quattro numeri in progressione geometrica di ragione 3 a partire dal 3: 3, 9, 27, 81, la cui somma è 120: allora le frazioni prese in numero di 120 superano il numero dato 4. Infatti le prime tre superano il triplo di $\frac{1}{3}$, cioè l'unità: le successive nove sono maggiori del triplo dell'aggregato di $\frac{1}{6}$, $\frac{1}{9}$, $\frac{1}{12}$, e cioè dell'aggregato di $\frac{1}{2}$, $\frac{1}{3}$, $\frac{1}{4}$; ma questo aggregato supera l'unità, come si è mostrato, quindi le nove successive superano l'unità. Infine, con la medesima dimostrazione le successive 27 ed 81 superano ognuna l'unità."

[28] Naturalmente per $C > 1$, dato che se $C \leq 1$ il risultato è ovvio.

Le Prime Ricerche di Pietro Mengoli

$$\frac{1}{Cn+B} > \frac{1}{C}\frac{1}{n+B}$$

e quest'ultima serie diverge in quanto per il corollario precedente diverge la serie $\frac{1}{n+B}$.

Una volta determinata la divergenza dela serie armonica, Mengoli procede all'esame della convergenza della serie degli inversi dei numeri triangolari

$$\sum_{n=2}^{\infty} \frac{2}{n(n+1)}$$

dimostrando in tre modi che la sua somma è 1. La prima dimostrazione consiste nell'osservare che

$$\sum_{n=2}^{k} \frac{2}{n(n+1)} = \frac{k-1}{k+1}$$

o nelle parole di Mengoli:

> aggregatae quotlibet a prima sunt aequales numero multitudinis ipsarum denominato per numerum binario maiorem, & propterea semper unitate sunt minores eo defectu, qui iuxta multitudinis additarum fractionum incrementum infra quamlibet assignatam magnitudinem diminuitur, & in infinitum evanescit.[29]

La seconda dimostrazione si fonda sulla formula

$$\frac{2}{2n(2n+1)} + \frac{2}{(2n+1)(2n+2)} = \frac{1}{2}\frac{2}{n(n+1)}$$

dalla quale, sommando a partire da $n=2$ e indicando con S la somma della serie, si ottiene:

$$S = \frac{1}{2}(S+1)$$

da cui $S = 1$. Infine, la terza dimostrazione si basa sul fatto che sommando successivamente 2, 4, 8, 16,... termini della serie data, si ottiene la serie geometrica di ragione $\frac{1}{2}$:

$$\frac{1}{2} + \frac{1}{4} + \frac{1}{8} + \frac{1}{16} + \cdots$$

che ha somma 1.

Il successo di tali dimostrazioni viene però subito a temperarsi a causa dell'impossibilità di ottenere la somma degli inversi dei quadrati:

---

[29] Ibidem: "la somma di un numero qualunque dalla prima è uguale al numero dei termini denominato da un numero maggiore di due, e dunque sono sempre minori dell'unità di un difetto tale, che all'aumentare del numero delle frazioni sommate diminuisce al di sotto di una qualsiasi grandezza assegnata, e all'infinito si annulla."

Ab hiuus fractionum dispositionis contemplatione faciliter expeditus, ad aliam progrediebar dispositionem, in qua singula unitates numeris quadratis denominantur. Haec speculatio fructus quidem laboris rependit, nondum tamen effecta est solvendo, sed ingenij ditioris postulat adminiculum, ut praecisam dispositionis, quam mihimetipsi proposui, summam valeat reportare.[30]

Di questa serie, Mengoli non parlerà più nel corso del volume, nemmeno per dimostrarne la convergenza; una dimostrazione questa non difficile nella struttura assiomatica dell'opera. D'altra parte il fatto stesso che egli rimandi ad altri la determinazione della somma *esatta* della serie in esame sta a testimoniare che la sua convergenza era da considerarsi acquisita.

Peraltro questo è un tratto caratteristico dell'elaborazione mengoliana, nella quale il problema centrale non è mai quello di dimostrare la convergenza (qui di una serie, altrove[31] di una successione), ma di trovare il valore a cui la serie converge. Un esempio di questo atteggiamento è costituito dal secondo metodo di sommazione della serie degli inversi dei numeri triangolari, nel quale, come abbiamo visto, la convergenza della serie è data per scontata, ed anzi è utilizzata proprio allo scopo di calcolarne la somma.

Il problema teorico della convergenza è dunque ignorato, o quanto meno taciuto, a tutto vantaggio del calcolo esplicito del limite. Se questo punto di vista non emerge esplicitamente nella dottrina delle serie, ciò è perché in esse, quanto meno in quelle che Mengoli prende in esame, tutte a termini positivi, c'è una nozione intuitiva di somma, e il problema è di determinare se essa è finita o infinita. Ed in effetti, la contrapposizione tra la serie armonica e quella degli inversi dei numeri triangolari non è a ben vedere quella tra serie convergenti e divergenti, ma piuttosto tra somma finita o infinita[32]. Questa caratteristica della teoria mengoliana diverrà molto più manifesta quando nella *Geometria Speciosa* passerà a considerare limiti di successioni, la cui convergenza verrà sempre supposta anche se, contrariamente a quanto avviene per le serie, essa in molti casi sarebbe tutta da dimostrare.

Con queste osservazioni possiamo considerare concluso l'esame dell'introduzione, e possiamo passare all'individuazione dell'apparato assiomatico che Mengoli pone in opera per trattare la convergenza e la divergenza delle serie. Prima però descriveremo brevemente i risultati principali, servendoci per non appesantire la trattazione di un formalismo moderno.

L'opera è divisa in tre libri, nei quali i risultati sono presentati in ordine crescente di difficoltà. Nel primo, vengono trattate le serie il cui termine generico è del tipo

$$\frac{1}{a_k a_{k+1}}$$

---

[30] Ibidem: "Una volta terminato facilmente lo studio di questa serie, procedevo ad un'altra, nella quale le unità erano denominate dai quadrati dei numeri. Questa ricerca ha dato qualche frutto, ma tuttavia non è ancora giunta alla soluzione, ma richiede il contributo di un ingegno superiore, affinché si giunga a trovare la somma precisa della serie che mi sono proposta."

[31] Nella *Geometria speciosa*.

[32] Più precisamente, come è chiaro dalla discussione che precede, tra la dimostrazione a priori della convergenza di una data serie e la verifica a posteriori che la sua somma è finita. Avendo chiarito questo punto, potremo parlare indifferentemente di serie convergenti o di somma finita.

Le Prime Ricerche di Pietro Mengoli

dove a_k è una progressione aritmetica: $a_k = Ck + B$. In particolare, Mengoli dimostra i seguenti teoremi:

Teorema 16:
$$\sum_{n=1}^{\infty} \frac{1}{n(n+1)} = 1$$

Teorema 24:
$$\sum_{n=0}^{\infty} \frac{1}{(n\alpha+1)[(n+1)\alpha+1]} = \frac{1}{\alpha}$$

Teorema 37:
$$\sum_{n=0}^{\infty} \frac{1}{(n\alpha+q)[(n+1)\alpha+q]} = \frac{1}{\alpha q}$$

Nel secondo libro è la volta delle serie
$$\sum_{n=1}^{\infty} \frac{1}{a_n a_{n+1} a_{n+2}},$$

per le quali vengono ottenuti, tra gli altri, i risultati seguenti:

Teorema 7:
$$\sum_{n=1}^{\infty} \frac{1}{n(n+1)(n+2)} = \frac{1}{4}$$

Teorema 14:
$$\sum_{n=0}^{\infty} \frac{1}{(2n+1)(2n+3)(2n+5)} = \frac{1}{12}$$

Teorema 21:
$$\sum_{n=0}^{\infty} \frac{1}{(n\alpha+1)[(n+1)\alpha+1][(n+2)\alpha+1]} = \frac{1}{2\alpha(\alpha+1)}$$

Teorema 25:
$$\sum_{n=0}^{\infty} \frac{1}{(n\alpha+q)[(n+1)\alpha+q][(n+2)\alpha+q]} = \frac{1}{2\alpha q(\alpha+q)}$$

Di gran lunga più complessa e più ampia è la struttura del terzo libro, dove i risultati precedenti vengono estesi a serie molto più generali di quelle contenute nei primi due. Più precisamente, si considerano serie del tipo:

$$\sum_{n=1}^{\infty} \frac{a_{n+s} - a_n}{a_n a_{n+1} \cdots a_{n+s}}$$

in particolare quando a_n è una progressione aritmetica $(a_n = Cn + B)$, nel qual caso il numeratore si riduce alla costante sC, e di conseguenza il termine generico è l'inverso del prodotto di s termini in progressione aritmetica.

Naturalmente, non è questione di dare una dimostrazione generale, valida per ogni s, una dimostrazione che è al di là delle possibilità della matematica di Mengoli e dei suoi contemporanei. D'altra parte, anche se le dimostrazioni mengoliane sono condotte in casi particolari (ad esempio per $s = 3$), la tecnica è del tutto generale, in modo che chi volesse adattarle ad altri valori di s non avrebbe che da compiere le evidenti modifiche del caso.

Il libro è diviso in due parti: dopo un risultato propedeutico (Teorema 1) vengono trattate dapprima (Teoremi 2–5) le serie in cui i termini a_k formano una successione aritmetica, e quindi serie generiche, in cui la successione a_k è supposta solamente crescente. Per quanto riguarda le serie del primo tipo, il risultato principale è dato dal Teorema 5: se $a_n = Cn + B$, allora

$$\sum_{n=1}^{\infty} \frac{1}{a_n a_{n+1} \cdots a_{n+s}} = \frac{1}{Cs} \frac{1}{a_1 a_2 \cdots a_s}$$

Quando invece la successione a_n è una generica successione crescente, il problema si biforca a seconda che essa tenda o meno all'infinito. Nel primo caso (Teorema 7) risulta[33]

$$\sum_{n=1}^{\infty} \frac{a_{n+s} - a_n}{a_n a_{n+1} \cdots a_{n+s}} = \frac{1}{a_1 a_2 \cdots a_s}$$

mentre nel secondo (Teorema 9) si ha:

$$\sum_{n=1}^{\infty} \frac{a_{n+s} - a_n}{a_n a_{n+1} \cdots a_{n+s}} = \frac{b^s - a_1 a_2 \cdots a_s}{b^s a_1 a_2 \cdots a_s}$$

dove si è indicato con b il "limite" della successione a_n. Infine, il Teorema 10 dimostra che se si ha

$$\lim_{n \to \infty} a_{nP+1} = \alpha_1$$

$$\lim_{n \to \infty} a_{nP+2} = \alpha_2$$

$$\dots$$

$$\lim_{n \to \infty} a_{nP} = \alpha_P$$

allora

---

[33]Si noti che questo risultato contiene come caso particolare il Teorema 5 enunciato sopra, dato che se a_n è una successione aritmetica si ha $a_{n+s} - a_n = sC$.

$$\sum_{n=1}^{\infty} \frac{a_{n+P} - a_n}{a_n a_{n+1} \cdots a_{n+P}} = \frac{a_1 a_2 \cdots a_P - \alpha_1 \alpha_2 \cdots \alpha_P}{a_1 a_2 \cdots a_P \alpha_1 \alpha_2 \cdots \alpha_P}$$

Poiché abbiamo parlato del limite di una successione, sarà bene fare alcune precisazioni. In effetti, Mengoli non dà qui una definizione precisa di limite di una successione, cosa che farà più tardi nella *Geometria speciosa*; si limita invece a distinguere, tra le "dispositiones continuae magnitudinum", le successioni crescenti di grandezze, quelle che "procedunt in infinitum" da quelle che "procedunt ab una ad alteram", e cioè che a partire dalla prima a_1 si avvicinano alla seconda b. Se poi si esaminano più da vicino le dimostrazioni, si riconoscerà che nel primo caso, quello cioè delle grandezze procedenti all'infinito, Mengoli fa uso del fatto che, dato comunque un numero G si possa trovare nella disposizione data un numero maggiore[34]; mentre nel secondo, delle grandezze che procedono da a a b, egli utilizza la proprietà che la somma di tutte le differenze $a_{i+1} - a_i$ sia uguale a $b - a^{35}$. Sono esattamente queste proprietà che permettono la dimostrazione dei risultati che abbiamo descritto.

Fin qui i risultati del *Novae quadraturae*, una serie di teoremi complessi e di non facile dimostrazione, specie se riferiti all'epoca in cui Mengoli scrive, nella quale in genere non si andava oltre la somma delle serie geometriche[36]. Ma l'interesse del volumetto non si esaurisce in questi risultati, per quanto importanti; un posto di altrettanta preminenza merita il metodo di dimostrazione di Mengoli, che per molto tempo resterà un esempio pressoché unico di rigore nella sommazione delle serie numeriche; ed in particolare l'apparato di definizioni e di assiomi che egli introduce per condurre a termine le sue dimostrazioni.

Si tratta di un corpus di dieci definizioni e due assiomi, con i quali Mengoli può trattare il tema delicato della convergenza senza fare appello in nessun caso alla manipolazione algebrica priva di giustificazioni.

Definitiones.

I. Differentiam duarum magnitudinum, quando prima excedit secundam, voco, excessum primae & secundae.

---

[34] *Novae Quadraturae*, pag. 114: ".. sit defectus F; & ut F ad unitatem denominatam per A, ita sit A ad G; & *inter numeros dispositos ab A, inveniatur C, numerus maior quam G*" (sia F il difetto, e come F all'unità denominata da A, così sia A a G; e tra i numeri disposti a partire da A, si trovi F maggiore di G). Il corsivo è mio. Si noterà come Mengoli usi qui il termine "numero" come sinonimo di "grandezza".

[35] Ivi, pag. 120: "... & quoniam ab A ad B sunt dispositae magnitudines infinitae; etiam differentiae in ea dispositione sunt infinitae; & *simul compositae sunt aequales uni differentiae extremarum*" (e poiché le grandezze disposte da A a B sono infinite, anche le differenze saranno infinite; e composte insieme sono uguali in totale alla differenza delle estreme). Il corsivo è mio.

[36] Ricordiamo che ancora nel 1673 Leibniz non crede che Mengoli sia riuscito a sommare le serie dei numeri figurati (triangolari, piramidali, ecc.) e suggerisce che possa trattarsi di somme finite. Si veda la lettera a Oldemburg del 26 aprile in *G. W. Leibniz sämtliche Schriften und Briefe*, Dritte Reihe, Erster Band, pag.88.

II. Quando vero prima deficit a secunda, voco, defectum primae, & secundae.[37]

III. Similes differentias, voco, tum excessus, tum defectus inter se.

IV. Dissimiles vero excessus defectibus.

V. Magnitudines Arithmetice dispositas, voco, quarum (sumptis continue binis quibuslibet) differentiae similes antecedentium, & consequentium sunt aequales.

VI. Magnitudines Harmonice dispositas, voco, quarum (sumptis continue ternis quibuslibet) prima se habet ad tertiam, ut differentia primae, & secundae ad similem differentiam secundae, & tertiae.[38]

VII. Unam magnitudinem altera denominatam, voco, quamlibet fractionem, in qua una magnitudo stat loco numeratoris, altera vero loco denominatoris.[39]

VIII. Differentia, & plana in aliqua dispositione, voco absolute, differentias, & plana magnitudinum, quae sunt continue conseguentes in illa dispositione, primae videlicet, & secundae; secundae, & tertiae; & sic deinceps usque ad ultimam, si dispositae sunt in aliqua multitudine; vel in infinitum, si dispositae concipiuntur infinitae.[40]

IX. Continuam magnitudinum dispositionem, voco, cum differentiae antecedentium, & consequentium sunt similes.[41]

X. Magnitudines dicuntur implere propositam extensionem, quando existentes infinitae sunt extensionis minoris proposita; vel quando existentes finitae, ita sunt minores proposita, ut una alia magnitudine adiecta in earundem ordine continuato proxima, fiant extensionis maioris proposita.[42]

Abbiamo qui una continua ricerca di precisione terminologica, che non lascia nulla di approssimativo o di ambiguo. Un esempio quanto mai caratteristico è dato dalla

---

[37] *Novae quadraturae*, pag 1. "Definizioni. 1. Chiamo *eccesso* della prima sulla seconda la differenza di due grandezze, quando la prima è maggiore della seconda. 2. Se invece la prima è minore della seconda, la chiamo *difetto* della prima dalla seconda". Si noti che Mengoli, seguendo peraltro l'uso comune del tempo, chiama differenza di due grandezze quello che per noi è il valore assoluto della differenza.

[38] Ivi, pag. 2. "3. Gli eccessi tra loro e i difetti tra loro si dicono *differenze simili*. 4. Gli eccessi e i difetti, *dissimili*. 5. Chiamo *disposte aritmeticamente* delle grandezze tali che, presene due coppie qualsiasi, le differenze simili delle antecedenti e delle conseguenti sono uguali. 6. Chiamo *disposte armonicamente* delle grandezze tali che, presene ad arbitrio tre consecutive, la prima stia alla terza come la differenza della prima e della seconda alla differenza simile della seconda e della terza."

[39] Ivi, pag. 4-5. "7. Data una qualsiasi frazione, in cui una prima grandezza sta al numeratore e una seconda al denominatore, chiamo la prima grandezza *denominata* dalla seconda."

[40] Ivi, pag. 5-6. "8. In una data successione, chiamo semplicemente *differenze epiani* le differenze e i piani delle grandezze che si susseguono nella detta successione; ad esempio della prima e della seconda, della seconda e della terza, e così via fino all'ultima, se le grandezze sono in numero finito, ovvero all'infinito, se si suppongono infinite."

[41] Ivi, pag. 9. "Chiamo *continua* una successione in cui le differenze delle antecedenti e delle conseguenti sono simili."

[42] Ivi, pag. 19. "Delle grandezze si dicono *riempire* una data estensione quando essendo infinite sono di estensione minore della data; ovvero quando prese in numero finito sono minori della data in modo tale, che aggiunta ad esse la prima grandezza che le segue nel loro ordinamento, risultino maggiori della data."

definizione V, nella quale si precisa che in una progressione aritmetica le differenze devono essere sì uguali, ma anche simili, e cioè o tutte di eccesso o tutte di difetto; in altre parole essa deve essere o crescente o decrescente.

Di particolare importanza è l'ultima definizione, che verrà usata frequentemente nelle dimostrazioni. In essa, come abbiamo già avuto occasione di osservare in altre circostanze, si assume implicitamente che un numero infinito di grandezze abbia sempre un'estensione; ovvero in termini moderni che una serie (a termini positivi) abbia sempre una somma. È questa l'unica ipotesi nascosta che sia possibile rintracciare nell'opera, un'assunzione strettamente legata con il problema della completezza dei numeri reali, e dunque al di là dell'orizzonte non solo del nostro autore, ma di tutto il diciassettesimo secolo.

Allo stesso ordine di idee è legato il primo (e in pratica unico) assioma della teoria mengoliana, che capovolge in un certo senso la definizione moderna di serie divergente:

> *Axioma primum.* Quando infinitae magnitudines infinitae sunt extensionis, possunt in aliqua multitudine sumi, ut superent quamlibet propositam extensionem.[43]

Meno essenziale sembra invece il ruolo del secondo assioma, che sembrerebbe posto allo scopo di garantire che se si esegue due volte la stessa somma si ottiene sempre lo stesso risultato:

> *Axioma secundum.* Quando infinitae magnitudines finitae sunt extensionis, & singulae magnitudines eaedem in infinitum concipiuntur in una, & altera extensione disponi, & aggregari, congruit una extensio alteri.[44]

Il gioco combinato della definizione X e dell'assioma 1 permette di condurre a termine le dimostrazioni di convergenza. Si comincerà per questo col dimostrare un risultato generale sulle serie convergenti:

> *Theorema 15.* Quando magnitudines a prima dispositae in infinitum, & aggregatae sunt extensionis finitae, sunt in aliqua multitudine a prima, quae implent propositam extensionem maiorem quidem prima, minorem tamen extensione omnium.[45]

---

[43] Ivi, pag. 18. "*Assioma primo* . Quando un numero infinito di grandezze sono di estensione infinita, se ne possono sempre prendere un numero finito, in modo da superare una qualsiasi estensione data."

[44] Ivi, pag. 19-20. "*Assioma secondo*. Quando infinite grandezze sono di estensione finita, e si suppone che le singole stesse grandezze siano disposte ed aggregate in una prima ed una seconda estensione, la prima estensione è congruente alla seconda." Questo assioma entra soltanto nella dimostrazioe del Teorema 15 del primo libro (vedi più oltre) e del Teorema 7 del terzo.

[45] Ivi, pag. 20. "*Teorema 15* . Quando delle grandezze, disposte dalla prima in infinito ed aggregate, sono di estensione finita, se ne possono prendere un numero finito dalla prima in modo da riempire una qualsiasi data estensione maggiore della prima grandezza e minore dell'estensione di tutte."

ovvero che se M è maggiore del primo termine a_1 e minore della somma A della serie, esiste un intero N tale che

$$\sum_{k=1}^{N} a_k < M \; ; \; \sum_{k=1}^{N+1} a_k > M.$$

La dimostrazione è semplice: si prendano abbastanza grandezze da *implere* M. Se fossero infinite la loro estensione C dovrebbe essere minore o uguale a M (qui Mengoli dice uguale ad M, probabilmente per un lapsus calami); per il secondo assioma, occupando le stesse grandezze sia l'estensione A che C, si dovrebbe avere $C = A$, e cioè la parte uguale al tutto. Ne segue che esse sono in numero finito, e dunque la tesi.

Si vede di qui la regione dell'alternativa nella definizione X; infatti essa permette di assumere che in ogni caso un certo numero (finito o infinito) di grandezze della serie riempiano la grandezza M, e quindi di ridursi a provare che esse sono in numero finito.

A questo punto sono messi in opera tutti gli ingredienti per condurre a termine le dimostrazioni dei vari teoremi. Per dare un'idea della tecnica dimostrativa, seguiremo Mengoli nella dimostrazione del Teorema 16 del primo libro:

Theorema 16. Unitates denominatae planis omnium numerorum ab unitate in infinitum dispositae, & aggregatae sunt aequales unitati.[46]

La dimostrazione passa per vari gradini. Si comincia col provare (Teorema 13) che si ha

$$\sum_{k=1}^{n} \frac{1}{k(k+1)} = \frac{n}{n+1} < 1.$$

Si conclude allora, utilizzando l'assioma 1, che la somma S della serie data è finita. Ora non può essere $S < 1$; infatti altrimenti, preso N in modo tale che risulti $S < N/(N+1)$ (problema 1) si avrebbe

$$\sum_{k=1}^{N} \frac{1}{k(k+1)} > S$$

D'altra parte non può essere neanche $S > 1$, poiché altrimenti si potrebbe prendere $M = 1$ nel Teorema 15, ed ottenere

$$\sum_{k=1}^{N+1} \frac{1}{k(k+1)} > 1$$

contro il Teorema 13. Ne consegue che $S = 1$, e dunque la tesi.

Il percorso dimostrativo che abbiamo delineato in un caso semplice, è tipico di tutta l'opera, che in tal modo assume, al di là dei risultati che vi sono contenuti, il carattere di una costruzione strutturata assiomaticamente e priva, se si eccettuano quelle connesse con la completezza della retta reale, di assunzioni implicite o di ipotesi

---

[46]Ivi, pag. 21. "*Teorema 16*. Le unità, denominate dai piani di tutti i numeri a partire da uno, disposte in infinito ed aggregate, sono uguali all'unità."

nascoste. Sarà questa una caratteristica costante dell'opera geometrica di Mengoli, che se da una parte varrà a garantirgli un posto nella storia della matematica come anticipatore di esigenze di rigore che si manifesteranno solo due secoli più tardi, non mancherà dall'altra di sovraccaricare il suo stile di un formalismo tanto personale quanto ingombrante, e in ultima analisi a nascondere anche quanto di interessante era contenuto nelle sue opere. Né d'altronde i suoi risultati sono così importanti da imporsi all'attenzione degli studiosi, che invece nella migliore delle ipotesi lasceranno le sue opere dopo un primo sguardo, spaventati spesso dalla stravaganza di un linguaggio troppo personale. Scriverà Barrow a Collins:

> his language is so uncouth and ambiguous, his definitions so many and obscure, that I think it were easier, toward the understanding any matter, to learn Arabic than his dialect. So that (beside that I do very much dislike such kind of writing, and hope very little from those that use it), having business enough [...] I can hardly allow leisure, and indeed have not patience enough to pierce into the depth of his obscurities. I see that he propounds many ordinary things involved in this way, but what he hath performed new I cannot guess.[47]

---

[47] I. Barrow a J. Collins, 1 febbraio 1666/67: "Il suo linguaggio è così complesso e ambiguo, le sue definizioni tante e così oscure, che credo sarebbe più facile, quanto a comprendere una qualche materia, imparare l'arabo piuttosto che il suo dialetto. Cosicché (a parte che questo modo di scrivere non mi piace affatto, e non spero granché da quelli che lo usano), avendo molto da fare [...] non posso perdere tempo, né ho abbastanza pazienza, per scavare nel profondo delle sue oscurità. Vedo che propone molte cose ordinarie raggirate in questo modo, ma cosa di nuovo possa aver compiuto, non posso indovinarlo."
S. J. Rigaud, *Correspondence of scientific men of the seventeenth century*, Rist. anastat. G. Olms, Hildesheim 1965, vol. II.

Composition and Fatou Limits

MAURICE HEINS   University of Maryland, College Park, Maryland, 20742 U.S.A.

1. It is a privilege to participate in the International Meeting Geometry and Complex Variables in Bologna being held in the context of the Nine Hundredth Anniversary of the founding of the University of Bologna – a central event in the history of Western Civilization – and to share in honoring the great achievements of the Bologna School of Mathematics. I wish to express my thanks to Professor Salvatore Coen for the kind invitation to participate and to present a paper.

2. Statements of results. Given F, a meromorphic function with domain $\Delta = \{|z| < 1\}$, let F^* denote the prolongation of F with domain the union of Δ and the set of points $\zeta \in C = \{|z| = 1\}$ at which F possesses a limit in each Stolz angle with vertex ζ, the value of $F^*(\zeta)$ being the limit in question. We term $F^*(\zeta)$ the *Fatou limit of F at ζ*. Let f denote an analytic function mapping Δ into itself, let g denote a nonconstant quotient of bounded analytic functions, each with domain Δ, and let $h = g \circ f$. The object of this paper is to relate the Fatou limits of h with those of f and g. The functions f^*, g^* and h^* have domains whose intersections with C have measure 2π. We shall establish the following theorem.

THEOREM. *The set E of $\zeta \in C$ such that $h^*(\zeta)$ and $g^*[f^*(\zeta)]$ are defined and satisfy the composition law*

$$h^*(\zeta) = g^*[f^*(\zeta)] \tag{2.1}$$

is a measurable set of measure 2π. The domain of f^* contains the domain of h^*. If g is bounded, the set E is the intersection of C and the domain of h^*.

The case where f is constant is immediate and will be put aside. The situation where g is replaced by a constant calls for obvious modifications. The details are routine and will be omitted. The proof of the theorem as stated, to be given in §3, will be based on the theorem of F. and M. Riesz [4] and the version of the Lindelöf sectorial limit theorem given on p. 105 of [2]. For one part of the proof the precursor form of the theorem of F. and M. Riesz given by Fatou [1] will suffice.

In §6 nonconstant functions f and g will be exhibited, where g is bounded, for which E is a proper subset of $\widetilde{E}$ = the set of $\zeta \in C$ such that $g^*[f^*(\zeta)]$ is defined. In fact, the construction may be so carried out that $\widetilde{E} - E$ contains a preassigned subset of C which has zero measure.

In §7 nonconstant functions f and g will be exhibited for which the domains of h^* and f^* contain a given nonempty subset of C of zero measure, $h^*(\zeta) = f^*(\zeta) = 1$ for the points ζ of the subset in question, but g^* is not defined at 1. Of course, this is not possible if g is bounded.

The composition of a harmonic function with domain Δ which is given by a Poisson-Lebesgue integral with an allowed function f was considered by Ohtsuka [3] who showed that the resulting function is given by a Poisson-Lebesgue integral. The work of Ohtsuka extends earlier work of Tsuji [5]. A counterpart of (1.1) for the boundary values of the composite function is not indicated in the cited paper of Ohtsuka.

The relation (1.1) may be applied to obtain Fatou composition results for the situation where g is replaced by the difference of two nonnegative harmonic functions with domain Δ. However, here the latter two assertions need not hold even when the harmonic function in question is bounded and not constant. Cf. §8.

3. Proof of Theorem. We start with the case where g is bounded. Given ζ, a point of C at which h has a Fatou limit, we introduce $\gamma(t) = f(\zeta t), 0 \leq t < 1$, and show that $\lim_{t \to 1} \gamma(t)$ exists. Suppose that this were not the case. If $\liminf_{t \to 1} |\gamma(t)| < 1$, there would exist strictly increasing sequences (s_k) and (t_k) of positive numbers, each with limit 1, which satisfy $s_k < t_k < s_{k+1}$, $k = 0, 1, \ldots$, and $\lim_{k \to \infty} \gamma(s_k) = a \in \Delta$, $\lim_{k \to \infty} \gamma(t_k) = b (\neq a) \in \Delta$. It follows that g would take the value $h^*(\zeta)$ at the points of an infinite subset of a compact subset of Δ. This would force g to be constant, contrary to assumption. There remains to be considered the case where $\lim_{t \to 1} |\gamma(t)| = 1$. Here g would have the Fatou limit $h^*(\zeta)$ almost everywhere on an arc of C having positive length. It would follow by the theorem of F. and M. Riesz (or even by the precursor form given by Fatou in his thesis [1, pp. 394-5] and noted by the Riesz brothers in their celebrated 1916 paper [4, pp. 27-8]) that g would be the constant taking the value $h^*(\zeta)$. This is not possible. Hence $\lim_{t \to 1} \gamma(t)$ exists, so that f has a Fatou limit at ζ. When $|f^*(\zeta)| < 1$, we have $h^*(\zeta) = g[f^*(\zeta)] = g^*[f^*(\zeta)]$. When $|f^*(\zeta)| = 1$, we conclude by the Lindelöf sectorial limit theorem in the formulation referred to above that $h^*(\zeta) = g^*[f^*(\zeta)]$ from the fact that $\lim_{t \to 1} g[f(t\zeta)]$ exists.

The assertions of the theorem are seen to hold for the case of a nonconstant bounded g.

There remains for consideration the unrestricted situation where g is a nonconstant quotient g_1/g_2 of bounded analytic functions g_1 and g_2 with domain Δ. We assume, as we may, that g_1 and g_2 are both nonconstant. Let $h_k = g_k \circ f$, $k = 1, 2$.

We consider the second assertion of the theorem. It suffices to consider the case where $c = \lim_{t \to 1} h(t\zeta)$ is finite, the remaining case being readily reduced to this one. The function $h_1 - ch_2$ has Fatou limit 0 at ζ. Further $g_1 - cg_2$ is not constant. The second assertion for the case of bounded g applies. We conclude that f has a Fatou limit at ζ. (We note that the existence of a radial limit of h at ζ is sufficient to draw the desired conclusion).

The first assertion for the present g may be established as follows. Let E_1 denote the intersection of C and the domain of h_1^*, let E_2 denote the subset of the intersection of C and the domain of h_2^* at the points of which h_2^* takes a nonzero value. By the Fatou limit theorem and the theorem of F. and M. Riesz the sets E_1 and E_2 have measure 2π and hence so does $E_1 \cap E_2$. For $\zeta \in E_1 \cap E_2$ we have

$$h^*(\zeta) = \frac{h_1^*(\zeta)}{h_2^*(\zeta)} = \frac{g_1^*[f^*(\zeta)]}{g_2^*[f^*(\zeta)]} = g^*[f^*(\zeta)]. \tag{3.1}$$

Since $E \supset E_1 \cap E_2$, the first assertion follows.
The theorem is established.

4. From this point on our concern will be with the exceptional set $C - E$. It would be a desideratum to have an exact characterization. This does not appear to be an easy question. For nonconstant f and g, the latter being bounded, the exceptional set is the set of $\zeta \in C$ at which h^* does not exists, so that in any case it is a $G_{\delta\sigma}$ of zero measure. We shall see in §6 that there exist such f and g with the property that the exceptional set contains an assigned subset of C of zero measure at the points ζ of which $g^*[f^*(\zeta)]$ is defined. Here one is led to the question of characterizing the set of ζ for which $g^*[f^*(\zeta)]$ is defined but $h^*(\zeta)$ is not.

5. Some auxiliary functions. The following functions will be used in the construction of the desired functions.

m. We define m as the restriction of the Möbius transformation $z \mapsto (1+z)/(1-z)$ to Δ.

a. We shall want a suitably normalized univalent conformal map of Δ onto a conveniently chosen subregion of Δ bounded by two distinct oricycles tangent to C at 1. To that end we introduce the map L of $\{\mathrm{Re}\, z > 0\}$ onto $\{1/2 < \mathrm{Re}\, z < 3/2\}$ given by

$$L(z) = \frac{i}{\pi} \mathrm{Log}\, z + 1. \tag{5.1}$$

Here «Log» denotes the principal logarithm. The function a is thereupon defined as $m^{-1} \circ L \circ m$. We note that its image is of the desired kind and that $\lim_{z \to 1} a(z) = 1$.

b. We next construct a Blaschke product that will be useful for producing one of the desired examples (v. §6). We first consider a product of the form

$$\frac{z - \tau}{z + \overline{\tau}} \cdot \frac{z - \overline{\tau}}{z + \tau}, \tag{5.2}$$

where $\mathrm{Re}\,\tau = 1$ and $\mathrm{Im}\,\tau > 0$. For z nonnegative real or ∞ the product (5.2) takes positive values and satisfies

$$\max \left| \frac{z - \tau}{z + \overline{\tau}} \cdot \frac{z - \overline{\tau}}{z + \tau} - 1 \right| = \frac{2}{|\tau| + 1}. \tag{5.3}$$

Let $\tau_k = 1 + i(k+1)^2$, $k = 0, 1, \ldots$. The infinite product

$$\prod_{k=0}^{\infty} \left(\frac{z - \tau_k}{z + \overline{\tau}_k} \cdot \frac{z - \overline{\tau}_k}{z + \tau_k} \right) \tag{5.4}$$

converges on $\{\operatorname{Re} z > 0\}$ to an analytic function b_1, which takes values of modulus less than 1, has simple zeros at the points τ_k and $\bar{\tau}_k$ but no other zeros, and has the sectorial limit 1 at ∞. We define b as $b_1 \circ m$ and note that it is a Blaschke product the zeros of which are $m^{-1}(\tau_k), m^{-1}(\bar{\tau}_k), k = 0, 1, \ldots$. We note that b has the Fatou limit 1 at 1.

c. It will be convenient to have available an analytic function that maps Δ into itself and has the limit 1 at each point of a given nonempty subset $\mathcal{E}$ of C where $\mathcal{E}$ is of measure zero. To that end we introduce χ_k, the characteristic function of an open subset O_k of C, where O_k contains $\mathcal{E}$ and has measure $< 2^{-k}$, $k = 0, 1, \ldots$. Let φ be given by

$$\varphi(z) = \frac{1}{2\pi} \int_0^{2\pi} \left[\sum_{k=0}^{\infty} \chi_k(e^{i\theta})\right] \frac{e^{i\theta} + z}{e^{i\theta} - z} d\theta, \quad |z| < 1. \tag{5.5}$$

The function φ is an analytic function on Δ with positive real part and has the limit ∞ at each point of $\mathcal{E}$. We define c as $m^{-1} \circ \varphi$ and observe that it has the stated properties.

d. The last auxiliary function to be introduced will be a quotient of Blaschke products having the property that it does not possess a radial limit at 1 but possesses the limit 1 at 1 outside of each Stolz sector with vertex 1. Such a function may be constructed as follows. We first put down as the denominator the Blaschke product given by

$$\prod_{k=0}^{\infty} \frac{\alpha_k - z}{1 - \alpha_k z}, \quad |z| < 1, \tag{5.6}$$

where $\alpha_k = 1 - 2^{-(k+1)}$, $k = 0, 1, \ldots$. The numerator is taken as a Blaschke product

$$\prod_{k=0}^{\infty} \frac{\beta_k - z}{1 - \beta_k z}, \quad |z| < 1, \tag{5.7}$$

where the β_k are introduced step-by-step, each β_k lying to the right of α_k and taken so close that the quotient of the Blaschke products has the stated limit property. The details are straightforward and will be omitted.

6. Examples concerning the set $\tilde{E} - E$ in the situation where g is bounded and not constant. For the first example we take f as the function a and g as the function b. By the construction of these functions each has Fatou limit 1 at 1. However, h does not possess a Fatou limit at 1, as we see on noting that h has zeros on $\{0 < x < 1\}$ clustering at 1 and using the theorem. Hence $1 \in \tilde{E} - E$.

We remark that $h_1 = f \circ g$ possess the Fatou limit 1 at 1. This says that h_1 has the Fatou fixed point 1. The function h_1 is an example of an analytic function mapping Δ into itself and having 1 as a Fatou fixed point which has the property that none of its iterates of order > 1 has a Fatou limit at 1, let alone 1 as a Fatou fixed point. It suffices to consider the identity $h_1 \circ h_1 = f \circ (g \circ f) \circ g$ to verify with the aid of the third statement of the theorem that $h_1 \circ h_1$ does not possess a Fatou limit at 1 and thereupon to proceed inductively.

We now exhibit an example where $\tilde{E} - E$ contains a given nonempty subset $\mathcal{E}$ of C which has zero measure. Here we propose $f = a \circ c$ and $g = b$. For $\eta \in \mathcal{E}$ the function f has the limit 1 at η. The function g has the Fatou limit 1 at $1 = f^*(\eta)$. Hence $\eta \in \tilde{E}$. To see that $h = g \circ f$ does not possess a Fatou limit at η we note that in the contrary case from $h = (b \circ a) \circ c$ it would follow by the third statement of the theorem that $b \circ a$ would possess a Fatou limit at $c^*(\eta) = 1$. This would contradict the property of $b \circ a$ established in the first paragraph of this section. Hence $\eta \in \tilde{E} - E$ so that $\mathcal{E} \subset \tilde{E} - E$.

7. An example concerning the situation where g is an allowed meromorphic function. Taking $g = d$ and $f = a \circ c$ we see that $h = g \circ f$ has the property that $\mathcal{E} \subset H - \widetilde{E}$ where H is the intersection of the domain of h^* and C. The contrast with the case where g is bounded is pronounced.

8. An example concerning the composition of a nonconstant bounded harmonic function with domain Δ and an allowed f. The following example shows that the latter two assertions of the theorem do not hold in this setting. Let Ω be the simply-connected region

$$\{|\operatorname{Re} z| < 2^{-1}, |\operatorname{Im} z| < 2^{-1}\} - \bigcup_{k=0}^{\infty} \{2^{-1} - 2^{-(k+1)} + (-1)^k ti : -2^{-1} < t \leq 2^{-2}\}, \quad (8.1)$$

let f be a Riemann mapping function for Ω and let u be defined by $u(z) = \operatorname{Re} z$, $|z| < 1$. Let η denote the point of C at which the cluster set of f is the segment $\{2^{-1} + ti : -2^{-1} \leq t \leq 2^{-1}\}$. It is readily verified that $\lim_{z \to \eta} u \circ f(z) = 2^{-1}$ but f does not have a radial limit at η.

REFERENCES

[1] P. Fatou, Séries trigonométriques et séries de Taylor, *Acta Math.* 30 (1906), 335-400.
[2] M. Heins, *Selected topics in the classical theory of functions of a complex variable*, Holt, Rinehart and Winston, New York, 1962.
[3] M. Ohtsuka, A theorem on the Poisson integral, *Proc. Japan Acad.* 22 (1946), No. 6, 195-197.
[4] F. and M. Riesz, Ueber die Randwerte einer analytischen Funktion, *C. R. du 4ième Congr. des Math. Scand. à Stockholm* (1916), 27-44.
[5] M. Tsuji, Theorems concerning Poisson integrals, *Japanese J. Math.* 7 (1930), 227-253.

The Bernstein Theorems for the Fantappiè Indicatrix and Their Applications to Mathematical Economics

G.M. HENKIN, A.A. SHANANIN Academy of Sciences U.S.S.R.

In production function theory there arose [13] investigation problem of the following integral transform of the Radon type:

$$\prod(p, p_0) = \int_{x \in R_+^n} (p_0 - px)_+ \mu(dx), p \in R_+^n, p_0 \in R_+^1, \tag{1}$$

where $\mu(dx)$ a non-negative measure concentrated in positive octant $R_+^n = \{x = (x_1, \ldots, x_n) \in R^n : x_j \geq 0, j = 1, 2, \ldots, n\}$,

$$(p_0 - px)_+ = \begin{cases} p_0 - px, & \text{if } p_0 - px \geq 0 \\ 0, & \text{if } p_0 - px < 0. \end{cases}$$

For applications (in mathematical economics) the following questions about transform (1) are very important:

1. Is the measure $\mu(dx)$ uniquely reconstructed through function $\prod(p, p_0), (p, p_0) \in R_+^n \times R_+^1$?
2. What are necessary and sufficient conditions for given function $\prod(p, p_0), (p, p_0) \in R_+^{n+1}$ in order to be the transform (1) of the positive measure $\mu(dx)$?
3. How to find the measure $\mu(dx)$ through function $\prod(p, p_0), (p, p_0) \in R_+^{n+1}$?

Investigation of these questions is reduced to study of the projective Fantappiè indicatrix properties [5] of the non-negative finite measure $\mu(dx)$ with a carrier in R_+^n:

$$\Phi(p, p_0) = \int_{x \in R_+^n} \frac{\mu(dx)}{p_0 + px}. \tag{2}$$

In fact, transforms (1), (2) are connected with the simple relation

$$\Phi(p, p_0) = \int_0^\infty \frac{1}{\tau + p_0} \frac{\partial^2 \prod(p, \tau)}{\partial \tau^2} d\tau, \; (p, p_0) \in R_+^{n+1}.$$

The finite measure $\mu(dx)$ with a carrier in R_+^n is an analytic functional. The Fantappiè indicatrix of this functional of the form

$$\Phi(w, 1) = \int_{x \in R_+^n} \frac{\mu(dx)}{1 + wx}$$

is defined and holomorphic in dual to $R_+^n \subset \mathbb{C}^n$ domain of the form

$$(R_+^n)_\mathbb{C}^* = \{w \in \mathbb{C}^n : \text{Im } w \in \text{con}(-\text{Re } w, R_+^n) \cup \text{con}(\text{Re } w, -R_+^n)\}, \tag{3}$$

where $\text{con}(u, \Gamma)$ denotes convex cone spanned on vector $u \in R^n$ and cone $\Gamma \subset R^n$. The domain $(R_+^n)_\mathbb{C}^*$ contains tube domain $\{w \in \mathbb{C}^n : \text{Re } w \in R_+^n\}$. By the Fantappiè-Martineau theorem (see [5], [6]) the functional $\mu(dx)$ (not necessary positive) is uniquely defined by values of the indicatrix $\Phi(w, 1)$ and its derivatives in any fixed point $w \in (R_+^n)_\mathbb{C}^*$. For non-negative measures with a compact *carrier* in R_+^n it is possible to obtain useful quantitative variant of this uniqueness theory.

Let indicatrix $\Phi(p, 1), p \in R_+^n$ correspond to the non-negative measure $\mu(dx)$ with a compact *carrier* X in R_+^n. We fix a point $q \in R_+^n$ and the natural number K. We consider a convex set $\mathcal{M}_{q,K}$ of non-negative measures $\tilde{\mu}(dx)$ with a *carrier* belonging to a carrier of the measure $\mu(dx)$, whose the Fantappiè indicatrixes $\tilde{\Phi}(p, 1)$ satisfy the equalities

$$\frac{\partial^K \tilde{\Phi}(q, 1)}{\partial p_1^{k_1} \ldots \partial p_n^{k_n}} = \frac{\partial^K \Phi(q, 1)}{\partial p_1^{k_1} \ldots \partial p_n^{k_n}} = m_k, \; |k| \leq K. \tag{4}$$

PROPOSITION 1. *For any two measures μ_1, μ_2 from $\mathcal{M}_{q,K}$ and any holomorphic function φ on the compact $X_\varepsilon = \{z \in \mathbb{C}^n : \inf_{x \in X} |x - z| \leq \varepsilon\}$ there takes place the inequality:*

$$\left| \int_{x \in X} \varphi(x)(\mu_1(dx) - \mu_2(dx)) \right| \leq \frac{\gamma_1}{\varepsilon^{2n}} \exp\{-\gamma_2 K \varepsilon\} \sup_{z \in X_\varepsilon} |\varphi(z)|,$$

where positive γ_1, γ_2 depends only on X, m_0 and q.

This proposition is a consequence of the classical Bernstein theorem [3] about polynomial approximations of real analytic functions. Another proof which gives precise estimates of γ_1, γ_2 is based on the Cauchy-Fantappiè-Leray formula [6] for φ on X_ε and on pluricomplex Schwarz lemma [16] for $\Phi - \tilde{\Phi}$.

With the help of the proposition 1 it is obtained the following well-interpreted answer to the question 1.

We fix the cone $\Gamma_N \subset R_+^n$ consisting of N rays.

THEOREM 1. *Let μ_1 and μ_2 be two positive measures with compact carriers in R_+^n of diameter d and masses not larger than m.*
Let transforms $\prod_1(p, p_0)$ and $\prod_2(p, p_0)$ of these measures of the form (1) coincide when $p \in \Gamma_N, p_0 > 0$. Then for the Fourier transforms $\hat{\mu}_1(\xi), \hat{\mu}_2(\xi)$ of these measures the following estimate is valid

$$|\hat{\mu}_1(\xi) - \hat{\mu}_2(\xi)| \leq \gamma_1 \exp\{-\gamma_2 \frac{N^{\frac{1}{n-1}}}{|\xi|}\}|\xi|^{2n},$$

where $\gamma_j = \gamma_j(n, m, d)$.

Theorem 1 is L_1-analogy of the Logan theorem [10] about reconstruction precision of the Fourier transform of L_2-function with a compact carrier through the finite number of röntgénográms.

In reply to the question 2 a main difficulty is connected with characterization of the Fantappiè indicatrixes of non-negative measures with a carrier in R_+^n. Following Bernstein (see [1], [4]) we call a function $\Phi(p, p_0)$ in R_+^{n+1} completely monotonic if $\Phi(p, p_0)$ is infinitely differentiable when $(p, p_0) \in R_+^n \times \text{int } R_+^1$ and for any multi-index $k = (k_0, k_1, \ldots, k_n)$ the following inequalities are held

$$(-1)^{|k|} \frac{\partial^{|k|} \Phi(p, p_0)}{\partial p_0^{k_0} \partial p_1^{k_1} \ldots \partial p_n^{k_n}} \geq 0. \qquad (5)$$

The following analogy of the Bernstein-Hausdorff theorem [1], [4] about characterization of the Laplace transforms of non-negative measures with a carrier in R_+^n is valid.

PROPOSITION 2. *A function $\Phi(p, p_0)$, $p \in R_+^n$, $p_0 > 0$ is the Fantappie indicatrix of the form (2) for some non-negative measure μ with finite power moments of any order iff*
a) Φ is completely monotonic in R_+^{n+1} and
b) $\Phi(\lambda p, \lambda p_0) = \lambda^{-1} \Phi(p, p_0)$, $\lambda > 0$, $p \in R_+^n$, $p_0 > 0$.

It follows from the proposition 2 that any function $\Phi(p, p_0)$ with properties a) and b) admits for any fixed $p_0 > 0$ analytic extension in p to the domain $(R_+^n)_{\mathbb{C}}^*$ of the form (3).

For applications it turned out to be very essential [7] that inequalities (5) are not independent. Particularly significantly this fact becomes apparent under a priori condition that a function Φ is real-analytic. Namely if a function Φ is real-analytic in convex domain $G \subset R^{n+1}$, and satisfies (5) in a point $(q, q_0) \in G$ then Φ satisfies the same inequalities (5) in domain $G \cap \{(p, p_0) \in R^{n+1} : p \leq q, p_0 \leq q_0\}$.

Contemporary variants (going back to S.N. Bernstein) of the theorem about separate analyticity [3], [9] make it possible to strengthen the proposition 2. For a vector $\xi = (\xi_1, \ldots, \xi_n)$ we set

$$D_\xi = \sum_{j=1}^n \xi_j \frac{\partial}{\partial p_j}.$$

PROPOSITION 3 ([7]). *In order that a function $\Phi(p, p_0)$, $p \in R_+^n$, $p_0 > 0$ would be represented in the form (2) where μ a non-negative measure with a compact carrier in R_+^n it is necessary and sufficient that*

a) $\lambda \Phi(\lambda p, \lambda p_0) = \Phi(p, p_0)$ for all $p > 0$, $p_0 > 0$, $\lambda > 0$ and in addition $\Phi(p, p_0)$ as a function in p belongs to $C^\infty(R_+^n)$ for fixed $p_0 > 0$,

b) for any fixed $p \in \text{int } R_+^n$ the function $\varphi_p(\lambda, p_0) = \Phi(\lambda \cdot p, p_0)$ is completely monotonic in λ and p_0 for $\lambda \geq -\varepsilon(p, p_0)$, $p_0 > 0$, where $\varepsilon(p, p_0) > 0$,

c) for any open cone Γ from R_+^n and some $q \in \Gamma$ and $q_0 > 0$ the following inequalities are held

$$(-1)^{k+m} D_{\xi^{(1)}} \ldots D_{\xi^{(k)}} \frac{\partial^m \Phi(q, q_0)}{\partial p_0^m} \geq 0, \qquad (6)$$

for any $\xi^{(1)}, \ldots, \xi^{(k)} \in \Gamma$, $k, m = 0, 1, 2, \ldots$.

On account of the proposition 3 it is obtained the following efficient in applications [7] reply to the question 2.

THEOREM 2 ([7]). *In order that a function* $\prod(p, p_0)$ *for* $p \in R_+^n$ *and* $p_0 > 0$ *would be represented in the form (1), where* μ *non-negative absolutely continuous in zero measure with a compact carrier* R_n^+ *it is necessary and sufficient that*

a) $\prod(p, p_0)$ *is a homogeneous of the first order convex function in* $R_+^n \times R_+^1$,

b) $\dfrac{\partial^2 \prod(p, p_0)}{\partial p_0^2} = 0$ *, if* $p_0 > \text{const}(p)$,

$$\lim_{p_0 \to +0} \prod(p, p_0) = \lim_{p_0 \to +0} \frac{\partial \prod}{\partial p_0}(p, p_0) = 0,$$

c) *for fixed* $p_0 > 0$ *the function*

$$\Phi(p, p_0) = 2 \int_0^\infty \frac{\prod(p, \tau)}{(\tau + p_0)^3} d\tau$$

belongs to $C^\infty(R_+^n)$ *and for any open convex cone* Γ *from* R_+^n *and some* $q \in \Gamma$ *and* $q_0 > 0$ *the inequalities (6) are held.*

Now we will discuss on opportunity of inversion of transforms (1), (2). In case when the Fantappiè indicatrix $\Phi(p, 1)$ is given in all dual domain (3) the corresponding inversion formula is obtained in works of Fantappiè and Martineau (see [5], [6]).

Peculiarity of this case consists in that the indicatrix $\Phi(p, p_0)$ is given only for $(p, p_0) \in R_+^{n+1}$.

Let for simplicity the measure $\mu(dx)$ have a continuous density with a carrier in $\text{int } R_+^n$, that is $\mu(dx) = b(x) dx$, where $b(x) \in C_0(R_+^n)$. Then the following inversion formula of the transform (2) is valid:

$$b\left(\frac{\alpha_1}{\alpha_0 q_1}, \ldots, \frac{\alpha_n}{\alpha_0 q_n}\right) =$$

$$= \lim_{k \to \infty} (-1)^{k(\alpha_0 + \ldots + \alpha_n) - 1} \frac{\Gamma(k(\alpha_0 + \ldots + \alpha_n)) q_1^{\alpha_1 k} \ldots q_n^{\alpha_n k}}{\Gamma(k\alpha_1 + 1) \ldots \Gamma(k\alpha_n + 1) \cdot \Gamma(k\alpha_0 - n)} \qquad (7)$$

$$\cdot \frac{\partial^{k(\alpha_0 + \ldots + \alpha_n) - 1} \Phi(q, 1)}{\partial p_0^{k\alpha_0 - 1} \partial p_1^{k\alpha_1} \ldots \partial p_n^{k\alpha_n}},$$

where $\Gamma(\cdot)$ the Euler function, $\alpha_0, \alpha_1, \ldots, \alpha_n$ natural numbers.

Besides formula (7) using derivatives (all orders) of the indicatrix $\Phi(p, p_0)$ for one fixed point $(q, 1)$ in order to reconstruct $b(x)$ we point to one integral inversion formula of the transform (2):

$$b(x) = \lim_{R \to \infty} \int_{p \in R_+^n} K_R(p_1 x_1, \ldots, p_n x_n) \Phi(p, 1) \, dp_1 \ldots dp_n, \tag{8}$$

where the limit is regarded in the sense of L_2-convergence in compacts of R_+^n,

$$K_R(u_1, \ldots, u_n) =$$

$$= \frac{1}{(2\pi)^n} \int_{\{\tau \in R^n : |\tau_j| \leq R, j=1,2,\ldots,n\}} u_1^{-i\tau_1} \ldots u_n^{-i\tau_n} \cdot$$

$$\cdot \frac{d\tau_1 \ldots d\tau_n}{\Gamma(1 - i\tau_1)\Gamma(1 - i\tau_2) \ldots \Gamma(1 - i\tau_n)\Gamma(i\tau_1 + \ldots + i\tau_n - n + 1)}.$$

Formulas (7), (8) may be regarded as multidimensional analogies of the Widder and the Paley-Wiener inversion formulas [11] for the Laplace and the Stieltjes one-dimensional transforms. These formulas are unstable from the computational point of view because the problem is not correct. Nevertheless the formula (8) can not be essentially improved. It implies from results [12] where optimal formula in related inversion problem of the Radon transform of finite function through incomplete data is obtained. Mentioned formulas do not take into account a priori restriction – non-negativity of the measure $\mu(dx)$. Using of this condition in one-dimensional case made it possible even in last age (P. Tchebycheff, A. Markov, T. Stieltjes) to obtain effective inversion algoritms of transform (2) which go under the name of the rational Padé approximations.

Very important unsolved problem is to find multidimensional analogies of the Padé approximations. We will make several remarks on this problem.

PROPOSITION 4. *Any homogeneous of the order* (-1) *rational completely monotonic function* $R(p, p_0), p \in R_+^n, p_0 > 0$ *is represented in the form*

$$R(p, p_0) = \sum_{j=1}^{N} \frac{c_j}{p_0 + px^{(j)}},$$

where $c_j > 0, x^{(j)} \in R_+^n, j = 1, 2, \ldots, N$.

This proposition shows that an approximation of indicatrixes (2) by rational completely monotonic indicatrixes is related to an approximation of the measure $\mu(dx)$ by non-negative measures $\mu_N(dx)$ concentrated in the finite number of points:

$$\mu_N(dx) = \sum_{j=1}^{N} c_j \delta(x - x^{(j)}) dx.$$

Let a function $\Phi(p, p_0)$ be completely monotonic in R_+^{n+1} and have homogeneity of the order (-1). We also fix a point $q \in R_+^n$ and the natural number K. We consider a convex set $\widetilde{\mathcal{M}}_{q,K}$ of all homogeneous of the order (-1) completely monotonic function $\widetilde{\Phi}(p, p_0)$ in R_+^{n+1} for which equalities (4) are valid in a point $q \in R_+^n$.

PROPOSITION 5. *Any extreme function in $\widetilde{M}_{q,K}$ represents a rational completely monotonic function of the order not exceeding C^n_{n+K}.*

This proposition is a consequence of proposition 2, 4 and classical results on moment problem [1], [2], [15].

The problem is to find constructively even if one of these extreme rational completely monotonic functions and to estimate minimal possible degree of such a function.

Consider now relation of described results to mathematical economics.

Attemps of mathematical description of technological structure of economics go back to works of V. Leontiev. In works of H. Houthakker and L. Johansen this direction received its development which made it possible to explain qualitatively many economic phenomena.

The following model of pure industry producing homogeneous outputs and using n kinds of production factors of current use (PFCU) is laid as a basis of technological description of economics. Suppose that in industry there are different technological production processes every of which requires input of n kinds of PCFU in given proportions. Then every technology is given by a vector $x = (x_1, \ldots, x_n)$ coefficients of PCFU input of per unit production output. Intensities of technologies use are restricted by available capacities in industry. Suppose that when building up capacities there is selection of technology by which this capacity is functioning. Then at any moment of time the capacities of industry turn out to be distributed according to technologies. Denote by μ the corresponding non-negative measure in R^n_+. Let $l = (l_1, \ldots, l_n)$ be a vector of PFCU going into disposal of this industry. Denote by $u(x)$ a coefficient of capacity load corresponding to technology x.

The production function $F(l)$ is called function which associates to the vector $l \geq 0$ maximum value of output $\int_{R^n_+} u(x)\mu(dx)$ under natural restrictions

$$\int_{R^n_+} xu(x)\mu(dx) \leq l,\ 0 \leq u(x) \leq 1.$$

In [13] it is supposed along with production function $F(l)$ to consider profit function $\prod(p, p_0)$ which is connected with $F(l)$ by the Legendre transform

$$\prod(p, p_0) = \max_{l \geq 0}[p_0 F(l) - pl], \qquad (9)$$

where $p_0 > 0$ is the price of output; $p = (p_1, \ldots, p_n) \geq 0$ – prices of PCFU used by the industry; $\prod(p, p_0)$ – summary profit of the industry. Profit function and capacity distribution on technologies turn out to be connected with integral transform of the Radon type (1).

In order to introduce the notion of production capacities for more complicated systems of industries using intermediate products and (or) producing inhomogeneous products it is necessary to consider problems of aggregation. It implies from formulas expressing agregated profit function through profit functions of pure industries [14] that they always satisfy conditions a), b) of theorem 2. However the condition c) of this theorem for agregated profit function may be broken. Interesting economic phenomena are related to this fact [7].

As the simplest application of theorem 2 we will point to conditions [7] of representation of production functions of the type CES, which are popular with econometrics, through capacity distribution or technologies. CES-function has the form

$$F(l_1, l_2) = (\beta_1 l_1^{-\rho} + \beta_2 l_2^{-\rho})^{-\gamma/\rho},$$

where $\beta_1 > 0, \beta_2 > 0, \rho \geq -1, 0 < \gamma \leq 1$.

Formula (9) makes it possible to compute profit function $\prod(p_1, p_2, p_0)$.

It follows from theorem 2 that to the funtion $F(l_1, l_2)$ corresponds some efficiency distribution of technologies if only $\rho > -1$ and $\gamma < 1$.

REFERENCES

[1] N.I. Akhiezer, *The classical moment problem and some related questions in analysis*, Hafner, New York, 1965.

[2] C. Berg, «The multidimensional moment problem and semigroups», *Proceedings of Symposia in Applied Mathematics*, 1987, vol. 37, 110-123.

[3] S. Bernstein, *Sur l'ordre de la meilleure approximation des fonctions continues par des polynomes de degré donné*, Bruxelles, 1912.

[4] S. Bochner, *Harmonic analysis and theory of probability*, Univ. Calif. Press, 1955.

[5] L. Fantappiè, «L'indicatrice proiettiva dei funzionali lineari e i prodotti funzionali proiettivi», *Annali di Mat.* 1943, v.20, serie 4.

[6] G.M. Henkin, Method of integral representation in complex analysis, in *Several Complex Variables I*, (Encycl. Math. Sc.) Springer-Verlag, 1989. (Trans. from Russian ed., Moscow, 1985).

[7] G.M. Henkin and A.A. Shananin, The Bernstein theorem and the Radon transform. Application in production function theory. *Computational cybernetics*, 1989, v. 157, (to appear in Russian).

[8] L. Johanson, *Production functions*, North Holland, 1972.

[9] J. Korevaar and J. Wiegenerinck, «A representation of mixed derivatives with an application to the edge-of-the-wedge theorem», *Proc. Nederl. Akad. Wetensch. Ser. A.*, 1985, v. 88, p. 77-86.

[10] B.F. Logan, The uncertainty principle in reconstructing functions from projections, *Duke Math. J.*, 1975, v. 42, N4, p. 661-706.

[11] R. Paley and N. Wiener, *Fourier transforms in complex domain*, Providence: AMS, 1934.

[12] V. Palamodov, A. Denisjuk, *Inversion des transformations de Radon d'après les donnés non complètes*, C.R. Acad. Sc. Paris, Série 1, 1988, t. 307, p. 181-183.

[13] A.A. Shananin, Investigation of one class of production (and profit) functions, appearing in macrodescription of econom. systems. *Journ. of computational mathem. and mathem. physics*, 1984, v.24, 1799-1811; 1985, v. 25, 53-65 (in Russian).

[14] A.A. Shananin, Agregative description systems of industries with the aim of product index, in *Mathem. modelling. Processes in the complex economics and ecological systems*, Moscow «Nauka», 1986, p. 106-147, (in Russian).

[15] V. Tchakaloff, Formules de cubatures mécaniques a coefficients non negatifs, *Bull. Sci. Math.*, Ser. 2, 1957, v. 81, N3, 123-134.

[16] P. Lelong, Notions capacitaires et fonctions de Green pluricomplexes dans espaces de Banach, *C.R. Acad. Sc. Paris*, Série 1, 1987, t. 305, p. 71-76.

Microlocal Analysis on Three Dimensional CR Manifolds

J. J. KOHN Department of Mathematics, Princeton University
Fine Hall, Princeton N. J. 08544, U.S.A.

The aim of this lecture is to illustrate how microlocal analysis is intimately connected with CR geometry. Here we will show it in the study of the operators $\bar{\partial}_b$ and the Szegö operator S_b. We will restrict our attention to three dimensional manifolds, which is somewhat anomalous (the $\bar{\partial}_b$-cohomology of $(0,1)$-forms has infinite dimension) at the same time CR structure in dimension three is particularly simple since locally it is determined by a single complex vector field and hence the ingrability condition is automatically satisfied.

DEFINITION: Let M be a three dimensional manifold. A CR structure on M is a subbundle of $T^{1,0}(M)$ of the complexified tangent bundle $CT(M)$ satisfying the following conditions:

(i) The fiber dimension of $T^{1,0}(M)$ is one.
(ii) $T^{1,0}(M) \cap \overline{T^{1,0}(M)} = \{0\}$, where $\overline{T^{1,0}(M)}$ denotes the complex conjugate of $T^{1,0}(M)$ and $\{0\}$ denotes the zero section in $CT(M)$.

A CR manifold consist of a manifold M together with a CR structure on M.

Set $T^{1,0}(M) = \overline{T}^{1,0}(M)$ let $\mathcal{B}^{0,1}(M)$ denote the $(0,1)$-forms i.e. section the dual bundle of $T^{0,1}(M)$. The exterior derivative induced by this bundle is denoted by $\bar{\partial}_b$, with $\bar{\partial}_b : C^\infty(M) \to \mathcal{B}^{0,1}(M)$. If L is a local section of $T^{1,0}(M)$ over an open set U and $u \in C^\infty(U)$ then $< \bar{\partial}_b u, \overline{L} > = \overline{L}(u)$. A CR function on M is a function u that satisfies $\bar{\partial}_b u = 0$.

If X is a complex manifold of dimension $n \geq 2$ and if $M \subset X$ such that for each $x \in M$ we have $\dim \mathbb{C}T_x(M) \cap T_x^{1,0}(X) = 1$ then the subbundle $\mathbb{C}T(M) \cap T^{1,0}(X)$ defines a CR structure on M which is called the induced CR structure. A CR manifold M is embeddable in a complex manifold M if there exists a C^∞ diffeomorphism $\varphi : M \to X$ such that the induced structure on $\varphi(M)$ coincides with $\varphi^*(T^{1,0}(M))$. Rossi (see [R]) found a remarkable example of a complex strongly pseudo-convex real analytic CR manifold M which is not embeddable in $\mathbb{C}^n$ for any n. Rossi's example can be described as follows. Let $S^3 \subset \mathbb{C}^2$ be the unit sphere given by $S^3 = \{(z_1, z_2) \in \mathbb{C}^2 \mid |z_1|^2 + |z_2|^2 = 1\}$. Let L be the tangent vector field on S^3 defined by

$$(1) \qquad L = \bar{z}_2 \frac{\partial}{\partial z_1} - \bar{z}_1 \frac{\partial}{\partial z_2}$$

for each small t define $L_t = L + t\bar{L}$. Then for t small L_t defines a CR structure on S^3 the resulting CR manifold is denoted by M_t. Then for $t \neq 0$ every CR function on M_t is even. If M_t were embeddable in $\mathbb{C}^n$ then the pullbacks $w_j \circ \varphi$ of the coordinate functions $w_1, \ldots, w_n$ would be CR functions that separate points. To continue this discussion we need to recall the notion of pseudo-convexity. In a three dimensional CR manifold each point has a neighborhood U on which there is a non-vanishing vector field L with values in $T^{1,0}(M)$. Then there exists a vector field T on U such that at each point of $x \in U$ the vectors $T_x, L_x, \bar{L}_x$ are linearly independent. We will choose T so that $\bar{T} = -T$. Then we express the commutator $[L, \bar{L}] = L\bar{L} - \bar{L}L$ by

$$(2) \qquad [L, \bar{L}] = \lambda T + aL + b\bar{L}.$$

DEFINITION: M is pseudo-convex if in some neighborhood of each point there exists a vector field T so that $\lambda \geq 0$, it is called strongly pseudo-convex if $\lambda > 0$ (note that both of these conditions are independent of the choice of L).

Note that the above M_t is strongly pseudo-convex, this is contrast to a theorem of Boutet de Monvel (see [B]) which asserts that if M is a strongly pseudo-convex compact CR manifold of dimension greater than three then it is embeddable in $\mathbb{C}^n$ for some n. Burns, in [Bu], analyzed Boutet de Monvel's proof and showed that if M is compact, strongly pseudo-convex and if $\dim M = 3$ then it is embeddable under the additional two conditions:

(a) The range of $\bar{\partial}_b$ is closed in L_2.
(b) There is a pseudo local operator Φ mapping the range of $\bar{\partial}_b$ into the domain of $\bar{\partial}_b \Phi = \bar{\partial}_b$.

In higher dimensions these two conditions are always satisfied see [K1]. In [K2] it is shown that in dimension three (a) implies (b) and in [K3] it is shown that (a) holds whenever M is a smooth boundary of a domain in a Stein manifold. The purpose of this lecture is to indicate how microlocal analysis is used to prove the above results and also how it is used to obtain Hölder estimates on CR manifolds of finite type.

REMARK: The above discussion concerns global embedding. The problem of local embedding is: given $x \in M$ does there exist a neighborhood U of x which is embeddable in $\mathbb{C}^n$. In the real analytic case local embeddability in $\mathbb{C}^n$, with $n = \frac{1}{2}(\dim M + 1)$, follows by use of the Cauchy-Kowalevski theorem, without any pseudo-convexity assumptions. In the C^∞ case Nirenberg (see [N]) found an example of a strongly pseudo-convex three dimensional CR manifold on which all local CR functions are constant and hence it is not locally embeddable. For $\dim M \geq 9$ Kuranishi in [Ku] proved that if M is strongly pseudo-convex then it is locally embeddable in $\mathbb{C}^n$ with $n = \frac{1}{2}(\dim M + 1)$. This result was generalized by Akahori in [A] to $\dim M = 7$. Whether it is true in dimension five is an open problem.

Now suppose that Ω is a bounded domain in a two dimensional complex manifold X. Suppose further that the boundary of Ω is a C^∞ pseudo-convex manifold, which we denote by M, and that there exists a strongly plurisubharmonic function in a neighborhood of M. (In case $X = \mathbb{C}^2$ we could take $|z_1|^2 + |z_2|^2$ to be such a function and if X is a Stein manifold the required plurisubharmonic function can be taken a $\sum_1^N |h_j|^2$ where $\{h_1, \ldots, h_N\}$ is a set of holomorphic functions so that for each $x \in M$ the vector space spanned by $\{dh_j(x)\}_{j=1,\ldots N}$ is two dimensional.) Under these assumptions the range of $\bar{\partial}_b$ in L_2 is closed. We will explain how microlocalization is used to prove this. The detailed proofs are given in [K3]. In case $X = \mathbb{C}^2$ another proof due to Boas and Shaw is given in [BS].

To start we fix L_2 inner products on functions and on $(0,1)$-forms, these norms are all equivalent so it does not matter which we choose. Let $\mathcal{H}_b$ denote the space of square-integrable functions on M which satisfy the equation $\bar{\partial}_b h = 0$ in the distribution sense. Closed range of $\bar{\partial}_b$ is equivalent to the following estimate: There exists $C > 0$ such that

(3) $$\|u\| \leq C\|\bar{\partial}_b u\|$$

for all u in the domain of $\bar{\partial}_b$ which are orthogonal to $\mathcal{H}_b$. Here $\|\ \|$ denotes the L_2 norm of functions and of $(0,1)$-forms. By the domain of $\bar{\partial}_b$ we mean the set of $u \in L_2$ such that the distribution $\bar{\partial}_b u$ is square integrable. The proof of (3) now proceeds as follows, we want to construct an operator Φ bounded in L_2 which maps the range of $\bar{\partial}_b$ into the domain of $\bar{\partial}_b$ such that

(4) $$\bar{\partial}_b \Phi \bar{\partial}_b = \bar{\partial}_b.$$

Once such a Φ is obtained then (3) follows since $u = \Phi\bar{\partial}_b u - S_b \Phi \bar{\partial}_b u$, where $S_b : L_2 \to \mathcal{H}_b$ is the orthogonal projection. Let α be a $(0,1)$-form on Ω such that $\alpha \wedge \bar{\partial}r\big| = \bar{\partial}_b u \wedge \bar{\partial}r$. Now if $\alpha \in H^s(\Omega)$, that is each component of α is in the L_2 Sobolev space $H^s(\Omega)$, then, combining the results of [KR] and [K4] we find that there exists a $(0,1)$-form $\alpha \in H^{s-1}(\Omega)$ such that $\sigma \wedge \bar{\partial}r\big|_M = 0$, $\bar{\partial}\sigma = \bar{\partial}\alpha$ and $\frac{\partial\sigma}{\partial \bar{z}_j} \in H^{s-1}(\Omega)$. Setting $\beta = \alpha - \sigma$, we have $\bar{\partial}\beta = 0$ thus (using again the results in [K4]) there exists $v \in H^{s-1}(\Omega)$ with v orthogonal to the space of holomorphic functions in $L_2(\Omega)$ and such that $\bar{\partial}v \in \beta$. Furthermore we have $v, \frac{\partial v}{\partial \bar{z}_j} \in H^{s-1}(\Omega)$. Now set $\Phi \bar{\partial}_b u = v\big|_M$, then clearly Φ satisfies (4). It remains to be shown that Φ is bounded. If $\bar{\partial}_b u \in L_2(M)$ the form α can be chosen to be in $H^{1/2}(\Omega)$ that is there exists a linear bounded operator taking the range of $\bar{\partial}_b$ into $H^{1/2}(\Omega)$. Also σ can be chosen to be the unique form satisfying $\bar{\partial}\sigma = \bar{\partial}\alpha$,

$\sigma \wedge \bar{\partial}r\big|_M = 0$ and σ orthogonal to all the forms θ such that $\bar{\partial}\theta = 0$ and $\theta \wedge \bar{\partial}r\big|_M = 0$. So the linear operator that takes $\bar{\partial}_b u$ to σ is a bounded operator from the range of $\bar{\partial}_b$ to the space $H^{1/2}(\Omega)$. Now, $v \in H^{1/2}(\Omega)$ and hence $v|_M \in H^{-1}(M)$ so that Φ is a bounded operator from the range of $\bar{\partial}_b$ to $H^{-1}(M)$. We want to prove that Φ is bounded in $L_2(M)$ and that is where we will need microlocalization.

REMARK: The spaces H^s used here are a variant of the usual Sobolev spaces (see [K3]) for which the restriction and extension theorems hold as stated.

In a neighborhood of a point $P \in M$ we choose coordinates x, y, t with origin P and vector fields L, $\bar{L}$, T such that $T = \frac{1}{\sqrt{-1}}\frac{\partial}{\partial t}$ and so that at the origin we have $L|_0 = \frac{1}{2}(\frac{\partial}{\partial x} - \sqrt{-1}\frac{\partial}{\partial y})$. Denote by ξ, η, τ the coordinates dual to x, y, t. A function u defined on M supported in small neighborhood of P can be split as $u \sim u^0 + u^+ + u^-$ where $\hat{u}^+$, the Fourier transform of u^+, is supported in a conical neighborhood of $(0,0,1)$, $\hat{u}^-$ is supported in a conical neighborhood of $(0,0,-1)$ and $\hat{u}^0$ is zero in the union of a conical neighborhood of $(0,0,1)$ and a conical neighborhood of $(0,0,-1)$. If u is defined on $\bar{\Omega}$, where U is a small neighborhood of P in X then we also split u as above. To do this we consider r the defining function of Ω, that is r is a C^∞ function in a neighborhood of M with $dr \neq 0$ and $r = 0$ on M. Now x, y, t, r are coordinates on U and so u is a function of these variables. To define the above splitting for fixed r we take the Fourier transform only with respect to the variables x, y, and t.

LEMMA. *Suppose* u, $\frac{\partial u}{\partial z_1}$, *and* $\frac{\partial u}{\partial z_2}$ *are in* $H^{s-1}(\Omega)$ *then* u^0 *and* u^+ *are in* $H^s(\Omega)$.

The idea of this proof follows, for details see [KS]. Given a first order differential operator F defined on $U \cap \bar{\Omega}$ then we will associate a first order pseudo-differential operator $F^\#$ which will act on functions supported in $U \cap M$. Let E be a second order elliptic differential operator on Ω which equals the laplacian $\Delta = -(\frac{\partial^2}{\partial \bar{z}_1 \partial z_1} + \frac{\partial^2}{\partial \bar{z}_2 \partial z_2})$ in $U \cap \bar{\Omega}$. If w is defined on M and supported in $U \cap M$ we let h be the solution of $Eh = 0$ such that $h|_M = w$. We define $F^\# w = (Fh)_M$.

We define the vector fields L_1, and L_2 by

(5)
$$L_1 = r_{z_2}\frac{\partial}{\partial z_1} - r_{z_1}\frac{\partial}{\partial z_2}$$
$$L_2 = \frac{1}{|r_{z_1}|^2 + |r_{z_2}|^2}(r_{\bar{z}_1}\frac{\partial}{\partial z_1} + r_{\bar{z}_2}\frac{\partial}{\partial z_2})$$

then $\frac{\partial}{\partial r} = \frac{1}{2}(L_2 + \bar{L}_2)$ and we set $T = \frac{1}{2}(L_2 - \bar{L}_2)$, if the coordinates and signs are chosen correctly this T agrees with the T defined above. We then have, ignoring first order terms,

$$\Delta \sim -\frac{\partial^2}{\partial r^2} + (T\prime)^2 - \frac{1}{2}(L_1\bar{L}_1 + \bar{L}_1 L_1)$$

so that $\sigma(\frac{\partial}{\partial r^2}^\#)$ is given by

(6)
$$\sigma(\frac{\partial^\#}{\partial r}) \sim \sqrt{\sigma(T)^2 + |\sigma(L_1)|^2}$$

and hence

(7)
$$\sigma(L_2^\#) \sim \sqrt{\sigma(T)^2 + |\sigma(L_1)|^2} + \sigma(T).$$

Now $\Delta u^+ \in H^{s-2}(\Delta)$ and $L_2 u^+ \in H^{s-1}(\Omega)$ hence denoting by u_b^+ the restriction of u^+ to M we have $L_2^\# u_b^+ \in H^{s-3/2}(M)$ then since by (7) the symbol of $L_2^\#$ is elliptic on the support of $\hat{u}_b^+$ we conclude that $u_b^+ \in H^{s-1/2}(M)$ and hence $u^0 \in H^s(\Omega)$. That $u^0 \in H^s(\Omega)$ follows from the fact that L_1 is elliptic on the support of $\hat{u}_b^0$. This concludes the outline of the proof of the lemma.

Returning to the discussion of the proof of (3), we apply the lemma to see that if $\alpha \in H^s(\Omega)$ then σ^0 and σ^+ are in $H^s(\Omega)$. It then follows by an extension of the arguments of [K4] (for details see [K3]) that v^0 and v^+ are in $H^s(\Omega)$. Applying the lemma again we conclude that, since v and $\frac{\partial v}{\partial \bar{z}_j}$ are in $H^{s-1}(\Omega)$ then $v^- \in H^s(\Omega)$. Combining these we see that $v \in H^s(\Omega)$ and have $v|_M \in H^{s-\frac{1}{2}}(M)$ so, since we can take $s = \frac{1}{2}$ we have $v|_M$ which shows that Φ is bounded in L_2 and concludes the proof of (3).

Now we turn to the problem of regularity of the solutions of

(8)
$$\bar{\partial}_b u = \alpha.$$

Note that if h is a CR function then $u + h$ is also a solution of (8). Since CR functions can be highly non-smooth (they are boundary values of holomorphic functions) we see that only special solutions of (8) are smooth. Here we will study the solution u which is orthogonal to the space $\mathcal{H}_b$ consisting of square integrable CR functions. Assuming that the range of $\bar{\partial}_b$ is closed we see, by standard Hilbert space theory, that the condition u orthogonal to $\mathcal{H}_b$ is equivalent to the existence of a $(0,1)$-form φ in the domain of the $\bar{\partial}_b^*$, the adjoint of $\bar{\partial}_b$, such that

(9)
$$\bar{\partial}_b^* \varphi = u.$$

The notion of finite type, defined below, (as in [K5]) generalizes strong pseudo-convexity and is the correct setting to prove that the solution of (8) satisfying (9) is smoother than α.

DEFINITION: $P \in M$ is of finite type if the non-vanishing vector field L on a neighborhood of P which has values in $T^{1,0}(M)$ has the property that for some k $(L\bar{L})^k \lambda\big|_P \neq 0$. This is equivalent to the Hörmander condition (see [H]) that the Lie algebra generated by $\operatorname{Re}(L)$ and $\operatorname{Im}(L)$ evaluated at P equals the tangent space at P.

For a pseudo-convex hypersurface M in $\mathbb{C}^2$ the finite type condition can be described as follows. Finite type of P is equivalent to the condition that there exists an integer m such that the order of contact with M of every complex analytic curve through P is less than m. We will discuss the following result (see [K2]).

THEOREM. *Suppose that M is a compact pseudo-convex three-dimensional compact CR manifold such that the range of $\bar{\partial}_b$ is closed in L_2. If α is in the range of $\bar{\partial}_b$ and* is in H^s in a neighborhood of P then the solution of (8) which satisfies (9) is in $H^{s+\varepsilon}$ in a neighborhood of P, the optimal ε here is $\frac{1}{m}$ where m is the type of P.

The above theorem amounts to the following estimate

(10) $$\|\zeta u\|_{s+\varepsilon} \leq C(\|\zeta\prime\overline{\partial}_b u\|_s + \|\overline{\partial}_b u\|),$$

for all u orthogonal to $\mathcal{H}_b$ whenever $\overline{\partial}_b u$ is in H^s in a neighborhood of $\text{supp}(\zeta\prime)$, where $\zeta\prime = 1$ in a neighborhood of $\text{supp}(\zeta)$. The points in $\text{supp}(\zeta\prime)$ are of type less than or equal to m.

Our starting point is the inequality

(11) $$\|u\|_\varepsilon^2 \leq C(\|Lu\|^2 + \|\overline{L}u\|^2 + \|u\|^2),$$

which holds for all u with support near P. This inequality with $\varepsilon < \frac{1}{m}$ (and C depending on ε) was proved in [**H**] for $\varepsilon = \frac{1}{m}$ the proof was obtained by Rothschild and Stein (see [**RS**]).

The operator $\overline{\partial}_b$ is locally equal to $\overline{L}$, where L is a suitable non-vanishing tangent vector of $(1,0)$ defined on a (possibly small) open set. We denote by $|T|$ a first order pseudo-differential operator with a smooth symbol defined in a neighborhood of P such that the symbol of $|T|$ denoted by $\sigma(x,y,t,\xi,\eta,\tau)$ equals $|\tau|$ when $|\tau| \geq 1$ and (ξ,η,τ) are in the union of two conical neighborhoods of $(0,0,1)$ and $(0,0,-1)$. Next we define the pseudo-differential operator A by

(12) $$A = -\frac{1}{2}(L\overline{L} + \overline{L}L) + \frac{1}{2}\lambda|T|.$$

Observe that A is also given by:

(13) $$A = \begin{cases} -L\overline{L} + \frac{1}{2}\lambda(|T|+T) & \mod(L,\overline{L}) \\ -\overline{L}L + \frac{1}{2}\lambda(|T|-T) & \mod(L,\overline{L}). \end{cases}$$

Hence

(14) $$Au^- \sim -L\overline{L}u^-$$

and

(15) $$Au^+ \sim -\overline{L}Lu^+.$$

From (10) it follows that

(16) $$\|u\|_\varepsilon^2 \leq C(|(Au,u)| + \|u\|^2).$$

By standard methods we can then conclude that, locally, A^{-1} maps H^s into $H^{s+2\varepsilon}$, furthermore LA^{-1}, $A^{-1}L$, $\overline{L}A^{-1}$, and $A^{-1}\overline{L}$ map H^s into $H^{s+\varepsilon}$, and L^2A^{-1}, $A^{-1}L^2$,

$LA^{-1}L$, $\overline{L}^2 A^{-1}$, $A^{-1}\overline{L}^2$, $\overline{L}A^{-1}\overline{L}$, $LA^{-1}\overline{L}$ and $\overline{L}A^{-1}L$ map H^s into H^s. From (8) we have $\overline{L}u^- \sim \alpha^-$ so that $L\overline{L}u^- \sim L\alpha^-$, applying (14) we have

(17) $$u^- \sim -A^{-1}L\alpha^-.$$

From (8) and (9) we have $\overline{L}L\varphi^+ \sim \alpha^+$ so that from (15) we have $\varphi^+ \sim -A^{-1}\alpha^+$ and since $u^+ \sim L\varphi^+$ we have

(18) $$u^+ \sim -LA^{-1}\alpha^+.$$

Finally, since $\overline{L}$ is elliptic on supp $\hat{u}^0$ we conclude that if $\alpha \in H^s$ locally then $u^0 \in H^{s+1}$. Putting all this together (and filling in a great many details) shows that $u \in H^{s+\varepsilon}$ as required.

COROLLARY: The Szegö projection operator $S_b : L_2(M) \to \mathcal{H}_b(M)$ which is the orthogonal projection preserves H^s near a point of finite type. That if $f \in L_s(M)$ and $f \in H^s$ in a neighborhood of P then $S_b(f)$ is also in H^s in a neighborhood of P.

The above is true since if $u \perp \mathcal{H}_b$ and $\overline{\partial}_b u = \overline{\partial}_b f$ then $S_b f = f - u$. Using (17) and (18) we have

$$u^- \simeq A^{-1}L\overline{L}f^-$$
$$u^+ \simeq -LA^{-1}\overline{L}f^+.$$

So the corollary follows from the mapping properties of A^{-1}.

Next we take up the question of Hölder estimates for the solution of (8) which satisfies (9). The above argument reduces the problem to the proof of analogous mapping properties of A^{-1} with respect to the Hölder norms. This task has been carried out by Fefferman and Kohn in [**FK1**] and by Christ in [**Ch1**]. Christ works with a version of A in which $|T|$ is defined to be the operator whose symbol equals $|\tau|$ everywhere. This operator has the advantage of being compatible both with euclidean geometry and with the geometry imposed by the operator $L\overline{L} + \overline{L}L$, however it has the disadvantage of having a having a non-smooth symbol. Here we will discuss briefly the methods used in [**FK1**]. Related results were obtained by Nagel, Rosay, Stein, and Wainger (see [**NRSW**]) by Catlin (see [**C**]) and by McNeal (see [**M**]).

To emphasize that A is real we set $X = \text{Re}(L)$, $Y = \text{Im}L$ and we have $A = -X^2 - Y^2 + \lambda|T|$. Hölder regularity for Hörmander operators, such as $X^s + Y^2$ was established by Rothschild and Stein in [**RS**]. The underlying geometry for such operators is analyzed in [**NSW**] and in [**C**]. In our case this underlying geometry is described as follows. Given P, Q in M we define a non-euclidian distance as the infimum of the length of all piecewise smooth path joining P and Q whose pieces are integral curves of X or of Y. We denote by $B(P,\gamma)$ the "ball" consisting of all Q whose non-euclidean distance to P is less than γ. Set $\delta(P,\gamma)$ equal to the euclidean distance of P to the boundary of $B(P,\gamma)$ and $\gamma(P,\delta)$ is defined implicitly by $\delta = \delta(P,\gamma(P,\delta))$. If P is of type m then

(20) $$C_1\delta^{1/2} \leq \gamma(P,\delta) \leq C_2\delta^{1/m}.$$

To prove the function is of Hölder class s we will use the following microlocal characterization. Let $\psi \in C_0^\infty(\mathbb{R}^3)$ such that $\mathrm{supp}(\psi) \subset \{\xi \in \mathbb{R}^3 \mid \tfrac{1}{2} < |\xi| < \tfrac{5}{2}\}$ such that $\psi(\xi) = 1$ if $1 < |\xi| < 2$.

For each $\delta \in (0, 1]$ define $\Gamma_\delta u$ by

$$\widehat{\Gamma_\delta u}(\xi) = \psi(\delta \xi) \hat{u}(\xi). \tag{21}$$

DEFINITION: The class of $LIP(s)$ consists of all compactly supported distributions u such that there exists $C > 0$ and

$$|\Gamma_\delta u(x)| \le C \delta^s \tag{22}$$

for all $x \in \mathbb{R}^3$ and $0 < \delta \le 1$.

If k is a non-negative integer and if $k < s < k+1$ then $LIP(s)$ consists of all functions with compact support that are in C^k and whose partials derivatives of order k are in the Lipschitz class $s - k$. $LIP(k)$ contains all compactly supported functions with bounded k partial derivatives. Our approach to studying the mapping properties of A^{-1} in these spaces it to establish estimates for u which satisfies $Au = f$ of the form

$$\|\Gamma_\delta u(x)\| \le C_N(\max|\tilde{\Gamma}_\delta f| \delta^{\frac{2}{m}} + \delta^N(\|u\| + \|f\|)) \tag{23}$$

$$\|\Gamma_\delta L u(x)\| + \|\Gamma_\delta \overline{L} u(x)\| \le C_N(\max|\tilde{\Gamma}_\delta f| \delta^{\frac{1}{m}} + \delta^N(\|u\| + \|f\|))$$

and

$$\|\Gamma_\delta L \overline{L} u(x)\| + \|\Gamma_\delta \overline{L} L u(x)\| \le C_N(\max|\tilde{\Gamma}_\delta f| + \delta^N(\|u\| + \|f\|))).$$

Here $\tilde{\Gamma}_\delta$ is the operator obtained by multiplying $\hat{u}(\xi)$ by $\tilde{\psi}(\delta \xi)$ where $\tilde{\psi} = 1$ in a neighborhood of $\mathrm{supp}\,\psi$ and is also supported in $\tfrac{1}{2} < \|\xi\| < \tfrac{5}{2}$. The integer m is the type of p and N is arbitrarily large. Similarly we obtain analogous estimates when u satisfies $Au = Lf$, $Au\overline{L}f$, and $Au = L\overline{L}f$.

The starting point for such estimates is an estimate of the form

$$\|\sigma_\delta \Gamma_\delta u\| \le C_N(\|\tilde{\sigma}_\delta \tilde{\Gamma}_\delta f\| \delta^{\frac{2}{m}} + \delta^N(\|u\| + \|f\|), \tag{24}$$

where $\sigma_\delta, \tilde{\sigma}_\delta \in C_0^\infty(B(P, \gamma(P, \delta)))$. The first step to obtain this estimate is to substitute $\sigma_\delta \Gamma_\delta u$ for u in (11). Once (24) is established there is still a major problem to overcome before deriving (23). Namely, in the localization the product of the volumes of the supports of $\psi(\delta \xi)$ and $B(P, \gamma(P, \delta))$ is unbounded (in fact greater than $C \delta^{-1/2}$). We refer the reader to [**FK1**] for details.

Finally we want to mention that M. Christ found a very interesting application of these estimates. He proved (see [**Ch2**]) that: if $\dim h = 3$, M is compact pseudoconvex, if every point of M is of finite type and if the range of $\overline{\partial}_b$ is closed then M is embeddable in $\mathbb{C}^n$ for some n.

REFERENCES

[A] T. Akahori, *A new approach to the local embedding theorem of CR-structure for $n \geq 4$*, Memoirs AMS **366**, 1987.

[BS] H. P. Boas and M.-C. Shaw, Sobolev estimates for the Lewy operator on weakly pseudo-convex boundaries, *Math. Ann.*, **274**, 1986, 221-231.

[B] L. Boutet de Monvel, Integration des equation de Cauchy-Riemann, Seminaire Goulaouie-Lions-Schwartz, Exré IX, 1974-1975.

[Bu] D. Burns, *Global behaviour of some tangential Cauchy-Riemann equations*, P.D.E. and Geometry Conference, Park City, Utah, 1977, Dekker, New York, 1979.

[C] D. Catlin, Estimates of invariant metrics on pseudo-convex domains of dimension two, preprint.

[Ch1] M. Christ, Regularity properties of the $\overline{\partial}_b$ equation on weakly-pseudo-convex CR manifolds of dimension 3, *Jour. Amer. Math. Soc.*, 1988, 587-664.

[Ch2] M. Christ, Embedding compact three-dimensional CR manifolds of finite type in $\mathbb{C}^n$, *Annals of Math.* **129** 1989, 195-213.

[FK1] C. L. Fefferman and J. J. Kohn, *Hölder estimates on domains of complex dimension two and on three dimensional CR manifolds*, Adv. in Math. **2** 1988, 223-303.

[FK2] C. L. Fefferman and J. J. Kohn, *Estimates of kernels on three-dimensional CR manifolds*, to appear in Rev. Math. Iberoam. 4, No. 3, 1988.

[FP] C. L. Fefferman and D. Phong, Subelliptic eigenvalue problems, in "Proceedings Conference on Harmonic Analysis, in Honor of Antoni Zygmund,", pp. 590-606.

[H] L. Hörmander, Hypoelliptic second order differential equations, *Acta Math.*, **119**, 1967, 147-171.

[K1] J. J. Kohn, *Boundaries of complex manifolds*, in "Proceedings Conference on Complex Analysis, Minneapolis, 1964," pp. 81-94, Springer-Verlag New York Berlin, 1965.

[K2] J. J. Kohn, Estimates for $\overline{\partial}_b$ on pseudo-convex CR manifolds, *Proc. Sympos. Pure Math.* **43**, 1985, 207-217.

[K3] J. J. Kohn, The range of the tangential Cauchy-Riemann operator, *Duke Math. J.* **53**, 1986, 525-545.

[K4]   J. J. Kohn, Global regularity for $\bar{\partial}$ on weakly pseudo-convex manifolds, *Trans. Amer. Math.* **181**, 1973, 272-292.

[K5]   J. J. Kohn, Boundary behaviour of $\bar{\partial}$ on weakly pseudo-convex manifolds of dimension two, *J. Differential Geom.* **6**, 1972, 523-542.

[KR]   J. J. Kohn and H. Rossi, On the extension of holomorphic functions from the boundary of a complex manfold, *Ann. of Math.* **28**, 1965, 451-472.

[Ku]   M. Kuranishi, Strongly pseudoconvex CR structures over small balls, *Annals of Math.* I, **115**, 1982, 451-500; II, **116**, 1982, 1-64; III, **116**, 1982, 249-330.

[M]   J. McNeal, Boundary behaviour of the Bergman kernel, *Duke Math. J.* 58, **2**, 1989, 499-512.

[N]   L. Nirenberg, Lectures on Linear Partial Differential Equations, CBMS No. 17, AMS, Providence, 1973, 10-12.

[NRSW] A. Nagel, J. P. Rosay, E. M. Stein, and S. Wainger, Estimates for the Bergman and Szegö kernels in $\mathbb{C}^2$, *Annals of Math.* **129**, 1989, 113-149.

[R]   H. Rossi, Attaching analytic spaces to a space along a pseudo-convex boundary, in "Proceedings, Conference on Complex Analysis," pp. 242-256, Springer-Verlag, New York/Berlin, 1965.

[RS]   L. P. Rothschild and E. M. Stein, "Hypoelliptic differential operators and nilpotent groups, *Acta Math.* **137**, 1976, 247-320.

Recent Contributions to the Calculus of Finite Differences: A Survey

D. LOEB* Université de Bordeaux I, Département des Mathématiques et de l'Informatique, 33405, Talence, France.

G.C. ROTA**, MIT Dept. of Mathematics Cambridge, MA 02139 USA

Abstract. We retrace the recent history of the Umbral Calculus. After studying the classic results concerning polynomial sequences of binomial type we generalize to a certain type of logarithmic series. Finally, we demonstrate numerous typical examples of our theory.

1. Polynomials of Binomial Type

Much of the calculus of finite differences leans upon certain analogies between two linear operators defined on functions $f(x)$ ranging over a field K of characteristic zero, and taking values in K:

1. The *derivative operator* D, defined as

$$\mathrm{D} f(x) = f'(x) ,$$

and

2. The *forward difference operator* Δ, defined as

$$\Delta f(x) = f(x + 1) - f(x) .$$

---

* Supported by a Chateaubriand Fellowship.
** Supported by National Science Foundation under Grant MCS-8104855

1.1. The Powers of x

The sequence of polynomials[1] $p_n(x) = x^n$ is related to the derivative operator by the elementary property

$$Dp_n(x) = np_{n-1}(x).\qquad(1)$$

Such sequences of polynomials have been called *Appel*. The sequence x^n is distinguished among all Appell sequences by the further property

$$D_n(0) = \delta_{n0}$$

for all $n \geq 0$.

Furthermore, $p_n(x) = x^n$ plays a key role in Taylor's theorem

$$f(x+a) = \sum_{n=0}^{\infty} p_n(a) \frac{[D^n f(x)]_{x=0}}{n!} = \sum_{n=0}^{\infty} \frac{a^n [D^n f(x)]_{x=}}{n!}$$

for all $a \in K$ and for all suitable functions $f(x)$.

In particular, we note that the sequence $p_n(x)$ is of *binomial type*. That is to say, it satisifies the *binomial theorem*

$$p_n(x+a) = \sum_{k=0}^{\infty} \binom{n}{k} p_k(a) p_{n-k}(x)\qquad(2)$$

for all $a \in K$ and all $n \geq 0$.

1.2. The Lower Factorial

Define another sequence of polynomials called the *lower factorial powers* as follows

$$p_0(x) = (x)_0 = 1,$$

and

$$p_n(x) = (x)_n = x(x-1)(x-2)\ldots(x-n+1)$$

for $n \geq 0$. It can be interpretted as the number of *injections* from an n-element set to an x-element set [9].

The lower factorial is related to the forward difference operator by the property

$$\Delta p_n(x) = np_{n-1}(x).$$

Such sequences are called *Sheffer sequences* for the forward difference operator. The lower factorial is distinguished among all Sheffer sequences by the further property

$$p_n(0) = \delta_{n0}$$

---

[1] When one speaks of a sequence of polynomials, one generally means a sequence $p_n(x)$ indexed by a nonnegative integer n such that the degree of each polynomial $p_n(x)$ is exactly equal to n.

Calculus of Finite Differences

for $n \geq 0$.

Furthermore, for suitable functions $f(x)$ and all constants $a \in K$, we have *Newton's formula*

$$f(x+a) = \sum_{n=0}^{\infty} (a)_n \frac{\Delta^n f(x)}{n!}.$$

In particular, *Vandermonde's identity* states that the sequence $p_n(x) = (x)_n$ is of binomial type:

$$(x+a)_n = \sum_{k=0}^{\infty} \binom{n}{k} (a)_k (x)_{n-k}.$$

1.3. The General Case

Our initial problem will be that of carrying this analogy as far as reasonably possible in order to make the parallel of D and Δ into an accidental special case.

To this end, we begin by classifying all sequences of polynomials of binomial type, that is, all sequences of polynomials $p_n(x)$ (with $\deg(p_n(x)) = n$) satisfying eq. 2

$$p_n(x+a) = \sum_{k=0}^{\infty} \binom{n}{k} p_k(a) p_{n-k}(x).$$

PROPOSITION 1. *A sequence of polynomials $p_n(x)$ of binomial type satisfies*

$$p_n(0) = \delta_{n0}$$

for all $n \geq 0$.

Proof. By eq. 2 with $n = a = 0$, $p_0(0) = 1$. Now, for n positive,

$$0 = p_n(x) - p_n(x) = \sum_{k=1}^{n} \binom{n}{k} p_k(0) p_{n-k}(x).$$

However, $p_k(x)$ is a basis for the vector space of polynomials, and in characteristic zero the binomial coefficients indicated are never zero, so $p_k(0) = 0$. ∎

Now, to each sequence of binomial type $p_n(x)$, we associate a linear operator Q defined by

$$Qp_n(x) = np_{n-1}(x). \qquad (3)$$

For the moment, the operator Q is defined only for polynomials; later (Section 3), we shall extend its domain to all more general types of functions.

By iteration, we obtain

$$Q^k p_n(x) = (n)_k p_{n-k}(x).$$

Recalling that $\binom{n}{k} = (n)_k / k!$ we can write the binomial identity (eq. 2) in the form

$$p_j(x+a) = \sum_{k=0}^{\infty} p_k(a) \frac{Q^k p_j(x)}{k!}$$

for $j \geq 0$.

Now, let $p(x)$ be any polynomial. Given its expansion $p(x) = \sum_{j=0}^{\infty} c_j p_j(x)$ in terms of the basis $p_j(x)$, we have

$$p(x+a) = \sum_{k=0}^{\infty} c_j p_j(x+a)$$

$$= \sum_{k=0}^{\infty} p_k(a) Q^k \left(\sum_{j=0}^{\infty} c_j p_j(x) \right) \quad (4)$$

$$= \sum_{k=0}^{\infty} p_k(a) \frac{Q^k}{k!} p(x) .$$

This calculation for all polynomials $p(x)$ can be recast into a more elegant form by introducing the *shift operator*

$$E^a f(x) = f(x+a) .$$

We then can write eq. 4 as the operator identity

$$E^a = \sum_{k=0}^{\infty} p_k(a) Q^k / k! \quad (5)$$

which we shall call the *Taylor's Formula for the sequence* $p_n(x)$. Hence, Newton's formula is nothing more than Taylor's formula for the lower factorials. Again, eq. 5 has been proven only for operations on polynomials for the time being; however, we shall see (Theorem 9) that its domain of validity is considerably greater.

The right side of eq. 5 obviously commutes with Q. Thus,

$$QE^a = E^a Q$$

for all constants a. Any linear operator T with the property that

$$TE^a = E^a T \quad (6)$$

for all scalars a will be said to be a *shift-invariant operator*. A linear shift-invariant operator is called a *delta operator* if its kernel is exactly the field of all constants K.

We have thus proved all of Theorem 1 and the hard part of Theorem 2.

THEOREM 1. *Let $p_n(x)$ be a sequence of polynomials of binomial type. Then there exists a unique delta operator such that*

$$Qp_n(x) = np_{n-1}(x)$$

for $n \geq 0$. ∎

Next, we have the converse.

Calculus of Finite Differences

THEOREM 2. *Let Q be a delta operator. Then there exists a unique sequence of polynomials of binomial type $p_n(x)$ satisfying*

$$Qp_n(x) = np_{n-1}(x)$$

for $N \geq 0$.

Proof. We can recursively define $p_n(x)$ subject to the requirements that
- $p_0(x) = 1$,
- For n positive, $p_n(x)$ is in the inverse image of Q on $np_{n-1}(x)$, and
- For n positive, $p_n(0) = 0$.

It remains only to show that the resulting sequence of polynomials obeys eq. 2. By the reasoning before Theorem 1, we again have eq. 4. Finally, setting $p(x) = p_n(x)$ yields the binomial theorem. ∎

Thus, we see that all of the features of the analogy between the operators D and Δ, and the polynomial sequences x^n and $(x)_n$ are shared by all other delta operators, and their corresponding sequences of binomial type.

Sequences of polynomials of binomial type are of frequent occurence in combinatorics,[2] probability, statistics, function theory and representation theory.

1.4 The Abel Polynomials

We note the following striking example of a sequence of polynomials of binomial type. Given a constant b, define the *Abel polynomials*[3] as follows

$$A_0(x; b) = 1$$

and for n positive

$$A_n(x; b) = x(x - nb)^{n-1}.$$

It is easily verified that, setting $Q = E^b D = D E^b$, we have $QA_n(x; b) = nA_{n-1}(x; b)$. Hence, Theorem 2 implies at once *Abel's identity*

$$(x + a)(x + a - nb)^{n-1} = \sum_{k=0}^{\infty} \binom{n}{k} a(a - kb)^{k-1} x(x - (n-k)b)^{n-k-1}.$$

Furthermore, Taylor's formula for the sequence of Abel polynomials gives

$$p(x + a) = \sum_{k=0}^{\infty} \frac{a(a - kb)^{k-1}}{k!} D <^k p(x + kb).$$

---

[2] In combinatorics, sequences of binomial type enumerate the number of functions from an n-element set to an x-element set enriched with some mathematical structure on each fiber. For details see [9].
[3] For example, with $b = -1$, the Abel polynomial $A_n(x; -1)$ counts the number of *reluctant functions* from an n-element set to an x-element set. A reluctant function is a function enriched with a rooted labelled forest on each fiber.

For example, for $p(x) = x^n$, we obtain

$$(x+a)^n = \sum_{k=0}^{\infty} \binom{n}{k} a(a-kb)^{k-1}(x+kb)^{n-k}$$

which is not *a priori* obvious.

In order to devise more examples, it will be helpful to first study shift-invariant operators in greater detail.

2. Shift-Invariant Operators

For convenience, we introduce the following bit of notation. For any function $f(x)$, we write

$$\langle f(x) \rangle_{(0)} = f(0) = [f(x)]_{x=0} . \tag{7}$$

The linear functional $\langle \rangle_{(0)}$ is called the *augmentation* or *evaluation at zero*.

Let $p_n(x)$ be a sequence of binomial type and let Q be its delta operator. Taylor's theorem for $p_n(x)$ can be written in terms of the augmentation (interchanging the roles of x and a)

$$E^a p(x) = \sum_{k=0}^{\infty} p_k(x) \frac{\langle E^a Q^k p(x) \rangle_{(0)}}{k!} .$$

Now, let T be any shift-invariant operator, and $p(x)$ any polynomial. We have

$$E^a T p(x) = T E^a p(x)$$

$$= \sum_{k=0}^{\infty} T p_k(x) \langle E^a Q^k p(x) \rangle_{(0)} / k! .$$

Again exchanging the role of x and a, therefore

$$T p(x+a) = \sum_{k=0}^{\infty} \frac{\langle E^a T p_k(x) \rangle_{(0)}}{k!} Q^k p(x) .$$

Setting $a = 0$, we obtain the operator identity

$$T = \sum_{k=0}^{\infty} c_k Q^k / k! \tag{8}$$

where $c_k = \langle T p_k(x) \rangle_{(0)}$.

Thus, we have proven

Calculus of Finite Differences

THEOREM 3. (Expansion Theorem) *Given a shift-invariant operator T and a delta operator Q associated to a sequence of binomial type $p_n(x)$ there is a unique expansion given by eq. 8.* ∎

Actually, some mild continuity conditions must be imposed, but we will refer the reader to the bibliography for such technical (and ultimately trivial) questions of topology.

The expansion theorem is to operators essentially what the generalized Taylor's theorem is to polynomials. Together, they allow the expansion of either operators or polynomials with respect to any basis.

The above result can be restated in more elightening terms. Let $K[[t]]$ be the ring of all formal series (sometimes called *Hurwitz series*) in the variable t with coefficients in K

$$f(t) = c_0 + c_1 t + c_2 t^2/2! + c_3 t^3/3! + \dots .$$

Then we have

THEOREM 4. (Isomorphism Theorem)

1. For every delta operator Q, one has the following isomorphism of the ring of formal power series $K[[t]]$ onto the ring of shift-invariant operators

$$f(t) \mapsto \sum_{k=0}^{\infty} c_k Q^k / k! = f(Q) . \qquad (9)$$

2. Furthemore, let ϕ be an isomorphism of the ring of formal power series with the ring of shift-invariant operators. Then there exists a delta operator Q such that the isomorphism ϕ is of the form eq. 9.

Proof. Part one is immediate from the Expansion Theorem. Part two is classical result concerning formal series where $Q = \phi(t)$. ∎

In particular for $Q = D$, we notice that all shift-invariant operators are formal power series in the derivative. That is, the ring of shift-invariant operators is $K[[D]]$. Moreover, delta operators $T = T(D)$ are formal power series in the derivative in which the coefficient of D^0 is zero, and the coefficient of D^1 is nonzero. Such series are known as *delta series*.

For example, the shift operator is given by the formal power series

$$E^a = \exp(aD) = \sum_{k=0}^{\infty} a^k D^k / k! .$$

As another example, let $K = \mathbb{C}$ be the complex field, and define the *Laguerre operator* by

$$Lf(x) = -\int_0^{\infty} e^{-y} f'(x+y) dy . \qquad (10)$$

The Laguerre operator is obviously a shift-invariant operator since

$$LE^a f(x) = -\int_0^{\infty} e^{-y} f'(x+a+y) dy = E^a Lf(x) .$$

Moreover,

$$\langle Lx^n \rangle_{(0)} = -\int_0^\infty e^{-y} n y^{n-1}\, dy$$

$$= [e^{-y} n y^{n-1}]_0^\infty - \int_0^\infty e^{-y} n(n-1) y^{n-2}\, dy$$

$$= -\int_0^\infty e^{-y} n(n-1) y^{n-2}\, dy$$

$$\vdots \quad \vdots$$

$$= -\int_0^\infty e^{-y} n!\, dy$$

$$= -n!,$$

and hence by the Expansion Theorem (Theorem 3), we infer that L is a delta operator and that

$$L = -D - D^2 - D^3 - \cdots$$
$$= D/(D - I) \tag{11}$$

where $I = D^0$ is the identity operator. We will later (Section 4.2) compute the sequence of binomial type associated with the Laguerre operator.

As another example, let W be the *Weierstrass operator* defined as

$$Wf(x) = \frac{1}{\sqrt{2\pi}} \int_{-\infty}^\infty f(y) \exp(-(x-y)^2/2)\, dy.$$

Again, we calculate $c_n = \langle Wx^n \rangle_{(0)}$. First, note that for n odd, the integrand in question is an odd function, so the integral from $-\infty$ to ∞ is zero. For $n = 0$, it is easier to compute c_0^2.

$$c_0^2 = \frac{1}{2\pi} \int_{-\infty}^\infty \int_{-\infty}^\infty \exp(-(x^2 + y^2)/2)\, dx\, dy$$

$$= \frac{1}{2\pi} \int_0^{2\pi} \int_0^\infty \exp(-r^2/2)\, r\, dr\, d\theta$$

$$= \int_0^\infty \exp(-r^2/2)\, r\, dr$$

$$= \int_0^\infty \exp(-u)\, du$$

$$= 1.$$

For n a positive even integer, we have the following recurrence for c_n via integration by parts

$$c_n = \frac{1}{\sqrt{2\pi}} \int_{-\infty}^\infty y^n \exp(-y^2/2)\, dy$$

$$= \frac{1}{\sqrt{2\pi}} \left(-[\exp(-y^2/2) y^{n-1}]_{-\infty}^\infty + (n-1) \int_{-\infty}^\infty y^{n-2} \exp(-y^2/2)\, dy \right)$$

$$= 0 + (n-1) c_{n-2}.$$

Calculus of Finite Differences

Thus, for n even,

$$c_n = (n-1)(n-3) \cdots 5 \cdot 3 \cdot 1 = \frac{n!}{(n/2)! \, 2^{n/2}}$$

(often denoted as $n!$) so that

$$W = \sum_{n=0}^{\infty} \frac{D^{2n}}{n! \, 2^n} = \exp(D^2/2) \,.$$

From the Isomorphism Theorem (Theorem 4), it follows that a shift-invariant operator over polynomials has a unique inverse if and only if $T_1 \neq 0$. For example, the Weierstrass operator W has a unique inverse.

Lastly, we compute the *Bernoulli operator*

$$Jf(x) = \int_x^{x+1} f(y)\, dy$$

where we have

$$\langle Jx^n \rangle_{(0)} = \int_0^1 y^n \, dy = \left[\frac{y^{n+1}}{n+1} \right]_0^1 = \frac{1}{n+1} \,.$$

Thus,

$$J = \sum_{k=0}^{\infty} \frac{D^k}{(k+1)!} = \frac{e^D - I}{D} = \frac{\Delta}{D} \,.$$

We close this section with a result which may be considered to be fundamental.

THEOREM 5. (Fundamental Theorem of the Calculus of Finite Differences) *Let $p_n(x)$ be the sequence of binomial type associated to a delta operator $Q(D)$. Let $Q^{(-1)}(t)$ be the inverse formal power series of $Q(t)$. (That is, suppose that $Q(Q^{(-1)}(t)) = Q^{(-1)}(Q(t)) = t$). Then the exponential generating function for $p_n(x)$ is*

$$\sum_{n=0}^{\infty} p_n(x) t^n / n! = \exp(x Q^{(-1)}(t)) \,.$$

Since this result is so important, we give two proofs. A third proof found in [9] is more powerful and applies equally to symmetric functions.

Proof 1. By Proposition 1 and eq. 3, the left hand side is characterized by the property

$$\langle Q(D)^k p(x,t) \rangle_{(0)} = t^k \,,$$

so it will suffice to verify the right hand side obeys this property equally. Let q_{nk} denote the coefficients of $Q(D)^k = \sum_{n=0}^{\infty} q_{nk} D^n / n!$. Then

$$\langle Q(D)^k \exp(xQ^{(-1)}(t)) \rangle_{(0)} = \sum_{n=0}^{\infty} \langle Q(D)^k x^n \rangle_{(0)} (Q^{(-1)}(t))^n / n!$$

$$= \sum_{n=0}^{\infty} q_{nk} (Q^{(-1)}(t))^n / n!$$

$$= Q(Q^{(-1)}(t))^k$$

$$= t^k \,. \qquad \blacksquare$$

Proof 2. By Taylor's Formula (eq. 5),

$$\exp(aD) = \sum_{k=0}^{\infty} p_k(a) Q^k / k! .$$

Substituting x for a, and $Q^{(-1)}(t)$ for D, we have the desired result. ∎

3. Extension to Logarithmic Series
3.1. The Harmonic Logarithms

We shall now extend the domain of every shift-invariant operator to a more general domain of formal series originating from the Hardy field [4] $\mathcal{L}$ called the Logarithmic Algebra consisting of all expansions of real functions in a neighborhood of infinity in terms of the monomials

$$\ell^\alpha = x^{\alpha_0} (\log x)^{\alpha_1} (\log \log x)^{\alpha_2} \ldots$$

for all vectors of integers [5] where only a finite [6] number of the α_i are different from zero.

Another characterization of $\mathcal{L}$ [6, Theorem 5.4A.4] is that $\mathcal{L}$ contains a dense subset which is the smallest proper *field* extensions of the ring of polynomials with real coefficients such that the derivative D is a derivation of the entire field, and the antiderivative D^{-1} is well defined up to a constant of integration.

$\mathcal{L}$ is equivalent to the set of all formal sums

$$\sum_\alpha b_\alpha \ell^\alpha$$

where the sum is over vectors of integers α with finite support, and for all nonnegative integers n, all integers $\alpha_0, \alpha_1, \ldots, \alpha_{n-1}$, and all integers β, there exists finitely many $\alpha_n > \beta$ such that there exists integers $\alpha_{n+1}, \alpha_{n+2}, \ldots$ only finitely many different from zero) such that $b_\alpha \neq 0$.

It will be noted that the derivative of such an expression is awkward

$$D\ell^\alpha = \alpha_0 \ell^{(\alpha_0-1,\alpha_1,\alpha_2,\alpha_3\ldots)} + \alpha_1 \ell^{(\alpha_0-1,\alpha_1-1,\alpha_2,\alpha_3\ldots)} + \alpha_2 \ell^{(\alpha_0-1,\alpha_1-1,\alpha_2,\alpha_3\ldots)} + \ldots .$$

Moreover, the formula for the antiderivative of such a function is not known in general, and even in the case $\alpha_0 \neq -1$ when it is known, it is given by a horrendous expression [6, p. 87]. Since the monomials ℓ^α are so unwieldy and do not even begin to compare to the powers of x in terms of ease and simplicity of calculation, we are forced to resort to another basis for the vector space $\mathcal{L}$ which as will be seen is the true *logarithmic analog* of the sequence x^n. This basis is called the *harmonic logarithm* $\lambda_n^\alpha(x)$.

---

[4] No knowledge of Hardy fields nor of asymptotic expansions is expected of the reader.
[5] We could achieve far greater generality by allowing α to be a vector of reals as in [7]. The definitons in [5] provide one with the basic tools to carry out these calculations.
[6] Since all the vectors we deal with here have finite support, we will usually adopt the convention of not writing the infinite sequence of zeroes they all end with. Thus, we write (1) for the vector $(1,0,0,0,\ldots)$.

Calculus of Finite Differences

For n an integer, and $\alpha = (\alpha_1, \alpha_2, \ldots)$ a vector of integers with only a finite number of nonzero entries, the *harmonic logarithm* of degree n and order α is defined by the series

$$\lambda_n^\alpha(x) = \sum_\rho \frac{s(-n, \rho_1)}{\lfloor -n \rfloor !} \left[\prod_{i=1}^\infty e_{\rho_i - \rho_{i+1}}(\alpha_i, \alpha_i, -1, \ldots, \alpha_i, -\rho_i + 1) \right] \ell^{(n),(\alpha-\rho)}$$

where
1. The *linear partition* $\rho = (\rho_1, \rho_2, \ldots)$ is a weakly decreasing sequence of nonnegative integers,
2. The *Roman factorial*[7] $\lfloor n \rfloor!$ is given by the formula

$$\lfloor n \rfloor ! = \begin{cases} n! & \text{for } n \text{ a nonnegative integer, and} \\ (-1)^{-n-1}/(-n-1)! & \text{for } n \text{ a negative integer,} \end{cases}$$

3. For any integer n negative, zero or positive, and for any nonnegative integer k, the *Stirling number* $s(n, k)$ is the coefficient of y^k in the Taylor series expansion of $(y)_n = \Gamma(y+1)/\Gamma(y-n+1)$, and
4. The *elementary symmetric function* $e_n(x_1, x_2, \ldots)$ is defined the generating function

$$\sum_{n=0}^\infty e_n(x_1, x_2, \ldots) y^n = \prod_{k=1}^\infty (1 + x_k y) .$$

It may be helpful to point out several important special cases.
1. The harmonic logarithms of degree zero are given by

$$\lambda_0^\alpha(x) = \ell^{(0), \alpha} = (\log x)^{\alpha_1} (\log \log x)^{\alpha_2} \ldots .$$

2. The harmonic logarithm of order $\alpha = (-1)$ and degree one $\lambda_1^{(-1)}(x)$ is the *logarithmic integral* $\text{li}(x) = \int_0^x dt / \log t$.
3. The harmonic logarithms of order $\alpha = (0)$ are given by

$$\lambda_n^{(0)}(x) = \begin{cases} x^n & \text{for } n \geq 0, \text{ and} \\ 0 & \text{for } n < 0. \end{cases}$$

4. The harmonic logarithms of order $\alpha = (1)$ are given by

$$\lambda_n^{(0)}(x) = \begin{cases} x^n \left(\log x - 1 - \frac{1}{2} - \ldots - \frac{1}{n} \right) & \text{for } n \geq 0, \text{ and} \\ x^n & \text{for } n < 0. \end{cases}$$

5. The harmonic logarithms of order $\alpha = (2)$ are given by

$$\lambda_n^{(0)}(x) = \begin{cases} x^n (\log x)^2 - x^n \left(\frac{2}{1} + \frac{2}{2} + \ldots + \frac{2}{n} \right) \\ + x^n \left[\frac{2}{2} \left(1 + \frac{1}{2}\right) + \ldots + \right. \\ \left. \frac{2}{n} \left(1 + \frac{1}{2} + \ldots + \frac{1}{n}\right) \right] & \text{for } n \geq 0, \text{ and} \\ 2 x^n \left[\log(x) - 1 - \frac{1}{2} - \ldots - \frac{1}{n-1} \right] . & \text{for } n < 0. \end{cases}$$

---

[7] After Steve Roman who rediscovered these numbers originally defined by Leibniz [15, 16, 17].

6. The harmonic logarithms of order $\alpha = (t)$ where t is a nonnegative integer[8] are given by the following formulas

$$\sum_{t=0}^{\infty} \lambda_n^{(t)}(x) z^t/t! = \lfloor n \rceil! x^{n+z} \Gamma(z+1)/\Gamma(z+n+1)$$

$$\lambda_n^{(t)}(x) = x^n \lfloor n \rceil! (xD)_{-n}(\log x)^t$$

$$= x^n \sum_{k=0}^{\infty} \lfloor n \rceil! (t)_k s(-n, k)(\log x)^{t-k}.$$

Now we can give the association between the derivative and the harmonic logarithms; the harmonic logarithms behave under derivation exactly like the powers of x once the ordinary factorial $n!$ is replaced by the Roman factorial $\lfloor n \rceil!$.

THEOREM 6. *For all integers* n, *and vectors of integers with finite support* α, *we have*

$$D\lambda_{n+1}^{\alpha}(x) = \lfloor n+1 \rceil \lambda_n^{\alpha}(x) \tag{12}$$

where

$$\lfloor n \rceil = \frac{\lfloor n \rceil!}{\lfloor n-1 \rceil!} = \begin{cases} n & \text{for } n \neq 0, \text{ and} \\ 1 & \text{for } n = 0. \end{cases}$$

More generally, for any nonnegative integer k,

$$D^k \lambda_{n+k}^{\alpha}(x) = \frac{\lfloor n+k \rceil!}{\lfloor n \rceil!} \lambda_n^{\alpha}(x).$$

Proof. By induction, it will suffice to demonstrate eq. 12. The following proof is a direct application of the recursion for the Stirling numbers. (See [11].)

$$D\lambda_{n+1}^{\alpha}(x)$$

$$= \sum_{\rho} \frac{s(-n-1, \rho_1)}{\lfloor -n-1 \rceil!} \left[\prod_{i=1}^{\infty} e_{\rho_i - \rho_{i+1}}(\alpha_i, \alpha_i - 1, \ldots, \alpha_i - \rho_i + 1) \right] D \ell^{(n+1),(\alpha-\rho)}$$

$$= \sum_{\rho} \frac{s(-n-1, \rho_1)}{\lfloor -n-1 \rceil!} \left[\prod_{i=1}^{\infty} e_{\rho_i - \rho_{i+1}}(\alpha_i, \alpha_i - 1, \ldots, \alpha_i - \rho_i + 1) \right]$$

---

[8] Our original paper [12] dealt exclusively with this case.

Calculus of Finite Differences

$$\times \left[(a+1)\ell^{(n),(\alpha-\rho)} + \sum_{k=1}^{\infty}(\alpha_k - \rho_k)\ell^{(n),(\alpha-\rho-\overbrace{(1,1,\ldots,1)}^{n})} \right]$$

$$= \sum_{\rho} \frac{(a+1)s(-n-1,\rho_1) + s(-n-1,\rho_1)}{\lfloor -n-1 \rceil !}$$

$$\times \left[\prod_{i=1}^{\infty} e_{\rho_i - \rho_{i+1}}(\alpha_i, \alpha_i - 1, \ldots, \alpha_i - \rho_i + 1) \right] \ell^{(n),(\alpha-\rho)}$$

$$= \sum_{\rho} \frac{s(-n,\rho_1)}{\lfloor -n-1 \rceil !} \left[\prod_{i=1}^{\infty} e_{\rho_i - \rho_{i+1}}(\alpha_i, \alpha_i - 1, \ldots, \alpha_i - \rho_i + 1) \right] \ell^{(n),(\alpha-\rho)}$$

$$= \lfloor n+1 \rceil \lambda_n^\alpha(x) \, . \qquad \blacksquare$$

Hence, if we denote by $\mathcal{L}^+$ the closure of the span of the harmonic logarithm $\lambda_n^\alpha(x)$ with $\alpha \neq 0$, then we notice that D restricted to $\mathcal{L}^+$ is a bijection. In particular, the antiderivative is given by

$$D^{-1} \lambda_n^\alpha(x) = \lambda_{n+1}^\alpha(x) / \lfloor n+1 \rceil \, .$$

We can thus characterize the harmonic logarithms by the relationship.

$$\lambda_n^\alpha(x) = \lfloor n \rceil ! D^{-n} \ell^{(0),\alpha} = \lfloor n \rceil ! D^{-n} \lambda_0^\alpha \, .$$

Now, let us dissect the logarithmic algebra even more finely. Let $\mathcal{L}^\alpha$ denote the logarithmic series involving $\lambda_n^\alpha(x)$ for all integers n and a fixed α. Evidently, $\mathcal{L}^+$ is the direct sum of the $\mathcal{L}^\alpha$ for $\alpha \neq (0)$.

THEOREM 7. (Roman Modules) *All of vector spaces $\mathcal{L}$ for $\alpha \neq (0)$ are naturally isomorphic as differential vector spaces.*

Proof. The isomorphism is given by the *skip operator*

$$\text{skip}_{\alpha\beta} \lambda_n^\beta(x) = \lambda_n^\alpha(x) \, . \qquad \blacksquare$$

In view of the preceeding theory, most of our calculations can be made in $\mathcal{L}^{(1)}$ which is called the *Roman module* after Steve Roman. The Roman module consists of series of the form

$$f(x) = \sum_{j=-\infty}^{-1} c_j x^j + \sum_{j=0}^{n} c_j x^j \left(\log x - 1 - \frac{1}{2} - \ldots \frac{1}{j} \right) = \sum_{j=-\infty}^{n} c_j \lambda_j^{(1)}(x) \, ;$$

that is, there are a finite number of terms in which $\log x$ appears, and an infinite series in inverse powers of x. Thus, the Roman module is fairly convenient for our calculations. All of our concrete examples will be drawn from the Roman module.

However, $\mathcal{L}^{(0)}$ is not isomorphic to the Roman module; it is isomorphic to the algebra of polynomials $\mathbf{C}[x]$.

3.2. Shift-Invariant Operators

A linear operator Q will be called *shift-invariant* if not only does it commute with the shift operator $E^a = \sum_{n=0}^{\infty} a^n D^n / n!$ for all a, but also that it commutes with the *skip operator* $\text{skip}_{\alpha\beta}$ from the logarithmic algebra $\mathcal{L}$ to $\mathcal{L}^\alpha$

$$\text{skip}_{\alpha\beta} \lambda_n^\gamma(x) = \delta_{\beta\gamma} \lambda_n^\alpha(x)$$

for $\beta \neq (0)$.

Any shift-invariant operator $\mathcal{L}$ is obviously shift-invariant (in the sense of eq. 6) when restricted to $\mathcal{L}^{(0)} = \mathbf{C}[x]$. Conversely, any shift-invariant operator Q on $\mathbf{C}[x]$ can be expanded as a formal series in the derivative $Q(D)$, and thus Q can be extended to all $\mathcal{L}$. However, are these the only shift-invariant operators which act on $\mathcal{L}$? On $\mathcal{L}^+$?

THEOREM 8. (Characterization of Shift-Invariant Operators)
  (1) *The algebra of shift-invariant operators on $\mathcal{L}^+$ is naturally isomorphic to the algebra of Laurent series[9] in the derivative $\mathbf{R}(D)$.*
  (2) *The algebra of shift-invariant operators on $\mathcal{L}$ is naturally isomorphic to the algebra of formal power series in the derivative $\mathbf{R}[[D]]$.*

Proof. It will suffice to prove part one, since the only Laurent series in the derivative which are well defined on $\mathcal{L}^{(0)}$ are those which involve no negative powers of the derivative; that is, members of the algebra of formal power series in the derivative.

Since the derivative commutes with $\text{skip}_{\alpha\beta}$ and E^a, all Laurent series in the derivative are shift-invariant operators on $\mathcal{L}^+$.

Conversely, suppose that Q is a shift-invariant operator. Q is determined by its actions on the harmonic logarithms of order α for any particular $\alpha \neq (0)$.

$$Q\lambda_n^\alpha(x) = \sum_{k=-\infty}^{n} c_{nk} \lambda_k^\alpha(x).$$

Next, since Q commutes with E^a for all a, it also commutes with D^n for all n. Thus,

$$\sum_{j=-\infty}^{k} c_{kj} \frac{\lfloor j \rfloor!}{\lfloor j - n \rfloor!} \lambda_{j-n}^\alpha(x) = D^n Q \lambda_k^\alpha(x)$$

$$= QD^n \lambda_k^\alpha(x)$$

$$= \frac{\lfloor k \rfloor!}{\lfloor k - n \rfloor!} \sum_{j=-\infty}^{k-n} c_{k-n,j} \lambda_j^\alpha(x).$$

---

[9] A *Laurent series* is a formal series of the form

$$f(x) = \sum_{i=n}^{\infty} c_i x^i$$

where n may be any integer positive or negative.

Equating coefficients of $\lambda_n^\alpha(x)$, and setting $n = k$, we have

$$c_{0j} = \left\lfloor \begin{matrix} j+k \\ k \end{matrix} \right\rceil c_{k,j+k}$$

where the *Roman coefficient* $\left\lfloor \begin{matrix} j \\ k \end{matrix} \right\rceil$ is defined to be $\lfloor j \rceil! / \lfloor k \rceil! \lfloor j - k \rceil!$.

Hence, Q is determined by the c_{0j}, and therefore equals the Laurent series $\sum_{j=-\infty}^{m} c_{0j} D^j / \lfloor j \rceil!$ where m is the largest j such that $c_{0j} \neq 0$. ∎

Since the algebra of Laurent series is a field, we immediately derive that all shift-invariant operators on $\mathcal{L}^+$ are invertible. Thus,

COROLLARY 1. (Differential Equations)

(1) *All linear differential equations with constant coefficients on $\mathcal{L}^+$ have a unique solution. That is, any equation of the form*

$$f(D)p(x) = q(x)$$

(where $q(x) \in \mathcal{L}^+$) has a unique solution $p(x) \in \mathcal{L}^+$.

(2) *All linear differential equations on $\mathcal{L}$ have solutions. Of these, one particular solution can be naturally chosen as the canonical one.*

Proof. We have just proven 1. The proof for 2 (see [7, §2.4.4A]) relies on the use of the projection maps $\text{skip}_{\alpha\beta}$. ∎

3.3. Binomial Theorem

Any shift-invariant operator acts on each level $\mathcal{L}^\alpha$ in the same way. For example, consider the logarithmic version of the binomial theorem.

$$E^a \lambda_n^\alpha(x) = \sum_{k=0}^{\infty} a^k D^k \lambda_n^\alpha(x)/k! = \sum_{k=0}^{\infty} \left\lfloor \begin{matrix} n \\ k \end{matrix} \right\rceil a^k \lambda_{n-k}^\alpha(x) . \qquad (13)$$

Several special cases are of particular interest. For $\alpha = (0)$, we have the standard binomial theorem. For $\alpha = (1)$, n positive, and $x = 1$, we have [6, p. 97]

$$(1+a)^n \log(1+a)$$
$$= ((1+a)^n - 1)\left(1 + \frac{1}{2} + \ldots + \frac{1}{n}\right) - na\left(1 + \frac{1}{2} + \ldots + \frac{1}{n-1}\right)$$
$$- \binom{n}{2} a^2 \left(1 + \frac{1}{2} + \ldots + \frac{1}{n-2}\right) - \ldots - \frac{3}{2}\binom{n}{n-2} a^{n-2} - na^{n-1}$$
$$+ a^{n+1} \left\lfloor \begin{matrix} n \\ -1 \end{matrix} \right\rceil + a^{n+2} \left\lfloor \begin{matrix} n \\ -2 \end{matrix} \right\rceil + \ldots .$$

Similarly, for $\alpha = (\overbrace{0,0,\ldots,0}^{k-1},1)$ and $n = 0$,

$$\overbrace{\log\log\ldots\log}^{k}(x+a)$$
$$= \overbrace{\log\log\ldots\log}^{k}(x) + \sum_{p_1 \geq p_2 \geq \cdots \geq p_k = 1} (-1)^{p_1+\ldots+p_{k-2}+p_{k-1}+1}$$
$$\times \left[\prod_{j=1}^{k-2} e_{p_j - p_{j+1}}(1,\ldots,p_j - 1)\right] \sum_{n=0}^{\infty} a^n(n-1)!\, s(-n,p_1)\, \ell^{(-n,p_1,\ldots,p_{k-1})}$$

This framework immediately gives rise to logarithmic versions of Taylor's theorem and the expansion theorem in terms of the logarithmic generalization of evaluation at zero or augmentation. The *augmentation*[10] of order α is defined to be the linear functional $\langle\rangle_\alpha$ from the logarithmic algebra $\mathcal{L}$ to the complex number $\mathbf{C}$ such that

$$\langle \lambda_n^\beta(x) \rangle_\alpha = \delta_{\alpha\beta} \delta_{n0} \, .$$

THEOREM 9. (Logarithmic Taylor's Theorem) *Let $p(x) \in \mathcal{L}^+$. Then we have the following expansion of $p(x)$ in terms of the harmonic logarithms.*

$$p(x) = \sum_{\alpha \neq (0)} \sum_{n=-\infty}^{\infty} \frac{\langle D^n p(x) \rangle_\alpha}{\lfloor n \rfloor!} \lambda_n^\alpha(x) \, . \tag{14}$$

Proof. It suffices to show that eq. 14 holds for a basis of $\mathcal{L}^+$. However, Theorem 6 states that eq. 14 holds for the basis of harmonic logarithms. ∎

THEOREM 10. (Logarithmic Expansion Theorem)
 (1) Let $Q(D)$ be a shift-invariant operator on $\mathcal{L}^+$. Then we have the following expansion of $Q(D)$ in terms of the powers of the derivative

$$Q(D) = \sum_{n=-\infty}^{\infty} \frac{\langle Q(D)\lambda_n^\alpha(x) \rangle_\alpha}{\lfloor n \rfloor!} D^n$$

where $\alpha \neq (0)$.
 (2) Let $Q(D)$ be a shift-invariant operator on $\mathcal{L}$. Then we have the following expansion of $Q(D)$ in terms of the powers of the derivative

$$Q(D) = \sum_{n=0}^{\infty} \frac{\langle Q(D)\lambda_n^\alpha(x) \rangle_\alpha}{n!} D^n$$

---

[10] The restriction of the augmentations of order (0) to $\mathcal{L}^{(0)}$ is the evaluation at zero mentioned in eq. 7. The other augmentations can profitably be rewritten in terms of the residue at zero.

Calculus of Finite Differences

where α is any vector of integers with finite support.

Proof. Theorem is immediate for $Q(D) = D^n$ from Theorem 6. ∎

We now see that the sequence of harmonic logariths $\lambda_n^\alpha(x)$ is in some sense (to be made clear later) of "binomial type", and that in this sense the harmonic logarithms are associated with the derivative operator.

3.4. The Logarithmic Lower Factorial

We can now extend to the logarithmic domain the analogy between the derivative operator D and the forward difference operator Δ which has been noted for polynomials.

Since $\mathcal{L}(0)$ is the algebra of polynomials, one would rightly expect that the logarithmic version $p_n^\alpha(x) = (x)_n^\alpha$ of the lower factorial function $p_n(x) = (x)_n$ would include the subsequence

$$p_n^{(0)}(x) = (x)_n^{(0)} = \begin{cases} (x)_n = x(x-1)\ldots(x-n+1) & \text{for } n \text{ nonnegative, and} \\ 0 & \text{for } n \text{ negative} \end{cases}$$

It has been known for a long time that, setting for n negative,

$$p_n(x) = (x)_n = 1/(x+1)(x+2)\ldots(x-n),$$

one has

$$\Delta p_n(x) = n(x)_{n-1}.$$

Given this preliminary information,[11] we guess that when n is negative, $1/(x+1)\ldots(x-n)$ is to the forward difference operator as x^n is to the derivative. Thus, we write

$$p_n^{(1)}(x) = (x)_n^{(1)} = 1/(x+1)(x+2)\ldots(x+n)$$

for n negative.

This leaves open the question of what is the forward difference version of the logarithm $\lambda_0^{(1)}(x) = \log x$ and of the harmonic logarithms

$$\lambda_n^{(1)}(x) = x^n \left(\log x - 1 - \frac{1}{2} - \ldots - \frac{1}{n}\right)$$

for n positive. In other words, what is the formal solution $p_0^{(1)}(x) = (x)_0^{(1)}$ of the difference equation $\Delta p(x) = 1/(x+1)$? This question was first formulated by Gauss; in the present context, it can be dealt with very easily. By Corollary 1, all such difference equations have unique solutions.

In fact, we can easily calculate $(x)_0^{(1)}$. Since the *Bernoulli numbers* B_n are defined as the coefficients of Δ^{-1}:

$$\frac{D}{e^D - 1} = \sum_{k=0}^{\infty} B_k D^k/k!,$$

---

[11] This information is only preliminary since without the residual theorem (Theorem 14) to pinpoint the exact value of $p_{-1}(x)$, the entire sequence might be off by a factor equal to all shift-invariant operator of degree zero $a_0 + a_1 D + a_2 D^2 + \ldots$.

we have

$$p_0^{(1)}(x) = x_0^{(1)} = \log(x+1) + B_1/(x+1) - B_2/2(x+1)^2 + B_3/3(x+1)^3 - \ldots .$$

Note that $(x-1)_0^{(1)} = \psi(x)$ is the *Gauss psi function*. In this domain, we verify that

$$\psi(x) = \Gamma'(x)/\Gamma(x)$$

as follows. Begin with the fundamental identity for the Gamma function,

$$\Gamma(x+1) = x\Gamma(x) \qquad (15)$$

Now, take the derivative of both sides.

$$\Gamma'(x+1) = \Gamma(x) + x\Gamma'(x) . \qquad (16)$$

Next, divide eq. (16) by eq. (15).

$$\frac{\Gamma'(x+1)}{\Gamma(x+1)} = \frac{1}{x} + \frac{\Gamma'(x)}{\Gamma(x)} .$$

Finally, note that

$$\frac{1}{x} = \Delta \frac{\Gamma'(x)}{\Gamma(x)} .$$

Thus, $\Gamma'(x)/\Gamma(x)$ equals $(x)_0^{(1)}$ possibly up to a constant.[12]

For the logarithmic lower factorial of order (1) and positive degree n, we simply employ the higher order Bernoulli numbers

$$p_n^{(1)}(x) = x_n^{(1)} = \sum_{k=1}^{\infty} B_{k,n+1} \begin{bmatrix} n \\ k \end{bmatrix} \lambda_{n-k}^{\alpha}(x+1)$$

where

$$\left(\frac{D}{e^D - 1}\right)^n = \sum_{k=0}^{\infty} B_{kn} D^k/k! .$$

Finally, to compute the logarithmic lower factorials of order $\alpha \neq (1)$, merely use the projection maps.

$$p_n^{\alpha}(x) = (x)_n^{\alpha} = \text{skip}_{\alpha,(1)}(x)_n^{(1)} .$$

To carry the analogy to the end, we note that by eq. 5 and part 2 of Theorem 10, we still have

$$E^a = \sum_{n=0}^{\infty} (a)_n \Delta^n/n! .$$

---

[12] In fact, no constant is needed.

Calculus of Finite Differences

From this, we have the logarithmic version of Vandermonde's identity

$$(x+a)_n^\alpha = \sum_{k=0}^{\infty} \left\lfloor \begin{matrix} n \\ k \end{matrix} \right\rceil (a)_k (x)_{n-k}^\alpha$$

for all integers n and all α. Similarly, we have the logarithmic analog of the Newton's formula

$$f(x) = \sum_{\alpha \neq (0)} \sum_{n=-\infty}^{\infty} \langle \Delta^n f(x) \rangle_\alpha (x)_n^\alpha / n! \ .$$

We note that this formula constitutes a substantial extension of Newton's formula. For example, if $f(x) = 1/x$, then we have for n negative [13]

$$\Delta^n \frac{1}{x} = \left\langle E^{-1} \Delta^n (x)_{-1}^{(1)} \right\rangle_{(1)}$$

$$= \frac{\left\langle E^{-1} (x)_{-n-1}^{(1)} \right\rangle_{(1)}}{(-n-1)!}$$

$$= \frac{\left\langle E^{-1} (x)_{-n-1}^{(0)} \right\rangle_{(0)}}{(-n-1)!}$$

$$= \left[(x)_{-n-1}^{(0)} / (-n-1)! \right]_{x=-1}$$

$$= (-1)(-2) \ldots (n) / (-n-1)!$$

$$= (-1)^{n+1} n \ .$$

Thus,

$$\frac{1}{x} = \sum_{n=-\infty}^{-1} \frac{(-1)^{n+1}}{\lfloor n \rfloor !} (x)_n^{(1)}$$

$$= \sum_{n=0}^{\infty} \frac{n!}{(x+1)(x+2) \ldots (x+n)}$$

(15)

(see [$3\frac{3}{4}$] for an alternate proof).

3.5. Logarithmic Sequences of Binomial Type

As we did with polynomials, we can now extend the interplay between the derivative D and the forward difference operator into a more general framework involving all delta operators.[14] Instead of sequences of polynomials $p_n(x)$, we have *logarithmic sequences* $p_n^\alpha(x)$ indexed

---

[13] Obviously, for n nonnegative, $\langle \Delta^n \left(\frac{1}{x}\right) \rangle_{(1)} = 0$.

[14] As before, a delta operator is a delta series in the derivative; that is a shift-invariant operator whose kernel is the set of constants – in this case $\mathbb{C}$.

by an integer n and a vector with finite support of integers α. Again we have a requirement that $p_n^\alpha(x)$ be of *degree* n, and also a new requirement that it be of *order* α; that is to say,

$$p_n^\alpha(x) = \sum_{-\infty}^{n} c_k^\alpha \lambda_k^\alpha(x)$$

with $c_n^\alpha \neq 0$. In particular, $p_n(0)(x)$ is a sequence of polynomials: one of each degree with $p_n^{(0)}(x) = 0$ for n negative. More generally, the $p_n^\alpha(x)$ each belong to $\mathcal{L}^\alpha$ and in fact form a basis for it.

The only two examples of logarithmic sequences which we have seen so far are the harmonic logarithm $\lambda_n^\alpha(x)$ and the logarithmic lower factorial $(x)_n^\alpha$. However, there are clearly many more; every sequence of polynomials can be extended in many ways into a logarithmic sequence. However, we are particularly interested in sequences of polynomials of binomial type. We seek a natural definition of a logarithmic sequence of binomial type such that every sequence of polynomials has one unique such extension. The following theorem is exactly what we are seeking.

THEOREM 11. (Logarithmic Sequences of Binomial Type) *Let $p_n(x)$ be a sequence of polynomials of binomial type. Then there exists a unique logarithmic sequence $p_n^\alpha(x)$ called the logarithmic sequence of binomial type which extendes $p_n(x)$*

$$p_n^{(0)}(x) = p_n(x) \text{ for all intergers } n,$$

obeys an analog of the binomial theorem

$$p_n^\alpha(x + a) = \sum_{k=0}^{\infty} \left\lfloor \begin{matrix} n \\ k \end{matrix} \right\rfloor p_k(a) p_{n-k}^\alpha(x) \tag{17}$$

for all integers n and constants a, and is invariant under the skip operator $\mathrm{skip}_{\alpha\beta}$ *for $\beta \neq (0)$; that is to say,*

$$\mathrm{skip}_{\alpha\beta} p_n^\gamma(x) = \delta_{\beta\gamma} p_n^\alpha(x). \tag{18}$$

Proof[15]. **Existence.** Let $f(D)$ be the delta operator associated with $p_n(x)$ and let $g(D)$ be its compositional inverse.[16] Now define $p_n^\alpha(x)$ to be what is known commonly as the *conjugate logarithmic sequence* for $g(D)$.

$$p_n^\alpha(x) = \sum_{k=-\infty}^{n} \frac{\langle g(D)^k \lambda_n^\alpha(x) \rangle_\alpha}{\lfloor k \rfloor!} \lambda_k^\alpha(x).$$

After briefly noting that $p_n^\alpha(x)$ is a well defined logarithmic sequence invariant under the skip operators, it will now suffice to prove it obeys eqs. 16 and 17. This will follow immediately from the following two lemmas.

---

[15] Proof ends page 260.
[16] They are compositional inverses in the sense of composition of formal power series in the derivative. That is, $f(g(D)) = D = g(f(D))$.

Calculus of Finite Differences

LEMMA 11.1. (Associated Logarithmic Sequences) *Let $p_n^\alpha(x)$ be the conjugate logarithmic sequence for the delta operator $g(D)$ whose compositional inverse is $f(D)$. Then $p_n^\alpha(x)$ is the associated logarithmic sequence for $f(D)$ in the sense that*

$$\langle p_n^\alpha(x) \rangle_\alpha = \delta_{n0} \qquad (19)$$

$$f(D) p_n^\alpha(x) = \lfloor n \rfloor p_{n-1}^\alpha(x) . \qquad (20)$$

Proof. Eq. 19 follows directly from the definition of a conjugate logarithmic sequence and the augmentation,

$$\langle p_n^\alpha(x) \rangle_\alpha = \langle \lambda_n^\alpha(x) \rangle / \lfloor 0 \rfloor! = \delta_{n0} .$$

For eq. 20, we must first write $f(D)$ in terms of D^n : $f(D) = \sum_{k=0}^\infty a_k D^k$. Then we proceed as follow

$$f(D) p_n^\alpha(x) = \sum_{k,j \geq 0} a_j \frac{\langle g(D)^k \lambda_n^\alpha(x) \rangle_\alpha}{\lfloor k-j \rfloor!} \lambda_{k-j}^\alpha(x)$$

$$= \sum_{k,j \geq 0} \frac{\langle a_j g(D)^{k+j} \lambda_n^\alpha(x) \rangle_\alpha}{\lfloor k \rfloor!} \lambda_k^\alpha(x)$$

$$= \sum_{k=0}^\infty \frac{\left\langle \left(\sum_{j=0}^\infty a_j g(D)^j \right) g(D)^k \lambda_n^\alpha(x) \right\rangle_\alpha}{\lfloor k \rfloor!} \lambda_k^\alpha(x)$$

$$= \sum_{k=0}^\infty \frac{\langle D g(D)^k \lambda_n^\alpha(x) \rangle_\alpha}{\lfloor k \rfloor!} \lambda_k^\alpha(x)$$

$$= \lfloor n \rfloor \sum_{k=0}^\infty \frac{\langle g(D)^k \lambda_{n-1}^\alpha(x) \rangle_\alpha}{\lfloor k \rfloor!} \lambda_k^\alpha(x)$$

$$= \lfloor n \rfloor p_{n-1}^\alpha(x) . \qquad \blacksquare$$

Note that in the case $\alpha = (0), p_n^{(0)}(0) = \delta_{n0}$ and $f(D) p_n^{(0)}(x) = n p_{n-1}^{(0)}(x)$, so that $p_n^{(0)}(x)$ is associated to the delta operator $f(D)$. Hence, we have eq. 16.

LEMMA 11.2. *Let $p_n^\alpha(x)$ be the associated logarithmic sequence for the delta operator $f(D)$. Then for all shift–invariant operators $h(D)$ we have*

$$h(D) p_n^\alpha(x) = \sum_{k=-\infty}^\infty \begin{bmatrix} n \\ k \end{bmatrix} \langle h(D) p_k^\alpha(x) \rangle_\alpha p_{n-k}^\alpha(x) .$$

Proof. It suffices to prove the lemma for $h(D) = f(D)^j$ since the powers of the delta operator $f(D)$ form a basis for the space of shift-invariant operators. However, by the definition of an associated logarithmic sequence, we immediately have $\langle f(D)^j p_k^\alpha(x) \rangle = \delta_{jk} \lfloor j \rfloor!$. Thus,

$$f(D)^j p_n^\alpha(x) = \frac{\lfloor n \rfloor!}{\lfloor n-j \rfloor!} \sum_{k=-\infty}^\infty \begin{bmatrix} n \\ k \end{bmatrix} \langle f(D)^j p_k^\alpha(x) \rangle_\alpha p_{n-k}^\alpha(x) . \qquad \blacksquare$$

Finally, eq. 17 follows from the lemma with $h(D) = E^a$.

Uniqueness. For theorem 11, it remains only to show the uniqueness of logarithmic sequences of binomial type. We will do so in two lemmas. The first shows that all logarithmic sequences of binomial type are associated logarithmic sequences, and the other shows that every delta operator has a unique associated logarithmic sequence. These two lemmas will complete the proof of Theorem 11.

LEMMA 11.3. *Let $p_n^\alpha(x)$ be a logarithmic sequence of binomial type; that is, let it obey eq. 17. Then there is a unique delta operator $f(D)$ such that $p_n^\alpha(x)$ is associated with $f(D)$. That is, $p_n^\alpha(x)$ obeys eqs. 19 and 20.*

Proof. Eq. 19 immediately follows from Proposition 1. Now, we *define* a linear operator Q by
$$Qp_n^\alpha(x) = \lfloor n \rfloor p_{n-1}^\alpha(x)$$
Eq. 20 is equivalent to the proposition that Q is a delta operator. Since Q obviously commutes with $\text{skip}_{\alpha\beta}$ for $\beta \neq (0)$, and the kernel of Q is clearly the set of real constants $\mathbb{R}$, it remains only to show that Q commutes with the shift operator E^a.

$$QE^a p_n^\alpha(x) = \sum_{k=0}^{\infty} \begin{bmatrix} n \\ k \end{bmatrix} p_k(a) Q p_{n-k}^\alpha(x)$$

$$= \sum_{k=0}^{\infty} \begin{bmatrix} n \\ k \end{bmatrix} \lfloor n-k \rfloor p_k(a) p_{n-k-1}^\alpha(x)$$

$$= \sum_{k=0}^{\infty} \begin{bmatrix} n-1 \\ k \end{bmatrix} \lfloor n \rfloor p_k(a) p_{n-k-1}^\alpha(x)$$

$$= E^a Q p_n^\alpha(x) . \quad \blacksquare$$

LEMMA 11.4. *Let $f(D)$ be a delta operator. Then there is exactly one logarithmic sequence $p_n^\alpha(x)$ associated with $f(D)$.*

Proof. By Taylor's theorem (Theorem 9), $p_n^\alpha(x)$ is determined by its augmentations $a_k = \langle D^k p_n^\alpha(x) \rangle_\alpha$. However, since $f(D)^k$ is a basis for the space of shift-invariant operators, the augmentations a_k are determined by the augmentations $b_k = \langle f(D)^k p_n^\alpha(x) \rangle_\alpha$ which we know are equal to $\lfloor n \rfloor ! \delta_{nk}$. $\blacksquare$

We also derive – in the same manner as Lemma 11.2 – generalizations of the logarithmic Taylor's theorem and expansion theorem.

THEOREM 12. (Logarithmic Taylor's Theorem) *Let $p_n^\alpha(x)$ be the logarithmic sequence associated with the delta operator $f(D)$. Then any logarithmic series $p(x) \in \mathcal{L}^+$ can be expanded*

$$p(x) = \sum_{\alpha \neq (0)} \sum_{n=-\infty}^{N_\alpha} \frac{\langle f(D)^n p(x) \rangle_\alpha}{\lfloor n \rfloor !} p_n^\alpha(x) . \quad \blacksquare$$

THEOREM 13. (Logarithmic Expansion Theorem) *Let $p_n^\alpha(x)$ be the logarithmic sequence associated with the delta operator $f(D)$, and let $\alpha \neq (0)$. Then any shift-invariant operator $g(D)$ can be expanded*

$$g(x) = \sum_{n=-\infty}^{\infty} \frac{\langle g(D) p_n^\alpha(x) \rangle_\alpha}{\lfloor n \rfloor !} f(D)^n . \quad \blacksquare$$

Calculus of Finite Differences

3.6. Explicit Formulas

We note that if any term of the logarithmic sequence of binomial type $p_n^{(1)}(x)$ associated with the delta operator $f(D)$ is known *a priori*, then all of the other terms can be computed in terms of it by appropriate use of the various powers of $f(D)$ and the projections $\text{skip}_{\alpha,(1)}$. In particular, we only need to determine the term $p_{-1}^{(1)}(x)$ which is called the *residual term*, and then

$$p_n^\alpha(x) = \text{skip}_{\alpha,(1)} f(D)^{-n-1} p_{-1}^{(1)}(x) / \lfloor n \rfloor! \,.$$

Thus, the following result is extremely fundamental,

THEOREM 14. (Residual Term) *Let $p_n^\alpha(x)$ be the logarithmic sequence associated with the delta operator $f(D)$, and let $\alpha \neq (0)$. Then*[17]

$$p_{-1}^{(1)}(x) = f'(D) \frac{1}{x} \,.$$

Proof. We will actually prove a seemingly stronger result

$$p_n^\alpha(x) = \lfloor n \rfloor! \, f'(D) f(D)^{-1-n} \lambda_{-1}^\alpha(x) \,. \tag{21}$$

Let $q_n^\alpha(x)$ denote the right hand side of eq. 21. Clearly, $q_n^\alpha(x)$ is a logarithmic sequence, so by Lemma 11.4, it will suffice to show that $q_n^\alpha(x)$ is associated with the delta operator $f(D)$. Of the two defining properties, eq. 20 is trivial to demonstrate.

$$f(D) q_n^\alpha(x) = \lfloor n \rfloor \lfloor n-1 \rfloor! \, f'(D) f(D)^{-n} \lambda_{-1}^\alpha(x) = \lfloor n \rfloor q_{n-1}^\alpha(x) \,.$$

To prove eq. 19, we first consider the case $n \neq 0$.

$$\langle q_n^\alpha(x) \rangle_\alpha = \langle \lfloor n \rfloor! \, f'(D) f(D)^{-1-n} \lambda_{-1}^\alpha(x) \rangle_\alpha$$

$$= \frac{\lfloor n \rfloor!}{-n} \langle (f(D)^{-n})' \lambda_{-1}^\alpha(x) \rangle_\alpha$$

Moreover, $\langle (f(D)^{-n})' \lambda_{-1}^\alpha(x) \rangle_\alpha$ is given by the coefficient of D^{-1} in $(f(D)^{-n})'$; however, this coefficient is always zero.[18] Thus, $\langle p_n^\alpha(x) \rangle_\alpha = 0$ for $n \neq 0$.

Whereas, for $n = 0$, we have

$$\langle p_0^\alpha(x) \rangle_\alpha = \langle f'(D) f(D)^{-1} \lambda_{-1}^\alpha(x) \rangle_\alpha \,.$$

However, $f'(D) f(D)^{-1}$ is equal to D^{-1} plus terms of higher degree, so

$$\langle p_0^\alpha(x) \rangle_\alpha = 1 \,.$$

∎

---

[17] The operator $f'(D)$ called the *Pincherle derivative* of $f(D)$ representes the operator $g(D)$ where $g(t) = df(t)/dt$. Thus, if $f(D) = D^n$, then $f'(D) = nD^{n-1}$. Another characterization of the Pincherle derivative will be given later. (Proposition 2)

[18] This theorem will be seen to be equivalent to the Lagrange inversion formula (Theorem 17). This observation here that the residue of the derivative of a formal power series is zero is the key step in all known proofs of the Lagrange inversion formula.

Theorem 14 allows us to derive a series of formulas – as explicit as possible – for the computation of sequences of polynomials of binomial type or logarithmic sequence of binomial type associated with a delta operator $f(D)$.

However, we require one last definition. We define the *Roman shift*[19] as the linear operator σ over the logarithmic algebra such that for all α

$$\sigma \lambda_n^\alpha(x) = \begin{cases} \lambda_{n+1}^\alpha(x) & \text{for } n \neq -1, \text{ and} \\ 0 & \text{for } n = -1. \end{cases}$$

The Roman shift is the logarithmic generalization of multiplication by x. In particular, note that for $n \neq -1$,

$$\sigma x^n = x^{n+1}.$$

We observe the following remarkable facts about the Roman shift.

PROPOSITION 2. (Pincherle Derivative) *For all logarithmic series $p(x) \in \mathcal{L}$,*

$$(D\sigma - \sigma D)p(x) = p(x) ; \qquad (22)$$

in other words,

$$D\sigma - \sigma D = I.$$

More generally, for all shift-invariant operators $f(D)$, we have the operator identity

$$f(D)\sigma - \sigma f(D) = f'(D) \qquad (23)$$

Proof. It will suffice to consider the case $f(D) = D^k$ and $\lambda_n^\alpha(x)$. When $k = 0$, eq. 23 is a triviality. Now, consider the case in which n is neither $k-1$ nor -1:

$$(D^k \sigma - \sigma D^k)\lambda_n^\alpha(x) = D^k \lambda_{n+1}^\alpha(x) - \frac{\lfloor n \rfloor!}{\lfloor n-k \rfloor!} \sigma \lambda_{n-k}^\alpha(x)$$

$$= \left(\frac{\lfloor n+1 \rfloor!}{\lfloor n-k+1 \rfloor!} - \frac{\lfloor n \rfloor!}{\lfloor n-k \rfloor!} \right) \lambda_{n-k+1}^\alpha(x)$$

$$= (\lfloor n+1 \rfloor - \lfloor n-k+1 \rfloor) \frac{\lfloor n \rfloor!}{\lfloor n-k+1 \rfloor!} \lambda_{n-k+1}^\alpha(x)$$

$$= kD^{k-1} \lambda_n^\alpha(x).$$

Next, consider the case in which $n = -1$ but $k \neq 0$:

$$(D^k \sigma - \sigma D^k) \lambda_{-1}^\alpha(x) = -\sigma D^k \lambda_{-1}^\alpha(x)$$

$$= -\frac{1}{\lfloor -k-1 \rfloor!} \sigma \lambda_{-k-1}^\alpha(x)$$

$$= -\lfloor -k \rfloor \frac{1}{\lfloor -k \rfloor!} \lambda_{-k}^\alpha(x)$$

$$= kD^{k-1} \lambda_{-1}^\alpha(x).$$

---

[19] Named after Steve Roman. Sometimes also called the *Standard Roman Shift*, or the *Roman shift for D* or the *Roman Shift for $\lambda_n^\alpha(x)$*.

Finally, consider the case in which $n = k - 1$:

$$(D^k\sigma - \sigma D^k)\lambda_n^\alpha(x) = D^k \lambda_{n+1}^\alpha(x) - \frac{\lfloor n \rfloor!}{\lfloor 1 \rfloor!}\sigma\lambda_{-1}^\alpha(x)$$
$$= \lfloor n+1 \rfloor! \lambda_0^\alpha(x)$$
$$= \lfloor n+1 \rfloor D^n \lambda_n^\alpha(x)$$
$$= k D^{k-1} \lambda_n^\alpha(x) .$$
∎

Thus, we see that the Roman shift gives a non-trivial extension of the classical commutation relation $Dx - xD = I$ of quantum mechanics in the logarithmic algebra. In effect, eq. 22 is really quite incredible, for Van Neuman has shown that for polynomials,[20] essentially the only operators (θ, ϕ) obeying the commutation relation

$$\theta\phi - \phi\theta = I$$

are $\theta = D$ and $\phi = x$. However, by eq. 22, we see that for logarithmic series, $\theta = D$ and $\phi = \sigma$ is yet another solution.

THEOREM 15. (Rodrigues) *Let $p(x)$ be a finite linear combination of the harmonic logarithms. Then for all real numbers a, we have the following formal identity*

$$(\exp(-a\sigma)D\exp(a\sigma))p(x) = (D - aI)p(x) .$$
∎

We can now state our main result.

THEOREM 16. *Let $f(D) = Dg(D)$ be the delta operator associated both with the logarithmic sequence $p_n^\alpha(x)$, and with the sequence of polynomials $p_n(x) = p_n^{(0)}(x)$. Then we have,*
  1. (Transfer Formula) *For all n,*

$$p_n^\alpha(x) = f'(D)g(D)^{-n-1}\lambda_n^\alpha(x) ,$$

  2. (Shift Formula) *For n neither 0 nor -1,*

$$p_n^\alpha(x) = \sigma g(D)^{-n}\lambda_{n-1}^\alpha(x) ,$$

and
  3. (Recurrence Formula) *For $n \neq 0$,*

$$p_n^\alpha(x) = \sigma f'(D)^{-1}p_{n-1}^\alpha(x) .$$

In particular, for the polynomial sequence $p_n(x)$, we have
  1. (Transfer Formula) *For all $n \geq 0$,*

$$p_n(x) = f'(D)g(D)^{-n-1}x^n ,$$

---

[20] Since then others have shown that this result holds for much more general sorts of functions although obviously not for logarithmic series.

2. (Shift Formula) *For all* $n > 0$,

$$p_n(x) = xg(D)^{-n}x^{n-1},$$

and

3. *(Recurrence Formula) For all* $n > 0$,

$$p_n(x) = xf'(D)_1 p_{n-1}(x).$$

Proof. (1) From eq. 21, we have for $\alpha \neq (0)$

$$p_n^\alpha(x) = \lfloor n \rfloor! f'(D)g(D)^{n+1} D^{-1-n}\lambda_{-1}^\alpha(x) = f'(D)g(D)^{n+1}\lambda_n^\alpha(x).$$

For $\alpha = (0)$, we derive the result from the case $\alpha = (1)$ as follows[21]

$$\begin{aligned}p_n(x) &= \mathrm{skip}_{(0),(1)} p_n^{(1)}(x) \\ &= \mathrm{skip}_{(0),(1)} f'(D)g(D)^{n+1}\lambda_n^{(1)}(x) \\ &= f'(D)g(D)^{n+1}\mathrm{skip}_{(0),(1)}\lambda_n^{(1)}(x) \\ &= f'(D)g(D)^{n+1}x^n.\end{aligned}$$

(2) For this part, we need the following routine calculation

$$f'(D)g(D)^{n+1} = g(D)^n - \frac{1}{n}(g(D)^n)'D.$$

Thus, by part (1), we have

$$p_n^\alpha(x) = g(D)^n \lambda_n^\alpha(x) - (g(D)^n)'\lambda_{n-1}^\alpha(x).$$

Finally, by Proposition 2, we have

$$\begin{aligned}p_n^\alpha(x) &= g(D)^n \lambda_n^\alpha(x) - g(D)^n \sigma \lambda_{n-1}^\alpha(x) + \sigma g(D)^n \lambda_{n-1}^\alpha(x) \\ &= g(D)^n \lambda_n^\alpha(x) - g(D)^n \lambda_n^\alpha(x) + \sigma g(D)^n \lambda_{n-1}^\alpha(x) \\ &= \sigma g(D)^n \lambda_{n-1}^\alpha(x).\end{aligned}$$

(3) This follows immediately from the substitute of the Transfer Formula

$$\lambda_{n-1}^\alpha(x) = f'(D)^{-1}g(D)^n p_{n-1}^\alpha(x)$$

in the Shift Formula

$$p_n^\alpha(x) = \sigma g(D)^{-n}\lambda_{n-1}^\alpha(x). \qquad \blacksquare$$

Actually, Theorem 16 can be generalized to cover the case in which the role of the harmonic logarithm $\lambda_n^\alpha(x)$ can be played by any logarithmic sequence of binomial type $p_n^\alpha(x)$.

---

[21] This is then the best possible example of a purely logarithmic proof of a result involving polynomials. It is derived from a formula (eq. 21) which does not even make sense for polynomials.

Calculus of Finite Differences 265

COROLLARY 2. *Let $f(D)$ and $g(D)$ be delta operators associated with the logarithmic sequences $p_n^\alpha(x)$ and $q_n^\alpha(x)$, and the sequences of polynomials $p_n(x) = p_n^{(0)}(x)$ and $q_n(x) = q_n^{(0)}(x)$. Then*

1. We have

$$p_n^\alpha(x) = \frac{f'(D)g(D)^{n+1}}{g'(D)f(D)^{n+1}} q_n^\alpha(x)$$

for all α and n, and in particular for $\alpha = (0)$ we have

$$p_n(x) = \frac{f'(D)g(D)^{n+1}}{g'(D)f(D)^{n+1}} q_n(x) ,$$

2. And for the polynomial sequences only,[22] we have

$$p_n(x) = x(g(D)^n f(D)^{-n}[x^{-1}q^n(x)]) .$$

Proof. Part one is immediate from the Transfer Formula. Part two results from the Shift Formula since the Roman shift of a polynomial is multiplication by x. ∎

3.7. Lagrange Inversion

The explicit formulas[23] of the preceeding section lead to an elegant proof of the Lagrange Inversion formula for the compositional inverse $f^{(-1)}(t)$ of a delta series $f(t)$.

THEOREM 17. (Lagrange Inversion) *Let $f(t)$ be a delta series, and $g(t)$ be any Laurent series of degree d, then*

$$g(f^{(-1)}(t)) = \sum_{k=d}^{\infty} \left\langle g(D)f'(D)f(D)^{-1-k}\frac{1}{x}\right\rangle_{(1)} t^k . \tag{24}$$

In particular,

$$f^{(-1)}(t)^n = \sum_{k=n}^{\infty} \left\langle D^n f'(D)f(D)^{-1-k}\frac{1}{x}\right\rangle_{(1)} t^k .$$

In other words, the coefficient of t^k in $f^{(-1)}(t)^n$ is the coefficient of t^{n-1} in $f'(t)f(t)^{-1-k}$.

Proof. It will suffice to demonstrate eq. 24. By eq. 21 formula,

$$\lfloor k \rfloor ! \left\langle g(D)p_k^{(1)}(x)\right\rangle_{(1)} = \left\langle g(D)f'(D)f(D)^{-1-k}\left(\frac{1}{x}\right)\right\rangle_{(1)}$$

where $p_k^\alpha(x)$ is the logarithmic sequence of binomial type associated with the delta operator $f(D)$. Now,

$$\langle f(D)^n p_k^\alpha(x)\rangle_\alpha = \delta_{nk}\lfloor n\rfloor ! = \langle D^n \lambda_k^\alpha(x)\rangle_\alpha ,$$

---

[22] The Roman shift is not in general invertible.
[23] In particular, eq. 21.

so in an augmentation of an operator acting on a $p_k^\alpha(x)$, one can substitute $f^{(-1)}(D)$ for D in the operator provided one also substitutes $\lambda_k^\alpha(x)$ for $p_k^\alpha(x)$. In particular,

$$\lfloor k \rfloor! \left\langle g(D) p_k^{(1)}(x) \right\rangle_{(1)} = \lfloor k \rfloor! \left\langle g(f^{(-1)}(D)) p_k^{(1)}(x) \right\rangle_{(1)}. \tag{25}$$

However, by the Expansion theorem (Theorem 10), $\lfloor k \rfloor! \left\langle g(f^{(-1)}(D)) p_k^{(1)}(x) \right\rangle_{(1)}$ is the coefficient of D^k in $g(f^{(-1)}(D))$. Putting all this information together, we have eq. 24. ∎

4. Examples

4.1. The Logarithmic Lower Factorial Sequence

We now return to our example of a logarithmic sequence of binomial type (Section 3.4.). The logarithmic lower factorial $(x)_n^\alpha$ is the logarithmic sequence of binomial type associated with the delta operator $f(D) = \Delta = e^D - I$.

By Theorem 14, to determine the residual series, we must first compute the Pincherle derivative

$$f'(D) = \Delta' = eD = E^1;$$

Thus, the residual series is

$$(x)_{-1}^{(1)} = E^1 \frac{1}{x} = \frac{1}{x+1}.$$

This completes our calculations of section 3.4., since all of the other terms $(x)_n^\alpha$ can then be computed via the formula

$$(x)_n^\alpha = \lfloor n \rfloor!^{-1} \mathrm{skip}_{\alpha,(1)} D^{-n-1} \left(\frac{1}{x+1} \right).$$

However, we should note that in some cases the recursion formula is much more useful

$$(x)_n^\alpha = \sigma(x-1)_{n-1}^\alpha.$$

In particular, for $\alpha = (0)$ and $n \geq 0$ or $\alpha = (1)$ and $n < 0$, this is equivalent to

$$(x)_n^\alpha = x(x-1)_{n-1}^\alpha.$$

In these cases, we immediately have

$$(x)_n = x(x-1)\ldots(x-n+1)$$

for n positive, and

$$(x)_n^{(1)} = \frac{1}{(x+1)(x+2)\ldots(x-n)}$$

for n negative.

4.2. The Laguerre Logarithmic Sequence

Our next example, the Laguerre logarithmic sequence $L_n^\alpha(x)$ and the Laguerre polynomials $L_n(x) = L_n^{(0)}(x)$, are the logarithmic sequence and polynomials sequence of binomial type associated with the Laguerre operator $f(D) = W$ defined by eq. 10 or eq. 11.

We have for all n and α by the Transfer formula,

$$L_n^\alpha(x) = f'(D)\left(\frac{D}{f(D)}\right)\lambda_n^\alpha(x)$$
$$= -(D-I)^{n-1}\lambda_n^\alpha(x) \qquad (26)$$
$$= \sum_{k=0}^\infty (-1)^{n+k}\left[\begin{array}{c}n-1\\k\end{array}\right]\frac{\lfloor n \rfloor!}{\lfloor n-k \rfloor!}\lambda_{n-k}^\alpha(x)$$

where $\binom{n}{k} = (n)_k/k!$. In particular, for $\alpha = (0)$ and n positive, we have

$$L_n(x) = \sum_{k=0}^{n-1}(-1)^{n+k}\left[\begin{array}{c}n-1\\k\end{array}\right]\frac{n!}{n-k!}x^{n-k} .$$

Note that for the Laguerre logarithmic sequence, we not only have

$$\mathrm{skip}_{(0),(1)} L_n^{(1)}(x) = L_n(x)$$

as required by eq. 18, but we also have

$$\mathrm{skip}_{(1),(0)} L_n(x) = L_n^{(1)}(x) \qquad (27)$$

for n positive. This a very special property[24]; it is only true of logarithmic sequences associated with delta operators of the form $aD/(D-b)$ for a,b real scalars ($a \neq 0$).

However,

$$L_0^{(1)} = \log x + \frac{1}{x} - \frac{1}{x^2} + \frac{2}{x^3} - \frac{6}{x^4} + \dots$$

which is not equal to $\mathrm{skip}_{(1),(0)} L_0(x) = \mathrm{skip}_{(1),(0)} 1 = \log x$.

From eq. 26 and the classical identity (for polynomials)

$$e^x D^n e^{-x} = (D-I)^n , \qquad (28)$$

we have the *Rodrigues formula* for the Laguerre sequence

$$L_n(x) = e^x D^{n-1} e^{-x} x^n .$$

---

[24] The logarithmic sequences of the form $a^n \lambda_n^\alpha(x)$ (for example the harmonic logarithm) are even more special since they obey eq. 27 even for $n = 0$. Nevertheless, there are no logarithmic sequences satisfying eq. 27 for any negative integers n.

4.3. The Abel Logarithmic Sequence

We continue our series of examples with the logarithmic generalization $A_n^\alpha(x; b)$ of the Abel polynomials $A_n(x; b)$ associated with the delta operator $f(D) = E^a D$ mentioned in Section 1.4.

By the shift formula, for n neither 0 nor 1, we have

$$A_n^\alpha(x) = \sigma \lambda_{n-1}^\alpha (x - nb) .$$

For $\alpha = (0)$ and $n > 1$, or $\alpha = (1)$ and $n < 0$, this immediately gives us

$$x(x - nb)^{n-1} .$$

In particular, the residual series is

$$A_{-1}^{(1)}(x) = x(x + b)^{-2} .$$

It remains now only to compute the terms of degre zero and one. The series of degree zero is computed via the transfer formula, and turns out be much simpler than might have been expected:

$$A_0^{(1)}(x) = (DE^b)' E^{-b} \log x$$
$$= (I + bD) \log x$$
$$= \log x + b/x .$$

As an application, we infer from the logarithmic binomial theorem (eq. 17)

$$A_0^\alpha(x + a) = \sum_{k=0}^\infty \left[\begin{matrix} 0 \\ k \end{matrix} \right] A_k(a; b) A_k^{(1)}(x)$$

the remarkable identity

$$\frac{b}{x+a} + \log(x + a) = \frac{b}{x} + \log x + \sum_{k=1}^\infty \frac{(-1)^{k+1} a(a - bk)^{k-1} x}{k(x + bk)^{k+1}} .$$

For example, we can substitute $a = 1, b = 2$, and $x = 5$, and compute the first twelve terms of both sides yielding equality to seven decimal places.

In general, the transfer formula gives us

$$A_n^\alpha(x) = E^{-nb}(I + bD) \lambda_n^\alpha(x)$$
$$= \lambda_n^\alpha(x - nb) + b \lfloor n \rfloor \lambda_{n-1}^\alpha (x - nb) .$$

For example, the term of degree one is given by

$$A_1^{(1)}(x) = x \log(x - a) + a - x .$$

Finally, consider the Taylor's theorem for the logarithmic Abel sequence (Theorem 12). Any $p(x) \in \mathcal{L}^+$ can be expanded in terms of the logarithmic Abel sequence as follows

$$p(x) = \sum_{\alpha \neq (0)}^{N\alpha} \sum_{n=-\infty}^{\infty} \frac{a_n^\alpha}{\lfloor n \rfloor!} A_n^\alpha(x)$$

Calculus of Finite Differences

where $a_n^\alpha = \langle E^{nb} D^n p(x) \rangle_\alpha$. In particular, for $p(x) = \log x$, $a_n^\alpha = 0$ for $\alpha \neq (1)$ or n positive, but for n nonpositive and $\alpha = (1)$ we have

$$a_n^\alpha = \langle E^{nb} D^n \log x \rangle_{(1)}$$
$$= \langle E^{nb} \lambda_{-n}^{(1)}(x)/(-n)! \rangle_{(1)}$$
$$= \langle E^{nb} \lambda_{-n}^{(0)}(x)/(-n)! \rangle_{(0)}$$
$$= \langle (x+nb)^{-n} \rangle_{(0)} /(-n)!$$
$$= (nb)^{-n}/(-n)! \ .$$

Thus,

$$\log x = \log x + \frac{b}{x} + \sum_{n=1}^\infty \frac{(-nb)^n}{n!\lfloor -n\rfloor!} A_{-n}^{(1)}(x)$$

$$-\frac{b}{x} = \sum_{n=1}^\infty \frac{(-nb)^n(-1)^{n-1}}{n} x(x+nb)^{-n-2} \qquad (29)$$

$$\frac{b}{x^2} = \sum_{n=1}^\infty \frac{(nb)^n}{n}(x+nb)^{-n-2}$$

Note that eq. 29 is the correct version of the calculations in [12, p. 105]. We thank Richard Askey for bringing our error to our attention.

4.4. The Logarithmic Ramey Sequences

Let $p_n^\alpha(x)$ and $p_n(x) = p_n^{(0)}(x)$ be the logarithmic and polynomial sequences of binomial type associated with a delta operator $f(D)$. The sequences $p_n^\alpha(x;b)$ and $p_n(x;b)$ associated with the delta operator $E^b f(D)$ are called the logarithmic and polynomial *Ramey sequences*[25] of $p_n(x)$ and $p_n^\alpha(x)$. (See Ramey's paper on polynomial Ramey sequences for $f(D)$). For example, if $f(D) = D$, the Ramey sequences are the Abel sequences dealt with in sections 1.4. and 4.3,

The following proposition is very useful in the calculation of Ramey sequences.

PROPOSITION 3. *Let $p_n^\alpha(x)$ and $p_n(x)$ be the Ramey sequences for the logarithmic and polynomial sequences of binomial type $q_n^\alpha(x)$ and $q_n(x) = q_n^{(0)}(x)$ for the delta operator $f(D)$.*

1. Then for n neither zero nor one, the explicit relationship between $p_n(x)$ and $q_n(x)$ is given by

$$p_n(x) = \frac{xq_n(x-nb)}{x-nb} \ .$$

2. *For all n and α, we have*

$$p_n^\alpha(x) = q_n^\alpha(x-nb) + \lfloor n \rfloor b f'(D)^{-1} q_{n-1}^\alpha(x-nb) \ .$$

---

[25] Also called the Abelization.

3. For n negative and $\alpha = (1)$, we have

$$p_n^{(1)}(x) = q_n^{(1)}(x - nb) + \lfloor n \rfloor bx^{-1} q_n^\alpha(x) .$$

Proof. (1) Immediate from part two of Corollary 2.
(2) Immediate from part one of Corollary 2.
(3) Since the Roman shift is multiplication by x in this context, we have by the Recurrence Formula

$$x^{-1} q_n^\alpha(x) = f'(D) q_{n-1}^\alpha(x) .$$

The conclusion now immediately follows from part two. ∎

For example, let us now compute the polynomial Ramey sequence for the lower factorial. This is the polynomial sequences $G_n(x)$ commonly called the *Gould polynomials* associated with the delta operator $D\Delta$. We have

$$G_n(x) = \frac{x(x - nb)_n}{x - nb} = x(x - nb - 1)_{n-1} .$$

5. Connection Constants

Let $p_n^\alpha(x)$ and $q_n^\alpha(x)$ be logarithmic sequences with

$$q_n^\alpha(x) = \sum_{k=-\infty}^{n} c_{nk}^\alpha \lambda_k^\alpha(x) . \tag{30}$$

Then the sequence

$$r_n^\alpha(x) = \sum_{k=-\infty}^{n} c_{nk}^\alpha p_k^\alpha(x)$$

is called the *umbral composition* of the logarithmic sequences $p_n^\alpha(x)$ and $q_n^\alpha(x)$. To denote umbral compostion, we write

$$r_n^\alpha(x) = q_n^\alpha(\mathbf{p}) .$$

Restricting out attention to polynomials, we see the if $p_n(x)$ and $q_n(x)$ with

$$q_n(x) = \sum_{k=0}^{n} c_{nk} x^k ,$$

then the umbral composition is uniquely defined, and is given by

$$q_n(\mathbf{p}) = \sum_{k=0}^{n} c_{nk} p_n(x) .$$

Calculus of Finite Differences

When $p_n^\alpha(x)$ and $q_n^\alpha(x)$ are of binomial type, then umbral compostion has a remarkable property.

THEOREM 18. *Let $p_n^\alpha(x)$ and $q_n^\alpha(x)$ be the logarithmic sequences of binomial type associated with the delta operators $f(D)$ and $g(D)$ respectively. Then their umbral composition $r_n^\alpha(x) = p_n^\alpha(\mathbf{q})$ is the logarithmic sequence of binomial type associated with the delta operator $f(g(D))$.*

Proof. We must show the $r_n^\alpha(x)$ obeys eqs. 19 and 20. Let the coefficients of $p_n^\alpha(x)$ and $f(D)$ be given by

$$p_n^\alpha(x) = \sum_{k=-\infty}^{n} c_{nk} \lambda_k^\alpha(x)$$

$$f(D) = \sum_{j=1}^{\infty} a_j D^j$$

We write c_{nk} since by eq. 18 the coefficients do not depend on α.

To prove eq. 19, we proceed as follows:

$$\langle r_n^\alpha(q) \rangle_\alpha = \left\langle \sum_{k=-\infty}^{n} c_{nk} q_k^\alpha(x) \right\rangle_\alpha$$

$$= \sum_{k=-\infty}^{n} c_{nk} \langle q_k^\alpha(x) \rangle_\alpha$$

$$= \sum_{k=-\infty}^{n} c_{nk} \delta_{0k}$$

$$= c_{n0}$$

$$= \langle p_n^\alpha(x) \rangle_\alpha$$

$$= \delta_{n0} .$$

Next, it remains to prove eq. 20.

$$f(g(D)) r_n^\alpha(x) = \sum_{j=1}^{\infty} \sum_{k=-\infty}^{n} a_j c_{nk} g(D)^j q_k^\alpha(x)$$

$$= \sum_{j=1}^{\infty} \sum_{k=-\infty}^{n} a_j c_{nk} \frac{\lfloor k \rfloor !}{\lfloor k-j \rfloor !} q_{k-j}^\alpha(x) .$$

However,

$$\lfloor n \rfloor p_{n-1}^\alpha(x) = f(D) p_n^\alpha(x)$$

$$= \sum_{j=1}^{\infty} \sum_{k=-\infty}^{n} a_j c_{nk} D^j \lambda_k^\alpha(x)$$

$$= \sum_{j=1}^{\infty} \sum_{k=-\infty}^{n} a_j c_{nk} \frac{\lfloor k \rfloor !}{\lfloor k-j \rfloor !} \lambda_{k-j}^\alpha(x) ,$$

so that

$$f(g(D))r_n^\alpha(x) = \lfloor n \rfloor r_{n-1}^\alpha(x) .$$ ∎

By restricting our attention to $\mathcal{L}^{(0)}$, we have the analogous result for polynomials.

COROLLARY 3. *Let $p_n(x)$ and $q_n(x)$ be the polynomial sequencese of binomial type associated with the delta operators $f(D)$ and $g(D)$ respectively. Then their umbral composition $r_n(x) = p_n(\mathbf{q})$ is the sequence of binomial type associated with the delta operator $f(g(D))$.* ∎

The preeceding theorem allows us to solve the *problem of connection constants*. That is, given two logarithmic sequence of binomial type $r_n^\alpha(x)$ and $q_n^\alpha(x)$ associated with the delta operators $g(D)$ and $h(D)$, we are to find constants[26] c_{nk} such that

$$r_n^\alpha(x) = \sum_{k=-\infty}^{n} c_{nk} q_k^\alpha(x) .$$

The preceeding theory shows that the constants c_{nk} are given by a logarithmic sequence $p_n^\alpha(x)$ as in eq. 30 associated with the delta operator $f(D) = h(g^{(-1)}(D))$. Of course, the same result holds true for polynomial sequences of binomial type.

In particular, given any logarithmic sequence of binomial type $p_n^\alpha(x)$ associated with the delta operator $f(D)$, there exists a unique inverse logarithmic sequence $\overline{p}_n^\alpha(x)$ associated with the delta operator $f^{(-1)}(D)$. In view of the logarithmic Taylor's theorem (Theorem 9),

$$\overline{p}_n^\alpha(x) = \sum_{k=-\infty}^{n} \frac{\langle D^k \overline{p}_n^\alpha(x) \rangle_\alpha}{\lfloor k \rfloor !} \lambda_k^\alpha(x) .$$

However, by the demonstration of eq. 25, $\overline{p}_n^\alpha(x)$ is the conjugate logarithmic sequence for the delta operator $f(D)$; that is,

$$\overline{p}_n^\alpha(x) = \sum_{k=-\infty}^{n} \frac{\langle f(D)^k \lambda_n^\alpha(x) \rangle_\alpha}{\lfloor k \rfloor !} \lambda_k^\alpha(x) .$$

Moreover, by further umbral compostion

$$\overline{p}_n^\alpha(\mathbf{q}) = \sum_{k=-\infty}^{n} \frac{\langle f(D)^k \lambda_n^\alpha(x) \rangle_\alpha}{\lfloor k \rfloor !} q_k^\alpha(x) .$$

Obviously, similar results hold for polynomials.

---

[26] Such constants exist since the two sequences are basis, and the coefficents do not depend on α by eq. 18 (see [$3\frac{1}{2}$] and [$3\frac{3}{4}$] for other techniques useful in calculating these connection constants).

6. Examples

6.1. From the Logarithmic Lower Factorial to the Harmonic Logarithm

To compute the connection constants expressing the harmonic logarithms $\lambda_n^\alpha(x)$ in terms of the logarithmic lower factorials $(x)_n^\alpha$ via the above techniques, we must first calculate the logarithmic generalization $\phi_n^\alpha(x)$ of the *exponential polynomials* $\phi_n(x)$ associated with the delta operator $\Delta^{-1} = \log(I + D)$ and which have been studied by Touchard [20] among others.

As noted above, this is the logarithmic conjugate sequence for forward difference operator Δ. That is,

$$\phi_n^\alpha(x) = \sum_{k=-\infty}^{n} \frac{\langle \Delta^k \lambda_n^\alpha(x) \rangle_\alpha}{\lfloor k \rceil !} \lambda_k^\alpha(x) .$$

In particular, for $\alpha = (0)$,

$$\phi_n(x) = \sum_{k=0}^{n} \frac{\langle \Delta^k x^n \rangle_\alpha}{\lfloor k \rceil !} x^k = \sum_{k=0}^{n} S(n, k) x^k$$

where $S(n, k)$ denote the *Stirling numbers of the second kind*. To calculate the residual series, we as usual employ Theorem 14,

$$\phi_{-1}^{(1)}(x) = \log(I + D)' x^{-1}$$

$$= \sum_{k=0}^{\infty} (-1)^k D^k x^{-1}$$

$$= \sum_{k=0}^{\infty} k! / x^{k+1}$$

which is equivalent via umbral composition to eq. 15.

Other terms of the logarithmic exponential sequence are most easily computed via the recurrence formula.

$$\phi_n^\alpha(x) = \sigma(\log(I + D)')^{-1} \phi_{n-1}^\alpha(x)$$
$$= \sigma(I + D) \phi_{n-1}^\alpha(x) .$$

For $\alpha = (0)$ and n positive, or $\alpha = (1)$ and n negative, we thus have by eq. 28

$$\phi_n^\alpha(x) = xe^{-x} D e^x \phi_{n-1}^\alpha(x)$$
$$= e^{-x}(xD) e^x \phi_{n-1}^\alpha(x) .$$

In particular, for polynomials, we have by induction

$$\phi_n(x) = e^{-x}(xD)^n e^x .$$

6.2. From the Logarithmic Lower Factorials to the Logarithmic Upper Factorials

The *Logarithmic upper factorial* $\langle x \rangle_n^\alpha$ is the logarithmic sequences associated with the backward difference operator $\nabla = I - E^{-D}$. In effect, it is the important special case of the Gould sequence with $b = -1$. Thus, for $\alpha = (0)$ and n positive, we have the sequence of polynomials

$$\langle x \rangle_n = \langle x \rangle_n^\alpha = x(x+1)\ldots(x+n-1),$$

and for $\alpha = (1)$ and n negative, we have

$$\langle x \rangle_n^{(1)} = \frac{1}{(x-1)(x-2)\ldots(x+n)}.$$

To express $(x)_n^\alpha$ in terms of $\langle x \rangle_n^\alpha$, we must first compute the logarithmic sequence of binomial type $p_n^\alpha(x)$ associated with the delta operator $f(D) = \Delta(\nabla^{(-1)})$.

$$\begin{aligned}
f(D) &= \Delta(\nabla^{(-1)}) \\
&= \exp(-\log(I-D)) - I \\
&= \frac{I}{I-D} - I \\
&= \frac{D}{I-D} \\
&= -W
\end{aligned}$$

where W is the Weierstrass operator. Hence, the logarithmic sequence $p_n^\alpha(x)$ is essential the logarithmic Laguerre sequence.

$$p_n^\alpha(x) = (-1)^n L_n^\alpha(x) = \sum_{k=0}^{\infty} (-1)^k \binom{n-1}{k} \frac{\lfloor n \rfloor !}{\lfloor n-k \rfloor !} \lambda_{n-k}^\alpha(x).$$

Thus,

$$(x)_n^\alpha = \sum_{k=0}^{\infty} (-1)^k \binom{n-1}{k} \frac{\lfloor n \rfloor !}{\lfloor n-k \rfloor !} \langle x \rangle_{n-k}^\alpha.$$

In particular, for $\alpha = (0)$ and n positive,

$$x(x-1)\ldots(x-n+1) = \sum_{k=0}^{n-1} (-1)^k \binom{n-1}{k} \frac{n!}{n-k!} x(x+1)\ldots(x+n-k-1)$$

a fact which was not at all obvious *a priori*. These connection constants $L_{nk} = (-1)^k \binom{n-1}{k} \frac{n!}{n-k!}$ are referred to as L_{nk} numbers.

6.3. From the Abel Series to the Harmonic Logarithms

To compute the harmonic logarithms $\lambda_n^\alpha(x)$ in terms of the logarithmic Abel sequence $A_n^\alpha(x; b)$ it will suffice to compute the logarithmic inverse Abel sequence $\mu_n^\alpha(x; b) = \overline{A}_n^\alpha(x; b)$ associated with the composition inverse of DE^b.

Calculus of Finite Differences

Unfortunately, there is no closed form expression for the compositional inverse of $D \exp(bD)$. Nevertheless, we can compute $\mu_n^\alpha(x)$ using the fact that it is the conjugate logarithmic sequence for DE^b.

$$\mu_n^\alpha(x) = \sum_{k=-\infty}^{n} \frac{\langle D^k E^{bk} \lambda_n^\alpha(x) \rangle_\alpha}{\lfloor k \rceil !} \lambda_k^\alpha(x) .$$

Now, $\langle D^k E^{bk} \lambda_n^\alpha(x) \rangle_\alpha / \lfloor n \rceil !$ is equal to $(bk)^{n-k} \lfloor n \rceil ! / \lfloor n-k \rceil !$. Thus,

$$\mu_n^\alpha(x; b) = \sum_{k=-\infty}^{n} \frac{(bk)^{n-k}}{\lfloor k \rceil !} \lambda_k^\alpha(x) .$$

And therefore

$$\lambda_n^\alpha(x) = \sum_{k=-\infty}^{n} \frac{(bk)^{n-k}}{\lfloor k \rceil !} A_k^\alpha(x; b) ,$$

or for $\alpha = (0)$ and n positive or $\alpha = (1)$ and n negative,

$$x^n = \sum_{k=-\infty}^{n} \frac{(bk)^{n-k}}{\lfloor k \rceil !} x(x - nb)^{n-1} .$$

References

[1] R. Askey, *Orthogonal Polynomials and Special Functions*, Regional Conference Series in Applied Mathematics, SIAM (1975).

[2] M. Barnabei, A. Brini, G. Nicoletti, Polynomial Sequences of Integral Type, *Journal of Mathematical Analysis and Its Applications*, **78** (1980), 598-617.

[3] N. Bourbaki, *Fonctions d'une Variable Rèelle*, Hermann, Paris, 1961.

[$3\frac{1}{2}$] E. Damiani, O. D'Antona, G. Naldi, L. Pavarino, *On the Connection Constants*, Rapporto Interno n. 54/89, Università di Milano, Dipartimento di Scienze dell'Informazione.

[$3\frac{3}{4}$] E. Damiani, O. D'Antona, D. Loeb, *The Complimentary Symmetric Function: Connection Constants Using Negative Sets*, to appear.

[4] Hardy, *Orders of Infinity*, Cambridge, University Press, 1910.

[5] D. Loeb, Series with General Exponents, *Journal of Mathematical Analysis*.

[6] D. Loeb, *The Iterated Logarithmic Algebra*, MIT Thesis, 1989.

[7] D. Loeb, The Iterated Logarithmic Algebra, *Advances in Mathematics*.

[8] D. Loeb, The Iterated Logarithmic Algebra II: Sheffer Sequences, *Journal of Mathematical Analysis*.

[9] D. Loeb, *Sequences of Symmetric Functions of Binomial Type*, Studies in Applied Mathematics.

[10] D. Loeb, A Generalization of the Binomial Coefficients, *Advances in Mathematics*.

[11] D. Loeb, A Generalization of the Stirling Numbers, *Journal of Discrete Mathematics*.

[12] D. Loeb, G.C. Rota, Formal Power Series of Logarithmic Type, *Advances in Mathematics*, **75** (1989) 1-118.

[13] I.G. Mcdonald, *Symmetric Functions and Hall Polynomials*, Oxford Mathematical Monographs, Claredon Press, Oxford, 1979.
[14] R. Mullin, G.C. Rota, On the Foundations of Combinatorial Theory III: Theory of Binomial Enumeration, *Graph Theory and Its Application*, (1970), 168-211.
[15] S. Roman, The Algebra of Formal Series, *Advances in Mathematics*, **31**, (1979), 309-329.
[16] S. Roman, The Algebra of Formal Series II: Sheffer Sequences, *Journal of Mathematical Analysis and Applications*, **74**, (1980), 120-143.
[17] S. Roman, *A Generalization of the Binomial Coefficients*. To appear.
[18] S. Roman, G.C. Rota, The Umbral Calculus, *Advances in Mathematics*, **27**, (1978), 95-188.
[19] G.C. Rota, D.Kahana and A. Odlyzko, Finite Operator Calculus, *Journal of Mathematical Analysis and Its Applications*, **42**, (1973).
[20] J. Touchard, Nombres Exponentiels et Nombres de Bernoulli, *Canadian Journal of Mathematics*, **8**, (1956), 305-320.

Local Differential Properties of Algebraic Surfaces

LUIGI MURACCHINI Dipartimento di Matematica, Università di Bologna

By means of local differential considerations (in the fields of euclidean, projective, affine and conformal geometries) one can associate, in invariantive way, to a non singular point P of an analytic complex surface Σ, an algebraic unidimensional variety V_P (abstractly an algebraic curve C_P) in general irreducible. This has been accomplished in many occasions by B. Segre [7] and by E. Bompiani [8] and exploited by them in their investigations on algebraic surfaces.

In the present brief communication I wish to expound shortly some results (see [1], [2]) all of which fall in the following same scheme: the reducibility of the variety V_P in components of a given kind (for every $P \in \Sigma$) allows the characterization of certain algebraic surfaces amongst the analytic complex surfaces. In later time interest for this kind of results and investigations and for the related classes of surfaces has been renewed (see [3],[4], [5]). Many results of the kind I shall expound suggest extensions and raise problems which, perhaps, may deserve attention but I shall not dwell on them. I shall confine myself to some typical results, leaving more detailed discussion to the papers quoted at the end.

A) in the field of euclidean geometry.

One has: the planes of the circles having in the point P of the complex analytic surface Σ 4–point contact (i.e. contact of order 3) envelop an algebraic cone of class 5 (having the tangent plane in P as plane of multiplicity 4). By duality one defines an algebraic curve C_P of order 5 with a point O of multiplicity 4.

It has been shown [1] that:

(a) C_P breaks down in a cubic curve C_P^* (with double point O) and two straight lines (different from the tangents to C_P^* in O) by O if, and only if, Σ is a *Dupin cyclide*. As it is known [6], the Dupin cyclide is enveloped in two different ways by a system of ∞^1 spheres (every sphere of one system touches all the spheres of the other system). The spheres touch the Dupin cyclide along circles which are curvature lines, and many other properties.

(b) if C_P breaks in a cubic curve C_P^* with double point O and the two straight lines tangent in O to the cubic (and only if), Σ is a *quadric* surface (not of revolution).

B) in the field of projective geometry.

One has: the planes of the conics having in P with Σ 6–point contact (contact of order 5) envelop an algebraic cone of class 9 (having the tangent plane of multiplicity 7). This gives an algebraic curve C_P of order 9 with a point O of multiplicity 7.

It has been shown [2] that:

(a) if, and only if, C_P breaks down in three straight lines by O and two cubic curves having double point in O (with the same tangents, different from the straight lines) Σ is a *cubic surface with three biplanar double points*. This surface is enveloped by three different systems of quadrics; every quadric of one of the three ∞^1 systems osculates the quadrics of the other two systems (two quadrics of the same system touch each other in two points). Every quadric touches Σ along a conic which is a line of Darboux of Σ (so the three systems of lines of Darboux are systems of conics).

(b) if, and only if, C_P breaks down in two irreducible components of orders 3 and 6 with, respectively, O of order 2 and 5 of multiplicity, Σ is a *surface of Steiner* (the so called *roman* surface, of order 4 with three double straight lines meeting in a triple point). Two of the three double lines can be coincident (*tacnodal type* of the Steiner surface); otherwise Σ is the *dual* surface of the Steiner surface (a cubic surface with four double conic points, known as the C. *Segre surface*).

(c) if, and oly if, C_P breaks in three irreducible components of orders 3,5 and 1 (a straight line) with O of multiplicity 2, 4 and the straight line by O, Σ is a Steiner's surface with the three double lines coincident (*oscnodal* type) or the dual surface.

C) in the field of conformal plane geometry (of Möbius).

The result concerns, in this case, analytic correspondences between the points of two euclidean planes and it is known that these correspondences can be considered abstractly as analytic surfaces.

One has: let T be an analytic *non conformal* correspondence between two euclidean planes π_1, π_2; consider a regular couple of correspondent points O_1, O_2. Consider the circles C_1 by O_1 in the plane π_1 and the circles C_2 osculating in O_2 the curves $T C_1$ in π_2, correspondent by T to the circles C_1 of π_1. Those circles C_2, which have in O_2 contact of order 3 with the relative curve $T C_1$, constitute an algebraic system ∞^2 having an image $C_{(O_1,O_2)}$: an algebraic curve of order 6 with a point of multiplicity 4.

It can be shown [9]: if T transforms the circles of a system of dimension 2 analytic in the plane π_2 in circles of the plane π_1, then the system is necessarily linear and the curve $C_{(O_1,O_2)}$ breaks down in a straight line and a curve of order 5 with a point of multiplicity [4]. Moreover the correspondence T is algebraic (2,2) (or (1,2) in particular cases).

I conjecture that the latter conclusion can be reached under the hypothesis of the breaking down of $C_{(O_1,O_2)}$ (in the way indicated) only.

D) in the field of affine geometry.

In this field I have only the preliminary result on which some final results can be conjectured as I shall say.

One has: the planes of the parables having in the point P of the analytic surface Σ a 5–point contact (contact of order 4) with the surface envelop an algebraic cone of class 6 (with the tangent plane in P of multiplicity 4). An algebraic curve C_P is so defined of order 6, with a quadruple point O, in general irreducible. It happens that C_P breaks down:

(a) if Σ is a central quadric, in two double straight linea by O and a conic (not containing O);

(b) if Σ is a paraboloid, in three double straight linea (two of them by O);

(c) if Σ is a Steiner surface of tacnodal or of oscnodal type, tangent to the plane at infinity along the double coincident straight lines, then C_P breaks down in two cubic curves with O double point.

I conjecture that the "only if" statements concerning (a), (b), (c) also hold; the results of section B) previously quoted support this conjecture.

REFERENCES

[1] Semin, F., *Sur les sections d'une surface surosculées par leurs cercles de courbure*, Revue de la Fac.des Sci.Université d'Istambul **19** (1954), 34–44.

[2] Muracchini L., *Sul contatto fra superficie e coniche. Una caratterizzazione differenziale della superficie romana di Steiner*, Annali di Matematica pura e appl.,(4) **44** (1957), 331–356.

[3] Griffiths, P. - Harris, J., *Algebraic geometry and local differential geometry*, Ann. Scient. Ec. Norm. Sup.,(4) **12** (1979), 355–432.

[4] Pinkall, U., *Dupin hypersurfaces*, Math. Ann **270** (1985), 427–440.

[5] Miyaoka, R., *Compact Dupin hypersurfaces with three principal curvatures*, Math. Z. **187** (1984), 433–452.

[6] Darboux, G., *Sur le contact des courbes et des surfaces*, Ann.Bull. des Sci. Math. **4** (1980), 348–384.

[7] Segre, B., *Opere scelte*, (cfr.the Introduction by E.Vesentini) **1** (1987).

[8] Bompiani, E., *Opere scelte vol.III Unione Matematica Italiana (cfr. specially: Nuovi enti geometrici: pseudoelementi differenziali*, Rend.Sem.Mat.Fis. Milano **33** (1963), 236–255).

[9] Muracchini, L., *Sulle trasformazioni puntuali analitiche che mutano circonferenze in circonferenze*, Atti Acc. Ist. Bologna, (9). **8** (1961), 1–19. Cfr. also.

  Segre, B., *Generalizzazione di un teorema di Beltrami*, Boll.U.M.I.,(3) **4** (1949), 16–22.

  Muracchini, L., *Sulla geometria differenziale conforme delle trasformazioni puntuali fra due piani*, Boll.U.M.I.,(3) **8** (1953), 252–258.

The Method of Holomorphic Waves and Its Applications to Partial Differential Equations

V. P. PALAMODOV University of Moscow, MGU, Leninskije gori, 119899 Moscow, USSR

1. The development of analysis of the past two centuries was closely related with problems of various wave equations: for example equations of acustic, electro-magnetic, elastic waves and so on. These problems were generalized in the theory of hyperbolic differential equations. Many prominent mathematicians brought contributions to this theory: Riemann, Poisson, Kirchhof, Volterra, Hadamard, Herglotz and others. At the beginning of thirties Luigi Fantappiè a disciple of Vito Volterra found a new approach to the Cauchy problem for partial differential equations based on some ideas of functional and complex analysis. To solve a Cauchy problem means in terms of functional analysis to construct a linear operator which maps a data to the solution. Supposing that all functions which occur are holomorphic this operator turns to be what L. Fantappiè called a «mixed analytic functional». In a serie of papers [1] Fantappiè proposed a lot of analytic methods to represent such a mixed functional in quadratures. His methods start from the notion of indicatrice of an analytic functional.

2. Let F be an analytic functional defined on C^n (which means in modern terms a linear continuous functional on the space of holomorphic functions). Its projective indicatrice is the holomorphic function p defined on an open set Ω of the dual space C^{*n} as

$$p(\xi) = F((1 - \xi z)^{-1})$$

The set Ω consists of covectors ξ such that the hyperplane $\xi \cdot z = 1$ does not meet a support of F. Fantappiè established an important representation of arbitrary holomorphic function f defined in the open polydisc $\Pi = \Pi_1 \times \ldots \times \Pi_n$ as a superposition of functions $(1 - \xi z)^{-n}$ [2].

This representation (as I understand it) can be written as follows:

$$f(z) = (2\pi i)^{-n} \int_{\partial^n \Pi} \frac{dw_1 \ldots dw_n}{w_1 \ldots w_n} f(w) \int_{\eta_i \geq 0, \sum \eta_i = 1} \frac{d\eta_1 \ldots d\eta_{n-1}}{(1 - \eta \cdot \frac{z}{w})^n} =$$

$$= (2\pi i)^{-n} \int_{\partial^n \Pi} f(w) \, dw \int_{\xi_i w_i \geq 0, \xi \cdot w = 1} \frac{\sigma_w}{(1 - \xi \cdot z)^n} \tag{1}$$

where $\partial^n \Pi$ is the skeleton of Π and $\sigma_w = d\xi/d(\xi \cdot w)$. The third term is simply another form of the second one with $\xi_i = \eta_i/w_i$. It ought to be mentioned some corollaries and developments of this formula:

i. The representation uses functions $h_\xi(z) = (1 - \xi \cdot z)^{-n}$ with the singular set equal to the complex hyperplane $\xi \cdot z = 1$ which is supporting to the compact $\overline{\Pi}$. In other terms f is represented as an integral over the «beam» of «plane waves» whose singular sets cover $C^n \backslash \Pi$. A «plane wave» means s function which depends on n variables $z_1, \ldots, z_n$ only through a linear combination $\xi \cdot z$ with some covector ξ. This representation can be generalized to the case of any bounded convex polycilindre Π.

In the frame of real analysis an idea of plane waves decomposition appeared later in F. John papers (see [3]) who interpreted the Radon's inversion formula [4] as a plane waves decomposition of the Dirac's delta-function.

ii. Any analytic functional F «supported on Π» can be recovered from its projective indicatrice with the formula

$$F(f) = (2\pi i)^{-n} \int_{\partial^n \Pi} f(w) \, dw \int_{\xi_i w_i \geq 0, \xi \cdot w = 1} q(\xi) \sigma_w, \tag{2}$$

where

$$q(\xi) = (-1)^{n-1}/(n-1)! \, D_t^{n-1}(t^{-1} p(\xi/t))|_{t=1} = F((1 - \xi z)^{-n})$$

iii. This formula can be also written in a cohomological form:

$$F(f) = \int_{\partial^n \Pi} f(w) \varphi(w) \, dw$$

where $\varphi(w)$ is the $(n-1)$-cocycle in $C^n \backslash \Pi$ which is given by the holomorphic function

$$\varphi(w) = \int_{\xi_i w_i \geq 0, \xi \cdot w = 1} q(\xi) \sigma_w$$

defined on a neighborhood of $C \backslash \Pi_1 \times \ldots \times C \backslash \Pi_n$ which closely related to the indicatrice. This approach to (1) which is implicit in Fantappiè's theory was developed later by A. Martineau [6] who identified analytic functionals on R^n with M. Sato's hyperfunctions with compact support.

iv. The transformation $F \mapsto p$ depends on the affine structure of C^n and therefore relates to harmonic analysis on this space, particularly to the theory of differential equations with constant coefficients. This relation was resourcefully exploited by Fantappiè to get some formulas of fundamental solutions of Cauchy problem for homogeneous operators [1]. In fact, for equations of hyperbolic type his methods give formulas using only quadratures in real domain. This formula is brought to an explicit expression for an operator of third order in three variables by Casulleras [5] a disciple of Fantappiè.

Later John [3] developing some ideas of Herglotz calculated in quadratures the fundamental solution for arbitrary equation of hyperbolic type with constant coefficients. In fact the method of plane waves decomposition is an alternative to the Fourier's methods and formulas (1), (2) can be considered as a parallel to the Fourier transformation. Fantappiè probably intended to compare both methods in his paper [7] devoted to a construction of a solution of parabolic equation with constant coefficients.

In the middle of thirthies S. Sobolev invented the method of generalized functions and applied it to a solution of the wave equation with variable coefficients. Fifteen year later L. Schwartz coupled Sobolev's ideas with the Fourier transformation in the theory of distributions. These powerful tools were used to get a big progress in the general theory of linear partial equations.

3. Now I expose some construction of [11] inspired by ideas Fantappiè and Martineau aimed to modify the method of plane waves and apply this modification to some problems for equations with variable coefficients.

DEFINITION 1. Let Γ be a compact set in R^n, V an open neighborhood of zero in C^n and $\xi \in C^n \backslash 0$.

I call a holomorphic wave in (Γ, V) with the direction ξ any holomorphic function $f = f(z)$ defined in $\Gamma + V_\xi$ where $V_\xi = V \cap C^n_\xi$ and

$$C^n_\xi = \{z \in C^n, \mathrm{Im}\,\xi z > 0\}$$

which satisfies for any compact set $K \subset C^n$ the inequality

$$|g(z)| \leq C \mathrm{dist}^{-q}(z, \partial(\Gamma + V_\xi)), \qquad z \in K \qquad (3)$$

with some constants q and C.

Because of (3) any holomorphic wave has its boundary value $b(g) \in D'(\Gamma)$. For any real ξ the distribution $b(g)$ is analytic along any hyperplane $\xi x = \mathrm{const}$.

PROPOSITON 1. If $\mathrm{Im}\,\xi \neq 0$ any holomorphic wave in (Γ, V) with the direction ξ is real analytic in $\mathrm{Int}\,\Gamma$.

PROOF. For any $x \in \mathrm{Int}\,\Gamma$ we can choose sufficiently small $z \in V_\xi$ such that $x' = x - z \in \Gamma$. We have $x = x' + z \in \Gamma + V_\xi$, hence g is analytic in a neighborhood of x.

EXAMPLES. 1. For any function or distribution f defined on an interval I of real line and a covector $\xi \in R^{*n} \backslash 0$ the function $g(x) = f(\xi x)$ of n variables is a plane wave with the direction ξ. If f is the boundary value of a holomorphic function f^+ defined in $I + V^+$ where V^+ is the intersection of upper halfplane with a neighborhood of zero then g is a holomorphic wave with the direction ξ. Particularly the function $(1 - \xi z)^{-n}$ which appears in Fantappiè's formula (1) is a holomorphic wave with the direction ξ and with the direction $-\xi$ (but for another Γ and V).

2. Let ρ be a complex-valued mesure on C^n such that

$$\int \exp(\gamma_\Gamma(\mathrm{Im}\,\xi))(|\xi| + 1)^{-q}|\rho| < \infty$$

for a compact set $\Gamma \subset R^n$ (γ_Γ is the supporting function) and for a constant q. Suppose that ρ vanishes outside of d-neighborhood of the ray $\zeta = t\xi$, $0 \le t < \infty$ for a constant d and a vector $\xi \ne 0$. Then the inverse Fourier Transform of the measure

$$u(x) = \int \exp(ix\xi)\rho$$

is a holomorphic wave in (Γ, C^n) with the direction ξ.

DEFINITION 2. We call a beam of holomorphic waves on (Γ, V) any mesurable family $g = \{g_\xi, \xi \in C^n, |\xi| = 1\}$ of holomorphic waves on (Γ, V) where ξ is the direction of g_ξ which satisfy (3) with some constant q and C which do not depend on ξ. For any beam g we define its integral boundary value as follows

$$\beta(g) := \int_{S^{2n-1}} b(g_\xi)\omega(\xi) \qquad (4)$$

where ω is the standart mesure on the sphere S^{2n-1}. This is a distribution on Γ. Let $K^0_{\Gamma,V}$ be the vector space of all beams of holomorphic waves on (Γ, V). Thus (4) defines a linear operator

$$\beta : K^0_{\Gamma,V} \to D'(\Gamma)$$

We shall say that a distribution $u \in \operatorname{Im}\beta$ has a representation by a beam of holomorphic waves. This is an analog of the representation (1). In fact, if the compact Π_i, $i = 1,\ldots,n$ in (1) tends to an interval $I_i \subset R$ then «passing to limit» in (1) we get a representation for distributions defined on $\Gamma = I_1 \times \ldots \times I_n$ by a beam of waves of the form $(1 - \xi z)^{-n}$.

Now a natural question arises: in what case the operator is surjective? The answer is positive at least if Γ is convex (see below theorem 1). It is easy to prove using the following sequences of arguments: to apply the Fourier transform to a distribution on Γ, to represent the result with a smooth complex-valued mesure on C^n, then to pass to polar coordinates on C^n in the inverse Fourier formula and to identify the ray integral with the boundary value of a holomorphic wave according to the example 2.

REMARK. For any slow increasing distribution u in R^n (i.e. $u \in S'(R^n)$) these arguments give a representation of u in the form (4) but with an integration over $S^{n-1} = S^{2n-1} \cap R^{*n}$. Particularly for $n = 1$ this representation goes back to the Carleman's idea of representation of a function as a sum of boundary values of analytic functions defined in upper and down halfplanes.

EXAMPLE 3. The following well known representation of the delta-function as a superposition of plane waves

$$\delta(x) = \frac{(n-1)!}{(-2\pi i)^n} \int_{S^{n-1}} \frac{\omega(\xi)}{(\xi x + i0)^n}$$

can be written in the form $\delta = \beta(w)$ where

$$w = \left\{ w_\xi(z) := \frac{(n-1)!}{(-2\pi i)^n}(\xi z + i0)^{-n} \right\}$$

is a beam of holomorphic waves in $(0, C^n)$.

Of course the map β is far from to be injective, therefore we need some information on Ker β. For this we construct a surjection in this space using the notion of a beam of holomorphic 2-waves on (Γ, V). This is a measurable family $g = \{g_{\xi,\eta}\}$ with two spherical parameters ξ, η where $g_{\xi,\eta}$ is a holomorphic function in $\Gamma + V_\xi + V_\eta$ which satisfies an inequality similar to (3) with some constants q and C which do not depend on ξ, η. The linear space of all beams of 2-waves on (Γ, V) I denote $K^1_{\Gamma,V}$. The formula

$$\mu(g)_\xi = \int g_{\xi,\eta}\omega(\eta) - \int g_{\eta,\xi}\omega(\eta)$$

defines a linear map

$$\mu : K^1_{\Gamma,V} \to K^0_{\Gamma,V}$$

which satisfies the relation $\beta\mu = 0$.

Because of Ker $\mu \neq 0$ we should continue this procedure. In this way we define for any integer $p \geq 0$ the space $K^p_{\Gamma,V}$ whose elements are called beams of holomorphic p-waves on (Γ, V). A beam of holomorphic p-waves is by a definition a measurable family $g = \{g_{\xi_0,\ldots,\xi_p}\}$ defined on $(S^{2n-1})^{p+1}$ where $g_{\xi_0,\ldots,\xi_p}$ is a holomorphic function in $\Gamma + V_{\xi_0} + \ldots + V_{\xi_p}$ sastisfying an inequality similar to (3). The map β is inserted in the sequence

$$\ldots \to K^{p+1}_{\Gamma,V} \xrightarrow{\mu} K^p_{\Gamma,V} \to \ldots \to K^1_{\Gamma,V} \xrightarrow{\mu} K^0_{\Gamma,V} \xrightarrow{\beta} D'(\Gamma) \to 0 \tag{5}$$

where μ is given by the following general formula

$$\mu(g)_{\xi_0,\ldots,\xi_p} := \sum_j (-1)^j \int_{S^{2n-1}} g_{\xi_0,\ldots,\xi_{j-1}\eta\xi_j,\ldots,\xi_p}\omega(\eta)$$

This is a complex of vector spaces since $\mu \cdot \mu = 0$.

PROPOSITION 2. For any $\varepsilon > 0$ there exists a neighborhood $V(\varepsilon)$ of zero in C^n such that any holomorphic p-wave $g_{\xi_0,\ldots,\xi_p}$ in (Γ, V) is holomorphic in $\Gamma + V(\varepsilon)$ provided there is a pair of indices $0 \leq i, j \leq p$ such that dist $(\xi_i, \xi_j) \geq \varepsilon$ where dist is the standart metric on the sphere.

For a proof we choose $V(\varepsilon)$ in such a way that $V_\xi + V_\eta \supset V(\varepsilon)$ for any ξ, η provided dist $(\xi, \eta) \geq \varepsilon$.

Now we formulate the main statement:

THEOREM 1 ([11]). For any convex compact set $\Gamma \subset R^n$ and any convex neighborhood V of zero in C^n the complex (5) is exact.

REMARK. Euristically speaking, (5) is a kind of resolvent of the space $D'(\Gamma)$ in a category of spaces of holomorphic functions which is in fact is not yet well-defined. However this exact sequence will play a part of resolvent in some applications to the theory of systems of differential equations.

4. We consider arbitrary system of linear differential equations

$$a(x, D)u(x) = 0 \tag{6}$$

with analytic coefficients on an real-analytic manifold X. First we remind some definitions. Let $\mathcal{D}$ be the sheaf on X of algebras of germs of linear differential operators on X with analytic coefficients. If k is an integer $\mathcal{D}^k$ means the free left $\mathcal{D}$-module of range k. Let $a(x, D)$ be a $t \times s$ matrix whose entries are sections of $\mathcal{D}$ over X. Then there is defined a morphism $a' : \mathcal{D}^t \to \mathcal{D}^s$ of left $\mathcal{D}$-modules: $b \mapsto b \cdot a$. The left $\mathcal{D}$-module $\mathcal{U} := \mathrm{Cok}\, a'$ is an invariant of (6). The sheaf $\mathcal{D}$ as a $\mathcal{O}_X$-module has a filtration $\{\mathcal{D}_m\}$ where $\mathcal{D}_m$ is the submodule of germs of operators of degree $\leq m$. We endow $\mathcal{U}$ with the filtration which is the image under the canonical surjection $\mathcal{D}^s \to \mathcal{U}$ of the filtration $\{\mathcal{D}_m^s\}$ and consider the graded module

$$\mathrm{gr}\mathcal{U} := \bigoplus_m \mathcal{U}_m/\mathcal{U}_{m-1}$$

of a finite type over the sheaf $\mathrm{gr}\mathcal{D}$ of graded commutative algebras.

DEFINITION 3. $\mathcal{U}$ satisfies the regularity condition of Spencer-Quillen if $\mathrm{gr}\mathcal{U}$ is locally free $\mathcal{O}_X$-module.

From now on we suppose that $\mathcal{U}$ is regular and put for any $x \in X$

$$\mathrm{gr}\mathcal{U}_x := \mathrm{gr}\mathcal{U} \bigotimes_{\mathcal{O}_{X,x}} C$$

This is a module of a finite type over the graded commutative algebra $\mathrm{gr}\, C[D] = \mathrm{gr}\mathcal{D}_x \otimes C$. Therefore $\mathrm{gr}\mathcal{U}_x$ defines a coherent algebraic sheaf over CP_{n-1} which we denote $\mathrm{Pr}\,\mathrm{gr}\mathcal{U}_x$.

DEFINITION 4. We call the characteristic set of (6) at x the algebraic set $\mathrm{Ch}\,(\mathcal{U}_x) := \mathrm{supp}(\mathrm{Pr}\,\mathrm{gr}\mathcal{U}_x)$. The union $\mathrm{Ch}\,(\mathcal{U}) := \bigcup_x \mathrm{Ch}\,(\mathcal{U}_x)$ (which is an analytic set in $CP(T^*(R^n))$) is called the characteristic variety of $\mathcal{U}$.

A system (6) is elliptic if $\mathrm{Ch}\,(\mathcal{U})$ has no real points.

THEOREM 2. Let a be a regular operator in a neighborhood of a point $y \in X$. For any neighborhood P of $\mathrm{Ch}\,(\mathcal{U}_y)$ in CP_{n-1} and any open set $Y \ni y$ there exists a compact set $\Gamma \subset Y$ such that $y \in \mathrm{Int}\,\Gamma$ and a neighborhood V of zero in C^n such that any distribution u which satisfies (6) in Y can be represented in Γ as a sum

$$u = \int_{p(\xi) \in P} u_\xi \omega(\xi) + u_0 \tag{7}$$

where $p : S^{2n-1} \to CP_{n-1}$ is the canonical projection, $\{u_\xi\}$ is a beam of holomorphic waves in (Γ, V) and u_0 is a holomorphic solution of (6) in a neighborhood of Γ.

Later we give a sketch of a proof and now we mention some corollaries.

5. COROLLARY 1. Let

$$a_1(x, D) w(x) = 0 \tag{8}$$

be the compatibility equation for (6) in a neighborhood Y of y. For any open set $P' \supset \mathrm{Ch}\,(\mathcal{U}_y) \cap RP_{n-1}$ there exist a compact neighboroood Γ of y and a neighborhood V of zero in C^n such that for any distribution w in Y satisfying (8) there exists a distribution u in Γ such that the equation

$$a(x, D) u(x) = w(x) + \int_{p(\xi) \in P'} w_\xi(x) \omega(\xi)$$

holds with some beam $\{w_\xi\}$ of holomorphic waves in (Γ, V).

To prove this corollary we apply theorem 2 to (8) and obtain an analytic solution w_0 of this system. Then we solve the equation $a(x, D) u(x) = w_0(x)$ using Cauchy-Kowalewskaja theorem (see for details for example [10]).

Method of Holomorphic Waves

REMARK. For any elliptic system (6) these arguments give a proof of local solvability without a reducing to a scalar elliptic equation. Theorem 2 with proposition 2 get also a proof of Petrowski's theorem on analiticity of solutions without any successive estimates of derivatives.

Now we describe an unicity property of solutions of arbitrary regular system.

DEFINITION 5. A smooth curve $C \subset X$ with ends y and z is called free relatively to $\mathcal{U}$ if there is no point $(x, \xi) \in \text{Ch}(\mathcal{U})$ such that $x \in C$ and ξ is orthogonal to the tangent vector t to C at x.

We call the $\mathcal{U}$-envelope of a free curve C the set of all points $x \in X$ such that there exists a homotopy C_τ, $0 \leq \tau \leq 1$ of free curves C_τ with ends y and z such that $C_0 = C$, $x \in C_1$.

COROLLARY 2. Any distribution-solution of (6) on X which vanishes in a neighborhood of a free curve C is equal to zero in $\mathcal{U}$-envelope of C.

When applied to a wave equation this corollary gives a well known unicity theorem for traces of solution on a time-like manifold (see for example [8]). For scalar equations this property was proved by John [9] in the frame of classical analysis. His method is to check that integrals of a solution with polynomial weights on C_τ are analytic functions of τ. In our case the analiticity of these integrals follows immediately from theorem 2. Our statement is also close to some results of Hörmander [12] and Kawai [13] where some relations between the boundary of support of a solution and the symbol of a scalar eqaution are established (see also [14] for other results of this kind).

6. Now we follow a sketch of a proof of theorem 2. We take a sequence of differential operators with analytic coefficients

$$a_0 = a, a_1, \ldots, a_n = 0$$

which give a local resolvent of left $\mathcal{D}$-module $\mathcal{U}$ on a neighborhood of y. This means that a_{k+1} gives a compatibility condition for a_k, $k = 0, 1, \ldots, n-1$. A coupling of this sequence with (5) gives the commutative diagramm

$$
\begin{array}{ccccccccc}
0 & & 0 & & & & & & \\
\uparrow & & \uparrow & & & & & & \\
K_{\Gamma,V}^n & \xrightarrow{\mu} & K_{\Gamma,V}^{n-1} & & & & & & \\
a_{n-1} \uparrow & & a_{n-1} \uparrow & & & & & & \\
K_{\Gamma,V}^n & \xrightarrow{\mu} & K_{\Gamma,V}^{n-1} & \to & \cdots & & & & \\
\uparrow & & \uparrow & & & & & & \\
\vdots & & \vdots & & & & & & \\
 & \cdots & \to & K_{\Gamma,V}^1 & \xrightarrow{\mu} & K_{\Gamma,V}^0 & \xrightarrow{\beta} & D'(\Gamma) & \to 0 \\
& & & a_0 \uparrow & & a_0 \uparrow & & a_0 \uparrow & \\
& & & K_{\Gamma,V}^1 & \xrightarrow{\mu} & K_{\Gamma,V}^0 & \xrightarrow{\beta} & D'(\Gamma) & \to 0 \\
\end{array}
\qquad (9)
$$

where we write simply $K_{\Gamma,V}^p$ instead of $[K_{\Gamma,V}^p]^s$ with some s and fix some small convex Γ and V. Taking a solution u of (6) we put it in the bottom right corner of (9). Then we start a «diagram chasing» with small modifications. For the first step we write $u = \beta(v^0)$ with some beam $v^0 \in K_{\Gamma,V}^0$ and consider the beam $a_0(v_0)$. Taking in account that $\beta a_0(v^0) = a_0 \beta(v^0) = 0$ we apply theorem 1 and find a beam $v^1 \in K_{\Gamma,V}^1$ such that $\mu(v^1) = a_0(v^0)$. Then we choose v^2 such that $\mu(v^2) = a_1(v^1)$ and so on. In this way we move to the left and up through the diagram. When we reach the top line we get a beam $v^{n-1} \in K^{n-1}_{\Gamma,V}$. Then to return to the start poin we should solve the equation

$$a_{n-1} w = v^{n-1} \tag{10}$$

This equation is formally solvable because of the compatibility equation $a_n v^{n-1} = 0$ is trivially satisfied ($a_n = 0$). The second term of (10) is a beam of holomorphic n-waves $v_{\xi_1,\ldots,\xi_n}$ in (Γ, V) and (10) means the family of equations

$$a_{n-1} w_{\xi_1,\ldots,\xi_n} = v_{\xi_1,\ldots,\xi_n} \tag{11}$$

LEMMA 1. For any (Γ, V) and $\varepsilon > 0$ there exists a convex compact neighborhood Γ_ε of y and a convex neighborhood of zero V_ε in C^n such that (11) has a holomorphic n-waves solution in $(\Gamma_\varepsilon, V_\varepsilon)$ provided either
    i) there exists i, $1 \leq i \leq n$ such that dist$(\xi_i, \mathrm{Ch}(\mathcal{U}_y)) \geq \varepsilon$
or
    ii) there exist i, j such that dist$(\xi_i, \xi_j) \geq \varepsilon$.

PROOF OF THE LEMMA. In the case i) we use proposition 1 and a combination of Cauchy-Kowalewskaya's theorem and the Duhamel's principle taking $\xi_i x$ as the time variable. In the case ii) we apply once more Cauchy-Kowalewskaya's theorem taking in account proposition 2.

Then we consider the beam w of holomorphic n-waves whose values $w_{\xi_1,\ldots,\xi_n}$ are defined by the lemma and equal to zero if neither i) nor ii) are satisfied. We put $u^{n-2} := v^{n-2} - \mu(w) \in K^{n-2}_{\Gamma_\varepsilon, V_\varepsilon}$. Because of (11) we have

$$a_{n-2} u_{\xi_1,\ldots,\xi_{n-1}} = 0$$

for any $\xi_1,\ldots,\xi_{n-1}$ satisfiing either i) or ii). This is the compatibility condition for the equation

$$a_{n-2} w_{\xi_1,\ldots,\xi_{n-1}} = u_{\xi_1,\ldots,\xi_{n-1}}$$

Using the same arguments as in lemma 1 we solve this equation for any $\xi_1,\ldots,\xi_{n-1}$ satisfying either i) or ii) and put $w_{\xi_1,\ldots,\xi_{n-1}} = 0$ in other cases. So we get a beam w' in $(\Gamma'_\varepsilon, V'_\varepsilon)$ for some small Γ'_ε and V'_ε. Then we define $u^{n-3} := v^{n-3} - \mu(w')$ and so on.

Finally we get a beam $u^0 \in K^0_{\Gamma_0, V_0}$ with some neighborhoods Γ_0 of y and V_0 of zero such that
    iii) $\beta(u^0) = u$,
    iv) $a_0 u^0_\xi = 0$ for any ξ satisfiing i).

LEMMA 2. There is a neighborhood U_0 of y such that u_ξ is analytic in U_0 provided i) holds for ξ.

This is an easy corollary of Cauchy-Kowalewskaja's and Holmgren's theorems. Now iii) and iv) imply theorem 2.

REFERENCES

[1] L. Fantappiè, La giustificazione del calcolo simbolico e le sue applicazioni all'integrazione delle equazioni a derivate parziali, *Mem. Reale Accademia d'Italia 1*, 1930.
Integrazione con quadrature dei sistemi a derivate parziali lineari e a coefficienti costanti in due variabili indipendenti, mediante il calcolo degli operatori lineari, *Rend. Circolo Matem. di Palermo* 57, 1933.
Integrazione in termini finiti di ogni sistema od equazione a derivate parziali, lineare e a coefficienti costanti, d'ordine qualunque. *Mem. Reale Accad. d'Ital. 8*, N13, 1937.

[2] L. Fantappiè, L'indicatrice proiettiva dei funzionali lineari e i prodotti funzionali proiettivi, *Annali di Mat.* 20, serie 4, 1943.

[3] F. John, *Plane waves and spherical means applied to partial differential equations*, New York 1955.

[4] J. Radon, Über die Bestimmung von Funktionen durch ihre Integralwerte länge gewisser Mannigfaltigkeiten, *Ber. Verh. Sächs Acad. Wiss. Leipzig Math.-Mat. kl.* 69 (1917), 262-277.

[5] J. Casulleras, *Aplication de la teoria de los funcionales analiticos a la resolution de un tipo de ecuaciones en derivadas parciales de 3 orden*, Collectanea Mathematica 1, fasc. 2, Barcelona, 1948.

[6] A. Martineau, Les hyperfonctions de M. Sato, *Sém. Bourbaki*, 1960-61, Exposé N 214.

[7] L. Fantappiè, Integrazione per quadrature dell'equazione parabolica generale a coefficienti costanti, *Rend. Accad. dei Lincei* 18, serie 6, sem. 2, 1933.

[8] R. Courant and D. Hilbert, *Methoden der Mathematischen Physic II*, Berlin, 1937.

[9] F. John, On linear differential equations with analytic coefficients, *Comm. Pure Appl. Math.* 2 (1949), 209-253.

[10] V.P. Palamodov, Differential operators in the class of convergent power series and the Weierstrass's preparation lemma, *Functional. analysis and its appl.* 2 (1968), N3, 58-69.

[11] V.P. Palamodov, The complex of holomorphic waves, *Trudy Sem. Petrovsk.* N1 (1975), 175-210; English translation: *Amer. Math. Soc. Transl.* (2), Vol. 122, 1984, 187-222.

[12] L. Hörmander, Uniqueness theorems and wave front sets for solutions of linear differential equations with analytic coefficients, *Comm. Pure Appl. Math.* 24 (1971), 671-704.

[13] M. Sato, T. Kawai and M. Kashiwara, *Hyperfunctions and pseudo-differential equations*, Springer Lecture Notes in Math. 287, (1973) 470-473.

[14] L. Hörmander, *The analysis of linear partial differential operators I*, Springer Verlag, 1983.

Ricerca Matematica e Vita Accademica a Bologna nel Secolo XVIII

LUIGI PEPE Dipartimento di Matematica, Università di Ferrara, via Machiavelli, 35 44100 - Ferrara, Italia.

Abstract. In this lecture the main figures of mathematicians in Bologna in the XVIIIth century are presented in their social and cultural context. Gabriele Manfredi (1681-1761) and Vincenzo Riccati (1707-1775) were not only the most important mathematicians, but also the teachers of other remarkable mathematicians as Gianfrancesco Malfatti, Gregorio Casali, Sebastiano Canterzani, Alfonso Bonfioli Malvezzi. Several Mathematicians were protagonists of the scientific life in this city from the foundation of the Istituto (Francesco Maria Zanotti (1692-1777) was secretary and president of the Accademia delle Scienze) to the republican triennium (1796-1799).

In questa conferenza si tesserà l'elogio di un secolo, il XVIII, spesso trascurato dalla storia delle matematiche, posto in ombra dai grandi protagonisti del secolo XVII (Galileo, Descartes, Leibniz, Newton, ecc.) e del secolo XIX (Cauchy, Galois, Riemann, Weierstrass, Cantor ecc.).

Conferenza tenuta il 22 marzo 1989 all'Università degli Studi di Bologna, in occasione del IX Centenario della fondazione.

Non sono mancati certo nel secolo XVIII eminenti matematici come Eulero e Lagrange, ma il loro nome non è legato alle scoperte di nuove teorie generali come il calcolo differenziale, la geometria analitica o la teoria dei gruppi. In altri casi è stata l'ampiezza degli orizzonti culturali a determinare una svalutazione dell'opera specialistica degli scienziati di questo secolo: probabilmente se d'Alembert non avesse scritto il *Discours preliminaire* dell'Encyclopédie sarebbe più noto per i suoi fondamentali contributi alla meccanica e alla fondazione della teoria delle equazioni differenziali alle derivate parziali.

Eppure il Settecento rappresenta uno dei periodi più luminosi per la cultura europea. Esso comincia con una grande rivoluzione scientifica, determinata dalla diffusione in Europa del calcolo differenziale ed integrale, della meccanica e dell'ottica di Newton ([6]) e termina con una grande rivoluzione politica e sociale che, grazie ai successi militari di Napoleone pone l'Europa di fronte a radicali riforme, sulle quali tanti intellettuali avevano scritto e alle quali la rivoluzione francese aveva dato concretezza.

L'interesse per la storia delle matematiche nel Settecento va ben al di là dei campi che oggi si fanno rientrare tra queste scienze e riguarda buona parte della storia della scienza e della cultura. Infatti l'assetto teorico, la pratica e l'organizzazione scientifica delle scienze matematizzate sono il vero modello epistemologico, se proprio si vuole usare questo termine, di quasi tutta l'attività scientifica nel Settecento, per il loro maggiore sviluppo teorico e per il successo del modello newtoniano dei principi matematici della filosofia naturale. E questo non può dirsi né per il secolo XVII, quando è ancora persistente l'impronta aristotelica, né per il secolo XIX, quando molte altre discipline come la chimica, la mineralogia hanno uno statuto epistemologico autonomo.

Nel Settecento Bologna è stata uno dei centri più importanti di cultura in Europa, molti scienziati bolognesi, nonostante i vincoli esercitati da una persistente censura ecclesiastica e da strutture universitarie invecchiate e di fatto irriformabili, fanno parte del grande circuito europeo senza i ritardi che si erano determinati nella seconda metà del Seicento e che si ripresenteranno in modo più grave per una buona metà del secolo XIX.

Il Settecento si caratterizza anche a Bologna con il fatto che sono intellettuali più attivi scientificamente che prendono anche le più importanti iniziative sul piano istituzionale e spesso emergono in funzioni di grande responsabilità: L.F. Marsigli, E. Manfredi, F.M. Zanotti, S. Canterzani, P. Lambertini.

E' interessante sottolineare anche come i maggiori scienziati nel Settecento a Bologna furono quasi tutti di origine bolognese, mentre molti scienziati di altri secoli provenivano da alte sedi e erano legati a Bologna principalmente dal loro insegnamento universitario: così era stato per Egnazio Danti e Girolamo Cardano, nel secolo XVI, per Giovanni Antonio Magini, Bonaventura Cavalieri e Giandomenico Cassini nel secolo XVII, così fu anche alla fine del secolo XIX per Cesare Arzelà, Federigo Enriques ecc.

Questo reclutamento locale di buon livello avveniva in concomitanza con una perdurante crisi dell'Università di Bologna, manifestatasi in modo molto grave nella seconda metà del secolo XVII, quando furono costretti all'emigrazione scienziati come Geminiano Montanari e Domenico Guglielmini.

Ma mentre le istituzioni culturali tradizionali erano luoghi di stagnazione culturale e quindi di sottocultura, sorgevano nuove forme di aggregazione culturale, grazie all'impegno di pochi individui il cui prestigio e il cui potere fu speso fino in fondo, contro i mille ostacoli frapposti dai ceti professionali privilegiati: nacquero così l'Accademia delle Scienze e l'Istituto, la cui origine e fortuna fu legata principalmente ai nomi di Luigi Ferdinando Marsigli e di Prospero Lambertini.

Per mettere un po' d'ordine nell'esposizione la articolerò, seguendo la cronologia, in quattro periodi.

Il primo dall'inizio del secolo al 1731 è segnato dalla diffusione del calcolo differenziale e dalla fondazione dell'Istituto. Il secondo va dalla pubblicazione del primo volume dei Commentari dell'Accademia delle scienze al 1773, anno della soppressione della Compagnia di Gesù. Il terzo copre la fine del secolo e termina con il periodo giacobino e napoleonico. Tre generazioni di intellettuali, legati da stretti vincoli di insegnamento e di vita scientifica in comune e sempre attenti alle maggiori novità della ricerca matematica europea si susseguono ed animano la vita scientifica bolognese.

LA DIFFUSIONE DEL CALCOLO DIFFERENZIALE E LA FONDAZIONE DELL'ISTITUTO

Un gruppo di giovani scienziati bolognesi, sostanzialmente autodidatti, si collocò in prima fila nella diffusione in Italia del calcolo differenziale di Leibinz (la quale precedette di più di un decennio quella del calcolo newtoniano delle flussioni).

Questo gruppo era costituito da Eustachio Manfredi (1674-1739), Gabriele Manfredi (1681-1761), Vittorio Francesco Stancari (1678-1709) e dal cesenate Giuseppe Verzaglia. ([14]) ([15])

Nel suo viaggio in Italia, Leibniz visitò due volte Bologna nell'aprile e nel dicembre del 1689. A Bologna Leibniz incontrò Domenico Guglielmini, con il quale rimase in corrispondenza epistolare. Guglielmini non aveva probabilmente conoscenze ed interessi nelle matematiche tali da consentirgli di impadronirsi del calcolo differenziale, le cui regole erano state pubblicate da Leibniz nel celebre articolo degli Acta Eruditorum del 1684 *Nova methodus pro maximis et minimis* ([7]) ([17]). Tuttavia proprio Guglielmini consigliò i giovani Manfredi, Stancari e Verzaglia ad apprendere la geometria cartesiana ed il calcolo differenziale, impresa alla quale i giovani scienziati bolognesi, uniti in un privato sodalizio, si accinsero con grande entusiasmo, come racconta Eustachio Manfredi nella bellissima introduzione delle *Schedae Mathematicae* di Vittorio Francesco Stancari (Bologna, 1713).

Il frutto più maturo delle ricerche analitiche dei giovani scienziati bolognesi fu rappresentato dalla pubblicazione del *De constructione aequationum differentialium primi gradus* (Bologna, 1707) di Gabriele Manfredi. Si tratta del primo libro nel mondo completamente dedicato alle equazioni differenziali: esso venne lodato da Leibniz e recensito sugli Acta Eruditorum. Il libro venne stampato a spese dell'autore per concorrere ad una cattedra dell'Università di Bologna, assegnata invece ad un modesto studioso: padre Ercole Corazzi.

Quanto al successo commerciale del volume ben ci illumina il seguente brano tratto da un'amara lettera, indirizzata dal Manfredi a Celestino Galiani il 27 agosto 1707 ([13]):

> Noi altri poveri disgraziati, a' quali la mala fortuna ha portato sotto gli occhi le cose algebriche et han svegliato nella mente l'abbominevole pensiero d'intenderle e la sventurata rissoluzione di coltivarle, come se avessimo con ciò formata la più enorme rissoluzione del mondo e che drittamente andasse a distruggere la Repubblica et a rovesciare le leggi del civile commercio, siamo non solamente beffati e mostrati a dito per la Città, ma perseguitati al più che si puote. E poco se ne manca che, nel passar per le strade, non ci corrano addosso e c'uccidano, tanta è l'animosità del volgo contro di noi che pure lasciamo vivere ogn'uno né facciamo male a chi che sia, anzi attendiamo a cosa che vuol essere di servizio per la Città e per le Repubbliche. Quando i più amorevoli che io abbia m'anno detto che ho stampato un libro che non intenderemo che quattro o sei in tutta Italia, m'anno detto la cosa meno disobbligante che mi sia peranco stata rinfacciata intorno al mio libro. E pensano bene d'avermi mostrata gran compassione a non dirmi di peggio et a darmi, come vedono loro, una staffilata coperta. Intanto il libro sta nella bottega, e non pure una copia se n'è per anche venduta, et io che mi ci sono spiantato, anderò col capello verde. Se avessi ristampato Palmerino d'Oliva o la storia del Comito de' Pulcini o che avessi posto insieme mille ragazzate e leggerezze toccanti la Madonna benedetta o san Giuseppe, et avessi loro posto in fronte, per titolo, Novena di S.Giuseppe, o pure Corona di Rose raccolte dal giardino del Cielo per inghindanone la Vergine Santissima, avrei a quest'ora vendute le mie divozioni e cavati i miei quattrini. Ma queste sono tutte seccaggini e bisogna aver pazienza, che ad ogni modo non vale il gridare.

Nel *De Constructione* il Manfredi si arresta di fronte alla soluzione dell'equazione differenziale omogenea (op. cit. p. 167):

$$(nx^2 - ny^2 - xy)\,dx + x^2\,dy = 0$$

Per questa e per tutte le equazioni omogenee di primo grado Manfredi diede il metodo, che porta il suo nome, per ricondurle ad equazioni a variabili separabili in una breve memoria pubblicata sul Giornale de' Letterati d'Italia (tomo XVIII, (1714) pp. 310-311). Ecco la sua formulazione:

> Sono queste equazioni tutte quelle nelle quali poste x, ed y, le due coordinate, le dimensioni dell'una aggiunte alle dimensioni dell'altra in ogni termine fanno una ugual somma, di sorte che non vi sia bisogno di supplire con quantità costanti le dimensioni, che ad esse mancassero in qualche termine ...
> In tutte queste equazioni posta $y = \dfrac{xz}{a}$ e $dy = \dfrac{xdz + zdx}{a}$ si giunge ad un'altra equazione ... in tal forma le indeterminate saranno separate co' loro differenziali.

Un altro risultato importante di Gabriele Manfredi riguarda l'integrazione dei trinomi:

$$\frac{dx}{x^{2m}+2nx^m+aa}$$

Esso è connesso ad un celebre problema proposto da Brook Taylor ai matematici dell'Europa continentale ([16]). Il manoscritto originale della memoria stampata sul Giornale de' Letterati d'Italia è ora tra le Carte di Alfonso Bonfioli Malvezzi, della Biblioteca dell'Archiginnasio di Bologna. ([1])

Nella prefazione del *De Constructione* si trova un riferimento a Luigi Ferdinando Marsigli (1658-1730). Generale al servizio dell'Imperatore d'Austria il Marsigli era stato allievo di Marcello Malpighi e di Geminiano Montanari e compagno di studi di Domenico Guglielmini. Rientrato a Bologna nel 1704 il Marsigli si diede interamente alle Scienze accogliendo nella sua casa l'*Accademia degli Inquieti* della quale lo Stancari e i Manfredi avevano fatto il riferimento dei loro studi scientifici. Il Marsigli si illuse in un primo momento di poter riformare gli ordinamenti dell'Università di Bologna per introdurvi le nuove discipline matematiche e sperimentali e in tal senso fece una proposta articolata nel 1709. Di fronte alle resistenze incontrate il Marsigli cominciò a coltivare l'idea della fondazione di un Istituto autonomo. L'*Istituto marsigliano* diventato celebre in tutta Europa per le sue collezioni scientifiche, accolse anche l'*Accademia degli Inquieti*, che prese il nome di Accademia delle Scienze dell'Istituto nonché la Specola per le osservazioni astronomiche. L'Istituto tenne la seduta inaugurale il 13 marzo 1714 ([5]), ([10]).

In cambio del suo disinteressato amore per la scienza (a cui diede personalmente anche significativi contributi riguardanti l'oceanografia, la storia naturale ecc.) il Marsigli ottenne tali e tante opposizioni dai suoi concittadini e in particolare dalla sua famiglia che fu costretto a lasciare Bologna e a vivere sotto falso nome per qualche tempo in Provenza.

Le controversie si avviarono ad una soluzione positiva nel 1726 solo per l'interessamento di Mons. Prospero Lambertini (1675-1758) illustre prelato bolognese che fu poi cardinale, arcivescovo di Bologna e infine papa Benedetto XIV. Ma ancora nel 1729 nel preparare per la stampa il primo volume dei *Commentarii*, che riferirono per tutto il secolo sulla vita scientifica dell'Istituto e dell'Accademia delle Scienze e pubblicarono numerose importanti memorie, Francesco Maria Zanotti esprimeva serissime preoccupazioni al suo amico Antonio Leprotti, archiatra pontificio, in una lettera pubblicata dal Bortolotti ([5] p. 157):

> Per lo che, fra tre o quattro mesi, io vo immaginando che il Tomo sarà ridotto a tale che bisognerà consegnarlo all'Inquisizione. Or questo è il luogo dove la prudenza mi pare essere il più grande imbarazzo del mondo, perché, a dirti il vero, non so come condurmi.
>
> Tu sai le vicende che ha sofferto il libro che stampa ora il Manfredi. Ora la mia istoria, contiene una compendiosa relazione di questo libro. Oltre a ciò io non ho potuto sfuggire, in altri luoghi, di esporre di qual sentenza fosse il Copernico, non ho potuto sfuggire di chiamare tal sentenza un'ipotesi ingegnosa, ed atta a spiegare i fenomeni della natura; non ho potuto sfuggire in qualche luogo di dire che alcuni filosofi hanno creduto che le bestie non sentano, e in somma, non ho potuto sfuggire di parlare con rispetto e con carità dei moderni.

Ora io temo forte che i Revisori, a ciascheduno di questi luoghi pretendano che io aggiunga qualche protesta, e, dove espongo quel che sente il Copernico, subito aggiunga che io però detesto il suo sistema, come un'eresia; dove dico che i moderni hanno dubitato se le bestie sentano, subito aggiunga che questo dubbio però è contrario alla Fede cattolica; e così temo che facendo lo stesso ora ad un luogo ora ad un altro, mi obblighino a riempire il libro di atti di Fede, intorno a certi articoli che non sono nel Credo e che io non sono tenuto a credere esplicitamente, e che, se per ventura, non fossero poi articoli, come è sentimento di tanti cattolici, noi faremmo ridere della nostra semplicità la stessa Chiesa Cattolica.

Solo due anni dopo, nel 1731 poteva uscire il sofferto primo volume dei Commentarii; nello stesso anno Prospero Lambertini divenne arcivescovo di Bologna.

GLI ANNI DI FRANCESCO MARIA ZANOTTI E VINCENZO RICCATI (1731-1773)

I sette volumi dei *De Bononiensi scientiarum et artium Instituto atque Academia commentarii* uscirono senza una periodicità fissa dal 1731 al 1791 (i tomi secondo e quinto sono divisi in due parti); il quinto tomo fu stampato nel 1767 ([18]). In quegli anni la vita accademica ruotava a Bologna intorno alla figura di Francesco Maria Zanotti, prima segretario, poi presidente dell'Accademia (dal 1766).

Francesco Maria Zanotti, nato a Bologna nel 1692 studiò algebra con V.F. Stancari "desideroso oltremodo che alcun giovane di alto ingegno vi si applicasse, il quale però niente sapesse delle scienze matematiche, volendo far prova di quanto innanzi potesse procedersi colla sola scorta dell'algebra".

Morto poco dopo lo Stancari, Zanotti studiò geometria con Eustachio Manfredi e Geminiano Rondelli e algebra con Gabriele Manfredi. Addottoratosi nel 1716 divenne pubblico lettore di filosofia nel 1718. Fu tra i primi a professare a Bologna la fisica cartesiana dei vortici e l'ottica newtoniana. Francesco Algarotti fu suo allievo. Nel 1723 Zanotti divenne segretario dell'Istituto licenziando per la stampa, come abbiamo visto nel 1731 il primo volume dei Commentarii. Nel 1732 rifiutava una cattedra a Padova fattagli offrire dal Morgagni.

Ingegno molto versatile lo Zanotti si occupò di varie questioni; tra i suoi scritti matematici ricordiamo: *De hyperbolicis quibusdam spatiis* (1746). *De elastris* (1747). *Sur les figures et les solides circonscrits au cercle et à la sphère* (Mém. Acad. Sci. 1748); il libro: *Della forza de' corpi che chiamano viva* (1752); *De separandis indeterminatis* (1755). I suoi scritti furono raccolti nelle *Opere* (1779-1802), in nove volumi in quarto.

Alcuni scritti dello Zanotti furono stampati a Napoli, dove egli era legato da amicizia con il matematico Niccolò di Martino; se si ricorda anche l'amicizia tra il Galiani e G. Manfredi possiamo far datare dall'inizio del Settecento una tradizione di buoni rapporti tra le Università di Napoli e di Bologna nel campo delle matematiche, destinata a durare nel tempo.

L'altro grande protagonista della vita matematica bolognese di quegli anni era Vincenzo Riccati (1707-1775). Vincenzo era giunto a Bologna per studiare nel Collegio dei Gesuiti da Castelfranco Veneto dove era nato dal conte Jacopo Riccati (1676-1754), dei cui grandi meriti di scienziato rimane testimonianza in tutti i trattati dell'analisi per aver dato il nome ad una celebre equazione differenziale. A Bologna, dopo un soggiorno romano, Vincenzo Riccati era tornato per insegnare, sempre nel collegio dei Gesuiti, fino alla soppressione dell'ordine nel 1773.

Riccati fu autore di pregevoli opere matematiche come: *Dialogo delle forze vive* (1749); *De usu motus tractorii in constructione aequationum differentialium* (1752); *De natura et proprietatibus quarundam curvarum* (1755); *De seriebus recipientibus summam generalem algebraicam, aut exponentialem commentarius* (1756); *Della integrazione della formula*

$$\frac{dz\sqrt{f+gzz}}{\sqrt{p+qzz}}$$

per mezzo degl'archi elittici ed iperbolici (1757); e della raccolta *Opusculorum ad res Physicas et Mathematicas pertinentium* t. I (1757), t. II (1762).

Il nome di Vincenzo Riccati, oltre allo studio delle funzioni iperboliche è più spesso legato al più ampio trattato pubblicato in Italia sulla geometria cartesiana ed il calcolo differenziale ed integrale (ivi comprese le equazioni differenziali e il calcolo delle variazioni): *Institutiones analyticae* (Bologna 1765-67), scritto in collaborazione con il suo allievo Girolamo Saladini (1731-1813).

Una parte notevole del Riccati-Saladini è dedicato allo studio delle curve piane tra le quali la lumaca di Pascal:

$$(x^2+y^2-2ax)^2 = l^2(x^2+y^2)$$

e le sestiche:

$$y^6+3x^2y^4+3x^4y^2+x^6-4a^2x^2y^2 = 0$$

$$4(x^2+y^2)^3+R^2y^2-4R^2(x^2+y^2)^2 = 0$$

Riccati aveva anche studiato le *trattrici*, ossia le curve per cui è costante la lunghezza della tangente, che possono essere descritte dall'equazione differenziale:

$$y^2+y^2\left(\frac{dx}{dy}\right)^2 = a^2$$

L'illustrazione seguente contiene la molteplici approvazioni a cui dovette sottostare le *Institutiones analyticae* e dà un'idea della persistente censura in pieno Settecento sulle opere dell'ingegno in Italia.

Vidit D. Johannes Maria Vidari Clericus Regularis Sancti Paulli, & in Ecclesia Metropolitana Bononiæ Pœnitentiarius pro Eminentissimo, & Reverendissimo Domino D. Vincentio Cardinali Malvetio Archiepiscopo Bononiæ, & S. R. I. Principe.

Die 26. Martii 1763.

A. R. P. Carolus Maria Offredi Ord. Theatinorum Pub. in Univ. Bonon. Professor, & S. Off. Revisor ordinarius videat pro S. O. & referat.

F. Seraphinus Maria Maccarinelli S. O. Bonon. Inq. Coadiutor.

30. Martii 1763.

Egregium opus inscriptum = *Institutiones Analyticæ collectæ a Vincentio Riccato Soc. Jesu, & Hieronymo Saladino Monacho Cœlestino, Tomus Primus* = de mandato Reverendissimi Patris Seraphini Mariæ Maccarinelli S. O. Bononiæ Inquisitoris Coadiutoris attente perlegi, nihilque in eo occurrit Fidei, aut bonis moribus contrarium: quapropter dignum censeo, ut publica luce donetur. In quorum fidem &c.
Ex Ædibus S Bartholomei Apost. Clericorum Regularium Bononiæ tertio Kal. Aprilis 1763.

D. Carolus Maria Offredi C. R. in Bononiensi Archigymn. Pub. S. T. Lector & S. O. Revisor Ord.

Die 31. Martii 1763.

Stante suprascripta attestatione.

INPRIMATUR

F. Seraphinus Maria Maccarinelli S. O. Bononiæ Inquisit. Coadiutor.

Riccati V., Saladini G., *Institutiones Analyticae, Tomus Primus*, Bononiae, MDCCLXV.

I due collegi tenuti dai Gesuiti: S. Francesco Saverio e S. Luigi e le loro scuole di S.Lucia erano in quei decenni il riferimento per quanti a Bologna desideravano dedicarsi agli studi scientifici assai più che l'Università, il cui ordinamento generale era bloccato, nonostante che, con lo stimolo dell'Istituto, qualche nuovo insegnamento venisse acceso. Prospero Lambertini, divenuto nel 1740 Papa Benedetto XIV non era favorevole ai Gesuiti e il suo pupillo Vincenzo Malvezzi (1715- 1775) creato cardinale e divenuto suo successore all'Arcivescovado di Bologna giocò un ruolo di primo piano nella soppressione dell'Ordine. Benedetto XIV, dopo aver anch'egli provato inutilmente a riformare l'Università, profuse il suo notevole impegno a favore della cultura a Bologna a vantaggio dell'Istituto: in particolare arricchì enormemente la Biblioteca, anche con un lascito personale, e creò nel 1745 nell'Accademia delle Scienze gli *Accademici Benedettini*, dotandoli di un trattamento economico stabile.

In quegli anni Bologna fu meta di giovani studiosi che vi si recavano da varie parti d'Italia per avviarsi allo studio delle scienze matematiche e naturali: basterà ricordare oltre a Girolamo Saladini, Giordano Riccati (1709-1790) e Gianfrancesco Malfatti (1731-1807); ma continuava anche un significativo reclutamento locale, da cui emersero per le scienze matematiche Sebastiano Canterzani (1734-1818), che sostituì lo Zanotti come segretario dell'Accademia e redattore dei Commentarii; Alfonso Bonfioli Malvezzi (1730-1804), Gregorio Casali (1721-1802); Petronio Caldani (1735-1808). Ma tutta questa generazione di intellettuali bolognesi, anche quelli che si dedicarono alla medicina e alle scienze naturali o alla storia e alla letteratura ebbero, principalmente ad opera di Vincenzo Riccati e Francesco Maria Zanotti, una solida formazione matematica; è il caso ad esempio dello storico e letterato Ludovico Savioli.

La soppressione dell'ordine dei Gesuiti, decisa da Clemente XIV nel 1773, fu tenacemente perseguita a Bologna dall'arcivescovo Vincenzo Malvezzi: a capo dei Gesuiti bolognesi si trovava allora un matematico di buon livello,corrispondentedell'Académie des sciences: Jacopo Belgrado (1704-1789), autore del trattato *De utriusque analyseos usu in re physica* (Parma, 1761-62 due volumi).

GLI ULTIMI DECENNI DEL SECOLO.

Tra i soci o corrispondenti bolognesi dell'Académie des Sciences uno solo appartiene a quest'ultimo periodo: Alfonso Bonfioli Malvezzi, divenuto nel 1773 corrispondente del Marchese di Condorcet. Gli altri bolognesi furono D. Guglielmini (1655 - 1711), G.B. Trionfetti (1656-1708), L.F. Marsigli (1658 -1730), E. Manfredi (1674 - 1739), F.M. Monti (1675 - 1754), F.M. Zanotti (1692 - 1777).

Anche se non si può non rilevare che i maggiori matematici che operarono a Bologna in questo secolo: Gabriele Manfredi e Vincenzo Riccati non furono accolti nell'Académie des Sciences, pure si deve costatare che dopo il Malvezzi a due soli bolognesi si sono aperte le porte dell'Académie: il chimico F. Malaguti (1802-1878), che lavorò lungamente in Francia ed il fisico A. Righi (1850-1920).

Alfonso Bonfioli Malvezzi, preferito dall'Académie des Sciences ai più attivi Canterzani e Casali, si era fatto conoscere personalmente da d'Alembert, Bossut e Condorcet grazie ad un lungo viaggio in Europa, compiuto tra il 1771 e il 1773 durante il quale egli soggiornò diverse settimane a Londra e alcuni mesi a Parigi. Di questo viaggio ci rimane una preziosa testimonianza recentemente pubblicata da S. Cardinali in un volume di scritti inediti del Malvezzi ([1]). Tra gli scritti inediti del Malvezzi, oltre la sua corrispondenza con Condorcet sul terremoto di Bologna del 1779, figura una memoria matematica (a cura di M.T. Borgato) diretta sotto forma di lettera, al suo compagno di studi Gianfrancesco Malfatti allora a Ferrara: *Sulle proprietà della curva cassiniana provate per la via del Calcolo* (1782). Il Malfatti aveva dedicato un trattato sintetico allo studio delle proprietà geometriche e meccaniche delle *ovali di Cassini*, cioè del luogo piano dei punti il prodotto delle cui distanze da due punti fissi è costante. Analiticamente le ovali di Cassini sono quartiche la cui equazione cartesiana è

$$(x^2 + y^2)^2 - 2a^2(x^2 - y^2) + a^4 - c^4 = 0$$

Un caso particolare delle ovali di Cassini si ha quanto a = c; si trova allora la *lemniscata di Bernoulli*:

$$(x^2 + y^2)^2 = 2a^2(x^2 - y^2)$$

curva dotata anche di interessanti proprietà meccaniche. Il Malvezzi si era occupato in prevalenza di meccanica e si era fatto conoscere con una memoria sul principio di Maupertuis che aveva incontrato l'approvazione di d'Alembert. Malfatti ebbe un'importante corrispondenza anche con Sebastiano Canterzani, dal 1766 segretario dell'Accademia delle scienze. Molti temi della ricerca matematica contemporanea vengono trattati in questa corrispondenza, solo recentemente pubblicata ([11]). Ad esempio riportiamo lo scambio di osservazioni dell'agosto 1779 relativa al calcolo delle variazioni. Scriveva il Malfatti:

> Gradirò moltissimo, che Ella dopo il suo ritorno in Città, mi mandi quelle sue carte, ove ha trascritto il metodo degl'isoperimetri dell'Eulero: e spero che quelle potranno bastarmi sicche non vi sia bisogno che Ella trascriva il resto. A proposito di questo io le voglio comunicare un mio dubbio, che anche dalla collina ove si trova, ella mi potrà sciogliere. Col metodo delle variazioni si risolve egregiamente questo Problema. Dato nella retta AH il punto A, e fuori della AH il punto C, determinare la curva minima che passa pei punti A, C.

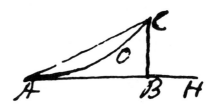

Ricerca Matematica a Bologna nel Secolo XVIII

Il metodo dà la linea retta, come debb'essere. Ciò posto, perché non sarà possibile sciogliere anche quest'altro Problema. Pei punti A, C far passare una curva AOC cosicché l'aja AOCB sia un minimo? Col metodo delle variazioni io trovo questa formula $\int dy dx = 0$, posto che varj la sola y, da cui non cavo niente di curva: eppure e' mi par possibilissimo, che tra tutte le curve ve n'abbia ad esser una, che costituisca l'area minima AOCB. Se non mi si prova evidentemente l'assurdità di questo Problema, perché non potrò io supporre qualche imperfezione nel metodo, sebbene io non sappia assegnarla? Che mi dic'ella? Parmi difficile, che essendo subito corsi i Geometri, per verificare il metodo anche colla sperienza, a cercare un minimo nella curva che passa per due punti dati, non abbiano poscia cercato la curva della minima aja, e non abbiano sviluppato la ragione, perché questa non si possa ottenere.

E il Canterziani precisava:

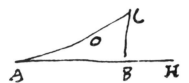

Il problema = dati due punti, uno A nell'asse AH, l'altro C fuori dell'asse medesimo, trovare la curva AOC, che dà l'aja AOCB minima = credo che sia troppo indeterminato, e più incerto, e vago, appunto come sarebbe il problema = sopra una data retta costruire il triangolo massimo, o minimo: il che posto non è meraviglia, se trattandolo col metodo inverso assoluto dei massimi, e dei minimi, non si cava niente di curva. Bisogna dunque limitar il problema obbligando la curva a qualche condizione: per esempio tra tutte le curve che hanno l'arco AOC d'una data lunghezza trovar quella, che dà l'aja AOCB massima, o minima. Allora il problema appartiene al metodo inverso relativo dei massimi, e dei minimi; e col calcolo delle variazioni Ella troverà il circolo, il quale dà l'aja AOCB massima. Parmi che in questo problema non possa aver luogo il minimo assoluto, il quale s'adempirebbe solamente colle due rette AB, BC che danno l'aja AOCB = 0.

Anche il conte Gregorio Casali, matematico e letterato tra i più stimati a Bologna nella seconda metà del secolo, fu tra i corrispondenti del Malfatti. Professore di architettura militare, segretario dal 1759 dell'Accademia Clementina, senatore di Bologna dal 1764, il Casali, che concluse la sua esistenza come Rettore dell'Università di Bologna tra il 1800 e il 1802, è figura che meriterebbe maggiori studi, anche per la sua produzione matematica, nella quale figura una bella memoria del 1757 intitolata: *De conicorum sectionum focis*, contenente lo studio di alcune curve dette *strofoidi* ([9]):

$$(x^2+y^2)(x-2a) + a^2 x = 0$$

Ma il Casali è forse più interessante per i rapporti tra scienza e letteratura, intellettuali e politica ([8]).

Merita infine almeno di essere citato Petronio Caldani (1735-1808) autore di una memoria, celebrata al suo tempo, che reca come titolo: *Della proporzione Bernoulliana fra il diametro e la circonferenza del circolo e i logaritmi* (Bologna, Dalla Volpe, 1782).

La proporzione bernoulliana è una formula che lega π ai logaritmi dei numeri negativi; essa trae origine da una memoria di Johann Bernoulli del 1702. Dalla scomposizione:

$$\frac{cdz}{z^2+a^2} = \frac{cdz}{2a\sqrt{-1}} \left(\frac{1}{z-a\sqrt{-1}} - \frac{1}{z+a\sqrt{-1}} \right)$$

Si ricava:

$$a\frac{\pi}{2} = \int_0^{+\infty} \frac{a^2 dz}{a^2+z^2} = \frac{a\sqrt{-1}}{2} \int_{-1}^{1} \frac{dt}{t} =$$

$$= \frac{a\sqrt{-1}}{2} \left[\log(+1) - \log(-1) \right] = \frac{a}{2} \frac{\log(-1)}{\sqrt{-1}}$$

IL TRIENNIO REPUBBLICANO.

A Bologna l'arrivo delle armate rivoluzionarie francesi, guidate dal generale Bonaparte, avvenne nel giugno del 1796; esso fu preceduto di poco dalla morte di due studenti Luigi Zamboni e G.B. De Rolandis, che avevano cospirato nel 1794 contro il governo pontificio.

Napoleone all'inizio seppe utilizzare il desiderio di autonomia da Roma dei bolognesi, conferendo al Senato il governo della città, e nelle mani del Gonfaloniere del Senato giurarono i professori dell'Università e i membri dell'Istituto; ma con l'aggregazione di Bologna alla Repubblica Cisalpina, avvenuta il 27 luglio 1797, l'Università di Bologna fu sottratta alla giurisdizione municipale e passò all'Amministrazione Centrale del Dipartimento del Reno, organo periferico del governo cisalpino, residente in Milano. Durante il governo cisalpino, pur tra le mille difficoltà politiche e militari e nonostante le spinte campanilistiche operanti in difesa della conservazione (delle quali si fece interprete anche l'uomo politico più influente che avesse allora Bologna: Antonio Aldini) si compì l'evento decisivo per la riforma dell'Università, invano perseguito in tutto il Settecento: la soppressione dei Collegi dottorali che conferivano i gradi accademici. Attraverso questa facoltà i Collegi, ben rappresentati da loro uomini all'interno delle magistrature cittadine, avevano impedito ogni reale riforma dell'ordinamento universitario che non fosse l'istituzione di qualche nuovo insegnamento.

La sorte dell'Università di Bologna e dell'Istituto delle Scienze fu argomento di discussione, e non di poco conto nelle Assemblee legislative della Repubblica Cisalpina.

Già nel negoziato che portò Bologna nella Cisalpina, una delle contropartite, per aver fissato in Milano la capitale della repubblica, fu il decreto, depositato dal generale Bonaparte, che fissava in Bologna l'*Istituto Nazionale* della Repubblica Cisalpina.

Tale Istituto era, secondo il modello della Costituzione della Repubblica francese dell'anno terzo, non solo essere un istituto di ricerca scientifica, ma doveva sovraintendere a tutta la pubblica istruzione.

Concretamente come l'*Institut* francese ereditava le strutture dell'Académie des sciences, l'*Istituto Nazionale* si doveva fondare sulla solida realtà organizzativa dell'Istituto marsigliano. La costruzione dell'Istituto nazionale a Bologna fu molto complessa e i primi trenta membri dell'Istituto poterono riunirsi per la prima volta solo l'8 gennaio 1803; la riunione avvenne a Bologna nei locali dell'antico Istituto ([2]).

Ugualmente complessa fu la riforma dell'Università di Bologna ([12]) che fu inserita nel *Piano generale degli Studi* elaborato da una Commissione per la Pubblica Istruzione, della quale facevano parte anche i matematici Gregorio Fontana e Lorenzo Mascheroni. Il *Piano* fu presentato al Gran Consiglio della Repubblica Cisalpina nella seduta del 24 luglio del 1798, e fu per un mese l'argomento principale di discussione dell'Assemblea.

Esso intanto proponeva il mantenimento delle Università di Bologna e di Pavia, chiamate Scuole di approvazione. Questa decisione era tutt'altro che scontata e segnava una chiara differenza della legislazione cisalpina dal modello francese (in Francia era stata abolita l'Università): essa incontrò vivaci reazioni soprattutto da parte di Vincenzo Dandolo, che oltre a farsi portavoce di istanze più "democratiche" cercava di difendere il mantenimento di fatto delle altre sedi Universitarie tradizionali (Ferrara, Modena ecc.) delle quali il piano prevedeva la soppressione; questo mantenimento doveva avvenire secondo il Dandolo con il potenziamento delle *scuole centrali* di cui il piano, seguendo in questo il modello francese, prevedeva la costituzione.

Concretamente l'Università di Bologna visse in quegli anni un'esistenza abbastanza stentata: furono istituite nuove cattedre ed altre vennero soppresse; e dodici docenti che, entro il 30 aprile 1798, non avevano giurato fedeltà alla Repubblica Cisalpina furono destituiti.

Il *piano della pubblica istruzione* si arenò; con la crisi della Repubblica Cisalpina e l'arrivo degli austro-russi, che entrarono a Bologna nel giugno 1799 per essere ricacciati dopo Marengo nel giugno 1800, si tentò la restaurazione dell'antica Università; cosa che, come la precedente riforma, riuscì solo parzialmente (e infatti gli antichi Collegi rimasero definitivamente sepolti).

Il secolo XVIII si chiuse a Bologna con la stampa nella Tipografia di S. Tommaso d'Aquino, anch'essa costituita con ingenti donazioni del Marsigli, della *Teoria generale delle equazioni* (1799) del modenese Paolo Ruffini (1765-1822), nella quale si dimostra l'irrisolubilità per radicali delle equazioni di grado superiore al quarto (finalmente le opere scientifiche di questo periodo non recano più i segni della censura e dell'inquisizione). Il Ruffini, come a Bologna il Canterzani, non aveva aderito al nuovo ordine napoleonico.

Altri matematici invece si impegnarono nella politica, che in quegli anni reclamava il primato: tra questi il conte Alfonso Bonfioli Malvezzi, membro della giunta per la prima costituzione di Bologna; il professore Giovanni Battista Guglielmini, membro attivo del Gran Consiglio della Repubblica Cisalpina (si occupò soprattutto dei problemi finanziari della repubblica) e inoltre il fisico Giovanni Aldini, fratello di Antonio e nipote di Luigi Galvani.

Vorrei concludere questa conferenza ricordando come l'interessante e complessa vicenda dell'Istituto nazionale nel periodo napoleonico è oggi nota nelle sue linee principali soprattutto grazie ai lavori del matematico Ettore Bortolotti (1866-1947), per molti anni professore nell'Università di Bologna. Egli seppe contrastare le interpretazioni riduttive di certa storiografia locale, dando un concreto esempio di come l'esperienza acquisita nella storia delle scienze, e con essa il rifiuto di punti di vista municipali, può essere decisiva anche nello studio di questioni più generali della storia della cultura e delle istituzioni ([2]), ([3]), ([4]), ([5]).

BIBLIOGRAFIA

[1] A. Bonfioli Malvezzi, *Viaggio in Europa e altri scritti*, a cura di S. Cardinali e L. Pepe. Ann. Univ. Ferrara, Sez. III, 1988.

[2] E. Bortolotti, *Materiali per la storia dell'Istituto Nazionale*, Estratto da Mem. Acc. Sci. Lett. Modena, Serie III, **12** (1915).

[3] E. Bortolotti, *L'école mathématique de Bologne. Aperçu historique*, Bologna, Zanichelli, 1928.

[4] E. Bortolotti, *L'Accademia delle Scienze di Bologna durante l'epoca napoleonica e la restaurazione pontificia*, Estratto da Rend. Acc. Sci., Istituto Bologna, cl. sci. fis. (1935-36).

[5] E. Bortolotti, *La storia della matematica nell'Università di Bologna*, Bologna, Zanichelli, 1947.

[6] P. Casini, *Newton e la coscienza europea*, Bologna, Il Mulino, 1983.

[7] M. Cavazza, *La corrispondenza inedita tra Leibniz, Domenico Guglielmini, Gabriele Manfredi*, Studi e Memorie per la Storia dell'Università di Bologna, Nuova Serie, **6**, (1987), pp. 51- 79.

[8] R. Cremante, W. Tega (a cura di), *Scienza e letteratura nella cultura italiana del Settecento*, Bologna, Il Mulino, 1984.

[9] G. Loria, *Curve piane speciali, algebriche e trascendenti*, Due volumi, Milano, Hoepli, 1930.

[10] *I Materiali dell'Istituto delle Scienze,* Bologna, Clueb, 1979.

[11] L. Miani, I. Ventura, S. Giuntini (a cura di), *Il carteggio Gianfrancesco Malfatti-Sebastiano Canterzani,* Boll. Storia Sci. Mat., **3**, n. 2 (1983).

[12] G. Natali, *L'Università degli studi di Bologna durante il periodo napoleonico (1796-1815)*, Atti e Memorie per la Storia dell'Università di Bologna N.S., **1** (1956) pp. 505-545.

[13] F. Palladino, *Tre lettere inedite di Gabriele Manfredi a Celestino Galiani sul calcolo infinitesimale*, Boll. Storia Sci. Mat., **4** (1984) fasc. 2, pp. 133-144.

[14] L. Pepe, *Il calcolo infinitesimale in Italia agli inizi del secolo XVIII*. Boll. Storia Sci. Mat., **1** n. 2 (1980) pp. 43-101.

15] L. Pepe, *Gabriele Manfredi (1681-1761) et la diffusion du calcul différentiel en Italie*. Studia Leibnitiana Supplementa XXVI (1986), pp. 79-87.

[16] L. Pepe, *Newton, le flussioni e la matematica italiana nel Settecento*. Atti del Convegno "Scienza e Immaginazione nella cultura inglese del Settecento" Gargnano 1985, Milano, Unicopli, 1987, pp. 309-319.

[17] A. Robinet, G. W. *Leibniz et la République des lettres de Bologne*. Studi e Memorie per la Storia dell'Università di Bologna. N.S., **6** (1987) pp. 3-49.

[18] W. Tega (a cura di), *Anatomie accademiche*, Due volumi, Bologna, Il Mulino, 1986-87.

Cauchy-Fantappiè Formules
in Multidimensional Complex Analysis

R. MICHAEL RANGE Department of Mathematics, State University of New York at Albany, Albany, New York 12222, USA

1. INTRODUCTION

To present a lecture on Cauchy-Fantappiè formulas at the occasion of the 9th centennial of the University of Bologna offers the opportunity to begin by remembering Luigi Fantappiè, who worked at the Università degli Studi di Bologna for several years during the 1930s.

Fantappiè was born in Viterbo, Italy, in 1901. After graduating from the Scuola Normale Superiore in Pisa in 1922, he was appointed assistant at the University of Rome. Here he met the two great masters Vito Volterra and Francesco Severi, who were very influential in shaping his interest in the foundations of the then still new theory of *Functional Analysis*. In particular, it was Severi who suggested to him to clarify the operation which assigns to a function its derivative at a given point. Fantappiè pursued this question in the complex analytic realm, thus starting a program to investigate the *theory of analytic functionals*, which was to occupy him throughout most of his scientific life. Had he chosen to investigate this question in the context of (real) differentiable functions, the theory of distributions might have been developed much earlier than it did!

Fantappiè's first appointment as professor was at the University of Palermo, followed by an appointment at the University of Bologna in 1932, though it appears that Fantappiè already worked in Bologna since the beginning of the 1930s. Around the mid 30s he moved to San Paolo, Brasil, while keeping his chair in Bologna. In 1939 he returned to Italy to assume the Chair in Analysis at the Istituto Nazionale di Alta Matematica in Rome, a position he held until his early death in 1956.

Two of Fantappiè's major memoirs on the theory of analytic functionals ([Fan], 19-247, and 249-416) were written during his years in Bologna, but he returned to this topic many times thereafter, either to rebuild and extend the foundations, as he did in 1940 (pp. 425-515) and in his deep 1943 memoir (pp. 515-626), or to develop applications of the theory.

Today, Luigi Fantappiè's name is widely known among complex analysts mainly in the context of the so-called «Cauchy-Fantappiè Forms», even though – it is fair to say – few mathematicians know of the underlying connection.

In this lecture I will first explain in simplest terms some of the basics of Fantappiè's theory of analytic functionals, and I will try to trace the tenuous path leading to the machinery of Cauchy-Fantappiè formulas as they are known today, the foundations of which were laid by Jean Leray in 1956. I will then discuss in broad outline some of the generalizations of the first formula of Cauchy-Fantappiè and some of the more recent applications, including a rather elementary solution of the «Levi problem», a result which deals with the characterization of domains of holomorphy by *local* (pseudo-) convexity properties. At the end I will discuss a new construction of Cauchy-Fantappiè forms with applications to estimates for solutions of the Cauchy-Riemann equations, which applies to pseudoconvex domains of finite type in dimension two, a topic at the present research frontier. In closing, I shall briefly return to Luigi Fantappiè.

Those interested in a detailed systematic exposition with complete proofs of most of the topics discussed here are referred to the book [Ran 2].

2. ANALYTIC FUNCTIONALS

Let us begin with the one variable case. Given a collection S of holomorphic (= analytic) functions, a *functional on* S is simply a map $F: S \to \mathbb{C}$. Incidentally, the corresponding *real* concept was first introduced by Vito Volterra in 1887 («Sopra le funzioni che dipendono da altre funzioni», Rend. Acc. Lincei #4, 1887), and this event marks as close as is possible the «birth» of *Functional Analysis*. F is said to be an *analytic functional* if S is *open* in an appropriate sense, and if for every holomorphic one-parameter family $\{f_t\} \subset S$ the function $t \to F[f_t]$ is holomorphic (cf. [Fan], pp. 35-36). If S is a linear space, one can consider *linear* analytic functionals in the obvious way. As an example, let us state one of the (elementary) results of the theory.

THEOREM. ([Fan], p. 52) *If F is a linear analytic functional, then for every holomorphic family $\{f_t\}$ one has*

$$\mathrm{d}/\mathrm{d}t(F[f_t]) = F[(\partial/\partial t)f_t].$$

Of course, these concepts can be extended to classes of functions holomorphic in several variables, and depending holomorphically on one or more parameters.

From today's perspective, all this may look like a rather elementary idea, but we should keep in mind that in the 1920s the idea of «functional», i.e. a function whose input variable is a function rather than a number, was still quite novel. More significantly, one of the main applications Fantappiè had in mind was the study of the general properties of the «rule» which gives the solution of a differential (or integral) equation in dependence of certain input data, like initial conditions at a point, or on hypersurfaces, or certain parameters. The scope of such a program is very vast indeed, even by today's standards! Conceptually, we are dealing with a higher level analogue of the step taken by mathematicians in the 19th century, when they shifted their interest from looking at individual functions to studying general properties of classes of functions.

One of the classical examples is the Cauchy integral formula

$$f \to (2\pi i)^{-1} \oint \frac{f(\xi)\,d\xi}{\xi - t},$$

which gives a representation for the «evaluation functional» at the point $t \in D$. In fact, Fantappiè recognized early on the fundamental role of this formula for his theory. This led him to introduce the *indicatrix* $I(F)$ of an analytic functional F, defined by

$$I(F)(t) = F[1/(\xi - t)],$$

with domain depending of course on F. The indicatrix determines a linear functional F completely; in fact one has

$$F[f] = (2\pi i)^{-1} \int_C I(F)(t) f(t)\,dt$$

for a suitable closed curve C ([Fan], p.60).

Properties of a linear analytic functional are thus reflected by properties of the corresponding indicatrix. For example, Fantappiè shows that the indicatrix $I(F)$ of an analytic functional F is *rational* if and only if F is a finite combination of differentiations, evaluation at a point, and multiplication by constants. Fantappiè also uses the indicatrix to develop a symbolic calculus to find solutions for general classes of *functional equations*. In particular, he shows how to solve the general second order linear differential equation

$$y'' + p_1(x) y' + p_2(x) y = 0 \tag{2.1}$$

by means of a single functional R, the *Riccati functional*, and explicit integrations ([Fan], 229-237).

Fantappiè's first investigations of the indicatrix in the case of *several* variables led him to consider the obvious generalization

$$I(F)(t_1, \ldots, t_n) = F\left[\prod_{j=1}^{n} 1/(\xi_j - t_j)\right]$$

([Fan], pp. 289-290, and again, ten years later, pp. 471-473). This indicatrix has the usual common drawback of the «one variable at a time» approach: it certainly is useful, but it fails to reflect the deeper properties of holomorphic functions of *several* variables. Fantappiè realized this, and in his deep 1943 paper ([Fan], 515-626) he introduced a new indicatrix, in which the variables were tied together in a more homogeneous way. In the process, he also shifted his point of view regarding the underlying space of definition for the indicatrix from a product of n spheres (the classical *Osgood space*) to the more natural complex n-dimensional projective space $P\mathbb{C}^n$. Considering that up to that time the theory of several complex variables was still largely «polydisc based» (domains of holomorphy were exhausted by analytic polyhedra which, following K. Oka, were embedded into higher dimensional polydiscs), Fantappiè's change in perspective has to be viewed as quite remarkable and original.

The new indicatrix $p(F)$, which Fantappiè called very appropriately *projective indicatrix*, is defined as follows:

$$p(F)(t_1, \ldots, t_n) = F[1/(1 + \sum \xi_j t_j)]$$

([Fan], 521-523). Fantappiè then obtains a remarkable formula written symbolically as

$$F[y] = \overset{\triangle}{p} \bullet \overset{\triangle}{y}, \qquad (2.2)$$

which expresses the value of the functional F on the function y in terms of the indicatrix $p = p(F)$ (see pp. 526-528).

The explicit expression for (2.2) involves a *characteristic function* χ and two integrations, something analogous to the convolution of p with χ and then with y. Fantappiè calls the operation on the right side of (2.2) the *projective functional product* of the functions p and y. A major part of the 1943 paper is devoted to a detailed study of this product and to finding effective ways to calculate it in special cases. Among the many formulas obtained (most of which are too complicated to write out here) is the following intriguing one. Suppose

$$p(z) = \sum p_\alpha z^\alpha \quad \text{and} \quad y(z) = \sum y_\alpha z^\alpha$$

are the power series expansions of p and y at the origin; then

$$\overset{\triangle}{p} \bullet \overset{\triangle}{y} = \sum_\alpha (-1)^{|\alpha|} \frac{\alpha!}{|\alpha|!} p_\alpha y_\alpha, \qquad (2.3)$$

when we are using the standard multiindex notion. In particular (2.3) shows that the projective functional product is commutative, something which is far from obvious from the relationship (2.2). Furthermore, this product is invariant under projective linear coordinate changes in one factor when the other factor is subjected to the corresponding dual transformation.

It is in these formulas for the projective functional product, which yield a representation of linear analytic functionals in terms of negative powers of linear (affine) functions, that one finds the key idea which is also present in the «Cauchy-Fantappiè Formula» discovered later by Leray.

Fantappiè gives a great deal of attention to the so-called *abeloid* linear analytic functionals, which are those whose projective indicatrix is a *rational* function. Fantappiè investigates them thoroughly, discusses their connection with algebraic functions and with abelian integrals, and – most importantly – he shows how they can be used to solve explicitly the Cauchy problem for arbitrary linear partial differential operators with constant coefficients, and with initial data on an arbitrary (analytic) non-characteristic hypersurface. This application is a several variable analogon of the solution of the 2nd order linear ordinary differential equation by means of the Riccati functional (see (2.1) above). I find the discovery of this relationship between functional analysis, algebraic geometry, and partial differential equations remarkable indeed, and it definitely gives evidence of the important role of the projective indicatrix.

Fantappiè's pioneering work on the solution of the Cauchy problem was carried on in 1956 by Leray [Ler 1], who was able to handle linear partial differential operators with *polynomial* coefficients by a method inspired by Fantappiè's, though casted in a much more abstract algebraic/topological framework which allowed for vast generalisations.

3. THE FIRST FORMULA OF CAUCHY-FANTAPPIÈ

Shortly after the work mentioned above, Leray discovered a new and very general integral representation formula for holomorphic functions of several variables, which shares a basic feature with Fantappiè's work, namely a representation of analytic objects by negative powers of linear functions. His result includes – as a special case – a new simple proof of a formula first discovered by Enzo Martinelli in 1938.

Leray's formula, originally stated for a convex region C in $\mathbb{C}^n$, is of a mixed algebraic/analytic/topological nature. Let X denote the dual space of $\mathbb{C}^n$, with duality written as $\xi \bullet x = \sum \xi_j x_j$ for $\xi \in X$, $x \in \mathbb{C}^n$, and set

$$\omega(x) = d x_1 \wedge \ldots \wedge d x_n, \qquad \omega^*(\xi) = \sum_{k=1}^{n}(-1)^{k-1}\xi_k d\xi_1 \wedge \ldots d\widehat{\xi_k} \ldots \wedge d\xi_n,$$

where $\widehat{}$ indicates that the term underneath is omitted.

THEOREM 1. [Ler 2] *Let the function f be holomorphic on the convex region $C \subset \mathbb{C}^n$ with C^1 boundary bC, and continuous on its closure. Then for all $z \in C$ one has*

$$f(z) = \frac{(n-1)!}{(2\pi i)^n} \int_{\alpha^*} \frac{f(x)}{[\xi \bullet (x-z)]^n} \omega^*(\xi) \wedge \omega(x), \tag{3.1}$$

where α^* is the cycle in $X \times \mathbb{C}^n$ consisting of those (ξ, x) with $x \in bC$ and such that ξ is zero on the tangent vectors to bC at x.

To be precise, α^* is a cycle and $\omega^*(\xi)/[\xi \bullet (x-z)]^n$ is a form on $PX \times \mathbb{C}^n$, where PX is the projective space over X, but by abuse of notation, we use the same symbols to denote their images in $X \times \mathbb{C}^n$. On α^*, ξ is thus a function of $x \in bC$, and the condition on α stated in the Theorem implies

$$\xi \bullet (x - z) \neq 0 \qquad \text{for } x \in bC \text{ and for all } z \in C. \tag{3.2}$$

So, after pulling back the integral under the map $x \to (\xi(x), x)$, one obtains the formula

$$f(z) = \frac{(n-1)!}{(2\pi i)^n} \int_{bC} \frac{f(x)}{[\xi \bullet (x-z)]^n} \omega^*(\xi) \wedge \omega(x). \tag{3.3}$$

In general, ξ will also depend on the point z.

Note that in case the dimension n is 1, the formula (3.3) reduces to the classical Cauchy integral formula. The special choice $\xi = \overline{x} - \overline{z}$ in (3.3), which obviously satisfies (3.2), leads to Martinelli's 1938 formula

$$f(z) = \frac{(n-1)!}{(2\pi i)^n} \int_{bC} \frac{f(x)}{|x-z|^{2n}} \omega^*(\overline{x} - \overline{z}) \wedge \omega(x) \tag{3.4}$$

(see [Mar]). The Martinelli integral formula (3.4), as well as Leray's formula (3.3) hold for arbitrary regions D with C^1 boundary bD, and not just for convex regions C.

The Martinelli integral representation formula (3.4) had been rediscovered independently a couple of years after Martinelli by S. Bochner (see the 1941 Princeton Ph.D. thesis of D.C. May, and Bochner's 1943 paper [Boc]), who derived it from Green's formula. For this reason (3.4) is now commonly referred to as the *Bochner-Martinelli Integral Formula*, although there is no question at all about Martinelli's priority.

Another important special case of (3.3) arises for convex domains C with C^2 boundary. Let r be a C^2 *defining function* for C, i.e., $C = \{x : r(x) < 0\}$ and $dr(x) \neq 0$ for $x \in bC$, and set

$$\xi = (\partial r/\partial x_1, \ldots, \partial r/\partial x_n).$$

Then (3.2) holds, and one obtains *Leray's Cauchy Integral Formula* for convex domains

$$f(z) = \frac{(n-1)!}{(2\pi i)^n} \int_{bC} \frac{f(x)}{\left(\sum_j \partial r/\partial x_j (x_j - z_j)\right)^n} \omega^*(\xi) \wedge (x). \qquad (3.5)$$

Notice that in contrast to (3.4) the kernel in (3.5) is holomorphic in z for $n > 1$ as well as for $n = 1$! (see (4.4) and (4.5) below for versions of (3.5) in different disguise).

It is in Leray's 1959 paper [Ler 3], in which detailed proofs of many of Leray's earlier announcements are given, that the formula (3.1) is – for the first time – called «*la première formule de Cauchy-Fantappiè*». Clearly this is an instance where the names attached to a result do not identify its author(s), but rather are intended to indicate a relationship to the work of others, which may be quite remote indeed, as it surely is with Cauchy in the case at hand. In my judgement Leray's choice of name reflects not only an appropriate historical perspective, but great modesty and generosity as well. In this context it should also be noted that Leray wrote his brief paper [Ler 2] during a stay in Rome in 1956, and that he and Fantappiè had friendly contacts during that period. And it was Fantappiè, as a member of the Accademia Nazionale dei Lincei, who presented Leray's note to the Accademia on May 12, 1956, barely two months before his sudden death on July 28, 1956.

4. THE KOPPELMAN FORMULAS

Before moving on to the next level of generality, let us make some changes, so as to be consistent with current simpler notation, as used for example in [Ran 2]. Recall that ξ is viewed as a function of x, and note that $d\xi_k = \partial\xi_k + \overline{\partial}\xi_k$. Since $\omega(x)$ is of type $(n,0)$ one obtains

$$\omega^*(\xi) \wedge \omega(x) = \left(\sum_{j=1}^n (-1)^{j-1} \xi_j \overline{\partial}\xi_1 \wedge \ldots \widehat{\overline{\partial}\xi_j} \ldots \wedge \overline{\partial}\xi_n\right) \wedge \omega(x) \qquad (4.1)$$

$$= \frac{(-1)^{n(n-1)/2}}{(n-1)!} \xi \wedge (\overline{\partial}\xi)^{n-1},$$

where in the second equation we have identified ξ with the $(1,0)$ form $\xi = \sum_j \xi_j \, dx_j$, which is consistent with the original interpretation of ξ as an element in the dual space X of $\mathbb{C}^n$. The exponent $(n-1)$ over $\overline{\partial}\xi$ indicates the $(n-1)$-fold exterior product of the $(1,1)$ form $\overline{\partial}\xi$. From now on we shall also replace x by ζ in all expressions. ζ will always be the variable of integration, and all differential operators act in ζ unless otherwise indicated. A simple computation then shows that

$$\frac{\xi \wedge (\overline{\partial}\xi)^{n-1}}{[\xi \bullet (\zeta - z)]^n} = W \wedge (\overline{\partial}W)^{n-1}, \quad \text{were we have set} \quad W = \frac{\xi}{[\xi \bullet (\zeta - z)]}$$

Cauchy-Fantappiè Formulas

The $(1,0)$ form $W = \sum_j w_j \, d\zeta_j$ obviously satisfies the relation

$$\sum_1^n w_j(\zeta_j - z_j) = 1. \tag{4.2}$$

A *generating form* for $z \in D$ is any $(1,0)$ form W defined on bD which satisfies (4.2). W will of course depend on z. The Cauchy-Fantappiè Formula (3.3) of Leray now takes the form

$$f(z) = \frac{1}{(2\pi i)^n} \int_{bD} f(\zeta) W \wedge (\bar{\partial} W)^{n-1} \tag{4.3}$$

for any C^1 generating form for z on bD and f holomorphic on D. [*Warning*: in passing from (3.3) to (4.3) the orientation of $\mathbb{C}^n$ has been changed, thus accounting for the loss of the factor $(-1)^{n(n-1)/2}$ in (4.1)!]

In particular, if D is convex with C^2 defining function r, Leray's Cauchy integral formula (3.5) takes the form

$$f(z) = \frac{1}{(2\pi i)^n} \int_{bD} f(\zeta) W^r \wedge (\bar{\partial} W^r)^{n-1}, \tag{4.4}$$

with generating form $W^r = \partial r(\zeta) / [\sum_j (\partial r / \partial \zeta_j)(\zeta_j - z_j)]$, or, equivalently,

$$f(z) = \frac{1}{(2\pi i)^n} \int_{bD} \frac{f(\zeta) \partial r \wedge (\bar{\partial} \partial r)^{n-1}}{[\sum_j (\partial r / \partial \zeta_j)(\zeta_j - z_j)]^n}. \tag{4.5}$$

We are now ready to proceed. In 1967 W. Koppelman rediscovered the integral representation formula (4.3) [Kop 1]. Shortly thereafter Koppelman introduced the appropriate generalization to integral representations of differential forms of type $(0, q)$, $0 \leq q \leq n$, which, in particular, include a version of (4.3) valid for arbitrary C^1 functions; the required correction term of course depends on $\bar{\partial} f$. To state these formulas, we follow Koppelman and introduce the *Cauchy-Fantappiè form* $\Omega_q(W)$ *of order* q generated by the $(1,0)$ form W:

$$\Omega_q(W) = \frac{(-1)^{q(q-1)/2}}{(2\pi i)^n} \binom{n-1}{q} W \wedge (\bar{\partial}_\zeta W)^{n-q-1} \wedge (\bar{\partial}_z W)^q. \tag{4.6}$$

Note that $B = \partial_\zeta \log |\zeta - z|^2$ is a generating form. The Cauchy-Fantappiè form $\Omega_0(B)$ generated by it is the *Bochner-Martinelli kernel*, i.e. the kernel which appears in the Bochner-Martinelli formula (3.4).

Koppelman's first formula, the so-called *Bochner-Martinelli-Koppelman Formula*, is as follows.

THEOREM 2. [Kop 2] *Let D be a domain with piecewise smooth C^1 boundary, and let f be a smooth $(0, q)$ form on the closure of D. Then the following representation holds for all $z \in D$:*

$$f(z) = \int_{bD} f \wedge \Omega_q(B) - \bar{\partial}_z \int_D f \wedge \Omega_{q-1}(B) - \int_D \bar{\partial} f \wedge \Omega_q(B).$$

The analog of Theorem 2 for arbitrary generating forms W is somewhat more complicated. Given W, we introduce the associated *homotopy form*

$$\hat{W}(\zeta, \lambda, z) = \lambda W + (1-\lambda) B \quad \text{for} \quad 0 \leq \lambda \leq 1,$$

and the form $\Omega_q(\hat{W})$ defined by (4.6), with $\hat{W}$ in place of W, and $\bar{\partial}$ replaced by the differential operator $\bar{\partial}_{\zeta,\lambda} = \bar{\partial}_\zeta + d_\lambda$ on the product manifold ($bD \times I$), where I is the unit interval $[0, 1]$. Then define the operator

$$T_q^W : f \to \int_{bD \times I} f \wedge \Omega_{q-1}(\hat{W}) - \int_D f \wedge \Omega_{q-1}(B), \tag{4.7}$$

which sends $(0, q)$ forms on $\overline{D}$ to $(0, q-1)$ forms on D.

Koppelman's second formula can now be stated as follows.

THEOREM 3. [Kop 2] *Let D have smooth C^1 boundary, and let f be a C^1 form on the closure of D. Suppose $W(\zeta, z)$ is a generating form on $bD \times I$ which is C^1 in $\zeta \in bD$. Then for $z \in D$ one has*

$$f(z) = \int_{bD} f \wedge \Omega_q(W) + \bar{\partial}_z(T_q^W(f)) + T_{q+1}^W(\bar{\partial}f). \tag{4.8}$$

Notice that for $q = 0$ the operator T_0 is identically zero (by default), and if f is holomorphic (i.e. $\bar{\partial}f = 0$), (4.8) is – modulo the notational changes introduced at the beginning of this section – exactly the Cauchy-Fantappiè Formula (3.3).

To be precise, Theorem 3 is not stated by Kopelman in the form given here. However, if one carries out the integration in $\lambda \in I$ in the first integral in (4.7) (something which can easily be done explicitly), one ends up with integrals over bD involving rather complicated kernels, resulting in an explicit version of formula (4.8) which follows directly from Propositions 3 and 4, and Theorem 5 in [Kop 2] by a straightforward application of Stokes' Theorem. Thus I feel it is justified to credit Theorem 3 above to Koppelman.

An untimely and tragic death prevented Koppelman from publishing the proofs of his results. Complete proofs for the case $q = 0$ where first published in 1969 by H. Grauert and I. Lieb [GrLi], and independently in 1970 by G.M. Henkin [Hen 2]. The case of arbitrary q was proved in 1970 by I. Lieb [Lie]. For the sake of accuracy let me add that the version of Theorem 3 stated here appears first in [Hen 2] for the case $q = 0$, and in 1973 in [RaSi] for arbitrary q.

Even though the auxiliary parameter λ in the integral in (4.7) may seem unnatural and involve perhaps unnecessary abstractions, it must be emphasized that it is the closed expression involved integration in λ which allows to take advantage most easily of the remarkable formal machinery of Cauchy-Fantappiè forms. The reader should compare the first detailed complete proof of Koppelman's second formula (4.8) in [Lie] to the proof of Theorem 3 given in [Ran 2, pp. 172-175].

5. THE GENERATING FORM OF HENKIN AND RAMIREZ

Theorem 3 is a very powerful general tool. Observe that if the generating form W in Theorem 3 can be chosen holomorphic in $z \in D$, then it immediately follows from (4.6) that $\Omega_q(W) \equiv 0$ for $q \geq 1$, and consequently (4.8) implies

$$f(z) = \bar{\partial}_z(T_q^W(f)) \tag{5.1}$$

whenever f is a $\bar{\partial}$-closed $(0,q)$ form on $\overline{D}$ and $q \geq 1$. Thus (5.1) gives an explicit solution formula for the $\bar{\partial}$-equation on the domain D, i.e. the equation central to multidimensional complex analysis!

A first immediate application occurs for a *convex* domain D. Here we have the generating form $W^r = \partial r(\zeta)/[\sum_j (\partial r/\partial \zeta_j)(\zeta_j - z_j)]$ (see (4.4)), which is indeed holomorphic in z. So (5.1), with the generating form W^r, gives what is perhaps the most elementary proof of the vanishing theorem

$$H^q(\overline{D}, \mathcal{O}) = 0 \quad \text{for } q \geq 1 \text{ for a convex domain } D. \tag{5.2}$$

Of course convex domains are far too special. Much more significant would be results for *strictly pseudoconvex* domains (their definition will be recalled at the beginning of the next section), since arbitrary pseudoconvex domains can be written as an increasing union of such domains.

The following result therefore plays – historically – a central role in the theory of Cauchy-Fantappiè Formulas. In hindsight, this is the result which led to the widespread interest in integral representations in modern multidimensional complex analysis during the past two decades.

THEOREM 4. *Let D be a bounded strictly pseudoconvex domain with smooth boundary. Then there exists a generating form*

$$W_D^{HR}(\zeta, z) \text{ defined on } bD \times \overline{D} - \{(\zeta, \zeta) : \zeta \in bD\}$$

which holomorphic in z and differentiable in ζ.

This result was proved first in 1969 by Henkin [Hen 1] in Moscow and, independently, by E. Ramirez [Ram] in Göttingen. The associated Cauchy-Fantappiè form $\Omega_0(W^{HR})$ is commonly known as the *Henkin-Ramirez kernel*. The original proofs made use of deep classical results of the global theory of several complex variables, including, in particular, the vanishing theorem stated in (5.2), but for the case of strictly pseudoconvex domain D. So, the proof of the vanishing theorem obtained by combining Theorem 3 and 4, as we had done for the case of convex domains, is indeed circular. «Technology» has improved since then, and today simpler proofs of Theorem 4, and of the resulting vanishing theorem are known, which make direct use of the theory of Cauchy-Fantappiè forms without using the classical several complex variable results (see [Ran 2] or [HeLe]).

Theorem 4 has been very influential. Of particular interest is the fact that the singularities of W^{HR}, which occur only for $z = \zeta \in bD$, are known explicitly. Thus the kernel $\Omega_0(W^{HR})$, as well as the operators T_q defined by W^{HR} (see (4.7)), can be estimated precisely in various norms, leading to many important applications and solutions of long outstanding problems far too numerous to be listed here. Instead, in the next section, we shall discuss another simple but far reaching application of Theorem 3, which is independent of Theorem 4.

6. THE SOLUTION OF THE LEVI PROBLEM

In 1910 E.E. Levi discovered that domains of holomorphy in dimension two must satisfy a local «complex convexity» condition, which has since become known as *pseudoconvexity*. If

$D \subset \mathbb{C}^n$ has C^2 boundary, and given an arbitrary C^2 defining function r near bD, this condition is characterized by

$$L_\zeta(r;t) = \sum_{j,k=1}^n \frac{\partial^2 r}{\partial \zeta_j \overline{\partial \zeta_k}}(\zeta) t_j \bar{t}_k \geq 0 \quad \text{for} \quad \zeta \in bD \quad \text{and} \tag{6.1}$$

$$\text{for all} \quad t \in \mathbb{C}^n \quad \text{with} \quad \sum_j (\partial r/\partial \zeta_j)(\zeta) t_j = 0. \tag{6.2}$$

If in (6.1) the *Levi form* $L_\zeta(r;t)$ is *positive* for all $\zeta \in bD$ and $t \neq 0$ which satisfy (6.2), the domain D is said to be *strictly pseudoconvex*.

Levi recognized in 1912 that if D is strictly pseudoconvex, then for every $\zeta \in bD$ there is a neighborhood U of ζ such that $U \cap D$ is a domain of holomorphy, but he could not answer the much deeper question whether a strictly pseudoconvex domain would be (globally) a domain of holomorphy. In essence, this question, which became known as the *Levi Problem*, asks whether domains of holomorphy can be characterized by *local* properties of the boundary. The Levi Problem remained unsolved for a long time. It was first solved affirmatively by K. Oka in 1942 in dimension two, and in 1953 by Oka, F. Norguet, and H. Bremermann, independently, in arbitrary dimension.

We shall now use Cauchy-Fantappiè forms to give an elementary solution of the Levi Problem.

It is elementary to show that if D is strictly pseudoconvex, one can choose the defining function to be *strictly plurisubharmonic*, i.e. the Levi form $L_\zeta(r;t)$ is positive definite for *all* t of $\mathbb{C}^n$. Assuming r to be so chosen, it then follows easily that the *Levi polynomial* of r defined by

$$F^{(r)}(\zeta, z) = \sum_{j=1}^n \frac{\partial r}{\partial \zeta_j}(\zeta)(\zeta_j - z_j) - \frac{1}{2} \sum_{j,k=1}^n \frac{\partial^2 r}{\partial \zeta_j \partial \zeta_k}(\zeta_j - z_j)(\zeta_k - z_k)$$

satisfies the estimate

$$2 \operatorname{Re} F^{(r)}(\zeta, z) \geq r(\zeta) - r(z) + c|\zeta - z|^2 \tag{6.3}$$

for ζ in a neighborhood of bD and $|\zeta - z| < \varepsilon$, for suitable (small) positive constants c and ε. Now choose a function $\chi(\zeta, z)$ which satisfies $0 \leq \chi \leq 1$, $\chi \equiv 1$ for $|\zeta - z| < \varepsilon/2$, and $\chi \equiv 0$ for $|\zeta - z| > 3\varepsilon/4$ and define

$$P_j(\zeta, z) = \chi \left(\frac{\partial r}{\partial \zeta_j}(\zeta) - \frac{1}{2} \sum_{k=1}^n \frac{\partial^2 r}{\partial \zeta_j \partial \zeta_k}(\zeta_k - z_k) \right) + (1 - \chi)(\overline{\zeta}_j - \overline{z}_j)$$

for $j = 1, \ldots, n$. It follows that if we set

$$\Phi(\zeta, z) = \sum_{j=1}^n P_j(\zeta, z)(\zeta_j - z_j),$$

then

$$L^r(\zeta, z) = \sum_{j=1}^n (P_j/\Phi) \, d\zeta_j \tag{6.4}$$

is a generating form on $bD \times D$ which is holomorphic in z for $|\zeta - z| < \varepsilon/2$. Therefore, the Cauchy-Fantappiè form $\Omega_1(L^\tau)$ generated by (6.4) vanishes identically on the set $\{(\zeta, z) : |\zeta - z| < \varepsilon/2\}$. Moreover, the estimate (6.3) and the construction of L^τ readily imply that

$$\Omega_1(L^\tau) \text{ extends smoothly to } bD \times D_\delta, \tag{6.5}$$

where $\delta > 0$ is sufficiently small, so that $D_\delta = \{z : r(z) < \delta\}$ is a strictly pseudoconvex neighborhood of the closure of D.

We now introduce the Banach space

$$Z_1 = Z_1(D) = \{f \in C^1_{0,1}(\overline{D}) : \overline{\partial} f = 0\},$$

with norm being the sum of the $C^1(\overline{D})$ norm of the coefficients of a form f. From the Koppelman formula (4.8) and from (6.5) it follows that the operator

$$E_1 : f \to \int_{bD} f(\zeta) \wedge \Omega(L^\tau)$$

defines a *compact* linear operator $E_1 : Z_1 \to Z_1$. By standard compact operator theory it follows that on the Banach space Z_1 the operator

$$\overline{\partial} T_1^{L^\tau} = \text{Identity} - E_1 \quad \text{(by (4.8))}$$

has closed image in Z_1 of *finite* codimension. By a classical argument going back to Grauert, this fact implies the solution of the Levi Problem, as follows.

What needs to be shown is that for an arbitrary fixed $\zeta \in bD$ there is a holomorphic function h on D which is singular at ζ. clearly the function $u(z) = 1/\Phi(\zeta, z)$ is smooth on $\overline{D} - \{\zeta\}$ with $u(z) \to \infty$ as $z \to \zeta$, and u is holomorphic near ζ; thus $\overline{\partial}(u^k)$ extends to a form in $Z_1(D_\delta)$ for δ sufficiently small, and for $k = 1, 2, \ldots$. By the finiteness statement above, applied to D_δ instead of the domain D itself, there are constants $c_1, \ldots, c_m, c_m \neq 0$, and a form $f \in Z_1$, such that

$$\sum_{k=1}^m c_k \overline{\partial} u^k = (\overline{\partial} T_1^{L^\tau})(f). \tag{6.6}$$

If we set $v = T_1^{L^\tau} f$, then $v \in C^1(D_\delta)$ – thus v is *bounded* on D – and it follows by (6.6) that $h = (\sum_k c_k u^k) - v$ is the required holomorphic function on D, since $h(z) \to \infty$ as $z \to \zeta$.

This solution of the Levi Problem was obtained in [Ran 1], where these techniques were carried farther to actually prove the solvability of the $\overline{\partial}$-equation on strictly pseudoconvex domains directly by integral operators, rather than the solvability up to *finitely* many obstructions discussed here. Solutions of the Levi Problem by similar methods were found, independently, by Harvey and Polking [HaPo] and by Henkin (see [HeLe]). The generating form L^τ had been introduced by Kerzman and Stein [KeSt], who however used it only in the representation of functions, i.e. for Cauchy-Fantappiè forms of order $q = 0$.

7. A GENERATING FORM AND ESTIMATES FOR $\bar{\partial}$ ON DOMAINS OF FINITE TYPE IN DIMENSION TWO

Until now, Cauchy-Fantappiè forms have mainly been used for convex and for strictly pseudoconvex domains. Already in 1972 J.J. Kohn [Koh] introduced a condition weaker than strict pseudoconvexity for domains in $\mathbb{C}^2$, which is called *finite type*, and he proved *subelliptic L^2 estimates* for the $\bar{\partial}$-Neumann problem on such domains. Loosely stated, a pseudoconvex domain with smooth boundary is of finite type if the Levi form, which is just a scalar function on bD when $D \subset \mathbb{C}^2$, vanishes at most to finite order in the complex tangential directions. In recent years a great deal of progress has been made in obtaining Hölder estimates for $\bar{\partial}$, etc., on domains of finite type by using techniques which are quite different from those discussed here (see the recent paper by C. Fefferman and J.J. Kohn [FeKo], which includes a broad overvew of this area).

Most recently I have been able to construct a generating form $W(\zeta, z)$ on $bD \times D$ on pseudoconvex domains of finite type in $\mathbb{C}^2$ which is holomorphic in z, and which satisfies the appropriate estimates, so that when combined with the methods of section 4, one obtains an integral solution operator for $\bar{\partial}$ which satisfies the right Hölder estimates (almost).

In order to state the results precisely, we need to introduce two fundamental invariants for domains of finite type, which are due to Nagel, Stein and Wainger (see [NSW]). These are the *nonisotropic distance function* $\rho(\zeta, z)$ defined on $\overline{D} \cap U \times \overline{D} \cap U$ for a suitable neighborhood U of bD (initially ρ was only defined on $bD \times bD$ – see [NRSW] for the extension), and the invariant $\Lambda(\zeta, \delta)$, defined for $\zeta \in bD$ and $\delta > 0$, which, in essence, measures the maximal «complex normal» radius of a polydisc $\subset D$ centered at a point in D on the interior normal at ζ in dependence of the «complex tangential» radius δ. If D is of finite type m at ζ, then there are constants C_1 and C_2 such that

$$C_1 \delta^m \leq \Lambda(\zeta, \delta) \leq C_2 \delta^2 \quad \text{for} \quad 0 < \delta \leq 1. \tag{7.1}$$

For the precise definitions we refer to [NRSW], and to [Cat] or [FeKo] for closely related work.

THEOREM 5. [Ran 3] *Let $D \subset \mathbb{C}^2$ be a smoothly bounded pseudoconvex domain of finite type. Then for any $\alpha > 0$ there are $\eta(\alpha) > 0$ with $\eta(\alpha) \to 0$ as $\alpha \to 0$, a constant C_α, and a generating form $W^\alpha(\zeta, z)$ defined on $bD \times D$ which satisfy the following properties:*

i) *For fixed $\zeta \in bD$, $W^\alpha(\zeta, z)$ is holomorphic in z on $\overline{D} - \{\zeta\}$;*

ii) *if $z \in D \cap U$ and $\delta = \rho(z, \zeta)$, then*

$$|W^\alpha(\zeta, z)| \leq C_\alpha \left[\frac{\delta^\alpha}{[\Lambda(\zeta, \delta)]^{1+\eta}} + \frac{|\zeta - z|^{1+\alpha}}{\delta[\Lambda(\zeta, \delta)]^{1+\eta}} \right]. \tag{7.2}$$

In brief outline, the proof involves the following ingredients. First, given $\zeta \in bD$, one constructs a domain $D_\zeta \supset \overline{D} - \{\zeta\}$ with the following properties: D_ζ is *pseudoconvex*, and

$$P(z, \gamma \delta) \subset D_\zeta \quad \text{for } z \text{ and } \delta \text{ as in ii) above,}$$

where $P(z, \gamma \delta)$ is the biholomorphic image of a polydisc centered at z of comples tangential radius $\gamma \delta$ and complex normal radius $\Lambda(\zeta, \gamma \delta)$ and γ is a constant independent of z and

ζ. The domain D_ζ does not have smooth boundary; its intersection with the complex normal looks like the complement of a wedge with vertex at ζ. Loosely speaking, D_ζ is an exterior analogon of the *admissible approach regions* in [NSW]. The construction of D_ζ is, essentially, a modification of a closely related result of Catlin [Cat], and it makes substantial use of Catlin's techniques. In the next step one uses the results of H. Skoda [Sko] to solve the division problem

$$\sum_{j=1}^{2} w_j^\alpha(\zeta,z)(\zeta_j - z_j) = 1 \quad \text{on} \quad D_\zeta, \tag{7.3}$$

with functions w_j^α holomorphic in $z \in D_\zeta$ which satisfy optimal weighted L^2 estimates, with weights depending on the choice of α and on ζ. The desired generating form is then $W^\alpha = \sum_j w_j^\alpha \, d\zeta_j$. Finally, in order to obtain from the weighted L^2 estimates the pointwise estimate (7.2) at $z \in D$, one uses a modification of Cauchy estimates on the «polydisc» $P(z, \gamma\delta)$.

In order to utilize this generating form W^α to construct integral solution operators for $\bar\partial$ as in section 4, one needs that W^α depends differentiably on $\zeta \in bD$. Apparently this is quite a delicate question, and the answer is still unknown. At this time I can only state a

CONJECTURE. *The generating form W^α given by Theorem 5 depends smoothly on ζ.*

In case of dimension two much less is in fact required than differentiability in ζ, and – modulo the weaker conjecture – one obtains the following result.

THEOREM 6. [Ran 3] *Let D be a smoothly bounded pseudoconvex domain in $\mathbb{C}^2$ of finite type m. Suppose that for every $\alpha > 0$ and $z \in D$ the generating form $W^\alpha(\zeta,z)$ given by Theorem 5 is a measurable function of ζ with respect to surface measure on bD. Then, for every $\varepsilon > 0$ there are $\alpha(\varepsilon) > 0$ with $\alpha \to 0$ as $\varepsilon \to 0$ and a constant A_ε, such that the solution operator T^α for $\bar\partial$ defined by formula (4.7) by using the generating form W^α satisfies the estimate*

$$|T^\alpha f(z) - T^\alpha f(z')| \leq A_\varepsilon |f|_{L^\infty(D)} |z-z'|^{\frac{1}{m}-\varepsilon} \tag{7.4}$$

for all $z, z' \in D$.

The existence of solutions to the equation $\bar\partial u = f$ on pseudoconvex domains of finite type m in $\mathbb{C}^2$ which satisfy the Hölder estimate (7.4) was proved first in 1987 by Fefferman and Kohn [FeKo] by using *microlocalization*. In fact, they obtain the optimal estimate (7.4) with $\varepsilon = 0$! It appears unlikely that the case $\varepsilon = 0$ can be obtained by the method discussed here. The reasons that $\varepsilon > 0$ is needed here is that in Theorem 5 the constants α and η are positive, which in turn is due to the fact that in Skoda's estimates for the solution of (7.3) it is required that $\alpha > 0$, and there does not seem to be any way to avoid that.

8. CONCLUDING REMARKS

We have come a long way from Luigi Fantappiè's theory of analytic functionals. Cauchy-Fantappiè formulas and forms are a lasting reminder of his work and vision, and I hope that this lecture will inspire some of today's mathematicians to browse through and perhaps study some of Fantappiè's original papers. The record shows that Fantappiè's work on analytic functionals and, in particular, the use of the symbolic calculus based on the indicatrix in the solution of partial differential equations, are a *complex* version, if not a precursor, of very general fundamental principles in (real) analysis, which are today used very widely – just to mention *distributions*, as *the* generalization of ordinary functions, and *pseudodifferential operators*, whose generalizations and amazing applications seem to find no end. Why Fantappiè's work has largely remained of the side lines of these developments, and why it did not bear fruits, is an intriguing question whose answer I must leave to others to find. But in view of today's great interest in non-linear analysis, for example in Partial Differential Equations and in Complex Dynamical Systems, going back to basic principles may not be all wrong, and perhaps somewhere in Fantappiè's work there is hidden the germ of an idea which, in the hands of the right master, might lead to new and far-reaching developments.

ACKNOWLEDGEMENTS

The author thanks J. Leray for kindly supplying relevant historical information. The work discussed in section 7 is supported by a grant from the National Science Foundation.

REFERENCES

[Boc]   S. Bochner, Analytic and meromorphic continuation by means of Green's formula, *Ann. of Math.* 44 (1944), 652-673.

[Cat]   D. Catlin, Estimates of invariant metrix on pseudoconve domains of dimension two, *Math. Zeit.* (to appear).

[Fan]   L. Fantappiè, *Opere Scelte*, Vol. 1 & 2, Unione Mat. Italiana 1973.

[FeKo] C. Fefferman and J. Kohn, Hölder estimates on domains of complex dimension two and on three dimensional CR manifolds, *Adv. in Math.* 69 (1988), 223-303.

[GrLi] H. Grauert and I. Lieb, Das Ramirezsche Integral und die Lösung der Gleichung $\bar{\partial} f = \alpha$ im Bereich der beschränkten Formen, *Rice Univ. Studies* 56 (1970), 29-50.

[Ha Po] R. Harvey and J. Polking, Fundamental solutions in complex analysis I: The Cauchy-Riemann operator, *Duke Math. J.* 46 (1979), 253-300.

[Hen]   G.M. Henkin,
  1. Integral representations of functions holomorphic in strictly pseudoconvex domains and some applications, *Math USSR Sb.* 7 (1969), 597-616.
  2. Integral representations of functions in strictly pseudoconvex domains and applications to the $\bar{\partial}$-problem, *Math. USSR Sb.* 11 (1970), 273-281.

[HeLe] G.M. Henkin and J. Leiterer, *Theory of Functions on Complex Manifolds*, Birkhäuser, Boston, Mass., 1984.

[KeSt] N. Kerzman and E.M. Stein, The Szegö kernel in terms of Cauchy-Fantappiè kernels, *Duke Math. J.* 45 (1978), 197-224.

[Koh]   J.J. Kohn, Boundary behavior of $\bar{\partial}$ on weakly pseudoconvex manifolds of dimension two, *J. Differential Geometry* 6 (1972), 523-542.

[Kop] W. Koppelman,
1. The Cauchy integral for functions of several complex variables, *Bull. Amer. Math. Soc.* 73 (1967), 373-377.
2. The Cauchy integral for differential forms, *Bull. Amer. Math. Soc.* 73 (1967), 554-556.

[Ler] J. Leray,
1. Le problème de Cauchy pour une équation linéaire à coefficients polynomiaux, *C.R. Acad. Sc. Paris* 242 (1956), 953-959.
2. Fonction de variable complexe: sa réprésentation comme somme de puissance négatives de fonctions linéaires, *Rend. Accad. Naz. Lincei*, ser. 8, 20 (1956), 589-590.
3. Le calcul différentiel et intégral sur une variété analytique complexe: Problème de Cauchy III, *Bull. Soc. Math. France* 87 (1959), 81-180.

[Lie] I. Lieb, Die Cauchy-Riemannschen Differentialgleichungen auf streng pseudokonvexen Gebieten, *Math. Ann.* 190 (1970), 6-44.

[Mar] E. Martinelli, Alcuni teoremi integrali per le funzioni analitiche di più variabili complesse, *Mem. della R. Accad. d'Italia* 9 (1938), 269-283.

[NRSW] A. Nagel, J.P. Rosay, E.M. Stein and S. Wainger, Estimates for the Bergman and Szegö kernels in $\mathbb{C}^2$; *Ann. of Math.* 129 (1989), 113-149.

[NSW] A. Nagel, E.M. Stein and S. Wainger, Boundary behavior of functions holomorphic in domains of finite type, *Proc. Nat. Acad. Sc. USA* 78 (1981), 6596-99.

[Ram] E. Ramirez, Ein Divisionsproblem und Randintegraldarstellungen in der komplexen Analysis, *Math. Ann.* 184 (1970), 172-187.

[Ran] R.M. Range,
1. An elementary integral solution operator for the Cauchy-Riemann equations on pseudoconvex domains in $\mathbb{C}^n$. *Trans. Amer. Math. Soc.* 274 (1982), 809-816.
2. *Holomorphic Functions and Integral Representations in Several Complex Variables*, Springer-Verlag, New York, N.Y., 1986.
3. A Cauchy-Fantappiè form and Hölder estimates for $\overline{\partial}$ on pseudoconvex domains of finite type in dimension two. (in preparation).

[RaSi] R.M. Range and Y.-T. Siu, Uniform estimates for the $\overline{\partial}$-equation on domains with piecewise smooth strictly pseudoconvex boundaries, *Math. Ann.* 206 (1973), 325-354.

[Sko] H. Skoda, Applications des techniques L^2 à la théorie des idéaux d'une algebre de fonctions holomorphes avec poids, *Ann. Sc. E.N.S.*, 4e ser., 5 (1972), 545-579.

Maria Gaetana Agnesi

J.H. SAMPSON The Johns Hopkins University, Baltimore, Maryland 21218, USA

In 1750 Maria Gaetana Agnesi was awarded the chair of Mathematics and Analytical Geometry at the University of Bologna, by Pope Benedetto XIV, with the full approbation of the Senate of Bologna, and it is thus a fitting occasion to recall that distinguished and gentle scholar whose remembrance has so faded with the passing years[1] that her name nowdays only evokes the cubic curve known as the Witch of Agnesi:

$$xxy + aa(y - a) = 0.$$

Agnesi never actually took up the duties of her chair, which she regarded as chiefly honourary, despite the urging of Beccari, president of the Accademia delle Scienze di Bologna, and of several other influential scholars. The surviving history of the Maria Gaetana Agnesi and her family is meager indeed[2]. Many spurious assertions have crept into it, amongst which one stating that her father was a professor in the University of Bologna. Nothing further from the truth. Nonetheless it was he who was responsible for starting his two favorite daughters, Maria Gaetana and Maria Teresa, along the road to early fame, Maria Gaetana as an erudite, and her slightly younger sister as a distinguished harpsichordist and composer. The family counted in fact twenty-one offspring (by three wives), several not surviving infancy and with little trace of the others.

---

[1] In fact the curve was known much earlier to Fermat and was named la Versiera by Grandi. Agnesi never made any special allusion to the curve, it being only one of dozens of often complex figures prepared by her and then engraved after her meticulously made drawings for her great book, *Istituzioni Analitiche ad uso della Gioventù Italiana*.
[2] From a lecture given on the sixth of June, 1988, at the Università degli Studi di Bologna on the occasion of the nine hundredth anniversary of its founding.

Not much has come down to us even about the two famous sisters. In her abundant writings and letters, conserved in the Biblioteca Ambrosiana in Milan, Maria Agnesi never sheds any light whatever upon what must have been a very complex and agitated household and family. Apart from the biography mentioned below and scattered brief epistolary references there survives very little but the testimony of the French nobleman and amateur historian Charles de Brosses, recounting, in a lengthy letter, an evening at a reception given by Don Pietro Agnesi. It is not even possible to establish a satisfactory concordance of noteworthy dates in her life. The chief sources which remain are the book *Elogio Storico di Maria Gaetana Agnesi*, by the close family friend Antonio Frisi, younger brother of the distinguished mathematician Paolo Frisi, and the cited sources in the Ambrosiana. Elsewhere (Maria Gaetana Agnesi, *Seminario di Geometria di Bologna*, (1988-1989) we have given a much more detailed account of Agnesi"s life, along with references and extracts.

The history of the Agnesi family, insofar as it concerns Maria Gaetana, was written in Milan. It is not even known whether or not she ever set foot in Bologna. The combined wealth of her father, Don Pietro, and his first wife, Anna Brivio, enabled the young family to purchase one of the finest houses in Milan, in via Pantano, along the Naviglio. The house offered superb resources for high-level entertainment, and the ultra-ambitious Don Pietro did not lose time in turning them to profit. His two eldest daughters were to become the starts of a long series of magnificent soirées, in which Don Pietro hoped to established Casa Agnesi as a worthy rival to the other highly fashionable intellectual salons of Milan. There is no doubt that he was in large measure successful. He was also able to acquire, by buying it, a small noble title, not hereditary, however.

Maria Gaetana was born in 1718, according to uncertain tradition in the house in Via Pantano. By the age of five she had a command of French, learnt from her nurse, and was fluent enough to impress a group of guests with a recitation. Thus was the frail vessel of her life sent forth on the swirling waters of fame. Her superior endowments were clearly visible to all who knew her.

She soon took to hanging around in the domestic classroom where her elder brother, Gaetano (a very popular name at the time), was being tutored in Latin. It was quickly apparent that she had learned the lessons much better than her struggling brother. And it was then that Don Pietro, in a very remarkable combination of perspicacity, generosity, and greed for renown, arranged for Maria Gaetana to study Latin methodically, and subsequently a variety of other subjects as well. He must also have embarked Maria Teresa on a course of musical studies, but no records of that have survived, except her fame as a performer. For its time, Don Pietro's decision was truly extraordinary. Even today it is not rare for girls in upper middle-class families to get nothing beyond a minimal high-school education. But for Don Pietro there was a Quid prop Quo which was to weigh heavily upon the young beneficiaries.

In 1727, at the age of nine years, Maria Gaetana received from her tutor, Father Gemelli, the difficult exercise of translating into Latin a lengtly discourse written by him, called the *Oratio*, propounding strong arguments favouring the education of women and girls. His young charge presented her translation verbatim in a reception in Casa Agnesi, speaking a Latin so pure and clear that she won instant recognition for her superior powers.

Maria Agnesi went on to the study of Greek, German and Hebrew, in all of which she excelled; then followed intensive studies of natural philosophy and mathematics, all of this taught by what appear to have been outstandingly mediocre pedants. But Maria Gaetana, endowed with a perfect memory and a profound sense of duty towards her father, was able to absorb vast amounts of very indigestible and stultifying stuff. Not until her lessons from Count Belloni and Father Rampinello did she have any sort of contact with modern topics or older ones in modern form. It is to be noted that all of her many tutors, with the exception of Count Belloni, were ecclesiastics. Constantly exposed to a sanctified atmosphere, Maria Gaetana grew up profoundly imbued with a sense of duty towards her conception of God, as well as (already noted) towards her father.

In Casa Agnesi all of her learning was immediately converted into marketable form. Don Pietro sponsored «Academies», i.e. gatherings of usually very distinguished guests in which his daughter Maria Gaetana displayed her talents and augmented his fame. Little is known about the frequency of such soirées or how they were conducted. But they were endoubtely lavish affairs, and we can assume that the one attended by Charles de Brosses must have been typical; Maria Agnesi and Maria Teresa were seated amidst the guests, who were invited to pose questions to Maria Gaetana on a very wide range of topics, such for example as the tides of the oceans. The questions were posed in Latin; Agnesi replied usually in the language of the guest, many of whom were foreign dignitaries. There is inescapable evidence of a considerable repetitiveness in those exchanges (the tides come in quite often), and it is hard to believe that the Academies really brought forth any original thought. Nevertheless, the nature of the gatherings supposes a very high degree of preparedness of the star, Maria Gaetana, and a very great facility with language and with the concepts that were then abroad. For example, she clearly had a good understanding of many of Newton's contributions (he was still alive during her early life). At these gatherings Maria Teresa gave concerts of contemporary pieces, such as those of Rameau, as well as her own compositions, and she also sang with a sweet voice. It is not hard to imagine the stress endured by those young people almost constantly on stage in their own house for more than twenty years.

The visit of De Brosses occured in 1739, the same year as the much awaited and minutely prepared visit of Maria Teresa d'Austria. The fame of the sisters Agnesi had by then gone well beyond the boundaries of Italy. But the strain of their filial duties was gathering, and Maria Gaetana, as De Brosses wrote, had expressed a desire to retire to a convent. It is not at all clear that she really had that intention; but when she expressed it to her father it obtained immediate results. Despite his bitter disappointment he had the intelligence to accede to her three conditions, which were to dress in a simple manner, to go to church whenever she desired to do so, and to leave forever the balls, the theatre and other profane divertissments. We can only conclude from this that she had endured a surfeit of balls and theatre – although no traces remain of such a mondaine existence – in addition of course to the frequent appearances as main attraction in Via Pantano.

Nevertheless receptions in Via Pantano continued until 1752, even if in subdued form. Nothing is available to inform us of the new format. But much more important was the fact that with her increased liberty Maria Gaetana was able to return to her partially interrupted studies. She maintained a very active scholarly correspondance, continuing over many years. In 1740 the famous Father Ramiro Rampinelli came to Casa Agnesi, and under his tutelage Maria Gaetana made very rapid progress in mathematics. It was then that she began her last scientific project, her *Istituzioni*, an undertaking which was to last for a decade.

Agnesi had published some minor works, including the *Oratio*, printed by Richini of Milan. As her *Istituzioni* neared completion Richini installed his presses in her new quarters in Via Pantano, and she undertook not only the execution of the figures but the training as well of Richini's compositors in the art of mathematical typography. The notation of the day differed in minor but typographically important aspects from more modern notation. For example, composite expression such as $x + 3y$, or $(x + 3y)$, were usually rendered as

$$\overparen{x + 3y},$$

which necessitated engraved symbols. Of course all figures had to be engraved.

The *Istituzioni* were finally published in 1748 in two large quarto volumes, each comprising around five hundred pages, elaborately dedicated to the Empress Maria Teresa. As the young authoress explains, the work did not begin as a treatise but rather as notes for herself and for the instruction of her younger brothers. Thus the work came to be written in Italian. As it progressed, however, it soon assumed the proportions of an encyclopedic treatise on the mathematics of the period. The first volume was devoted to algebra and analytic geometry, the second to the infinitesimal calculus and differential equations. It is remarkable that there are no applications whatever to mechanics, despite the fact that the authoress was a true Newtonian.

Agnesi sent copies of her work to several important people, including the Empress Maria Teresa, Pope Benedetto, members of the French Academy of Sciences, etc. The *Istituzioni* very quickly enlarged her already considerable fame, and praise came from many quarters. The Empress did not reply directly to the warm dedication, but the she did send the authoress, throught the intermediary of the Governer Pallavicini, a precious jewelled box of crystal and a ring of diamonds. Agnesi later sold the box to support her charitable works. The pope, as we have seen, rewarded her effort by her appointment to the University of Bologna.

It is difficult to assess the importance of the *Istituzioni*, which was Agnesi's only published scientific work. It had no originality, but it was widely praised for its clarity and organization. It was translated into English, and the French Academy of Sciences caused the second volume to be translated into French. Neverthless scientific and historical references to it are exceedingly rare, and it seems to have been as quickly forgotten as was its author. Euler's *Introductio in analysin infinitorum* was published in the same 1748, and soon his famous trilogy would dominate the scene. But the first volume, like Agnesi's, is devoted to algebra; the second did not appear until 1755, the third in 1768. For a time Agnesi's book was by far the best available.

The last of Don Pietro's Academies took place in 1752, attended by many notables, including the Governator Count Gian Luca Pallavicini. The next morning Don Pietro went to the ducal palace to thank the Governor for his august presence. The governor however at once turned to the rumors abroad accusing Don Pietro of keeping his daughters practically sequestered, failing in his paternal duty of getting them married. Don Pietro left in a burst of anger which surely would have resulted in a duel, had the severe strain of the event not led quickly to his death, a «fortunoso accidente», says the Canon Frisi.

Thus liberated, Maria Teresa, already thirty-two years of age, lost no time in getting married, not even respecting the six-month period of mourning prescribed by the Empress, finding just three sufficient.

After publication of the *Istituzioni* Maria Gaetana maintained desultory scientific contacts for some time. In 1750 Paolo Frisi asked her to examine his first scientific article, *De Figura et Magnetudine Telluris*, probably with the hope of getting her help in having it published. She refused to comment on the work, pleading a series of migraine headaches as the reason. However, a letter to Antonio Frisi shows that she felt that some parts of the paper were in conflict with the narrow orthodoxy which she seems to have learned from Father Rampinelli. Religion had an important place in her life since her recovery from a severe illness in childhood. Upon the death of her father the weight of paternal authority gave way to that of religious authority, and Agnesi turned to a life of complete self-abnegation, devoting her entire soul and energy to works of charity, mostly in the care of the sick and aged. In 1771 a hospice for the aged called the Luogo Pio Trivulzi was founded by Prince Trivulzi, and she was asked to become its directrice. In 1783 she herself moved there and remained until her death in 1799. From Frisi's moving account, Agnesi's later years must have been passed very sadly under the crushing burden of failing health and the infirmities of age.

Acknowledgement. I would like here to express my great gratitude to Professor Salvatore Coen, to the Dipartimento de Matematica of the University of Bologna, and to the Consiglio Nazionale delle Ricerche.

The original article by the same title, containing details, references and extracts, will appear in the *SEMINARI di GEOMETRIA* of the Università degli Studi di Bologna.

An Extension to Fantappiè's Theory of Analytic Functionals

DANIELE C. STRUPPA[*] Dept. of Mathematics, Univ. of Calabria, 87036 Arcavacata di Rende (CS), Italy and Dept. of Math. Sciences, George Mason Univ., Fairfax, VA 22030, U.S.A.

Dedicated to
the University of Bologna,
on «her» 900-th birthday.

0. INTRODUCTION

In this paper we discuss the links between some new results in the theory of infinite order differential operators on hyperfunctions [18], and some very classical theories of Fantappiè [8], [9]. More specifically, while it is well known the link between Fantappiè's analytic functionals and hyperfunction theory (see e.g. [35], for a rather large section on this topic), it is not so widespread the knowledge that the same algebraic and topological techniques which have led to hyperfunctions, can be used to provide some interesting extensions of Fantappiè's theory.

As it is well known, an analytic functional according to Fantappiè is (see sections 1 and 2 for the precise details) an element of the «dual»[1] space of the space of functions which are holomorphic is some region G of $\mathbb{CP}^1$ (or, more generally, of $\mathbb{CP}^n$) and the most striking result which Fantappiè obtained was a «duality» theorem [9] which associates, in a bijective correspondence, analytic functionals to holomorphic functions in $\mathbb{CP}^1 \backslash G$, vanishing at infinity (the so called *indicatrix* of a functional).

Work partially supported by the Matsumae International Foundation.
[*] The author is a member of G.N.S.A.G.A. of the Italian C.N.R.
[1] We must use quotation marks, as a duality in modern terms was never actually given by Fantappiè.

As far as we know, the first author to point out the possibility of substituting analytic functions with C^∞ solutions of systems of differential equations more general than the Cauchy-Riemann was Fichera who, in his beautiful [11], makes this remarkably precise suggestion, with the aim of constructing a similar theory, in which at least the duality theorem could be preserved. As Fichera pointed out recently [12], a partial progress was obtained, at the times, by M. Carafa, who replaced $\overline{\partial}$ by Δ_2 (his results are unfortunately unpublished, and the untimely death of their author prevents us from being more specific). We would like to point out that Fichera's brilliant suggestion is even more so if one considers that it was made in 1957, when thinking of holomorphic functions as (simply) C^∞ solutions of an overdetermined system of linear differential equations was not so much in fashion: this point of view has then become relatively commonplace, starting with the fundamental works of Ehrenpreis which culminated in his monograph [7]. But the italian school of analytic functionals never obtained any significant result on this subject (partially in view of the resistance of a part of the italian academy towards Fantappiè's approach), and things seemed to fade out. On the other hand, in a totally independent development, Sato created, in the years 1958-59, his magnificent theory of hyperfunctions [31] which (in a sense which can be made clear [37]) provides a rigorous setting for Fantappiè's analytic functionals, even though Sato never explicitly mentions his work (the motivations behind Sato's approach were of a completely different kind, namely concerning dispersion relations in quantum field theory [17]). The mathematician who first understood this link is Martineau who, in February 1961, explained the first «obscure» Sato's papers[2] to the european mathematicians in a Séminaire Bourbaki [22], where an abstract form of Fantappiè's duality theorem is given. Martineau's keen interest in Fantappiè's work is also clearly stated in his [27] from which we quote the relevant passages «... la théorie générale des espaces vectoriels topologiques doit son plein succès à la théorie des distributions de Schwartz ... Une application intéressante a été la reintroduction par G. Köthe et les deux Silva, de la théorie de Fantappié dans le cadre de l'analyse fonctionelle moderne ... Pour $n = 1$ une partie essentielle de la théorie des fonctionelles analytiques est l'interprétation de l'indicatrice de Fantappié. Par exemple, le dual de l'espace des fonctions holomorphes au voisinage d'un compact de $\mathbb{C}$ s'identifie à un quotient de l'espace des fonctions holomorphes dans le complémentaire de ce compact. Perfectionnant pour ce but précis les hypothèses de la dualité de Serre, j'ai montré, sous certaines hypothèses de cohomologie que, pour $n > 1$, ce dual pouvait être identifié à un espace de cohomologie défini sur CK, ou à un espace de cohomologie à support dans K [22]. Ce dernier résultat ... apparaît comme fondement dans la théorie des hyperfunctions de Sato ... Dans le cas connexe, au moins, il est possible de remplacer l'espace de cohomologie par un espace de fonctions holomorphes définies sur l'ensemble des hyperplans qui ne rencontrent pas K [24], [25]. On fonde ainsi la théorie de l'indicatrice projective de Fantappié ... ».

Proceeding along the lines described above by Martineau, the japanese mathematician Komatsu obtained in 1966 (but the result was widely distributed only in the lecture notes of the 1971 Katata conference[3] [20]) a general statement of the duality theorem, in which the Cauchy-Riemann system is replaced by a more general one (a more complete account should also mention the strongly related notion of P-functional, introduced and developed by Bengel in [2], [3] and [4]; for a related result, see our Corollary 5.1).

---

[2] As Kashiwara, Kawai and Kimura put it in [17]: «... the nature of the atmosphere in which Sato worked during the early stages of microfunction theory, when things were foggy for everyone except Sato ...».
[3] The proceedings of this conference were significantly dedicated to Martineau, whose fatal sickness prevented him from attending the conference itself.

In view of some recently obtained new results [18], we are now able to provide a further generalization of Komatsu's work, and we now can deal with a rather large class of infinite order differential operators with constant coefficients. This seems to be a significant improvement, as we are able to work out a very general and simple theory for such operators.

Let us now give a brief description of the organization of this paper, which aims at being somehow self-contained (aim which hopefully justifies its lengthy sections). The first two sections are devoted to a short description of those points of Fantappiè's theory which are of interest to us; this is done in an explicit way, since most of Fantappiè's works are in the italian language, which has probably created a barrier between them and the largest mathematical community (with exceptions, of course!); in particular section 1 describes Fantappiè's functionals in one complex variable, and section 2 describes those in several variables. We then deal (section 3) with the magnificent foundations which Martineau gave to Fantappiè's theory: in particular we describe those points of Martineau's works which deal with the precise cohomological definitions of the various indicatrices introduced by Fantappiè, and which therefore paved the way to meaningful generalizations. In section 4 we shall sketch the proof of Komatsu's result, indicating in which precise sense it can be viewed as an extension of Fantapppiè's theory. Even though this section does not contain any new theorems, we feel it has some interest in what it underscores a unifying point of view which does not seem to be widely known.

Finally, in section 5, we provide a new cohomological extension of Fantappiè's theory. The main result is probably a duality theorem (theorem 5.8) which generalizes the early representation theorems of Fantappiè (as well as its Komatsu's extension), but the section contains also a rather complete study of the cohomological properties of systems of infinite order linear differential equations with constant coefficients, which was not possible before the appearance of [18]. We feel that this section might as well be fruitfully generalized, and we point out some possible directions.

We hope that this paper will partially compensate the treatment which, in Italy, Fantappiè's theories have received, and we dedicate our effort to the University of Bologna (where Fantappiè actively worked during the first years of his career) which is now celebrating the 900-th birthday. Mille di questi giorni!!

Acknowledgements. I wish to express my deep gratitude to the Matsumae International Foundation, for its very generous support and to the people at the R.I.M.S. of Kyoto University, for their warm hospitality during the period in which this work was carried out. I also wish to thank Prof. S. Coen for inviting me to the celebrations to honor the University of Bologna and for asking me to contribute this work. Finally, a special word should be said about Mrs. Kay O'Grady who patiently and skillfully translated my handwriting into this beautiful typescript.

1. FANTAPPIÈ, I

In these first two sections we sketch very shortly the theory that L. Fantappiè created in the years 1924-1953, and which (it seems to us) is not quite common knowledge nowadays. In particular we will describe the final form of the theory, as it appears in [9], [30], and we refer the reader interested in its developments to [35], [37], where also the motivations for Fantappiè's work are discussed. Let us start, in this section, by talking about the case of analytic functionals in one variable (for which objects the theory was more complete).

DEFINITION 1.1. Let Ω be a (not necessarily connected) open subset of $\mathbb{C} \cup \{\infty\} = \mathbb{C}\mathbb{P}^1$ and let f be C^∞ in Ω and such that $\dfrac{\partial f}{\partial \bar{z}}(z) = 0$ for any $z \in \Omega$. Then f will be called a *locally analytic* function.

REMARK 1.1. A locally analytic function is therefore a section over Ω of the sheaf $\mathcal{O}$ of germs of holomorphic functions, and a section $s_2 \in \Gamma(\Omega_2, \mathcal{O})$ extending $s_1 \in \Gamma(\Omega_1, \mathcal{O})$, $\Omega_1 \subsetneq \Omega_2$ will be regarded as different from s_1.

REMARK 1.2. As the reader will notice, we use here (as long as no confusion arises), Fantappiè's original terminology.

REMARK 1.3. Let $s_1 \in \Gamma(\Omega_1, \mathcal{O})$, $s_2 \in \Gamma(\Omega_2, \mathcal{O})$, $\Omega_1 \subsetneq \Omega_2$, $s_1 = s_{2|\Omega_1}$. We then say that s_2 is a *continuation* of s_1; note that s_2 need not be an *analytic* continuations of s_1, as the following trivial example shows: set $\Omega_2 = \Omega_1 \cup \Omega_1'$, Ω_1 and Ω_1' both non empty open subsets of $\mathbb{C}\mathbb{P}^1$. Define $s_1(z) = z^3$ on Ω_1 but $s_2(z) = z^3$ on Ω_1 and $s_2(z) = z^4$ on Ω_1'. Then s_2 is a non analytic continuation of s_1.

DEFINITION 1.2. A locally analytic function $f \in \Gamma(\Omega, \mathcal{O})$ is said to be *ultraregular* if $f(\infty) = 0$, in case the point at infinity belongs to Ω. The function $f = 0$ defined on $\Omega = \mathbb{C}\mathbb{P}^1$, however, will not be considered to be ultraregular. To denote ultraregular functions we shall use the symbol $\Gamma^\infty(\Omega, \mathcal{O})$.

DEFINITION 1.3. Let $s_0 \in \Gamma^\infty(\Omega, \mathcal{O})$, let $A \subset \Omega$ be a closed set, and $\sigma > 0$ a positive real number. With the symbol (A, σ) we shall denote the «neighborhood» of s_0 given by all functions $s \in \Gamma^\infty(U(A), \mathcal{O})$, for $U(A)$ any open neighborhood of A, for which $|s(z) - s_0(z)| < \sigma$, $z \in A$.

REMARK 1.4. Any neighborhood (A, σ) of s_0 contains all continuations of s_0 itself.

DEFINITION 1.4. A *linear neighborhood* (A) of $s_0 \in \Gamma^\infty(\Omega, \mathcal{O})$, $A \subset \Omega$ closed, is defined as the set $\bigcup_{U(A)} (\Gamma^\infty(U(A), \mathcal{O}))$, or, equivalently, $\bigcup_{\sigma > 0} (A, \sigma)$, with $U(A)$ all possible open neighborhoods of A.

DEFINITION 1.5. The set of all ultraregular functions of one variable, endowed with the natural topology induced by the system of neighborhoods defined above, is a topological space, which will be denoted by $S^{(1)}$.

REMARK 1.5. Sum of two elements is $S^{(1)}$ is, clearly, not always possible, so that no vector space structure can be attributed to $S^{(1)}$.

DEFINITION 1.6. An open subset of $S^{(1)}$ will be called a *functional region* (or a *region*, for short).

REMARK 1.6. All sets of the form (A) or (A, σ) are regions according to Definition 1.6.

DEFINITION 1.7. A *linear region* is a region of $S^{(1)}$ which is closed with respect to the $\mathbb{C}$-linear combinations of its elements (in Fantappiè's point of view, the sum of a finite number of functions is defined only for those z's which belong to the intersection of the regions of $\mathbb{C}\mathbb{P}^1$ where each function is defined; therefore, in order for such a sum to exists, it is necessary and sufficient that this intersection be non empty).

These definitions can actually be simplified in view of the following important characterization of functional regions, due to Fantappiè.

THEOREM 1.1. *Let R be a linear functional region. The intersection of all the regions where the functions of R are defined is a proper, non empty, closed subset S of $\mathbb{CP}^1$, and each ultraregular function on A belongs to R. Conversely, the set of all functions which are ultraregular on a proper, non empty closed subset A of $\mathbb{CP}^1$ forms a linear functional region.*

REMARK 1.7. This theorem says that one can define a bijective correspondence between linear regions R and their characteristic sets A given by Theorem 1.1, and every region R can be shown to coincide with the neighborhood (A) of each of its functions; more precisely, Theorem 1.1 enables us to *define* a linear region as the set (A) of all functions which are ultraregular on the proper non empty closed subset A of $\mathbb{CP}^1$. Remember that, from a modern point of view, $(A) = \bigcup_{\substack{\Omega \supset A \\ \Omega \text{ open}}} \Gamma^\infty(\Omega, \mathcal{O})$, as given in Definition 1.4. The analytic functionals of Fantappiè will be seen to be introduced as elements of the dual space of (A), for A given. However this way of defining an object, which is quite natural for us, today, was not at all so, back in the twenties, so that the «continuity», which has to be built in the concept of functional, was suggested to Fantappiè by some well known results of Poincarè on analytic dependence of solutions of analytic Cauchy problems for partial differential equations. One therefore needs the notion of *analytic line* (see [35], [37] for the development of such notion in the early Fantappiè's works).

DEFINITION 1.8. A function $y(z, \zeta)$ of two complex variables z, ζ is said to be an *analytic line* in $S^{(1)}$ if:

i) for each ζ_0 in a region $\Omega \subset \mathbb{CP}^1$, the function $y_0(z) := y(z, \zeta_0)$ is ultraregular in a region $M(\zeta_0) \subset \mathbb{CP}^1$;
ii) for each z in a region $\Omega' \subset \mathbb{CP}^1$, the function $y(z, \zeta)$ is holomorphic in ζ;
iii) the set $I(\zeta) = \mathbb{CP}^1 - M(\zeta)$ is a continuous function ζ, i.e.: for any $y_0 \in \Omega$ and for any $\varepsilon > 0$, there exists an open set $U(\zeta_0) \subset \Omega$ such that, for any $\zeta \in U(\zeta_0)$, it is

$$d(I(\zeta), I(\zeta_0)) := \sup_{x \in I(\zeta)} d(x, I(\zeta_0)) < \varepsilon$$

REMARK 1.8. Some relevant examples of analytic lines can be given immediately (and the reader will soon realize their importance) by considering

$$y(z, \zeta) = \frac{1}{\zeta - z}$$

and

$$y(z, \zeta) = \frac{1}{1 - \zeta z}.$$

We can now give the definition of analytic functional, according to Fantappiè (just a reminder of the fact that *modern* analytic functionals are not quite the same thing).

DEFINITION 1.9. Let R be a linear region of $S^{(1)}$. A $\mathbb{C}$-linear map $F : R \to \mathbb{C}$ is said to be a *locally analytic linear functional* if:

i) given y_0, a continuation of y_1, then $F[y_0] = F[y_1]$;

ii) if an analytic line $y = y(z,\zeta)$, defined for $\zeta \in \Omega$, intersects, for $\zeta \in \Omega' \subset \Omega$, the region R, then the value of F on this line is, as a function of ζ,

$$f(\zeta) := F_z[y(z,\zeta)], \quad {}^4$$

a locally analytic function in Ω'.

REMARK 1.9. Among the many different examples, let us mention two very simple ones: differentiation and integration.

i) Consider the functional D which associates to each function $y(z)$ its derivative at a chosen point z_0, i.e.

$$D[y] = y'(z_0).$$

In this case the functional D is defined in the linear region R of all functions which are ultraregular in z_0, i.e. $R = \mathcal{O}_{z_0}$ is the space of germs holomorphic functions at z_0, and the characteristic set associated to R is $A = \{z_0\}$.

ii) For $a, b \in \mathbb{C}$, consider the segment $[a,b]$ joining them, and the analytic functional I defined by integrating along $[a,b]$, i.e.

$$I[y] = \int_a^b y(z)\,dz,$$

which we shall consider as defined on the linear region R of all functions which are ultraregular in some neighborhood of $[a,b]$. In this case, as it is probably clear, the characteristic set will be $A = [a,b]$.

Before we can get to the main results of Fantappiè's theory, we still need to mention a couple of basic facts which will be useful in the sequel.

THEOREM 1.2. *Let F be a linear analytic functional, defined on a linear region (A), and let $L = \{y(z,\zeta)\}$ be an analytic line which intersects (A). Then, one can take the derivative with respect to ζ inside the functional, i.e.*

$$\frac{d}{d\zeta} F_z[y(z,\zeta)] = F\left[\frac{\partial y(z,\zeta)}{\partial \zeta}\right],$$

and, by iteration,

$$\frac{d^n}{d\zeta^n} F_z[y(z,\zeta)] = F_z\left[\frac{\partial^n y(z,\zeta)}{\partial \zeta^n}\right]$$

A related theorem which will be needed in the sequel, and which we quote without proof, is the following:

---

[4] Here and in the sequel, the subscript z indicates that F acts on $y(z,\zeta)$ as a function of z, regarding therefore ζ as a parameter.

Fantappiè's Theory of Analytic Functionals

THEOREM 1.3. *Let F be an analytic functional defined in the region (A), $L = \{y(z,\zeta)\}$ an analytic line which intersects (A), and Ω the region (in $\mathbb{CP}^1$) of all values of ζ for which the line L intersects (A). Then for every rectifiable path $\gamma \subset \Omega$, it is*

$$\int_\gamma F[y(z,\zeta)]\,d\zeta = F\left[\int_\gamma y(z,\zeta)\,d\zeta\right].$$

A first question which one might ask, is how to compute the value of a functional on each of the functions where it is defined. The answer to this question can be provided through the introduction of the *antisymmetric indicatrix*

REMARK 1.10. Consider the following function:

$$y(z,\zeta) = \begin{cases} \dfrac{1}{\zeta - z} & \text{for } \zeta \neq \infty, \quad z \neq \zeta \\ 0 & \text{for } \zeta = \infty, \quad z \neq \infty \end{cases}$$

It is immediately seen that $y(z,\zeta)$ is an analytic line according to Definition 1.8, and, most important, *it enters every linear region (A)*.

Therefore, the following definition is meaningful:

DEFINITION 1.10. The function

$$u(\zeta) = F_z\left[\frac{1}{\zeta - z}\right]$$

is called the *antysimmetric indicatrix* of F.

REMARK 1.11. It is readily seen that if F is defined on (A), then $u(\zeta)$ is locally analytic in $B = \mathbb{CP}^1 - A$, and (if $\infty \notin A$) it is ultraregular, since $u(\infty) = 0$.

REMARK 1.12. If $0 \in B$, one has, for any integer $n \geq 0$

$$\frac{1}{n!}u^{(n)}(0) = -F\left[\frac{1}{z^{n+1}}\right]$$

or more generally

$$\frac{1}{n!}u^{(n)}(\zeta) = -F\left[\frac{1}{(z-\zeta)^{n+1}}\right].$$

REMARK 1.13. We can easily compute the indicatrix for the functionals D and I of Remark 1.9.

i) Let $D[y] = y'(z_0)$. Then its indicatrix will be given by

$$u(\zeta) = D\left[\frac{1}{\zeta - z}\right] = \frac{1}{(\zeta - z_0)^2}.$$

ii) Let $I[y] = \int_a^b y(z)\,dt$. Then its indicatrix will be given by

$$u(\zeta) = I\left[\frac{1}{\zeta - z}\right] = \int_a^b \frac{dt}{\zeta - z} = -[\log(z - \zeta)]_a^b = \log\frac{a - \zeta}{b - \zeta}, \qquad u(\infty) = 0.$$

Note that u is a locally analytic function on $\mathbb{CP}^1 - [a,b]$ and, despite the appearance of a logarithm, it is *not* multivalued, since the request $u(\infty) = 0$ chooses a sheet on the Riemann surface on which $u(\zeta)$ would otherwise be naturally defined.

REMARK 1.4. In an analogous way, Fantappiè also defines a *symmetric indicatrix*, by taking

$$y(z,\zeta) = \begin{cases} 1/(1-z\zeta) & \text{for } \zeta \neq \infty \\ 0 & \text{for } \zeta = \infty, \quad z \neq 0 \\ \text{undefined} & \text{for } \zeta = \infty, \quad z = 0 \end{cases}$$

and then applying F to it; therefore such an indicatrix is the locally analytic function

$$w(\zeta) = F_z\left[\frac{1}{1-z\zeta}\right] = \frac{1}{\zeta}u\left(\frac{1}{\zeta}\right).$$

We leave to the reader to check the analogues of Remarks 1.11, 1.12 and 1.13 for the simmetric indicatrix.

DEFINITION 1.11. Let A, B be two closed disjoint subsets of $\mathbb{C}\mathbb{P}^1$. A *separating curve* for (A, B) is any closed rectifiable curve γ which keeps A and B on its two different sides.

REMARK 1.15. If A and B are, respectively, the sets where two ultraregular functions u, v are undefined, then the value of the integral

$$\int_\gamma u(z)v(z)\,dz$$

does not depend on the choice of the separating curve γ (this is indeed, [30], the reason for the notion of ultraregularity).

We can now state and prove the main result of the theory (this is, actually, a reductive statement, and has to be interpreted in the sense that the next theorem is actually, the keystone for Fantappiè's theory and its applications).

THEOREM 1.4. *Let F be an analytic functional defined on (A), and let $y(z) \in (A)$ be any function, whose singular set (points where y is singular, or even undefined) we denote by B. Let now γ be a separating curve for (A, B) and denote by u the antisymmetric indicatrix of F. Then the following representation formula holds:*

$$F[y(z)] = \frac{1}{2\pi i}\int_\gamma u(\zeta)y(\zeta)\,d\zeta \qquad (1)$$

Proof: First note that we take γ such that it includes A in its interior, but, on the other hand, it is completely contained in a region where y is ultraregular (the existence of such a γ is proved, e.g., in [9]). Next, it is clear that y is a continuation of the (locally analytic) function $\tilde{y}$, defined inside γ by the formula

$$\tilde{y}(z) = \frac{1}{2\pi i}\int_\gamma \frac{y(\zeta)}{\zeta - z}\,d\zeta;$$

it is immediate to observe that $\tilde{y} \in (A)$, and therefore:

$$F[y(z)] = F[\tilde{y}(z)] = F\left[\frac{1}{2\pi i}\int_\gamma \frac{y(\zeta)}{\zeta - z}\,d\zeta\right].$$

By Theorem 1.3 we get

$$F[y(z)] = \ldots = \frac{1}{2\pi i}\int_\gamma F\left[\frac{y(\zeta)}{\zeta - z}\right]d\zeta = \frac{1}{2\pi i}\int_\gamma F\left[\frac{1}{\zeta - z}\right]y(\zeta)\,d\zeta =$$
$$= \frac{1}{2\pi i}\int_\gamma u(\zeta)y(\zeta)\,d\zeta.$$
□

It is not useless to point out the interest of the theorem we just proved. Indeed, Remark 1.11 shows that we can associate to every linear functional F, a locally analytic ultraregular function u; conversely, Theorem 1.4 shows that, for each locally analytic ultraregular function u, one can construct (via (1)) a linear functional F, whose antisymmetric indicatrix is exactly u (this last result follows from a standard use of Cauchy's integral formula), namely:

THEOREM 1.5. *There is a one to one correspondence (induced by* (1)*) between linear locally analytic functionals and ultraregular locally analytic functions.*

2. FANTAPPIÈ, II

In this section, we briefly discuss the case of analytic functionals in several variables. In this case, as it was clearly pointed out in [23], the theory of Fantappiè was far from being complete, in view of the different topological structure of singularities which several complex variables functions have, when compared with one complex variable functions. In particular, it may be of some interest to remark that in [23] (but also in previous works by Fantappiè) it is precisely mentioned the well known theorem of Hartogs on the removability of compact singularities; on the other hand, it is now widely known (since the basic work of Ehrenpreis in 1960, see e.g. [73]) that such a theorem is not peculiar to holomorphic functions, but it holds for other spaces of C^∞ solutions of suitably overdetermined systems (holomorphic functions arising when the Cauchy-Riemann system is considered); the large number of results in this direction (the interested reader may consult the recent work [36], where a rather complete research has been carried out on generalized Hartogs' phenomena) is therefore in complete agreement with Fichera's suggestion which we mentioned in section 0.

In the case of n variables, the approach followed by Fantappiè consisted in considering locally analytic functions in $\mathbb{C}\mathbb{P}^n$, which, again, had to be ultraregular. We will not spend any time in describing the development of Fantappiè's theory (see e.g. [35]), but we will simply say that most of the initial definitions and theory (which we have just described for one complex variable) go through to the case of several complex variables[5], the first problems arising exactly when trying to extend Theorems 1.4 and 1.5. Formally, at least, things go smoothly, and the antisymmetric indicatrix of an analytic functional is defined as follows:

DEFINITION 2.1. Let F be an analytic functional defined on a linear functional region (A) (in this case A is a closed non empty proper subset of $\mathbb{C}\mathbb{P}^n$). *Assume* that the *analytic*

---

[5] In this case $S^{(n)}$ is the space of ultraregular locally analytic functions, and, for some $A \subsetneq \mathbb{C}\mathbb{P}^n$, $A \neq \emptyset$, any linear region of $S^{(n)}$ can be written as (A).

variety[6]

$$y(z,\zeta) = y(z_1,\ldots,z_n;\zeta_1,\ldots,\zeta_n) := \frac{1}{(\zeta_1 - z_1)\cdot(\zeta_2 - z_2)\cdot\ldots\cdot(\zeta_n - z_n)} \qquad (2)$$

intersects (A). Then we define the *antisymmetric indicatrix* of F by

$$u(\zeta) = u(\zeta_1,\ldots,\zeta_n) := F_z[y(z,\zeta)].$$

REMARK 2.1. The most crucial difference with the case in which $n = 1$ is contained in the fact that, for $n > 1$, $y(z,\zeta)$ need not necessarily intersect (A), and so, in general, we are not even allowed to define the antisymmetric indicatrix. Thus, for example, the «test» function y given by (2) is singular on a certain number of hyperplanes, and therefore the antisymmetric indicatrix is not well defined if the characteristic set A of the linear region (A) where F is defined, intersects all hyperplanes. Generally speaking, in order for $y(z,\zeta)$ to belong to (A) it is necessary and sufficient that each ζ_i does not belong to the projection A_i of A on the i-th axis, and therefore one is bound to consider linear functionals F defined on linear regions (A) whose characteristic set $A \subset \mathbb{CP}^n$ is bounded.

Still, even when the antisymmetric indicatrix of a linear functional is defined (as it happens when A is a compact of $\mathbb{C}^n$, or of $\mathbb{R}^n \subset \mathbb{C}^n$, for example), we do not always have theorems like Theorems 1.4 and/or 1.5. The closest thing we have is the following consequence of Cauchy's formula in several variables:

THEOREM 2.1. *Let* $A = A_1 \times \ldots \times A_n$, *with each* A_i *a compact in* $\mathbb{C}$ *(hence* $A \subset \mathbb{CP}^n$ *is all contained in a ball of finite radius and centered at the origin), and let* F *be a linear locally analytic functional defined on* (A), *whose antisymmetric indicatrix we indicate by* $u(\zeta_1,\ldots,\zeta_n)$. *Then, for every function* $y \in (A)$, *we have the following representation formula*

$$F[y(z_1,\ldots,z_n)] = \left(\frac{1}{2\pi i}\right)^n \int_{\gamma_1} d\zeta_1 \int_{\gamma_2} d\zeta_2 \ldots \int_{\gamma_n} u(\zeta_1,\ldots,\zeta_n) y(\zeta_1,\ldots,\zeta_n) d\zeta_n, \qquad (3)$$

where each curve γ_i *has to contain at its interior the closed set* A;

REMARK 2.2. The representation formula (3) actually works even if A is not of the form $A_1 \times \ldots \times A_n$, in which case, however, $A \subset A_1 \times A_2 \times \ldots \times A_n$, i.e. $(A) \supset (A_1 \times \ldots \times A_n)$, and (3) *only* holds for those y which belong to $(A_1 \times \ldots \times A_n)$, the reason for this being obvious if one considers how the γ_i's have to be chosen.

REMARK 2.3. Another interesting phenomenon arises from the consideration of formula (3), when $n > 1$ of course. Indeed, integration in (3) takes place on a closed n-dimensional subvariety $\gamma = \gamma_1 \times \ldots \times \gamma_n$ of $\mathbb{CP}^n$ or, better, of the Segre variety $V_{2n} \subset \mathbb{R}^{n(n+2)}$ which gives a suitable real model of $\mathbb{CP}^n$ [8]. Actually, if I_u and I_y represent, respectively, the subsets of V_{2n} where u and y are not defined, one can replace, in (3), γ with any other n-dimensional variety γ' as long as γ' is homologous to γ relatively to $V_{2n}\setminus(I_u \cup I_y)$; γ' is then such that it is homologous to zero in $V_{2n}\setminus I_u$ or $V_{2n}\setminus I_y$, and any such variety will therefore be called a *separating variety* for I_u and I_y, even though (unlike what happens when $n = 1$) such a variety does not divide any longer V_{2n} in two separate regions.

---

[6] Obvious generalization of the notion of analytic line (see Definition 1.8).

REMARK 2.4. Even more interesting is the fact that, according to this interpretation of (3), most analytic functionals turn out to be multivalued, even though their indicatrices are single valued, as it follows by taking two different separating varietes γ and γ' which are not homologous with respect to $V_{2n}\setminus(I_u \cup I_y)$. This can be easily obtained by moving y along an analytic variety, so that the final separating variety γ' will not be homologous to γ. An explanation, in modern terms, of this unpleasant phenomenon will be given (through Martineau's works) in section 3, where it will be clear that it is somehow misleading to try to interpretate antisymmetric indicatrices (for $n > 1$) in terms of holomorphic functions, when a more sophisticated approach is necessary.

REMARK 2.5. In complete analogy with the theory of antisymmetric indicatrices, one can develop the theory of symmetric indicatrices, on which, however, we need not spend any further time.

Much more interesting (in itself and in view of its modern developments) is the introduction of a new kind of indicatrix, much more suited to the projective nature of the problems to be studied.

DEFINITION 2.2. Let F be an analytic functional defined on a linear functional region (A), which is intersected by the analytic variety

$$y(z,\zeta) = (1 + \zeta_1 z_1 + \zeta_2 z_2 + \ldots + \zeta_n z_n)^{-1};$$

then we define the *projective indicatrix* of F by

$$p(\zeta) := F_z[y(z,\zeta)].$$

REMARK 2.6. The discovery of this indicatrix (one of the higlights of Fantappiè's theory in several variables) was somehow delayed by the circumstance that, for $n = 1$, $p(\zeta) = w(-\zeta)$.

REMARK 2.7. Besides its applications, the projective indicatrix clearly stands out for conceptual simplicity (in comparison with the antisymmetric one); indeed, such an indicatrix will provide the value of F for those functions whose poles are confined to a single hyperplane; moreover, the projective indicatrix is independent on the choice of the coordinate system, a circumstance which clearly does not hold for the two other indicatrices.

REMARK 2.8. A final important point is the fact that a representation theorem for $F[y]$ can be obtained with the use of the projective indicatrix; such a theorem is quite complicated, so that we shall not state it here, and we refer the reader to [8] for its statement and proof; we shall only point out that Fantappiè never succeeded in perfecting this theorem into a duality statement (as Theorems 1.4 and 1.5), and this task was later carried out by Martineau in [25], but see also [22] and [23].

3. MARTINEAU'S FOUNDATIONS

In this section we shall try to describe Martineau's precise treatment of Fantappiè's theory; in particular we shall focus on Martineau's foundations of the theory of the antisymmetric indicatrix, which will culminate in Theorems 2.1 and 2.2. We ought to point out that Martineau also got very relevant results on the projective indicatrix (whose theory, on the other hand, Fantappiè had left somehow incomplete [10]) and these results may be susceptible of further refinements with the modern techniques described in sections 4 and 5; still, in this paper, we have decided not to discuss these aspects on which, however, we plan to return in the near future (the reader may also refer to [37]).

Before we state the main results of Martineau, a few words (details will appear in [37]) are in order on what happened from, say, 1941 (when Fantappiè's [8] appeared) to 1961 (when Martineau's [22] was delivered); on one hand, the theory of topological vector spaces (as envisaged in [6]) became increasingly significant in mathematical analysis, leading to the creation and success of Schwartz's distribution theory and therefore diminishing the appeal of the more hand-made methods of Fantappiè, but, on the other hand, some students of Fantappiè attempted a merging of these modern techniques with their teacher's ones; the most successful was, undoubtedly J.S. e Silva who, in his 1948 doctorate thesis [33], made Fantappiè's construction into a modern one, with the help of topological vector spaces. This work was, unfortunately, not completely correct, but it had, nonetheless, a very good effect on the whole, as more straight foundations for the theory of analytic foundations were laid down by A. Grothendieck [14] and G. Köthe [21] whose works were directly inspired by Silva's [33]. It was in this very active framework (mostly concerned with one-dimensional situations, though!) that Martineau developed his very far reaching theory, where algebraic techniques (such as cohomological theories) were coming to help the analysts (a preview of algebraic analysis!). Since topology has to play a relevant role (while completely absent in section 1), let us recall some well known topological definitions:

DEFINITIONS 3.1. If $\Omega \subset \mathbb{C}^n$ (here and in most places in the sequel, $\mathbb{C}^n$ might as well be replaced by any Stein variety), we denote by $H(\Omega)$ the space $\Gamma(\Omega, \mathcal{O})$ of holomorphic functions on Ω, endowed with the *topology of uniform convergence on the compact subsets of* Ω.

REMARK 3.1. $H(\Omega)$, with this topology, is a *nuclear Fréchet space*.

DEFINITION 3.2. For $K \subset \mathbb{C}^n$ a compact set, one sets

$$H(K) := \mathop{\text{ind lim}}_{\substack{\Omega \supset K \\ \Omega \text{ open}}} H(\Omega),$$

where the inductive limit has to be taken in the category of locally convex topological vector spaces, i.e. $H(K)$ has the finest locally convex topology which makes continuous all the restriction maps $H(\Omega) \to H(K)$.

REMARK 3.2. $H(K)$, with this topology, is a *complete nuclear (DF)-space* (hence an (LF)-space as well).

DEFINITION 3.3. The strong dual space $H'(K)$ of $H(K)$ is a *Fréchet nuclear* space whose elements will be called *linear analytic functionals*, carried by K.

REMARK 3.3. From a set theoretical point of view, $(A) = H(A)$, whenever $A \subset \mathbb{C}^n$ is a compact set, and a functional defined on (A) belongs to $H'(A)$. Of course, the structure we are describing in this section *has* a topology.

DEFINITION 3.4. A compact set $K \subset \mathbb{C}^n$ is said to be *almost convex* if:

i) $H^i(K, \mathcal{O}) = 0$ for $i > 0$,

and

ii) K has a Stein neighborhood.

REMARK 3.4. Condition ii) is automatically satisfied if $K \subseteq \mathbb{R}^n$, thanks to a theorem by Grauert [13], and, of course, for any K when $n = 1$. In condition i), of course, $\mathcal{O}$ denotes the sheaf of germs of holomorphic functions, while the H^i are the usual cohomology groups with coefficients in a sheaf.

We are therefore in the condition of giving the abstract version of Fantappiè's duality theorem, following [22]:

THEOREM 3.1. *Let $K \subseteq \mathbb{C}^n$, $n \geq 2$, be almost convex, and let Ω^j denote the sheaf of germs of holomorphic j-forms on $\mathbb{C}^n$. Then*

$$H'(K) \cong H^{n-1}(\mathbb{C}^n \backslash K, \Omega^n),$$

the isomorphism being a topological one.

Proof. The proof is, essentially, a corollary of Serre's duality lemma [32], and employs (by now) standard cohomological techniques. First one starts writing the cohomology sequence with compact supports of the pair $(\mathbb{C}^n, K)$ (the reader may consult [13] for all cohomological terminology);

$$\ldots \to H_c^i(\mathbb{C}^n \backslash K, \mathcal{O}) \to H_c^i(\mathbb{C}^n, \mathcal{O}) \to H_c^i(K, \mathcal{O}) \to H_c^{i+1}(\mathbb{C}^n \backslash K, \mathcal{O}) \to \ldots; \quad (4)$$

$\mathbb{C}^n$ being Stein, one has $H_c^i(\mathbb{C}^n, \mathcal{O}) = 0$, for any $i \neq n$, while $H_c^{i+1}(K, \mathcal{O}) = H^i(K, \mathcal{O})$, which vanishes for any $i > 0$ (by the almost convexity of K). We conclude that, for $i \neq 1, n$

$$H_c^i(\mathbb{C}^n \backslash K, \mathcal{O}) = 0,$$

while, for $i = 0$ in (4), and $H(K) = H^0(K, \mathcal{O})$

$$0 \to H(K) \to H_c^1(\mathbb{C}^n \backslash K, \mathcal{O}) \to 0$$

(notice that this would fail for $n = 1$).

We therefore obtain

$$H(K) \cong H_c^1(\mathbb{C}^n \backslash K, \mathcal{O}),$$

and the theorem will be proved if we succeed in showing that $H_c^1(\mathbb{C}^n \backslash K, \mathcal{O})$ and $H^{n-1}(\mathbb{C}^n \backslash K, \Omega^n)$ are in duality. This, however, is an immediate consequence of Serre's duality lemma, and the proof is complete. □

REMARK 3.5. Given an analytic functional F, one can therefore associate to it an $(n-1)$-cocycle of a covering of $\mathbb{C}^n \backslash K$, with values in Ω^n; such a cocycle is what Martineau calls (and for a very good reason which we shall see in a moment) the *Fantappiè indicatrix* of the functional F.

Before we give a more concrete description of Theorem 3.1 (a description which will clarify the connections between Theorem 2.1 and Theorem 3.1), we point out a simple, interesting cohomological corollary of Theorem 3.1, which shows the relations with hyperfunctions theory:

THEOREM 3.2. Let $K \subseteq \mathbb{C}^n$, $n \geq 1$, be almost convex; then

$$H'(K) \cong H^n(\mathbb{C}^n, \mathbb{C}^n \setminus K; \Omega^n) \qquad (5)$$

Proof. For $n \geq 2$, the isomorphism (5) follows immediately from Theorem 3.1 and from the exact sequence of relative cohomology; indeed one has

$$\ldots \to H^{n-1}(\mathbb{C}^n, \Omega^n) \to H^{n-1}(\mathbb{C}^n \setminus K, \Omega^n) \to H^n(\mathbb{C}^n, \mathbb{C}^n \setminus K; \Omega^n) \to H^n(\mathbb{C}^n, \Omega^n) \to \ldots,$$

but since $H^{n-1}(\mathbb{C}^n, \Omega^n) = H^n(\mathbb{C}^n, \Omega^n) = 0$, the thesis follows immediately. As for the case $n = 1$, this cannot be done via Theorem 3.1 (which is false for $n = 1$), but an elemetary proof can be given ([21], [33]) by noticing that each $[\omega] \in H^1(\mathbb{C}, \mathbb{C} \setminus K; \Omega^1)$ defines a continuous linear functional on $H(K)$ by setting

$$[\omega](f) = \int_\gamma f(z)\omega(z)$$

where $f \in H(K)$ is defined in a small neighborhood U of K, and γ is a rectifiable closed curve in U which encircles K once and clockwise: notice that $[\omega](f)$ does not depend on the arbitrary choice of γ and of ω in $[\omega]$. Conversely if $F \in H'(K)$, then $\omega(z) = \frac{1}{2\pi i} F_\zeta \left(\frac{1}{\zeta - z} \right) dz$ belongs to $H^1(\mathbb{C}, \mathbb{C} \setminus K; \Omega^1)$ and $[\omega](f) = F(f)$ for any function $f \in H(K)$. □

REMARK 3.6. If $K \subseteq \mathbb{R}^n$, then $H^n(\mathbb{C}^n, \mathbb{C}^n \setminus K; \Omega^n)$ is nothing but the space $\mathcal{B}_K(\mathbb{R}^n)$ of hyperfunctions on $\mathbb{R}^n$, whose support is contained in K.

REMARK 3.7. Given $F \in H'(K)$, the element in $H^n(\mathbb{C}^n, \mathbb{C}^n \setminus K; \Omega^n)$ that Theorem 3.2 associates to F is called, by Martineau, the *Sato indicatrix* of F. At least when $n = 1$, its link with section 1 is clearly understood; as for the case of $n \geq 2$, we shall see in a second its meaning. One can therefore claim that Theorem 3.2 (and its companion Theorem 3.1) are indeed the correct formalization of Fantappiè's duality theorems.

We shall now follow Martineau's [25] to describe more clearly in which way the «abstract» Theorems 3.1 and 3.2 relate to the more concrete Fantappiè's ones. This analysis will also help us in understanding the concrete meaning of the generalizations we shall deal with.

DEFINITION 3.5. Let $A \subseteq \mathbb{CP}^n$. Its *projective complement* is the set $C^*A \subseteq (\mathbb{CP}^n)^*$ which parametrizes the complex hyperplanes of $\mathbb{CP}^n$ which do not meet A.

DEEFINITION 3.6. A set $A \subseteq \mathbb{CP}^n$ is said to be *linearly convex* if $A = C^*C^*A$.

REMARK 3.8. If $A \subseteq \mathbb{CP}^1$, then $C^*A = \mathbb{CP}^1 \setminus A$.

REMARK 3.9. It can be shown (see [26], Theorem 4) that if $n \geq 2$, any real compact convex set K is linearly convex. This clearly provides us with a large number of examples for what follows.

At it is easily shown, see [25], if $K \subset \mathbb{C}^n$ is linearly convex, then it is also almost convex, and therefore,

$$H'(K) = H^{n-1}(\mathbb{C}^n \setminus K, \Omega^n);$$

we now try to interpretate (5). Let then ξ be a hyperplane in $\mathbb{C}^n$, and consider the closed set $K_\xi = \bigcup_{\{\xi+a\} \cap K \neq 0} \{\xi + a\}$ and its complement $\Omega_\xi = \mathbb{C}^n \backslash K_\xi$.

By the linear convexity of K one deduces that the family of the Ω_ξ naturally covers $\mathbb{C}^n \backslash K$, and therefore every element in $H^{n-1}(\mathbb{C}^n \backslash K, \Omega^n)$ can be represented using the (Leray) covering $\{\Omega_\xi\}$, i.e. by giving on each $\Omega_{0,\ldots,n-1} = \Omega_{\xi_0} \cap \Omega_{\xi_1} \cap \ldots \cap \Omega_{\xi_{n-1}}$ a holomorphic n-form

$$\varphi_{0,\ldots,n-1}(z)\, dz_1 \wedge \ldots \wedge dz_n$$

such that, for every $z \in \Omega_{0,\ldots,n}$, the cocycle identity

$$\sum_{h=0}^{n}(-1)^h \varphi_{0,\ldots,\hat{h},\ldots,n}(z) = 0$$

holds. In line with Fantappiè findings, one would expect that some interest could be attached to the notion of «vanishing at infinity» (in accordance with the role played in section 1 by ultraregularity). Indeed, the following basic result is true:

THEOREM 3.3. a) *If an $(n-1)$ cocycle associated to the covering $\{\Omega_\xi\}$ of $\mathbb{C}^n \backslash K$ is a coboundary, and if it vanishes at infinity, then it vanishes identically.*

b) *In each $(n-1)$-cohomology class as above, there is always a cocycle which vanishes at infinity.*

Proof. Only a sketch of the proof will suffice, for our purposes, as the reader may find the details in [25]. First, to prove a), assume that $K_1, \ldots, K_n$ are compact subsets of $\mathbb{C}^n$, and that $\Omega_{\xi_0} = \{z \in \mathbb{C}^n : z \notin K_1\}, \ldots, \Omega_{\xi_{n-1}} = \{z \in \mathbb{C}^n : z \notin K_n\}$. Take a cocycle φ as in the hypothesis: since φ vanishes at infinity, for any neighborhood $U = U_1 \times \ldots \times U_n$ in $\mathbb{C}^n$ of $K_1 \times \ldots \times K_n$ we can find rectifiable closed curves $\gamma_1, \ldots, \gamma_n$ ($\gamma_i \subseteq U_i$), such that, for z suitable,

$$\varphi_{0,\ldots,n-1}(z) = \left(\frac{1}{2\pi i}\right)^n \int_{\gamma_1} \ldots \int_{\gamma_n} \frac{\varphi_{0,\ldots,n-1}(\zeta)\, d\zeta_1 \wedge \ldots \wedge d\zeta_n}{(z_1 - \zeta_1) \cdot \ldots \cdot (z_n - \zeta_n)}; \tag{6}$$

but as φ is a coboundary, we have

$$\varphi_{0,\ldots,n-1}(\zeta) = \sum_{h=0}^{n-1}(-1)^h \varphi_h(\zeta), \tag{7}$$

where φ_h is holomorphic on z such that $z_i \notin K_i (i \neq h + 1)$; from (6) and (7), integrating in ζ_{h+1}, we get

$$\int_{\gamma_1} \ldots \int_{\gamma_n} \frac{\varphi_h(\zeta)\, d\zeta_1 \wedge \ldots \wedge d\zeta_n}{(z_1 - \zeta_1) \cdot \ldots \cdot (z_n - \zeta_n)} = 0,$$

which proves a). As far as b) is concerned, we take K_i to be the projection of K on the i-th coordinate and, as before, $\Omega_{\xi_h} = \{z \in \mathbb{C}^n : z_{h+1} \notin K_{h+1}\}$. Then if $F \in H'(K_1 \times \ldots \times K_n)$, one can define the Fantappiè indicatrix

$$\varphi(z) = \left(\frac{1}{2\pi i}\right)^n F_\zeta \left(\frac{1}{(\zeta_1 - z_1) \cdot \ldots \cdot (\zeta_n - z_n)}\right)$$

which turns out to be holomorphic in $\Omega_{0,\ldots,n-1}$, and vanishing at infinity; it is then possible to show that $\{\varphi(z)\,\mathrm{d}z_1 \wedge \ldots \wedge \mathrm{d}z_n\}$ is a cocycle which represents F in $H^{n-1}(\mathbb{C}^n\backslash K, \Omega^n)$, and which has the required vanishing properties. □

The precise topological version of the theorem which we want to state, needs some extra definitions. To this purpose we consider $\Gamma^\infty(\Omega_{0,\ldots,n-1}, \mathcal{O})$ (see section 1) which we endow with the topology of uniform convergence on compact subsets, and we take the topological product

$$\prod_{(\xi_0,\ldots,\xi_{n-1})} \Gamma^\infty(\Omega_{0,\ldots,n-1}, \mathcal{O});$$

in it we consider the closed subspace defined by the relations

$$\sum_h (-1)^h \varphi_{i_0,\ldots,\hat{i}_h,\ldots,i_n} = 0 \quad \text{on each} \quad \Omega_{i_1,\ldots,i_n};$$

this enables us to define a nuclear Fréchet topology on the space of $(n-1)$-cocycles which vanish at the origin, and a simple application of Theorem 3.3 provides the required formalization of Fantappiè's representation results:

THEOREM 3.4. *Let* $K \subset \mathbb{C}^n$ *be a linearly convex compact set. Put, for* ξ *any hyperplane through the origin of* $\mathbb{C}^n$,

$$\Omega_\xi = \{z \in \mathbb{C}^n : (z + \xi) \cap K = \emptyset\};$$

then $H'(K)$ *is naturally isomorphic to the topological vector space of the* $(n-1)$*-cocycles of the coverings* $\{\Omega_\xi\}$ *of* $\mathbb{C}^n\backslash K$, *with values in the sheaf of germs of n-holomorphic forms vanishing at infinity.*

Theorems 3.1 and 3.2 (as well as their «concrete» version, Theorem 3.4) give a correct foundation for the theory of Fantappiè's antisymmetric indicatrix; before getting into Komatsu's extension and into our own ones, we would like to remind the reader that Martineau succeeded in founding the theory of Fantappiè's projective indicatrix as well (see e.g. [25]): further details on this aspect, as well as a more precise historical account of these developments, will be found in [37], since dwelling into these aspects would take us far away from our main purpose.

4. KOMATSU'S GENERALIZATION

Let us now introduce some algebraic machinery which will be useful both to describe Komatsu's work [10] and to introduce our own contribution. To begin with, let $\mathcal{A}, \mathcal{B}, \mathcal{D}, \mathcal{E}$ denote respectively the sheaves of real analytic functions, hyperfunctions (see [17], e.g.) distributions and infinitely differentiable function on $\mathbb{R}^n$; by $\mathcal{S}$ we shall denote anyone of these sheaves, (or possibly the sheaf $\mathcal{O}$ over $\mathbb{C}^n$) when making a statement which holds for any one of them. If now $P(D)$ is any $r_1 \times r_0$ matrix of *linear differential operators with constants coefficients*, we can use it to define a sheaf homomorphism

$$P(D) : \mathcal{S}^{r_0} \to \mathcal{S}^{r_1},$$

whose kernel we denote by S^P, i.e. S^P is the sheaf of solutions u in S^{r_0} of the homogeneous system of equations

$$P(D)u = 0.$$

Denote now by $\mathbb{C}[X] = \mathbb{C}[X_1, \ldots, X_n]$ the ring of $\mathbb{C}$-polynomials in n indeterminates; for $Q(X)$ any matrix with entries in $\mathbb{C}[X]$, we denote by $Q'(X)$ its adjoint matrix ${}^t Q(-X)$.

If we consider a system $P(D)$ as before and we replace $-i\partial/\partial x_j$ by X_j we get a matrix $P(X)$ whose adjoint $P'(X)$ defines a $\mathbb{C}[X]$-homomorphism

$$P'(X) : \mathbb{C}[X]^{r_1} \to \mathbb{C}[X]^{r_0},$$

whose cokernel we denote by $M' = \mathbb{C}[X]^{r_0}/P'(X)\mathbb{C}[X]^{r_1}$ (M' is a finitely generated $\mathbb{C}[X]$-module, and this procedure can be seen as the starting point for modern algebraic analysis [17]). It was shown by Palamodov [29] that M' admits a free resolution

$$0 \leftarrow M' \leftarrow \mathbb{C}[X]^{r_0} \overset{P'(X)}{\leftarrow} \mathbb{C}[X]^{r_1} \overset{P'_1(X)}{\leftarrow} \mathbb{C}[X]^{r_2} \leftarrow \ldots$$
$$\ldots \leftarrow \mathbb{C}[X]^{r_{n-1}} \overset{P'_{n-1}(X)}{\leftarrow} \mathbb{C}[X]^{r_n} \leftarrow 0 \quad (8)$$

which terminates for some $m \leq n$ (for a concrete construction of (8), one may use the so called generalized Koszul complex, as described in [5], [19], [34]). Let us now take W a convex open set in $\mathbb{R}^n$ (or $\mathbb{C}^n$) or even a convex compact set in $\mathbb{R}^n$ (or $\mathbb{C}^n$); one has the following basic result due to Ehrenpreis, Malgrange, Hörmander, Palamodov, Komatsu, Harvey (see [20] for references):

THEOREM 4.1. *The sequence*

$$0 \to S^P(W) \to S(W)^{r_0} \overset{P(D)}{\to} S(W)^{r_1} \overset{P_1(D)}{\to} \ldots \overset{P_{n-1}(D)}{\to} S(W)^{r_n} \to 0$$

is exact.

An immediate consequence is the equally important

THEOREM 4.2. *The sequence*

$$0 \to S^P \to S^{r_0} \overset{P(D)}{\to} S^{r_1} \overset{P_1(D)}{\to} S^{r_2} \to \ldots \to S^{r_{n-1}} \overset{P_{n-1}(D)}{\to} S^{r_n} \to 0$$

is a resolution of S^P.

Before we can state Komatsu's main results, we need to introduce some further notations and propositions concerning relative cohomology groups. For S^P the sheaf defined before, one can consider $K \subset \mathbb{R}^n$ compact, and the long exact sequence associated to the pair $(\mathbb{R}^n, K)$, i.e.

$$0 \to H^0(\mathbb{R}^n, \mathbb{R}^n \backslash K; S^P) \to H^0(\mathbb{R}^n, S^P) \to H^0(\mathbb{R}^n \backslash K, S^P) \to$$
$$\to H^1(\mathbb{R}^n, \mathbb{R}^n \backslash K; S^P) \to H^1(\mathbb{R}^n, S^P) \to H^1(\mathbb{R}^n \backslash K, S^P) \to \ldots \quad (9)$$

The interesting result (consequence of Theorem 4.2, essentially) is the fact that (9) can be actually decomposed into short exact sequences, namely

THEOREM 4.3. *With the notations adopted previously, the following sequences are exact:*

$$0 \to H^0(\mathbb{R}^n, \mathbb{R}^n\backslash K; S^P) \to H^0(\mathbb{R}^n, S^P) \to H^0(\mathbb{R}^n\backslash K, S^P) \to H^1(\mathbb{R}^n, \mathbb{R}^n\backslash K; S^P) \to 0$$

and

$$0 \to H^p(\mathbb{R}^n, S^P) \to H^p(\mathbb{R}^n\backslash K, S^P) \to H^{p+1}(\mathbb{R}^n, \mathbb{R}^n\backslash K, S^P) \to 0 \qquad p \geq 1.$$

It is therefore of no little interest to discover conditions which may imply the vanishing of $H^p(\mathbb{R}^n, \mathbb{R}^n\backslash K, S^P)$; indeed, when $p = 0$, such a vanishing is equivalent (in view of Theorem 4.3) to the unique continuation property of solutions of $P(D)u = 0$, while for $p = 1$ it is equivalent to the existence of continuation and for $p \geq 2$, such a vanishing provides the isomorphism

$$H^p(\mathbb{R}^n, S^P) \cong H^p(\mathbb{R}^n\backslash K, S^P).$$

To determine such conditions, we are led to the introduction of the $\mathbb{C}[X]$-modules $\text{Ext}^p(M, \mathbb{C}[X])$ as follows: take the dual sequence of (8), namely

$$0 \to \mathbb{C}[X]^{r_0} \xrightarrow{P(X)} \mathbb{C}[X]^{r_1} \xrightarrow{P_1(X)} \mathbb{C}[X]^{r_2} \to \ldots \xrightarrow{P_{m-1}(X)} \mathbb{C}[X]^{r_m} \to 0,$$

which is obviously semi-exact. Now $\text{Ext}^p(M, \mathbb{C}[X])$ is, by definition, the p-th cohomology group of this complex (notice that, therefore, $\text{Ext}^0(M, \mathbb{C}[X])$ always vanishes if $r_0 = 1$ and $P \neq 0$).

The vanishing of the Ext modules has several interesting consequences, which are now well known, and which we list here for the reader convenience[7]

THEOREM 4.4.

$$\text{Ext}^0(M, \mathbb{C}[X]) = 0 \Leftrightarrow H^0_c(\mathbb{R}^n, \mathcal{B}^P) = 0 \Leftrightarrow H^0_c(\mathbb{R}^n, \mathcal{D}'^P) \Leftrightarrow$$

$$\Leftrightarrow H^0_c(\mathbb{R}^n, \mathcal{E}^P) = 0 \Leftrightarrow H^0(\mathbb{R}^n, \mathbb{R}^n\backslash\{0\}; \mathcal{B}^P) = 0 \Leftrightarrow H^0(\mathbb{R}^n, \mathbb{R}^n\backslash\{0\}; \mathcal{D}'^P) = 0$$

THEOREM 4.5. *For $p \geq 1$, and K bounded and convex in $\mathbb{R}^n$,*

$$\text{Ext}^p(M, \mathbb{C}[X]) = 0 \Leftrightarrow H^p(\mathbb{R}^n, \mathbb{R}^n\backslash K; \mathcal{B}^P) = 0 \Leftrightarrow H^p(\mathbb{R}^n, \mathbb{R}^n\backslash\{0\}; \mathcal{B}^P) = 0$$

THEOREM 4.6. *For $p \geq 1$, and K bounded convex and open in $\mathbb{R}^n$,*

$$\text{Ext}^p(M, \mathbb{C}[X]) = 0 \Leftrightarrow H^p(\mathbb{R}^n, \mathbb{R}^n\backslash K; \mathcal{D}'^P) = 0 \Leftrightarrow H^p(\mathbb{R}^n, \mathbb{R}^n\backslash K; \mathcal{E}^P) = 0$$

REMARK 4.1. In Theorems 4.5 and 4.6, the result is that the vanishing of the Ext modules are equivalent to the vanishing of the appropriate cohomology groups, *for all* bounded convex sets K.

We can now state what, in a sense, is Komatsu's answer to Fichera's original question.

---

[7] The results are not always stated in their most general form.

Fantappiè's Theory of Analytic Functionals 347

THEOREM 4.7. *Let K be a bounded open set in $\mathbb{R}^n$, let $P(D)$ be a system such that $\text{Ext}^p(M, \mathbb{C}[X]) = 0$ for $p = 0, 1, \ldots, m-1$, where m is the length of its free resolution (8), and let $Q(D) = P'_{m-1}(D)$, $Q_p(D) = P'_{m-p-1}(D)$. Suppose, moreover, that, either*

$$\dim H^p(\mathbb{R}^n, \mathbb{R}^n \setminus K; \mathcal{D}'^P) \leq \aleph_0, \quad p = 1, \ldots, m$$

or

$$\dim H^{m-p}(K, \mathcal{E}^Q) < +\infty \quad p = 0, 1, \ldots, m-1.$$

Then $H^p(\mathbb{R}^n, \mathbb{R}^n \setminus K; \mathcal{D}'^P)$ and $H^{m-p}(K, \mathcal{E}^Q)$ are respectively a (DSF)-space and an (FS)-space, and, for $p = 0, 1, \ldots, m$, they are dual to each other.

Proof. In view of Theorem 4.2, we have the following resolutions of $\mathcal{D}'^P$ and $\mathcal{E}^Q$, respectively:

$$0 \to \mathcal{D}'^P \to \mathcal{D}'^{r_0} \xrightarrow{P(D)} \mathcal{D}'^{r_1} \xrightarrow{P_1(D)} \ldots \xrightarrow{P_{m-1}(D)} \mathcal{D}'^{r_m} \to 0 \quad (10)$$

and

$$0 \to \mathcal{E}^Q \to \mathcal{E}^{r_m} \xrightarrow{Q(D)} \mathcal{E}^{r_{m-1}} \to \ldots \xrightarrow{Q_{m-2}(D)} \mathcal{E}^{r_1} \xrightarrow{Q_{m-1}(D)} \mathcal{E}^{r_0} \to 0. \quad (11)$$

Since (10) is a soft resolution of $\mathcal{D}'^P$ we have [13] that the groups $H^p(\mathbb{R}^n, \mathbb{R}^n \setminus K; \mathcal{D}'^P)$ are the cohomology groups of the complex (with obvious notations)

$$0 \to \mathcal{D}'_K(\mathbb{R}^n)^{r_0} \xrightarrow{P(D)} \mathcal{D}'_K(\mathbb{R}^n)^{r_1} \to \ldots \xrightarrow{P_{m-1}(D)} \mathcal{D}'_K(\mathbb{R}^n)^{r_m} \to 0. \quad (12)$$

since $\mathcal{D}'_K(\mathbb{R}^n)$ is a (DSF)-space, and since the open mapping theorem holds for the strong dual of reflexive Fréchet spaces, one can use the Schwartz lemma [32] together with the countability hypothesis, to show that the maps $P_j(D)$ have closed ranges; similarly, the groups $H^{m-p}(K, \mathcal{E}^Q)$ are the cohomology groups of the complex

$$0 \to \mathcal{E}(K)^{r_m} \xrightarrow{Q(D)} \ldots \to \mathcal{E}(K)^{r_1} \xrightarrow{Q_{m-1}(D)} \mathcal{E}(K)^{r_0} \to 0$$

and, if $\dim H^{m-p}(K, \mathcal{E}^Q) < +\infty$, again by the Schwartz lemma (and since each $\mathcal{E}(K)$ is an (FS)-space) we get that the ranges of $Q_j(D)$ are closed. Finally, since the sequences (12) and (13) are dual one another, we conclude with Serre's lemma [32] to get the desired result. □

REMARK 4.2. In view of the vanishing of the Ext modules we immediately get (from Theorems 2.8 and 2.10) that if K is also convex, then most relative cohomology groups vanish and what one gets is exactly

$$H^m(\mathbb{R}^n, \mathbb{R}^n \setminus K; \mathcal{D}'^P) \cong (H^0(K, \mathcal{E}^Q))'. \quad (14)$$

This is, indeed, the generalization of Fantappiè's theory; in fact, for $P = \bar{\partial} = \{\partial/\partial \bar{z}_j\}$, $\mathbb{R}^{2n} = \mathbb{C}^n$, $Q = -\bar{\partial}$, $m = n$, we deduce that (in hypotheses of finiteness of cohomology), $H^n(\mathbb{R}^{2n}, \mathbb{R}^{2n} \setminus K; \mathcal{O}) = \mathcal{B}_K(\mathbb{R}^n) = H'(K)$, which is just Theorem 3.2. Therefore, if we take a system Q and the sheaf of its C^∞ solutions $\mathcal{E}^Q$, and we look at the dual of the space $\mathcal{E}^Q(K)$, we still get a cohomology space, whose elements we can regard as *indicatrices* for our more general functionals.

REMARK 4.3. The topological isomorphism (14) (which might be referred to as the Fantappiè-Fichera-Martineau-Komatsu duality theorem) can actually be endowed of a more refined meaning, at least in some instances. Suppose, for example, that $H^m(\mathbb{R}^n, \mathcal{D}'^P)$ vanishes (this condition is, actually, a condition on the system P, which is verified, e.g., by the Cauchy-Riemann system) then Komatsu's result, together with the second part of Theorem 4.4, gives (for $m \geq 2$ and therefore $n \geq 2$)

$$[\mathcal{E}^Q(K)]' \simeq H^{m-1}(\mathbb{R}^n \backslash K, \mathcal{D}'^P),$$

which is the exact counterpart of Martineau's characterization of Fantappiè's antisymmetric indicatrix. Finally, in case we look at $\mathbb{C} = \mathbb{R}^2$, $m = 1$, we have an isomorphism of the form

$$[\mathcal{E}^Q(K)]' \simeq H^m(\mathbb{C}, \mathbb{C}\backslash K; \mathcal{D}'^P) = H^1(\mathbb{C}, \mathbb{C}\backslash K; \mathcal{D}'^P);$$

however, in view of the vanishing of Ext^0, and of the first part of Theorem 4.3, one gets a nice characterization of our generalized functionals, which is very much similar to the one which arises in the case of analytic functionals, namely

$$[\mathcal{E}^Q(K)]' \simeq \frac{H^0(\mathbb{C}\backslash K, \mathcal{D}'^P)}{H^0(\mathbb{C}, \mathcal{D}'^P)}.$$

REMARK 4.4. Let us conclude this section by pointing out that Theorem 4.7 can be also formulated in the framework of the sheaves $\mathcal{B}$ and $\mathcal{A}$ (instead of $\mathcal{D}'$ and $\mathcal{E}$), with a proof which follows exactly the lines of the proof of Theorem 4.7. We shall come back to this point in the next section.

5. INFINITE ORDER OPERATORS

In this last section we extend Theorem 4.7 to a more general situation, namely to the case in which $P(D)$ represents a suitable system of infinite order linear differential operators with constant coefficients. The theory of such operators is now quite well developed (mainly thanks to the work of the Japanese school following Sato) and has proved his usefulness in many instances (let us just point out [1] and the references therein); for this reason we have thought it may not be useless to investigate to which extent the theory of Fantappiè can be generalized to include solutions to suitable systems of such operators (of which, once again, the Cauchy-Riemann system is a simple example). As it will be clear, we follow the main ideas contained in section 4, but the technical problems which we have to face are much more complicated (e.g. we shall see that the polynomial ring $\mathbb{C}[X]$ has to be replaced by the ring A of entire functions of exponential minimal type) and we exploit here some recently developed results [18]. Of course, as it is well known, differential operators of infinite order act as sheaf homomorphisms on the sheaves $\mathcal{O}, \mathcal{B}$ and $\mathcal{A}$, which, therefore, will be the only sheaves to be considered in this section.

Let us start by giving some basic definitions: we denote by A the ring of entire functions of exponential minimal type, i.e.

$$A = \{f \in H(\mathbb{C}^n) : \text{ for any } \varepsilon > 0 \text{ there exists } C_\varepsilon > 0 \text{ such that } |f(z)| \leq C_\varepsilon \exp(\varepsilon|z|)\}.$$

REMARK 5.1. A is endowed with a natural (FS)-space topology which makes it topologically isomorphic, via the Fourier transform, to the strong dual of the space $\mathcal{O}(\{0\})$ of germs of holomorphic functions at the origin, i.e., with our notations,

$$A = \mathcal{F}(\mathcal{O}(\{0\})') = \mathcal{F}(H'(0)).$$

On the other hand, we denote by $\hat{A}$ the ring of linear differential operators of infinite order with constant coefficients, defined on $\mathbb{C}^n$; the same relation which links $\mathbb{C}[X]$ to finite order linear differential operators with constant coefficients, can be established between A and $\hat{A}$. Therefore, for $f(\zeta)$ in $A(\zeta = (\zeta_1, \ldots, \zeta_n) \in \mathbb{C}^n)$, we find an operator $f(D)$ in $\hat{A}$ by replacing ζ_j with $D_j = \partial/\partial z_j$ and viceversa. From this point of view, the operator $f(D)$ may be considered as the convolution operator determined by the inverse Fourier transform of $f(\zeta)$ (of course, we are dealing with a very special convolution operator, whose support is at the origin).

In the sequel we shall look to systems of such operators in which only one unknown appears (even though, as we shall poin out later, we can think of a natural way of going to the most general case), i.e. we take $\vec{f}(D) = (f_1(D), \ldots, f_N(D))$, $1 \leq N \leq n$, and we consider the sheaf homomorphism

$$\vec{f}(D) : S \to S^N,$$

where S is either A, B or $\mathcal{O}$. As in section 4, we shall denote by the symbol $S^{\vec{f}}$ the sheaf of solutions of the homogeneous system

$$\vec{f}(D)u = 0,$$

i.e.

$$f_1(D)u = f_2(D)u = \ldots = f_N(D)u = 0.$$

As we said previously, a key tool is the Hilbert[8] syzygy theorem (namely sequence (8)), but it can be easily shown that such a result will not necessarily hold in the ring A. This consideration led Kawai and the author to study, in [18], which conditions might be imposed to restore such a crucial result: let us briefly summarize our analysis, which is carried out in detail in [18], sections 2 and 3.

DEFINITION 5.1. Let $\vec{f} = (f_1, \ldots, f_N)$, $f_j \in A$, be such that the characteristic variety

$$V = \{z \in \mathbb{C}^n : f_1(z) = \ldots = f_N(z) = 0\}$$

is a complete intersection $(n-N)$-dimensional variety. We say that $\vec{f}$ is A-*slowly decreasing* if there exists a family $\mathcal{L} = \{L\}$ of N-dimensional affine complex spaces with

$$\bigcup_{L \in \mathcal{L}} L = \mathbb{C}^n,$$

---

[8] The name of Hilbert is attached to this result, which in his most general form is due to Palamodov, as Hilbert proved an earlier version for homogeneous modules.

and there are constants $C_1, C_2 > 0$ and two sequences of positive numbers $\delta_n \to 0$ and A_n, $\{A_n\}$ being increasing, such that

i) for each $L \in \mathcal{L}$, the set

$$S_L(\vec{f}, n) = \{z \in L : |f(z)| < A_n \exp(-\delta_n |z|)\}$$

has relatively compact connected components;

ii) if z_1, z_2 are two points in the closure of the same component of some $S_L(\vec{f}, n)$, then

$$|z_1| \leq C_1 |z_2| + C_2.$$

REMARK 5.2. For $N = 1$, one can easily verify that any $f \in A$ satisfies Definition 5.1, i.e. it is A-slowly decreasing. This implies a fairly well known fact, see e.g. [16], which is the surjectivity of $f(D) : H(0) \to H(0)$ or equivalently the closure of any principal ideal in A. Definition 5.1 arose when trying to find systems for which similar closure of ideals properties might hold; we should mention that our search was clearly directed by the basic paper of Berenstein and Taylor [38], where these problems are addressed, though in a different setting.

DEFINITION 5.2. Let Ω be an open set in $\mathbb{C}^n$; we say that Ω is *good* if there is a connected component G of some $S_L(\vec{f}, n)$ such that, for some A_θ, $\theta > 0$,

$$\Omega = \{z \in \mathbb{C}^n : \text{ there exists } \zeta \in G \text{ such that } |z - \zeta| < A_\theta \exp(-\theta |z|)\}.$$

DEFINITION 5.3. When we fix $\theta, A_\theta, \delta_n, A_n$, the family $\mathcal{G}$ of all good sets which one finds is called a *good family* of open sets. If we now increase δ_n, θ and decrease A_θ, A_n, we get a new good family $\mathcal{G}'$, which we call a *good refinement* of $\mathcal{G}$, and a natural map $\rho : \mathcal{G}' \to \mathcal{G}$.

DEFINITION 5.4. A family $\mathcal{L} = \{L\}$ as in Definition 5.1 is said to be *almost parallel* if, for every good family $\mathcal{G}$, there exists a good refinement $\mathcal{G}'$ of $\mathcal{G}$ such that

$$\forall \Omega_1, \Omega_2 \in \mathcal{G}', \quad \Omega_1 \cap \Omega_2 \neq \emptyset \Rightarrow \overline{\Omega}_1 \cap \overline{\Omega}_2 \subset \rho(\Omega_1) \cap \rho(\Omega_2).$$

The reason for all these complicated (and apparently unexplicable!) definitions is the following basic result, whose complete proof the reader may find in [18].

THEOREM 5.1. Let $f = (f_1, \ldots, f_N)$, $1 \leq N \leq n$ be A-slowly decreasing with respect to some almost parallel family on lines $\mathcal{L}$. Then:

i) the ideal $I = Af_1 + \ldots + Af_N$ is closed in A;
ii) the A-module $M = A/Af_1 + \ldots + Af_N$ admits a free Koszul complex resolution

$$0 \leftarrow M \leftarrow A \xleftarrow{F_0} A^N \xleftarrow{F_1} \ldots \leftarrow A^{\binom{N}{N-1}} \xleftarrow{F_{N-1}} A \leftarrow 0. \tag{15}$$

REMARK 5.3. Let us describe more explicitely the maps in (15), at least in the simple case in which $N = 3$. The detailed general construction can be found, e.g., in [18]. When $N = 3$, resolution (15) becomes

$$0 \leftarrow \frac{A}{Af_1 + Af_2 + Af_3} \leftarrow A \xleftarrow{F_0} A^3 \xleftarrow{F_1} A^3 \xleftarrow{F_2} A \leftarrow 0$$

Then, obviously, the map $A \to \dfrac{A}{Af_1 + Af_2 + Af_3}$ is the usual quotient map and nothing need to be said. Also no comment is necessary on $F_0 : A^3 \to A$ which act as follows:

$$F_0(g_1, g_2, g_3) = g_1 f_1 + g_2 f_2 + g_3 f_3.$$

Hence, in particular $F_0(A^3)$ vanishes in M. On the other hand we now want to define $F_1 : A^3 \to A^3$; in this case F_1 actually acts on the following spaces

$$F_1 : \wedge \binom{3}{2} \mathbb{C}^3 \otimes A \to \wedge \binom{3}{1} \mathbb{C}^3 \otimes A,$$

and it is given by

$$F_1(h_{12}, h_{13}, h_{23}) = (h_{12} f_2 + h_{13} f_3, -h_{12} f_1 + h_{23} f_3, -h_{13} f_1 - h_{23} f_2);$$

it is clear that $F_0 \circ F_1 = 0$. Finally we have

$$F_2 : \wedge \binom{3}{3} \mathbb{C}^3 \otimes A \to \wedge \binom{3}{2} \mathbb{C}^3 \otimes A,$$

i.e.

$$F_2 : A \to A^3,$$

which is defined as follows:

$$F_2(k) = (kf_3, -kf_2, kf_1);$$

once again, it is immediately seen that $F_1 \circ F_2 = 0$ and that F_2 is injective; notice, on the other hand, that Theorem 5.1 shows (in ii)) that under suitable hypotheses, this semiexact sequence is actually exact.

REMARK 5.4. As it was pointed out in [18], to which we refer the reader for more details, there are plenty of examples of N-tuples which satisfy the complicated requirement of being A-slowly decreasing. Many of them can actually be constructed by employing standard techniques from the theory of infinite products. To be explicit, we shall just point out that, in $\mathbb{C}^n$, the N-tuple, $1 \leq N \leq n$, given by

$$f_1(z_1, \ldots, z_n) = \operatorname{Ch}\sqrt{z_1}, \ \ldots, \ f_N(z_1, \ldots, z_n) = \operatorname{Ch}\sqrt{z_N}$$

(where, of course, $\operatorname{Ch}\sqrt{z_j} := \sum_{n=0}^{+\infty} \dfrac{z_j^n}{(2n)!}$) is A-slowly decreasing.

By recalling that the Fourier transform is an isomorphism between A and $\hat{A}$, and observing that $\vec{f}(z)$ is A-slowly decreasing, we can translate Theorem 5.1 in the following statement, in which we actually write F_0 to denote $F_0(D)$, and similarly F_j to denote $F_j(D)$:

THEOREM 5.2. *Let* $\vec{f} = (f_1, \ldots, f_N)$, $1 \leq N \leq n$ *be A-slowly decreasing (with respects to some almost parallel family of lines). Then :*

i) *the ideal* $\hat{I} = \hat{A}f_1(D) + \ldots + \hat{A}f_N(D)$ *is closed in* $\hat{A}$;
ii) *the* $\hat{A}$*-module* $\hat{M} = \hat{A}/\hat{A}f_1(D) + \ldots + \hat{A}f_N(D)$ *admits a free Koszul complex resolution*

$$0 \leftarrow \hat{M} \leftarrow \hat{A} \overset{F_0}{\leftarrow} \hat{A}^N \overset{F_1}{\leftarrow} \ldots \leftarrow \hat{A}^{\binom{N}{N-1}} \overset{F_{N-1}}{\leftarrow} \hat{A} \leftarrow 0.$$

Analogues of Theorems 4.1, 4.2 may as well be readily obtained.

THEOREM 5.3. *If* Ω *is any open convex set in* $\mathbb{C}^n$, *and if* $\vec{f}$ *is A-slowly decreasing, then*

$$0 \to H'(\Omega) \overset{{}^t F_{N-1}}{\to} H'(\Omega)^N \to \ldots \to H'(\Omega)^N \overset{{}^t \vec{f}(D)}{\to} H'(\Omega)$$

is exact, and the image of ${}^t\vec{f}(D)$ *is closed in* $H'(\Omega)$.

Proof. This result is essentially contained in [18], where the case of more general convolution operators is dealt with (so that a more restrictive slowly decreasing condition is needed). We will not repeat here the rather long argument which is required, and we simply point out that the whole proof is based on a division theorem in the space $\hat{H}'(\Omega)$. □

THEOREM 5.4. *With the hypotheses above, the following sequences is exact, where* $S = \mathcal{O}$, $\mathcal{A}$ *or* $\mathcal{B}$.

$$0 \to S(\Omega) \to S(\Omega) \overset{\vec{f}(D)}{\to} S(\Omega)^N \overset{F_1}{\to} \ldots S(\Omega)^N \overset{F_{N-1}}{\to} S(\Omega) \to 0.$$

Proof. When $S = \mathcal{O}$, the result is an immediate consequence of Theorem 5.3 and of Schwartz's lemma. Restricting $\mathcal{O}$ to $\mathbb{R}^n$, and taking $\Omega \subseteq \mathbb{R}^n$, we get the result for the sheaf $\mathcal{A}$. Finally, to prove the theorem for $S = \mathcal{B}$, one takes a convex open set V in $\mathbb{C}^n$ such that $V \cap \mathbb{R}^n = \Omega$, and represents $\mathcal{B}(\Omega)$ by means of $H^n(\mathcal{V}, \mathcal{V}'; \mathcal{O})$, where

$$\mathcal{V} = \{V_0, \ldots, V_n\}, \; \mathcal{V}' = \{V_1, \ldots, V_n\} \quad \text{with} \quad V_0 = V, \; V_j = \{z \in V : \text{Im } z_j \neq 0\}.$$

Since this theorem has been proved for $S = \mathcal{O}$ on each $V_{i_0, \ldots, i_p}$, we immediately get the thesis. □

An immediate corollary of this last result is

THEOREM 5.5. *If* $\vec{f}$ *is A-slowly decreasing, and* S *is any sheaf among* $\mathcal{O}, \mathcal{A}, \mathcal{B}$, *then*

$$0 \to S^{\vec{f}} \to S \overset{\vec{f}(D)}{\to} S^N \overset{F_1}{\to} \ldots \to S^N \overset{F_{N-1}}{\to} S \to 0$$

is a resolution (flabby if $S = \mathcal{B}$ *) of the sheaf* $S^{\vec{f}}$ *of solutions of* $\vec{f}(D)u = 0$.

Proof. Simply notice that convex sets form a fundamental system of neighborhoods at any point. □

Let us finally point out a significant consequence of Theorems 5.1 and 5.2, concerning the vanishing of some Ext groups (this, too, is proved in detail in [18]):

THEOREM 5.6. *If $\vec{f} = (f_1,\ldots,f_N)$ is A-slowly decreasing then we have:*

$$\mathrm{Ext}^j_{\tilde{A}}(\hat{M},\mathcal{O}) = 0, \quad \text{for} \quad j \geq 1$$

and

$$\mathrm{Ext}^j_{\tilde{A}}(\hat{M},\mathcal{B}_{\{0\}}) = 0, \quad \text{for} \quad j = 0,\ldots,N-1.$$

This result can be directly applied to be study of hyperfunctions solutions of $\vec{f}(D)u = 0$, just as we did in section 4: namely

THEOREM 5.7. *If $\mathrm{Ext}^0_{\tilde{A}}(\hat{M},\mathcal{B}_{\{0\}}) = 0$, then we have the unique continuation property for solutions of $\hat{f}(D)u = 0$. More generally, if $\mathrm{Ext}^j_{\tilde{A}}(\hat{M},\mathcal{B}_{\{0\}}) = \mathrm{Ext}^{j+1}_{\tilde{A}}(\hat{M},\mathcal{B}_{\{0\}}) = 0$, then, for every open subset U of $\mathbb{R}^n$ that contains the origin*

$$\mathrm{Ext}^j_{\tilde{A}}(\hat{M},\mathcal{B}(U)) \cong \mathrm{Ext}^j_{\tilde{A}}(\hat{M},\mathcal{B}(U\setminus\{0\})).$$

Proof. Indeed, from the flabbiness of the sheaf $\mathcal{B}$ of hyperfunctions, [31], it follows that

$$0 \to \mathrm{Ext}^0_{\tilde{A}}(\hat{M},\mathcal{B}_{\{0\}}) \to \mathrm{Ext}^0_{\tilde{A}}(\hat{M},\mathcal{B}(U)) \to$$
$$\to \mathrm{Ext}^0_{\tilde{A}}(\hat{M},\mathcal{B}(U\setminus\{0\})) \to \mathrm{Ext}^1_{\tilde{A}}(\hat{M},\mathcal{B}_{\{0\}}) \to \ldots$$

is exact for any open subset U as in the statement of the theorem. The result is now obvious.

□

REMARK 5.5. This result was exploited in [18] to prove a new generalization of Hartogs' removable singularities theorem.

REMARK 5.6. In every preceding statment, one can replace $\{0\}$ with any compact set $K \subset \mathbb{R}^n$, provided a slightly stronger A_K-slowly decreasing condition is requested: see [18] for details.

We can finally prove our improvement on Komatsu's results which, as we already noticed many times, goes in the direction of extending Fantappiè's theory:

THEOREM 5.8. *Let $\vec{f}(D) = (f_1(D),\ldots,f_N(D))$ be such that $\vec{f}$ is A-slowly decreasing. Let K be a compact set in $\mathbb{R}^n$ such that either*

$$\dim H^p(\mathbb{R}^n,\mathbb{R}^n\setminus K;\mathcal{B}^{\vec{f}}) < +\infty, \quad p = 1,2,\ldots,N \quad \text{or, with} \quad g =^t F_{n-1}(D), \quad (16)$$
$$\dim H^{N-p}(K,\mathcal{A}) \leq \aleph_0, \quad p = 0,1,\ldots,N-1. \quad (17)$$

Then the cohomology groups $H^p(\mathbb{R}^n,\mathbb{R}^n\setminus K;\mathcal{B}^{\vec{f}})$ and $H^{N-p}(K,\mathcal{A}^g)$, with natural (FS)-space and (DFS)-space topologies, are strong dual spaces one another for $p = 0,1,\ldots,N$. More precisely, all $H^p(\mathbb{R}^n,\mathbb{R}^n\setminus K;\mathcal{B}^{\vec{f}})$ vanish for $p < N$ and

$$H^N(\mathbb{R}^n,\mathbb{R}^n\setminus K;\mathcal{B}^{\vec{f}}) = (\mathcal{A}^g(K))' \quad (18)$$

Proof. In view of Theorem 5.3, the groups $H^p(\mathbb{R}^n, \mathbb{R}^n\setminus K; \mathcal{B}^{\vec{f}})$ are the cohomology groups of the complex

$$0 \to \mathcal{B}_K(\mathbb{R}^n) \xrightarrow{\vec{f}(D)} \mathcal{B}_K(\mathbb{R}^n)^N \to \ldots \to \mathbb{B}_K(\mathbb{R}^n) \to 0, \qquad (19)$$

and since each $\mathcal{B}_K(\mathbb{R}^n)^{\binom{N}{i}}$ is an (FS)-space, all the linear maps in (19) have closed range if (16) holds true (just Schwartz's lemma). On the other hand, the groups $H^{N-p}(K, \mathcal{A}^g)$ are the cohomology groups of the complex

$$0 \to \mathcal{A}(K) \xrightarrow{g} \mathcal{A}(K)^N \to \ldots \to \mathcal{A}(K) \to 0, \qquad (20)$$

which is the dual complex to (19). As before, the maps in (20) have closed range if (17) holds, in view again of Schwartz's lemma. Finally the theorem follows from Serre's lemma. The last part of the theorem is just a consequence of Theorems 5.4 and 5.5. □

REMARK 5.7. Theorem 5.8 greatly generalizes the first results of Fantappiè; we are now talking about hyperfunctions and about infinite order differential equations, while the natural environment for Fantappiè's work was with the Cauchy-Riemann system for C^∞ functions. Still we would like to point out two further possible generalization of our own result, which, we feel confident, can be successfully attacked. The first, and simplest, consists in replacing our system $\vec{f}(D)$, where only one unknown is allowed, by a «general» rectangular system. In this case, all the theory described in this section has to be built anew, and even [18] would have to be entirely written. Still closure of modules theorems, in case of entire functions with growth conditions, are worked out in detail in [34], where a «slowly decreasing» condition for rectangular systems is described and effectively applied. Therefore, it would probably be only a matter of patience to recover a result like Theorem 5.8 for these more general systems. A second, more delicate, extension, relate to the case of *variable coefficients* differential operators (either of finite or of infinite order). In this case, closure of ideals theorems have not yet been worked out, and most of the work has really to be done.

Let us conclude our paper by pointing out one other application of Theorem 5.8, which relates to Bengel's P-functionals.

DEFINITION 5.5. A system $\vec{f}$ of infinite order partial differential operators is said to be *elliptic* if

$$\mathcal{B}^{\vec{f}} = \mathcal{A}^{\vec{f}}.$$

In case $\vec{f}$ is of finite order, the theory of such operators is rather well known (see [20] and the references therein), while if $\vec{f}$ is of infinite order but $N = 1$ (i.e. we have a single operator) the elliptic case is studied in detail in [16]. We do not wish to carry out any detailed study of elliptic systems, except for the following simple observation: when $\vec{f}$ is elliptic, and always under the slowly decreasing hypotheses, Theorem 5.5 actually provides a flabby resolution for $\mathcal{A}^{\vec{f}}$ and one gets:

COROLLARY 5.1. If $\vec{f}$ is $\mathcal{A}$-slowly decreasing and elliptic, then (with the usual notations)

$$H^N(\mathbb{R}^n, \mathbb{R}^n\setminus K; \mathcal{A}^{\vec{f}}) = (\mathcal{A}^g(K))'.$$

Proof. Just apply the ellipticity condition to (18). □

REMARK 5.8. Let us point out the relations of Corollary 5.1 with the well known duality theorem of Grothendieck for compactly supported Bengel P-functionals, [4]. Indeed, if $P(D)$ is a single elliptic operator (with constants coefficients, and of finite order) in $\mathbb{R}^{n+1}$, Bengel defined [2], [3] the P-functionals on an open set $\Omega \subseteq \mathbb{R}^n$ to be the elements of the first relative cohomology group $H^1(V, V\backslash\Omega; \mathcal{A}^{P'(D)})$, where V is an open set in $\mathbb{R}^{n+1}$, containing Ω as a relatively closed set. Since the flabby resolution for $\mathcal{A}^{P(D)}$ has now length one, it is possible to prove the following duality:

$$(\mathcal{A}^{P(D)}(K))' = H^1(\mathbb{R}^{n+1}, \mathbb{R}^{n+1}\backslash K; \mathcal{A}^{P'(D)}). \tag{21}$$

When $P(D)$ is still elliptic, but of infinite order, (21) is still true, as it was shown by Kawai in [16]; henceforth, we can think of our Corollary 5.1 as an extension of the Bengel-Kawai duality, to the case of systems of infinite order differential operators.

REFERENCES

[1] T. Aoki, M. Kashiwara and T. Kawai, On a class of linear differential operators of infinite order with finite index, *Adv. in Math.* 62 (1986), 155-168.

[2] G. Bengel, Sur une extension de la théorie des hyperfonctions, *C.R. Acad. Sci. Paris* 262 (1966), 499-501.

[3] G. Bengel, Régularité des solutions hyperfonctions d'une équation elliptique, *C.R. Acad. Sci. Paris* 262 (1966) 569-570.

[4] G. Bengel, Das Weyl' sche Lemma in der Theorie der Hyperfunktionen, *Math. Z.* 96 (1967), 373-392.

[5] D. Buchsbaum, A generalized Koszul complex, I, *Trans. Amer. Math. Soc.* 111 (1964), 183-196.

[6] J. Dieudonnè and L. Schwartz, La dualité dans les espaces (F) et (LF), *Ann. Inst. Fourier* 1 (1949), 61-101.

[7] L. Ehrenpreis, *Fourier Analysis in Several Complex Variables*, New York, 1970.

[8] L. Fantappiè, Nuovi fondamenti della teoria dei funzionali analitici, *Mem. Acc. d'Italia* 12 (1941).

[9] L. Fantappiè, *Teoria de los funcionales analiticos y sus aplicaciones*, Barcelona, 1943.

[10] L. Fantappiè, Su un'espressione generale dei funzionali lineari mediante le funzioni «para-analitiche» di più variabili, *Rend. Sem. Mat. Padova* 22 (1953), 1-10.

[11] G. Fichera, La vita matematica di Luigi Fantappiè, *Rend. Sem. Mat. Roma* 16 (1957), 143-160.

[12] G. Fichera, *personal communication*, 1987.

[13] R. Godement, *Topologie Algèbrique et Théorie des Faisceaux*, Paris, 1958.

[14] H. Grauert, On Levi's problem and the imbedding of real analytic manifolds, *Ann. of Math.* 68 (1958), 460-472.

[15] A. Grothendieck, Sur certain espaces de fonctions holomorphes, I, *J. Reine Angew. Math.* 192 (1953), 34-64.

[16] T. Kawai, On the theory of Fourier hyperfunctions and its applications to partial differential equations with constant coefficients, *J. Fac. Sci. Univ. Tokyo* 17 (1970), 467-517.

[17] M. Kashiwara, T. Kawai and T. Kimura, *Introduction to Algebraic Analysis*, Princeton, 1986.

[18] T. Kaway and D.C. Struppa, *Existence theorems for holomorphic solutions of infinite order differential equations*, Int. J. Math. 1 (1990), to appear.
[19] J. Kelleher and B.A. Taylor, *On finitely generated modules over some rings of analytic functions*, unpublished (1971).
[20] H. Komatsu, Relative cohomology of sheaves of solutions of differential equations, *Springer Lecture Notes in Mathematics 287 (1973), 192-261.*
[21] G. Köthe, Dualität in der Funktionentheorie, J. Reine Angew. Math 181 (1953), 30-49.
[22] A. Martineau, Les hyperfonctions de M. Sato, *Séminaire Bourbaki* 244 (1960), 1-13.
[23] A. Martineau, Indicatrices des fonctions analytiques et inversion de la transformation de Fourier-Borel par la transformation de Laplace, *C.R. Acad. Sci. Paris* 255 (1962), 1845-1847.
[24] A. Martineau, Equation différentielles d'ordre infini, in *Séminaire sur les équations aux dérivees II, College de France* 94 (1954-66), 49-110 and 95 (1965-66), 109-154.
[25] A. Martineau, Sur la topologie des espaces de fonctions holomorphes, *Math. Ann.* 163 (1966), 62-88.
[26] A. Martineau, Sur la notation d'ensemble fortement linéellment convex, *Anais. Acad. Brasil. Ciênc.* 40 (1968), 427-435.
[27] A. Martineau, Travaux de A. Martineau en analyse fonctionelle, (1969), in *Oeuvres de André Martineau*, Paris, 1977, 15-20.
[28] A. Meril and D.C. Struppa, Convolutors in spaces of holomorphic functions, *Springer lecture Notes in Mathematics* 1276 (1987), 253-275.
[29] V.P. Palamodov, *Linear Differential Operators with Constant Coefficients*, Berlin-Heidelberg-New York, 1970.
[30] F. Pellegrino, La Théorie des fonctionelles analytiques et ses applications, in *Problèmes concrets d'analyse fonctionnelle*, by P. Levy, Paris, 1951.
[31] M. Sato, On a generalization of the concept of function, *Proc. Jap. Acad.* 34 (1958), 126-130 and 604-608.
[32] J.P. Serre, Un théorème de dualité, *Comm. Math. Helv.* 29 (1955), 9-26.
[33] J.S. e Silva, As funçoes analiticas e a Analise functional, *Port. Math.* 9 (1950), 1-130.
[34] D.C. Struppa, The fundamental principle for systems of convolution equations, *Mem. Amer. Math. Soc.* 273 (1988).
[35] D.C. Struppa, Luigi Fantappiè e la teoria dei funzionali analitici, in *Atti del Convegno «La matematica Italiana tra le due guerre mondiali», Milano-Gargnano, ottobre 1986*, Pitagora, Bologna, 1987.
[36] D.C. Struppa, The first eighty years of Hartogs' theorem, *Sem. Geom. Dip. Mat. Univ. Bologna*, 1987.
[37] D.C. Struppa and C. Turrini, in preparation.
[38] B.A. Taylor and C.A. Berenstein, Interpolation problems in $\mathbb{C}^n$ with applications to harmonic analysis, *J. An. Math.* 38 (1980), 188-254.

On B. Segre and the Theory of Polar Varieties

BERNARD TEISSIER, Département des Mathématiques et de l'Informatique, Ecole normale supérieure, 45 rue d'Ulm, 75230, Paris

It is an honour for me to lecture at the University of Bologna, especially since two great algebraic geometers, Beppo Levi and Beniamino Segre, whose work on singularities has had lasting influence, were Professors here.

Both Beppo Levi (see [14]) and Beniamino Segre made use of polar varieties of projective varieties in their studies of resolution of singularities of surfaces. Segre also introduced the Segre classes, which are similar in spirit to the polar classes, replacing tangent spaces by secant lines, and provide numerical invariants of embeddings. He used them to give an alternate construction of the characteristic classes of non singular projective varieties that had been defined by Todd using polar classes.

1. Local polar varieties

Let me begin by recalling the more modern definition of *local* polar varieties, according to [23], [11], [25],[6]. Consider a diagram of complex analytic spaces and maps, all assumed to be "sufficiently small" representatives of germs

$$\begin{array}{c} X \hookrightarrow S \times \mathbf{C}^N \\ \downarrow f \swarrow \\ S \end{array}$$

assume that there is a dense open analytic subset $X° \subset X$ such that the restriction of f to $X°$ is flat with nonsingular fibers purely of dimension d. Let us now fix an integer k, $o \leq k \leq d$ and take a linear projection $p: \mathbf{C}^N \to \mathbf{C}^{d-k+1}$. The closure $P_k(f;p)$ in X of the critical locus of the restriction $Id_S \times p|X°$ is by definition the polar variety of f associated to the given installation in $S \times \mathbf{C}^N$ and the projection p. One usually considers the polar varieties associated to " sufficiently general" projections.

One finds it helpful to build the relative conormal space $C_f(X)$ of X in $S \times \mathbf{C}^N$, which is the closure in $S \times \mathbf{C}^N \times \check{\mathbf{P}}^{N-1}$ of the set of couples (x, H) where $x \in X^\circ$ and H is a direction of hyperplane in $\mathbf{C}^N$ tangent at x to the fiber $f^{-1}(f(x))$. One then has the following diagram:

$$\begin{array}{ccc} C_f(X) & \hookrightarrow & S \times \mathbf{C}^N \times \check{\mathbf{P}}^{N-1} \\ \downarrow \kappa_f & & \downarrow \\ X & \hookrightarrow & S \times \mathbf{C}^N \\ \downarrow f & \swarrow & \\ S & & \end{array}$$

If one denotes by $\lambda_f: C_f(X) \to \check{\mathbf{P}}^{N-1}$ the natural projection, and by $L^{d-k} \subset \check{\mathbf{P}}^{N-1}$ the set of hyperplanes containing ker p, then one has for a sufficiently general projection p the set theoretic equality

$$\kappa_f(\lambda_f^{-1}(L^{d-k})) = P_k(f; p) .$$

However it is certain that one should also study the total family, parametrized by the projections, of the polar varieties of a given type.

In the special case where S is a point and X is the cone in $\mathbf{C}^N$ on a reduced projective variety in $\mathbf{P}^{N-1}$, one recovers the classical theory of polar loci.

Originally in [11] and [23], the definition made use of the relative Nash modification of X (see [26], [25], p.417) instead of the relative conormal space; one can then in a completely analogous way also define more generally polar varieties corresponding to a general Schubert cycle associated to a set of incidence dimensions $\mathbf{a} = (a_1, \ldots, a_d)$ and a given flag of linear subspaces. Since they depend upon the flag, what we get is actually a system of subvarieties of X parametrized by a flag manifold. In order to have a well-defined object it was deemed in the classical projective and «absolute» case to be necessary to introduce an equivalence relation on the set of algebraic varieties such that at least for «almost all» flags the corresponding polar varieties are all equivalent. Motivated by the search for numerical invariants, Segre, Severi and Todd considered rational equivalence classes, which, however, forget all but the simplest geometric characters (such as dimension and degree) of an algebraic variety.

The theory of equisingularity allows one to try and preserve much more and still have well-defined objects, since for instance, for «almost all» flags, the corresponding polar varieties are all equisingular. In particular, at least the topology of a general polar variety is well-defined.

The purpose of this lecture is to try to draw attention by examples to the importance of the geometric viewpoint in the theory of polar varieties, and to ask a few questions connected with this viewpoint. Here, geometric is meant as opposed to the cohomological, or cyclist, or numerical, viewpoint.

The geometry of the local polar varieties associated to a germ of a complex space or of a complex morphism contains a wealth of information about this germ, and since polar varieties are of lower dimension they can be used for inductions. They can also be used to define numerical invariants considerably more subtle than those arising from the cyclist viewpoint. Since we take a local viewpoint, it is necessary to check that the objects which we define are independant of the choices of local embeddings and coordinates. All the geometric situations considered will be local unless otherwise specified.

THEOREM - *Given a complex analytic morphism* $f: (X, x) \to (S, s)$ *satisfying the conditions above, the Whitney-equisingularity type of a general polar variety* $P_k(f; D_{d-k+1})$ *depends only upon the analytic type of the map-germ* f *at* x.

The proof is essentially contained in ([25], Chap.4), and let me first give it in the case of a complex analytic map-germ $f:(\mathbf{C}^2, 0) \to (\mathbf{C}, 0)$.

In this case, after a choice of coordinates x, y, given a linear projection $\ell: \mathbf{C}^2 \to \mathbf{C}$, the corresponding polar curve is the zero set of the 2-form $df \wedge d\ell$; il $\ell = x + by$, then an equation for our polar is $\frac{\partial f}{\partial y} - b\frac{\partial f}{\partial x} = 0$. By the general results on equisingularity we may assume, if $\ell_0 = x$ is a sufficiently general projection, that the family of polar curves is an equisingular family of plane curves as b varies near 0. So for $\ell = \ell_0$ we get the equisingularity type of the general polar curve *in the coordinates* x, y. Now to prove that this equisingularity type is independant of what we chose to call linear coordinates, it is sufficient to consider a one-parameter family of projections of the form $\ell_0 + vg(v; x, y)$ with $g(v; x, y) \in (x, y)^2 \mathbf{C}\{v, x, y\}$, so that we get a family of curves with equation $df \wedge (dx + vdg) = 0$. Instead of considering the corresponding determinant, we consider the associated linear system

$$\frac{\partial f}{\partial x}X + \frac{\partial f}{\partial y}Y = 0$$

$$(1 + v\frac{\partial g}{\partial x})X + v\frac{\partial g}{\partial y}Y = 0$$

which defines locally near the point $(0, 0, (0:1)) \in \mathbf{C} \times \mathbf{C}^2 \times \check{\mathbf{P}}^1$ an analytic subspace $Z \subset \mathbf{C} \times \mathbf{C}^2 \times \check{\mathbf{P}}^1$ such that its image in $\mathbf{C} \times \mathbf{C}^2$ is the family of our polar curves. Note that the first equation in independant of v, and defines therefore the product by $\mathbf{C}$ of a family of curves Z_1 in $\mathbf{C}^2 \times \check{\mathbf{P}}^1$. This family of curves is actually isomorphic with the family of polar curves as $b = -\frac{X}{Y}$ varies, so it is equisingular, and therefore so is its product with $\mathbf{C}$; the product $\mathbf{C} \times \check{\mathbf{P}}^1$ is the singular locus of Z_1, along which Z_1 is Whitney equisingular. Now we remark that the second equation, upon taking $v = 0$, describes a non singular hypersurface $Z_2(0)$ which is *transversal* in $\mathbf{C}^2 \times \check{\mathbf{P}}^1$ to the surface $\mathbf{C} \times \check{\mathbf{P}}^1$. Now it is a basic property of Whitney equisingularity that all sections by nonsingular spaces of a given dimension transversal to a stratum form themselves an equisingular family, and as v varies, we have here precisely a family of sections of $Z_1(0)$ by non singular hypersurfaces $Z_2(v)$, transversal for $v = 0$ and therefore also for v sufficiently small. Finally we see that Z is a family of curves equisingular along $\mathbf{C} \times 0$. It remains to see that for each value of v, the curve $Z(v)$ is isomorphic with the zero set of $df \wedge (dx + vdg)$, but this is obvious. The same proof works if we multiply the first equation by f, and shows that in fact the *equisingularity class* of the union of f and the polar curve depends only on the *analytic class of the germ* f. Exactly the same proof works for the general polar variety of dimension d of a complex analytic mapping $f: (\mathbf{C}^{d+1}, 0) \to (\mathbf{C}, 0)$.

The general case is somewhat more delicate, but the idea is the same. It suffices to study the following situation: a one-parameter analytic family of maps $p^*(v): \mathbf{C}^N \to \mathbf{C}^{d-k+1}$ such that $p(0)$ is a general linear projection and the corresponding map $P^*(v) = Id_{\mathbf{C} \times S} \times p^*: \mathbf{C} \times S \times \mathbf{C}^N \to \mathbf{C} \times \mathbf{C}^{d-k+1}$. Consider the graph embedding of P^*

$$\begin{array}{ccc} \mathbf{C} \times S \times \mathbf{C}^N & \hookrightarrow & \mathbf{C} \times S \times \mathbf{C}^{d-k+1} \times \mathbf{C}^N \\ \downarrow P^* & \swarrow & \\ \mathbf{C} \times S \times \mathbf{C}^{d-k+1} & & \end{array}$$

and the product diagram

$$\begin{array}{ccccc} \mathbf{C} & \times & C_f(X) & \hookrightarrow & \mathbf{C} \times S \times \mathbf{C}^N \times \check{\mathbf{P}}^{N-1} \\ & & \downarrow Id_{\mathbf{C}} \times \kappa_f & & \downarrow \\ \mathbf{C} & \times & X & \hookrightarrow & \mathbf{C} \times S \times \mathbf{C}^N \\ & & \downarrow Id_{\mathbf{C}} \times f & \swarrow & \\ \mathbf{C} & \times & S & & \end{array}$$

We choose as in [25], Chap.4, a Whitney stratification C_α of $C_f(X)$ compatible with the inverse image of the singular locus of X and with the inverse images by κ_f of the strata of a Whitney stratification of $P_k(f,p)$. Now we consider the intersection $\tilde{P} = C \times C_f(X) \cap C_{p*}(C \times S \times C^N)$ in $C \times S \times C^N \times \check{P}^{N-1}$ and its natural projection $\mathcal{P} \to C$.

Since we assume that the original projection $p(0) = p$ is a general linear projection, it is easy to verify that for each sufficiently small v, the image of $\tilde{\mathcal{P}}(v)$ is the polar variety of f with respect to the analytic projection $p(v)$, and the image of $\tilde{\mathcal{P}}$ in $C \times S \times C^N$ is the total space $\mathcal{P}$ of the family of polar varieties of f as v varies. Now it is no longer true, as in the curve case, that the fibers $\tilde{\mathcal{P}}(v)$ are isomorphic to the polar varieties, but one knows enough about the map $\tilde{\mathcal{P}}$ in $C \times S \times C^N \to \mathcal{P}$ to show the following:

If we choose carefully as above an equisingular stratification of $C_f(X)$, and denote by S_0 the largest stratum contained in $\kappa_f^{-1}(0)$, the product $C \times S_0$ is a stratum of an equisingular stratification of $C \times C_f(X)$. The intersection with $C_{p*}(C \times S \times C^N)$ is transversal for $v = 0$, so $C \times S_0 \cap (C^N \times L^{d-k})$ is a stratum of an equisingular stratification of $\tilde{\mathcal{P}}$.

Using the characterization of Whitney stratifications by properties of the Auréole (see [12], corollaire 2.2.4.1) instead of the characterization of equimultiplicity as in ([25], p.428) one then deduces from this that $C \times 0$ is a stratum of a Whitney stratification of $\mathcal{P}$, which proves the theorem.

This proof is given only partially because it could be simplified if one had the answer to the following

Question 1. Let $C(X) \subset C^N \times \check{P}^{N-1}$ be the conormal space of $X \subset C^N$. Let $(X_\alpha)_{\alpha \in A}$ be the canonical Whitney stratification of X. Can one describe a *purely Lagrangean* method to construct the Lagrangean variety $\cup_{\alpha \in A} C(X_\alpha)$. That is, without going down to stratify X, but using for example higher microlocalizations.

This question is in a way a geometric version of the following question, which Thom calls his *Dream of Youth*:

Question 2. (Thom) Given an ideal $I \subset C\{z_1,\ldots,z_n\}$ defining a reduced equidimensional germ $X \subset C^N$, let

$$X = F_0 \supset F_1 \supset \cdots \supset F_r \supset \emptyset$$

be the canonical Whitney filtration of X (near 0), characterized (see [25], Chap.6) by the fact that the connected components of the differences $F_i \setminus F_{i+1}$ are the strata of the minimal Whitney stratification, and let $I = I_0 \subset I_1 \subset \cdots \subset I_r \subset C\{z_1,\ldots,z_n\}$ be the corresponding sequence of ideals.

The problem is: find a way using Jacobian extensions of ideals (i.e, by adding suitable Jacobian minors of generators), to generate the sequence (I_j) from the ideal I.

I do not know how to answer this exact question (which I have taken the liberty to complexify and slightly adapt from the original) but using the results of [25] I can give (see [27]) a similar result but I have to use not only Jacobian extensions, but also the residuation $(J : K) = \{g \in C\{z_1,\ldots,z_n\}/gK \subset J\}$ of ideals, choices of special system of generators, and «generic» choices of coordinates.
Here is a question which is in a way intermediate between these two:

Question 3. Given a projective variety $V \subset P^N$, consider the canonical Whitney stratification (W_α) of its dual variety $\check{V} \subset \check{P}^N$. How can one build from equations of V the totality of the duals $\check{W}_\alpha \subset P^N$?

As an exercise, given a projective plane curve, try to build from its equation those of all the special lines associated with it : inflection tangents, double tangents, etc.., only by using Jacobian extensions, etc.., without dualizing.

2. The contact of the relative polar varieties

In fact this study was begun over a hundred years ago, in the case of plane curves, by H.J. Stephen Smith in [20], in connection with the study of Plücker formulas. Then the italian geometers, for whom the contact was best expressed in terms of common infinitely near points, tried to find the infinitely near points common to a curve and its general polar. Since the infinitely near points of the polar depend upon the analytic type of the singularity and not only on its tree of infinitely near points (that is, its equisingularity class), this viewpoint led essentially to frustration, at least until the very recent results of Casas ([1]) concerning the polar curve of a plane curve singularity which is «generic» among those belonging to a given equisingularity class.

However, there are some specific results at least in the case of hypersurfaces with isolated singularities, to which we will come back.

For the purpose of resolution of singularities, however, this fundamental impossibility to determine the equisingularity type of polar varieties (absolute or relative) from just the equisingularity type of the given germ of variety or map may not hamper the use of polar varieties as providers of inductive steps. Here is the prime example, after Segre [19] :

2.1. Contact of polar varieties and resolution of singularities

Given a projective hypersurface $X \in \mathbf{P}^N$ with equation $F(X_1, \ldots, X_{N+1}) = 0$, let $\xi = (\xi_1, \ldots, \xi_{N+1})$ be a point of $\mathbf{P}^N$, then the polar hypersurface of X with respect to the point ξ is the hypersurface with equation $\sum_1^{N+1} \xi_i \frac{\partial F}{\partial X_i} = 0$; it is the projective hypersurface corresponding to therelative polar variety $P_1(F; p)$ associated to the map $F: \mathbf{C}^{N+1} \to \mathbf{C}$ and the projection $p: \mathbf{C}^{N+1} \to \mathbf{C}^N$ corresponding to ξ.

In [19], Segre proposed to correct in the case of surfaces an attempt of Derwidue in [2] (see Zariski's Math. Review [31]) to resolve singularities of projective varieties, and convince the reader that one could after corrections extract a proof at least in the case of surfaces in characteristic zero (which at that time was already known by the work of Zariski). The main steps of the argument are as follows:
1) a computation of the strict transform of the polar curve under blowing up of a non singular center along which the given hypersurface is equimultiple, *at a point where the multiplicity has not decreased* and the constatation that at such a point the strict transform of a general polar is a polar of multiplicity $m-1$ of the strict transform.
2) In the case of surfaces, one can by blowing up points reduce to the case where the equimultiplicity locus is nonsingular, and by 1), one sees that it is sufficient to show that one can make the multiplicity of a general polar decrease by a finite number of blowing ups along non singular equimultiplicity loci. One thus reduces to the case where the given hypersurface has singularities of multiplicity two only.
3) A direct computational treatment of this last case.

Let $f(y_1, \ldots, y_t, z_1, \ldots, z_n) = 0$ be a local equation for a hypersurface in $\mathbf{C}^N$, where

$N = n+t$. Let assume that $f(y_1,\ldots,y_t,0,\ldots,0) = 0$ and consider locally around 0 the blowing up $\pi\colon Z \to \mathbf{C}^N$ of the subspace of $\mathbf{C}^N$ defined by the ideal $(z_1,\ldots,z_n)$. In a typical chart of Z we have coordinates $y_1',\ldots,y_t',z_1',\ldots,z_n'$ and the map π described by

$$y_j \circ \pi = y_j' \quad \text{for } 1 \leq j \leq t$$

$$z_1 \circ \pi = z_1'$$

$$z_i \circ \pi = z_i' z_1' \quad \text{for } i \neq 1.$$

The strict transform of f by π is decribed in the same chart by

$$f'(y_1',\ldots,y_t',z_1',\ldots,z_n') = z_1'^{-m} f(y_1',\ldots,y_t',z_1',z_2'z_1',\ldots,z_n'z_1').$$

Let us now compute the composition with π of the partial derivatives

$$\frac{\partial f}{\partial z_1} \circ \pi = m z_1'^{m-1} f' - z_1'^m \frac{\partial f'}{\partial z_1'} - \sum_{i \geq 2} z_i' z_1'^m \frac{\partial f'}{\partial z_i'}$$

$$\frac{\partial f}{\partial z_i} \circ \pi = z_1'^{m-1} \frac{\partial f'}{\partial z_i'} \quad (2 \leq i \leq n)$$

$$\frac{\partial f}{\partial y_s} \circ \pi = z_1'^m \frac{\partial f'}{\partial y_s'} \quad (1 \leq s \leq t)$$

$$f \circ \pi = z_1'^m f'.$$

So that if we take an element of the ideal $(f, \frac{\partial f}{\partial z_1}, \ldots, \frac{\partial f}{\partial z_n}, \frac{\partial f}{\partial y_1}, \ldots, \frac{\partial f}{\partial y_t})$, say $g = af + \sum_1^n b_i \frac{\partial f}{\partial z_i} + \sum_1^t c_s \frac{\partial f}{\partial y_s}$, we find the identity

$$z_1'^{-(m-1)} g \circ \pi = ((a \circ \pi)z_1' + mb_1 \circ \pi) f' - (b_1 \circ \pi) z_1' \frac{\partial f'}{\partial z_1'}$$

$$+ \sum_{i=2}^n (b_i \circ \pi - (b_1 \circ \pi) z_1' z_i') \frac{\partial f'}{\partial z_i'} + \sum_{s=1}^t (c_s \circ \pi) z_1' \frac{\partial f'}{\partial y_s'}.$$

So if g is of order $m-1$ at 0, its strict transform is in the ideal $(f', \frac{\partial f'}{\partial z_1'}, \ldots, \frac{\partial f'}{\partial z_n'}, \frac{\partial f'}{\partial y_1'}, \ldots, \frac{\partial f'}{\partial y_t'})$.

There are several observations to make:
1) The behaviour of g is actually governed by that of $\sum_1^n b_i \frac{\partial f}{\partial z_i} + \sum_1^t c_s \frac{\partial f}{\partial y_s}$, and one may as well call polar variety the hypersurface defined by $g = 0$ at least in the case where a and the b's and c's are constants.
2) If g is of the form

$$g = \sum_2^n b_i \frac{\partial f}{\partial z_i} + \sum_1^t c_s \frac{\partial f}{\partial y_s} \quad \text{with } b_i, c_s \in \mathbf{C}$$

then

$$z_1'^{-(m-1)} g' = \sum_2^n b_i \frac{\partial f'}{\partial z_i'} + \sum_1^t c_s z_1' \frac{\partial f'}{\partial y_s'}$$

so we have a very similar expression for the strict transform.

3) If f is equimultiple along Y and the multiplicity of f' at the point x' with coordinates $z'_1 = \ldots = z'_n = y'_1 = \ldots = y'_t = 0$ is again $m - 1$, then the exceptional divisor $z'_1 = 0$ is *transversal* to $f' = 0$ in Z at x'.

4) If f is equimultiple along Y, then so is an element

$$g = af + \sum_1^n b_i \frac{\partial f}{\partial z_i} + \sum_1^t c_s \frac{\partial f}{\partial y_s}$$

which is of multiplicity $m - 1$ at the point 0.

These observations result from straightforward computations, and they are part of the stock-in-trade of all resolvers of singularities. In particular they have been vastly generalized by Hironaka.

Another important fact discovered by Segre (in the projective situation; it has been rediscovered in the local situation several times since) is that in the blowing-up $\pi: X' \to X$ of X along a non singular center Y such that X is equimultiple along Y, the multiplicity of X' at any of its points x' is at most equal to the multiplicity of X at $\pi(x')$.

Segre uses this to show that (in modern language), the problem to prove that a permissible succession of blowing ups along isolated m-uple points and nonsingular m-uple curves is necessarily finite can be reduced to the case of a hypersurface of multiplicity 2. The solution of this problem implies resolution, and the proof in the case where $m = 2$ is fairly easy. Segre uses a case-by-case approach.

The idea of the reduction is that in view of the preceding results, as long as the multiplicity does not drop, the multiplicity of a general polar, which is $m - 1$, does not drop either. Of course in higher dimensions, the choice of centers of blowing-ups becomes a major problem. In any case, Segre's work suggests that the use of the polar hypersurface may offer some alternative to the «maximal contact» of Hironaka. In some sense, the general polar hypersurface of f has «maximal contact» at 0 with $f = 0$ among those of multiplicity $m - 1$. In the case of locally irreducible curves one can make this really precise using a result of Merle and the beautiful theory of maximal contact of singular curves of Monique Lejeune-Jalabert (see [13] and [16]). Notice also that if you iterate the polar curve operation for an irreducible plane curve until you get a non singular curve, this curve does have maximal contact at 0. It is a good exercise at this point to prove resolution of singularities of plane curves using Segre's method.

2.2. Contact of a polar curve of a germ of holomorphic map $f: (\mathbf{C}^{n+1}, 0) \to (\mathbf{C}, 0)$ with the hypersurface $f^{-1}(0)$

This is the only case where we know something precise. This contact is the subject of [24],[26], [28] and is decribed by the *Jacobian Newton polygon*, which is constructed as follows:

Let $f \in \mathbf{C}\{z_0, \ldots, z_n\}$ describe our map. The relative polar curve of the map with respect to a linear projection $\ell: \mathbf{C}^{n+1} \to \mathbf{C}^n$ is the zero set of the 2-form $df \wedge d\ell$ and if ℓ is sufficiently general, it is a germ of a reduced curve $P_n(f, \ell)$.

I now decompose this germ into its irreducible components $\Gamma_i, 1 \leq i \leq r$. By the definition of polar varieties, none of these components can be included in the hypersurface $f = 0$ (if it were the case, the component would have to be included in the critical locus of f), and each has a multiplicity at 0, which we denote by m_q, and an intersection number at 0 with $f = 0$, necessarily $\geq m_q$, and which we denote by $e_q + m_q$. From these numbers we construct the *contact polygon* at 0 of the hypersurface $f = 0$ and the curve $P_n(f,\ell)$ as follows : we construct a Newton polygon by adding the elementary Newton polygons

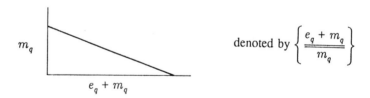

We get a Newton polygon, for example

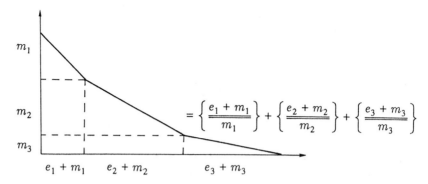

Because of the theorem of transversality of relative polar varieties of [24], th.1 (see also [30], [25] Chap.4 and [5] for generalizations) which states that when the base S is a point or a nonsingular curve the tangent cones at 0 of the polar varieties are transversal to the kernels of the corresponding projections, this abstract Newton polygon is actually the Newton polygon of the image in $\mathbf{C}^2$ with coordinates (t_1, t_2) of the polar curve which is the closure of the critical locus outside 0 of the map $(f, \ell): \mathbf{C}^{n+1} \to \mathbf{C}^2$. This Jacobian Newton polygon was introduced in [24] in the case where f has an isolated singularity; it refines in that case the ratio $\frac{\mu^{(n+1)} + \mu^{(n)}}{\mu^{(n)}}$, where $\mu^{(n+1)}$ is the Milnor number of the hypersurface and $\mu^{(i)}$ is the Milnor number of its intersection with a general i-dimensional plane through the origin.

But it has turned out that the slopes of its sides, often called the polar invariants of f, play a basic role in several *a priori* unrelated questions about the singularities of f :

1) the poles of the meromorphic map from $\mathbf{C}^2$ to the space of distributions in $\mathbf{C}^{n+1}$ obtained by analytic continuation from the map defined for $\mathrm{Re}\, s \gg 0, \mathrm{Re}\, t \gg 0$ by

$$(s, t) \mapsto (\phi \mapsto \int_{\mathbf{C}^N} \ell^s f^t \phi)$$

where ℓ is a sufficiently general linear form and ϕ a differentiable function with compact support. Work of Sabbah and Loeser has shown that the poles of this meromorphic map lie on lines in $\mathbf{C}^2$ with slopes given by those of the Jacobian Newton polygon. They actually show much more, see [15], [18].

2) The best possible exponents θ_1 and θ_2 in the Łojasiewicz inequalities (near 0)

$$\|\mathrm{grad} f(z)\| > C \|z\|^{\theta_1} \text{ and } \|\mathrm{grad} f(z)\| > C' |f(z)|^{\theta_2}.$$

3) The different rates at which in a 1-parameter Morsification $f_v = f + v\ell$, where ℓ is a general linear form, the Morse critical points $c_i(v)$ of f_v tend to 0 as v tends to 0, i.e the best possible exponents in the inequalities

$$\|c_i(v)\| \leq C_i |v|^{r_i}.$$

The point here is that we have the

THEOREM ([24]). *In a Whitney-equisingular family of hypersurfaces $f(v; z_1, \ldots, z_{n+1})$ with isolated singularity at 0 (for each v), the Jacobian Newton polygon is constant.*

For irreducible germs of plane curves Merle has given in [16] the expression for the Jacobian Newton polygon in terms of the Puiseux exponents and it turns out that in this case the Jacobian Newton polygon *determines* the Puiseux exponents and therefore the topology of the curve. In fact much more is true: for each of the Puiseux exponents of an irreducible plane curve X, the general polar curve has a bunch of irreducible components whose Puiseux expansion coincides (up to a suitable ramification of the parameter) with the Puiseux expansion of X *up to* (but excluding) that exponent. So the polar really has very high contact with the curve.

In the reducible case, some components of the polar curve may not have high contact with the curve, but except in very special cases some components do have high contact. There is a general result in this direction:

Fact 1. Let us assume that $f: \mathbf{C}^{n+1} \to \mathbf{C}$ has an isolated singularity at 0, and that the general polar curve is not tangent to $f = 0$ at 0 (that is, no tangent line to the curve is in the tangent cone of $f = 0$ at 0). Since, by the results of [29] and [24], the multiplicity of the polar curve is $\mu^{(n)}$ and its intersection multiplicity with $f = 0$ is $\mu^{(n+1)} + \mu^{(n)}$, while the multiplicity at 0 of $f = 0$ is $\mu^{(1)} + 1$, transversality is equivalent to the equality $\mu^{(n+1)} = \mu^{(1)} \mu^{(n)}$. By the results of ([29], Chap.2), this implies that the blowing-up of the origin resolves the singularity of $f = 0$, or if one prefers, that $f = 0$ is equisingular with its tangent cone at 0.

Fact 2. If we now take a hypersurface, say $f(v; x, y, z) = 0$ in $\mathbf{C}^4$, which has a nonsingular one dimensional singular locus along the v axis, we can think of it as a one parameter family of hypersurfaces with isolated singularity at the origin. If the general polar curve of f is empty, if the sections $v = v_0$ of the general polar surface have constant contact polygon with $f(v_0; x, y, z)$ as v_0 varies, then according to [24], the μ^* sequence is constant along the v axis and according to a theorem of Laufer in [8], our hypersurface has a simultaneous resolution of singularities along the v axis.

In view of Segre's proof, these two facts suggest the

Problem 5. Study the behaviour of the polar varieties of codimension > 1 under equimultiple blowing-up with non singular center.

There is also a similar problem for the absolute polar varieties; here the references for motivation are [9] and [20].

In fact, (see [28], [16] [18]) one can put both the absolute and the relative case in the same frame, and generalize the Jacobian Newton polygon as follows: Consider an analytic map $F: X \to \mathbf{C}^2$; choose coordinates t_1, t_2 on $\mathbf{C}^2$ and set $f_1 = t_1 \circ F, f_2 = t_2 \circ F$. Then the strict critical locus $S(F)$ of F is by definition the closure in X of the set of nonsingular points of X where the fibers of f_1 and f_2 are not both non singular and meeting transversally. The Newton polygon in the coordinates t_1, t_2 is a generalization of the Jacobian Newton polygon and we call it the Jacobian Newton polygon of f_1, f_2 on X.

Question 6. Is it true that in an equisingular deformation of X and F this Jacobian Newton polygon is constant ?

(The definition of equisingularity here is left to the reader).

The recent results of Gaffney in [3] and [4] make it plausible that, at least when X is a complete intersection wih isolated singularity, a proof can be given along the lines of [24].

REFERENCES

[1] E. Casas-Alvero : Infinitely near singularities and singularities of polar curves, *Preprint, Universitat de Barcelona, dep. Alg. i Geom.*1988

[2] L. Derwidué: Le problème de la réduction des singularités de variétés algébriques, *Math. Annalen* **123** (1951), 302-330.

[3] Terence Gaffney : Integral closure of modules and Whitney equisingularity, *preprint, Northeastern University*

[4] Terence Gaffney : Multiplicities and equisingularity of ICIS germs, *preprint, Northeastern University*

[5] J.P.G. Henry et M. Merle: Limites d'espaces tangents et transversalité de variétés polaires, *Algebraic Geometry, Proceedings, La Rábida, 1981*, Springer L.N.M., No.961, 189-199

[6] J.P.G. Henry, M. Merle et C. Sabbah: Sur la condition de Thom stricte pour un morphisme analytique complexe, *Ann. Sci. E.N.S.*, **17**, 1984, 227-268.

[7] Patrick Du Val: Benjamino Segre, *Bull. London Math. Soc.,* **11** (1979), 215-235

[8] Henry Laufer: Strong simultaneous resolution for surface singularities, *preprint,* Stony Brook, 1984

[9] Dũng Tráng Lê: Exposants polaires et résolution des surfaces, *Communication au 3ème Congrès des mathématiciens vietnamiens, 1985.*

[10] Dũng Tráng Lê, Françoise Michel, Claude Weber, Courbes polaires et topologie des courbes planes. A paraître dans *Annales scientifiques de l'E.N.S.*

[11] Dũng Tráng Lê et B. Teissier: Variétés polaires locales et classes de Chern des variétés singulières, *Annals of Math.*, **114**, 1981, 457-491.

[12] Dũng TrángLê et B. Teissier: Limites d'espaces tangents en géométrie analytique, *Commentarii Math. Helv.*, **63**, 4, 1988, 540-578.

[13] Monique Lejeune-Jalabert , *Thèse* Paris 7,1973.

[14] Beppo Levi : Sur la résolution des points singuliers des surfaces algébriques, *Comptes rendus Ac. Sci. Paris*, **134**, 1902, p.224.

[15] François Loeser : Fonctions dzeta locales d'Igusa à plusieurs variables, intégration dans les fibres, et discriminants, *Ann. Sci. E.N.S.*, **22**, 3,1989, 435-471,

[16] Michel Merle: Invariants polaires des courbes planes, *Inventiones Math.*, **41**, 1977, 103-111.

[17] Françoise Michel, Claude Weber, *Livre à paraître sur la topologie des courbes planes.*

[18] Claude Sabbah : Proximité évanescente 1: Structure polaire des D-modules (en collaboration avec F. Castro). *Compositio Math.*, **62**, 1987, 283-328.

[19] Beniamino Segre: Sullo scioglimento delle singolarità delle varietà algebriche, *Ann. di Mat.* **4**, 33, 5-48.

[20] H.J. Stephen Smith : On the higher singularities of plane curves, *Proc. London Math. Soc.*, **5**, 1873, 153-182.

[21] Mark Spivakovsky: Resolution of singularities by normalized Nash blowing-up, To appear, *Annals of Math.*
[22] Bernard Teissier: *appendice à* [32].
[23] Bernard Teissier: Variétés polaires locales et conditions de Whitney, *C.R.A.S., Paris* **280**, 1980.
[24] Bernard Teissier : Variétés polaires 1: invariants polaires des singularités d'hypersurfaces, *Inv. Math.* **40**, 1977, 267-292
[25] Bernard Teissier : Variétés polaires 2: multiplicités polaires, sections planes, et conditions de Whitney. *Algebraic Geometry, Proceedings, La Rábida, 1981* Springer L.N.M., No.961 (1982), 314-491.
[26] Bernard Teissier: The hunting of invariants in the geometry of discriminants. *real and complex singularities*, Sijthoff and Noordhoof 1977.
[27] Bernard Teissier: Analytic singularities: topology, algebra, algebraization, *Stone Lectures at Northeastern*, May1988, in preparation.
[28] Bernard Teissier : Polyèdre de Newton Jacobien et équisingularité, *Séminaire sur les singularités*, Publications de Paris 7, 1980.
[29] Bernard Teissier: Cycles évanescents, sections planes, et conditions de Whitney, "Singularités à Cargèse, *Asterisque* No.7-8, S.M.F. 1973.
[30] Bernard Teissier: Variétés polaires locales: Quelques résultats, Journées complexes, Nancy 1980. *Publications de l'Institut Elie Cartan*, 1981.
[31] Oscar Zariski: *Math. Reviews*, **13**, No.1, Jan. 1952, p.67.
[32] Oscar Zariski: *Modules de branches planes*, Cours au centre de Mathématiques de l'Ecole Polytechnique, 1973. Réedition Hermann 1986.

Pseudoconformal Invariants and Differential Geometry

GIUSEPPE TOMASSINI Scuola Normale Superiore, Piazza dei Cavalieri 7, 56126 PISA

0. INTRODUCTION

1. One of the more interesting problems in the geometric theory of the holomorphic functions of several complex variables is to study the real subvarieties of $\mathbb{C}^n, n \geq 2$ with respect to the pseudoconformal transformations. This problem, arised by Poincaré ([5]), has motivated many scientific papers starting with Poincaré himself.

Without any intention to be exhaustive I shall restrict myself to mention the contributions to this problem given by B. Segre ([6], [7]), E. Cartan ([1]), N. Tanaka ([9]) up to the paper of Chern and Moser ([2]).

In '78 Webster published a paper ([10]) in which the problem to construct pseudoconformal invariants was considered from the point of view of the riemannian metric ds^2 on a strictly pseudoconvex real hypersurface M and a metric connection D in $H(M)$, the bundle of the complex tangent hyperplanes of M.

This metric is not invariant with respect to a pseudo-conformal transformation of M but it provides an useful tool to construct pseudoconformal invariants. A natural question therefore arises to investigate the relationship between the «metric properties of M» and its CR-structure. In this scheme of things in a joint paper with G. Gigante ([3]) we have considered a deformation a $\{M_t\}$ of $M = M_0$ which preserves the «Webster metric» up to the second order and we have called such a deformation an *infinitesimal bending* of M.

This situation, recall the classical «rigidity problem» for a surface in $\mathbb{R}^3$ and an evocative exposition of the main results and problems in this context can be found in Ch. XII of treatise of Spivak ([8]).

To an infinitesimal bending of M we can associate a vector field ζ defined on M, the «velocity» of $\{M_t\}$ at $t = 0$ and the main result of the paper, that I would like to discuss here briefly is that on suitable hypothesis the components of ζ are CR-functions.

2. Before going into the precise statements we give an account of the Webster approach.

Let $M \subset \mathbb{C}^{n+1}$ be a C^∞-smooth hypersurface defined by $\sigma = 0$. Then $\theta = i\partial\sigma_{|M}$ is a global real form and for every point $p \in M$ we can find an open neighborhood $U \subset M$ of p and n differential 1-forms on $U, \theta^1, \ldots, \theta^n$ such that $\{\theta, \theta^\alpha, \overline{\theta}^\alpha\}_{1 \le \alpha \le n}$ form a base of complex differential 1-forms. For instance, if $z^1, \ldots, z^n, w$ denote complex coordinates in $\mathbb{C}^{n+1}$ and $\partial\rho/\partial w\,(p) \neq 0$, we can take $\theta^\alpha = dz^\alpha_{|M}$, $1 \le \alpha \le n$. Put $\theta^{\overline{\alpha}} = \overline{\theta}^\alpha$, $1 \le \alpha \le n$. If $\{\theta, \theta^\alpha, \overline{\theta}^\alpha\}_{1 \le \alpha \le n}$ is another base on U' and $U \cap U' \neq \emptyset$, then we have

$$\theta^{\alpha'} = \sum_{\alpha=1}^{n} A^{\alpha'}_\alpha \theta^\alpha + B^{\alpha'}\theta. \qquad (*)$$

In this way we get an atlas $\mathcal{U} = \{U_j, \theta^\alpha_{(j)}, \theta^{\overline{\alpha}}_{(j)}\}$, $1 \le \alpha \le n$, $j \in \Lambda$ and M, equipped with $\mathcal{U}$ is called a CR-hypersuface of $\mathbb{C}^{n+1}$ (with the induced CR-structure).

We also use the notation $M = \{U_j, \theta, \theta^\alpha, \theta^{\overline{\alpha}}\}$ (where θ is the global form $i\partial\rho_{|M}$ and $\theta^\alpha = \theta^\alpha_{(j)}, \ldots$).

A CR-map (or a *pseudoconformal map*) f between two CR-hypersurfaces $M = \{U_j, \theta, \theta^\alpha, \theta^{\overline{\alpha}}\}$, $M' = \{U'_j, \theta', \theta^{\alpha'}, \theta^{\overline{\alpha}'}\}$ is a map $f : M \to M'$ such that

$$f^*\theta' = \lambda\theta \qquad (1)$$

$$f^*\theta^{\alpha'} = \sum_{\alpha=1}^{n} a^{\alpha'}_\alpha \theta^\alpha + b^{\alpha'}\theta; \qquad (2)$$

in other words, f sends a local CR-function on M' into a local CR-function on M.

Let $M = \{U_j, \theta, \theta^\alpha, \overline{\theta}^\alpha\}$ be a CR-hypersurface defined by $\rho = 0$. Then, for every $p \in M$ we have

$$d\theta(p) = -i\partial\overline{\partial}\rho_{|H_p(M)} = -i\sum_{\alpha,\beta=1}^{n} g_{\alpha\overline{\beta}}\theta^\alpha \wedge \theta^{\overline{\beta}} + \sum_{\alpha=1}^{n}(\eta_\alpha \theta^\alpha + \eta_{\overline{\alpha}}\theta^{\overline{\alpha}}) \wedge \theta$$

and the hermitian matrix $(g_{\alpha\overline{\beta}})$ is non singular positive definite whenever M is strictly pseudoconvex. In this case we can find a new atlas $\{U_j, \theta, \theta^{\alpha'}, \theta^{\overline{\alpha}'}\}$ for which

$$d\theta = -i\sum_{\alpha'=1}^{n}\theta^{\alpha'} \wedge \theta^{\overline{\alpha}'}. \qquad (3)$$

We call it a *reduced atlas* and M a *reduced CR-hypersurface*. In particular, in $(*)$, we have $B^{\alpha'} = 0$, $1 \le \alpha' \le n$, and $\sum_{\alpha'=1}^{n} A^{\alpha'}_\alpha \overline{A^{\alpha'}_\beta} = \delta_{\alpha\beta}$.

Now assume that M is a strictly pseudoconvex reduced CR-hypersurface; then we can define a metric on M by the following formula

$$ds^2 = \theta \otimes \theta + \operatorname{Re}\sum_{\alpha=1}^{n}\theta^\alpha \otimes \theta^{\overline{\alpha}}$$

([3]).

We call it the «Webster metric» of M (with respect to a chosen defining function ρ). This metric depends on ρ in the following way: if $\rho' = \lambda\rho$, $\lambda > 0$, one has $\theta' = \lambda\theta$, $\theta^{\alpha'} = \sqrt{\lambda}\theta^\alpha$ and consequently

$$ds'^2 = \lambda^2 \theta \otimes \theta + \lambda \operatorname{Re} \sum_{\alpha=1}^{n} \theta^\alpha \otimes \theta^{\overline{\alpha}}.$$

Another metric, which is independent on ρ, is given by

$$d\sigma^2 = \frac{\theta \otimes \theta}{|\partial \rho|^2} + \frac{1}{|\partial \rho|} \operatorname{Re} \sum_{\alpha=1}^{n} \theta^\alpha \otimes \theta^{\overline{\alpha}}.$$

If M' is another strictly pseudoconvex CR-hypersurface and ds'^2 the Webster metric of M', then a CR-isometry $f : M \to M'$ is called a *pseudohermitian transformation* (and it can be proved ([4]) that a CR-transformation $f : M \to M'$ is a pseudohermitian transformation if and only if $f^*\theta' = \theta$).

Starting from (3), we obtain by differentiation the «structure equations»

$$d\theta = -i \sum_{\alpha=1}^{n} \theta^\alpha \wedge \theta^{\overline{\alpha}} \tag{3}$$

$$d\theta^\alpha = \sum_{\beta=1}^{n} \psi_\beta^\alpha \wedge \theta^\beta + \theta \wedge \tau^\alpha \tag{4}$$

$$\psi_\alpha^\beta + \overline{\psi}_\alpha^\beta = 0 \tag{5}$$

where the $-\psi_\alpha^\beta$ are the «connection forms» on $H(M)$ and the τ_α are the «torsion forms» for this connection ([10]). In particular, for $n = 1$ (i.e. $\dim_{\mathbb{R}} M = 3$) we have

$$d\theta = -i\theta^1 \wedge \theta^{\overline{1}} \tag{3'}$$

$$d\theta^1 = \psi \wedge \theta^1 + \nu \theta \wedge \theta^{\overline{1}} \tag{4'}$$

$$\psi + \overline{\psi} = 0 \tag{5'}$$

where $\nu\theta^{\overline{1}}$ is the torsion form. The curvature tensor and the pseudohermitian curvature are defined by the formulas

$$d\psi = R_{1\overline{1}\,1\overline{1}}\theta^1 \wedge \theta^{\overline{1}} \mod \theta \tag{6}$$

$$R = -\frac{1}{2}R_{1\overline{1}\,1\overline{1}}. \tag{7}$$

From now on we shall dealing with strictly pseudoconvex reduced CR-hypersurfaces M defined by $\rho = 0$ equipped with the Webster metric ds^2.

To emphasize the fact that ds^2 is depending on ρ we use notation (M, ρ, ds^2).

1. INFINITESIMAL BENDINGS AND RIGIDITY THEOREMS

1. Let M be a C^∞-smooth hypersurface of $\mathbb{C}^{n+1}$, $n \leq 1$. A C^∞-*family of hypersurfaces* through M is a pair $(M \times I, \Phi)$ where $\Phi : M \times I \to \mathbb{C}^{n+1}$ is a C^∞-smooth map such that

(i) Φ is of maximum rank.
(ii) $\Phi_t = \Phi_{|M\times\{t\}}$ is a diffeomorphism between $M\times\{t\}$ and a hypersurfaces M_t of $\mathbb{C}^{n+1}$.
(iii) $\Phi_o = \mathrm{id}_M$.

We use the notation $\{M_t\}_{t\in I}$ instead of $(M\times I, \Phi)$. To a family $\{M_t\}_{t\in I}$ we can associate a field ζ of $(1,0)$-vectors on $M = M_o$ in the following way: let $z^j = z^j(u)$, $1\leq j\leq n$, $w = w(u)$, $u = (u^1,\ldots,u^{2n+1})$ be a parametric representation of M in a neighborhood of p; then, locally at p, the family $\{M_t\}_{t\in I}$ is defined by $z_j = z^j(u,t)$, $1\leq j\leq n$, $w = w(u,t)$ (where $z^j(u,o) = z^j(u)$, $w(u,o) = w(u)$) and ζ is then given by the formula

$$\zeta = \sum_{j=1}^n \frac{\partial z^j}{\partial t}(u,o)\frac{\partial}{\partial z^j} + \frac{\partial w}{\partial t}(u,o)\frac{\partial}{\partial w}$$

ζ is called the *velocity* of $\{M_t\}_{t\in I}$ at $t = 0$.

Now suppose that M is strictly pseudoconvex and that $\bigcup_{j\in J} M_j$ where M_j is open in M and for every j there is a C^∞-function $\rho_{j,t} = \rho_j(.,t)$ on $\Phi(M_j\times t)$ such that $M_{j,t} = \{\rho_{j,t} = 0\}$. Let $ds^2_{j,t}$ be the Webster metric on $(M_j, \rho_{j,t})$. We say that $\mathcal{M} = \{M_{j,t}, \rho_{j,t}, ds^2_{j,t}\}$ is an *infinitesimal bending* of M if for every j we have

$$ds^2_{j,t} = ds^2_{j,o} + O(t^2).$$

For example suppose that X is a holomorphic vector field on a neighborhood of M. Then X generates a 1-parameter local group of holomorphic transformations $\{\Phi_t\}$ and consequently an infinitesimal bending $\{M_t, \rho\Phi_t^{-1}\}$ (for which $ds^2_t = ds^2_o$).

We say that M is *rigid* (with respect to infinitesimal bendings) if the velocity vector field ζ of an infinitesimal bending of M is given by $X_{|M}$ where X is a holomorphic vector field on a neighborhood of M.

2. The rigidity results that we have proved in [3] are contained in the following two theorems:

THEOREM 1. *If* $\dim_\mathbb{R} M > 3$, *then* ζ *is a CR-vector field.*

THEOREM 2. *Assume that M is compact and let* $\mathcal{M} = \{M_{j,t}, \rho_{j,t}, ds^2_{j,t}\}$ *be an infinitesimal bending of M. If M has no torsion and its pseudohermitian curvature is positive for every j, then ζ is a CR-vector field.*

The proof of theorem 1 is purely algebraic whereas in the situation of theorem 2 the main tool for the proof is a maximum principle which clarifies the role played by the curvature of M ([3]).

COROLLARY 3. *Assume that M is the real analytic. Then, under the hypothesis of theorem 1 or theorem 2, M is rigid. In particular the unit sphere S^{2n+1} is rigid.*

COROLLARY 4. *Let Y be an infinitesimal isometry of M. Then, under the hypothesis of theorem 1 or theorem 2, Y is a CR-infinitesimal automorphisms.*

REFERENCES

[1] E. Cartan, «Sur la géometrie pseudo-conforme des hypersurfaces de l'espace de deux variables complexes» I, II *Ann. Mat. Pura e Appl.* 11 (1932) 17-90 and *Ann. Sc. Nor. Sup. Pisa Sc. Fis. Mat.* 1(1932) 333-354.

[2] S.S. Chern and J. Moser, «Real hypersurfaces in complex manifolds», *Acta Math.* 133 (1974) 219-271.

[3] G. Gigante and G. Tomassini, Infinitesimal bendings of real hypersurfaces (to appear in *B.U.M.I.*).

[4] H. Jacobowitz, «Mappings between CR-manifolds», *Ann. of Mat. St.* 100 (1981).

[5] H. Poincaré, «Les fonctions analytiques de deux variables et la représentation conforme», *Rend. Circ. Mat. Palermo t.* 23 (1927) 91.

[6] B. Segre, «Questioni geometriche legate colla teoria delle funzioni di due variabili complesse», *Rend. del Sem. Mat. della R. Univ. di Roma* (2) serie 2 (1931) 59-107.

[7] B. Segre, «Intorno al problema di Poincaré della rappresentazione pseudoconforme», *Rend. R. Acc. Naz. Lincei* (6) 13 (1931) 676-683.

[8] M. Spivak, *A comprehensive introduction to differential geometry* Vol. 5.

[9] W. Tanaka, On the pseudoconformal geometry of hypersurfaces of n complex variables, *J. Math. Soc. Japan* 14 (1962) 397-429.

[10] S.M. Webster, Pseudo-hermitan structures on a real hypersurface, *J. Diff. Geom.* 13 (1978) 25-41.

Giuseppe Vitali and the Mathematical Research at Bologna

A.Vaz FERREIRA, Dipartimento di Matematica, University of Bologna, Piazza di Porta San Donato 5, I 40127 Bologna, Italy

0. Vitali's research papers on Analysis were recently reprinted and edited by the Unione Matematica Italiana as a book: Vitali [0] of References. This volume (to which I have also myself collaborated) contains moreover the complete list of Vitali's mathematical publications, an accurate biography (with a general outlook of Vitali's results), a catalogue of writings concerning Giuseppe Vitali, and a collection of letters, that makes up an important historical document, collected by M.T. Borgato and L. Pepe.

Vitali [0] constitues the main source for my talk; henceforth it will be refered to simply as *Opere*.

On page 31 of *Opere* are mentioned older scientific biographies or commemorations of Vitali by Ettore Bortolotti, G. Lampariello, S. Pincherle, C. Severini, A. Tonolo, F. Tricomi.

The recent papers on Vitali's life and work are the following ones:
   a) L. Pepe [1], the biography included in *Opere*;
   b) L. Pepe [2], dedicated to the teaching activity of Vitali;
   c) L. Pepe [3] and
   d) T. Viola [1] that focus papers on Real Analysis;
   e) M.T. Borgato and A. Vaz Ferreira [1] which concern is the period of Vitali's career after world War I, when Vitali became a University professor and dedicated himself mainly to Differential Geometry.

Vitali's life and scientific achievements are closely connected with Bologna and its University. My aim in this talk is just to enlighten this subject. In the spirit of these Conferences and, to complete somehow the picture furnished by the above mentioned writings on Vitali, I will particulary pay attention in this context to Vitali's papers on Complex Analysis.

As my talk has an historical character, I will try to avoid the risk of absolute appraisals by relativizing the discourse. A certain number of quotations and some aside notes will help me.

1. Vitali was born at Ravenna, a town rich of historical remembrances and monuments, the 24th August 1875. He was the first of five sons of Domenico Vitali, a railwayman. In 1895 Vitali entered the University of Bologna and attended this university during the academic years 1895-96, 1896-97 as a student of the Course in Mathematics. His professors of Geometry and Analysis were Enriques and Arzelà who thought highly of their pupil, and helped him to obtain a study grant which enabled him to enter the Scuola Normale Superiore at Pisa. Thus, Vitali attended this school in the academic years 1897-1898, 1898-99; Dini, Bertini and Bianchi were his teachers, and Vitali will obtain honourably his degree (Laurea in Matematica) with a dissertation on functions on Riemann surfaces under the direction of Bianchi.

Vitali had at the Scuola Normale Superiore as a fellow Guido Fubini, an exceptionally brilliant student, endowed with a charming personality. A close long-lasting friendship will tie these two deep mathematicians and will influence the mathematical thinking of both.

The first mathematical papers of Vitali are dated 1900. It is of interest to observe that Vitali [1] was written under invitation of Enriques in order to include it in his celebrated «Questioni riguardanti le matematiche elementari» the later title of the work.

During the academic years 1899-1900, 1900-1901 Vitali was Dini's assistant and Vitali took advantage of this to write his beautiful habilitation dissertation «Sopra le equazioni di Appell del 2° ordine e loro equazioni integrali», presented March 1902, also written under the direction of Bianchi. The contents of that dissertation is developed in two papers Vitali [4], Vitali [5] written in the years 1901, 1902.

Vitali [2] contains a simple nice proposition about holomorphic functions f of one variable for which $\lim f^{(n)}(a)$ exists at a given point a in the domain of f. Vitali [3] concerns the problem of construction on a Riemann surface of a function holomorphic with given singularities, the object of Vitali's dissertation for his «Laurea in Matematica» degree.

Vitali [4], Vitali [5] have their source in the important memoir Appell [1], more specifically in its second supplement «Sur une classe d'équations différentielles linéaires à coefficients algébriques». Here Appell introduces a class of fuchsian differential equations and develops sketchily some remarks about them, opening thereby the problem of the study of such equations which are called by Vitali Appell equations. What Appell calls in his memoir «fonctions à multiplicateurs» on a compact Riemann surface R gives rise naturally to Appell equations of the first order; the classes of that equations can be viewed as a principal homogeneous space with base Pic(R) having as a group $\Gamma(R, \Omega^1)$ [1].

Vitali's long, high-level, study (about 70 pages) is a geometrical study of equations of second order; here Vitali concerns himself with monodromy and determines the equations with a given group (i.e. a given class in $H^1(R, GL_2(\mathbb{C}))$).

Vitali's papers have as a premise a honest knowledge of a part of the classical treatise by Appell - Goursat on algebraic functions and the older treatises of Neumann and Briot on abelian integrals and abelian functions.

Although more questions remained open, such an interesting research was not pursued further by Vitali and by Appell (let me quote from a letter of Appell to Vitali (cf. *Opere*, p. 425): «Je suis maintenant bien loin de toutes ces questions et je ne m'occupe plus guère que de mécanique»).

In our days the subject blossoms anew; cf. e.g. Deligne [1] where equations of the 2nd order are paid much attention. Vitali's papers seem to be not known to the specialists in the field.

By the beginning of XX century also Bianchi and Fubini were interested in fuchsian equa-

---

[1] We are using the usual sheaf-theoretical notation, cf. Gunning [1].

tions and discontinuous groups; the curious postcard of Fubini to Vitali dated «Catania, 13 giugno 1903», cf. *Opere* pag. 427-8, seems to suggest the continuation of Vitali's interest for such subjects, but other writings of Vitali on this are not known.

2. At the beginning of the century assistant-professors were badly payed. On the other hand Vitali's family had insufficient means. Thus Vitali in 1902, after his habilitation, felt himself obliged to seek for a (better - payed!) position as a teacher of secondary schools. Actually Vitali will teach in secondary schools until 1922! First some months at Sassari, next two years at Voghera, then for 18 years at Genova (Liceo «Colombo»), cf. Pepe [1, 2].

Vitali, as a mathematician, was well known in Italy and abroad, so that the fact that for more than twenty years any italian university could not give a place to Vitali was scandalous[2].

Let us return to the beginning of the century! It is clear that Vitali finished his university studies as a specialist in a fruitful and difficult field of complex function theory which can also be considered as a subject belonging to Complex Analytic Geometry. His fellow and friend Fubini[3], by the same period, was mostly concerned with discontinuous groups and differential-geometric subjects.

On the other hand it is evident that students at Scuola Normale Superiore received certainly an excellent training in classical real analysis and stimulating ideas: Dini's «Fondamenti per la teorica delle funzioni di variabile reale», Pisa, 1878 was an authoritative, influential treatise that had also a german translation in 1892. The teaching of Dini (1845-1918) was a reference point for all a generation of celebrated, or simply good mathematicians: Arzelà (1847-1912), Pincherle (1853-1936), Ricci Curbastro (1853-1925), Bianchi (1856-1928), Volterra (1860-1940), Ascoli (1863-1896), Somigliana (1860-1955), Lauricella (1867-1913). The influence of Dini on the successive generation of pupils Severini (1872-1951), Enriques (1871-1946), Vitali (1875-1932), Fubini (1879-1943), E.E. Levi (1883-1917), ... was necessarily deep.

An abstract logical appraisal could suggest a natural continuous evolution from Dini's work to Vitali's work through Ascoli, Arzelà and Volterra's writings on real function theory. Historically the matter appears to be less simple.

The debt of Vitali towards Arzelà will be one of the subjects of my talk. For what concerns Dini and the generation to which belongs Vitali a few words are perhaps necessary. Although the integration of functions «che diventano infinite in infiniti punti» was one of Dini's concerns, it is sure that Dini did not understand the importance of Lebesgue's integration theory; his position changed only in the last years of his life, after Perron[4] had defined the integral of a function without the use of the new concept of a measure. Dini did not consider the new analysis too much important for the development of Science! For a testimony of historical interest cf. Fubini [1], cf. also *Opere* pag. 8, 9[5].

A similar generational break took place in France (cf. A.E. Taylor - P. Dugac in Revue d'histoire des sciences *34* (1981, p. 149-169). Lebesgue writes that Darboux (1842-1917) «ne

---

[2] Let us quote from G. Sansone's *Algebristi, analisti, geometri-differenzialisti, meccanici e fisici-matematici ex normalisti del periodo 1860-1929* (Scuola Norm. Sup. Pisa 1977, p. 41-42): Vito Volterra, forse nel 1922, ebbe occasione di incontrare a Parigi il Lebesgue. Il Lebesgue chiese a Volterra notizie del Vitali e informato che il Vitali insegnava matematica in un liceo a Genova, un po' meravigliato rispose «mi rallegro che l'Italia abbia la possibilità di tenere all'insegnamento liceale matematici come il Vitali».

[3] Cf. Fubini's *Opere Scelte*, Cremonese ed., Roma 1962, p. 3-5.

[4] Cf. Sitzungsber. d. Heidelberger Ak., V a (1914), No. 14.

[5] Moreover in his exposition of Lebesgue integral Dini does not take in consideration the work of Vitali, cf. Dini's *Lezioni di Analisi Infinitesimale*, Pisa 1915.

s'intéressa guère à mes mémoires sur l'intégration». In spite of the interest of Picard (1856-1941) for the evolution of the concepts in Analysis[6], he expressed on the celebrated dissertation of Baire (1874-1932) the opinion that the author «nous paraît avoir une tournure d'esprit favorable à l'étude de ces questions qui sont à la frontière de la mathématique et de la philosophie et qui sont aujourd' hui fort en honneur. Nous ne lui conseillerons pas d'ailleurs de s'y cantonner exclusivement...». It goes without saying that Baire worked in Italy in 1898 with Volterra and his dissertation appeared in the *Annali di Mat. pura ed appl.* (serie 3, *3* (1899), p. 1-123) under the title «Sur les fonctions de variables réelles». The director of the Annali was just Dini.

If Baire's text could seem rather abstract (for those times), to the dissertation of Lebesgue (1875-1941) «Intégrale. Longueur. Aire», an eloquent title!, also printed on *Annali di Mat. pura ed appl.* (ser. 3, (*7* (1902), p. 231-359), were not reserved better appraisals, cf. Fubini [1], loc. cit.: «I lavori di Lebesgue non erano considerati molto importanti da tutti i matematici ....Quando Lebesgue scrisse la sua tesi, Picard si rivolse al Dini, allora Direttore degli «Annali di Matematica», dicendogli che aveva una buona tesi di uno dei suoi allievi, che questa tesi studiava i fondamenti del Calcolo, che tali fondamenti erano studiati principalmente in Italia, e che perciò sarebbe stato meglio che la tesi fosse pubblicata sugli «Annali». Io venni a conoscenza di questa lettera, e subito pensai che Picard non apprezzasse molto questo genere di ricerche. Neppure il Dini era convinto dell'importanza della tesi di Lebesgue ...».

Let us quote also the testimony of Struik (cf. D.J. Struik [1], ch. III, §3): at Leiden University in 1916 people don't talk about Lebesgue; at Göttingen by 1925 Lebesgue was not generally appraised, so that Struik could know the importance of Lebesgue's work only when Wiener (1894-1964) met him there.

As a matter of fact, there is evidence that most mathematicians of the generation of Arzelà, Pincherle, Bianchi, could not grasp the importance of the new ideas; of this generation Volterra and Hadamard (1865-1963) had surely a more open view.

Within such a setting, the new analysis was created by Borel (1871-1956), Baire (1874-1932), Lebesgue (1875-1941) and Vitali!

Very likely, the interest of Fubini (himself an incredibly versatile mathematician) for the new ideas began only by the end of 1903; let us quote from his letters to Vitali: «Ti prego instantissimamente di volermi dire dove io potrei apprendere in modo rapido la teoria degli insiemi, o anche soltanto i primi rudimenti di essa» (cf. *Opere*, p. 433), and (what is more interesting for us to note) we have in a postcard from the 8th January, 1905, «Ho poi riconosciuta errata la dimostrazione da me data del tuo meraviglioso teorema sugli integrali definiti. Tu dovresti vedere la nuova definizione di Lebesgue di 4 o 5 anni fa di integrale negli Annali di Matem., e le definizioni di Borel di limite generalizzato: chi sa quante funzioni non integrabili o differenziabili diverrebbero tali, così come tante serie divergenti divengono coi metodi di Borel convergenti.

Se conosci questi studi, dimmene quanto ti sembra» (cf. *Opere*, p. 450). It is also of interest to read the postcards from 1st June, 1905, 12th June, 1905 on p. 454-455 of *Opere*.

Let us quote also from Fubini [1]: «Beninteso, nei miei studi non ero un rivoluzionario, e perciò, almeno in principio, mi accontentai degli integrali di Riemann...». Indeed, if we exclude two minor notes from 1905, the first important results of Fubini in real function theory are from 1907, 1912[7].

---

[6] Cf. Picard's *L'évolution de l'idée de fonction pendant le XIX siècle* (1899), in *Discours et notices*, Paris 1936, p. 189-225.

[7] Cf. Fubini's *Opere scelte*, papers No. 53, 68, 69, 71, 73, 74, 75, 77 of the list, p. 6, 7. The 1906 paper on Dirichlet Principle treats really another subject.

Lastly, the opinion of Bianchi on such a subject is illuminated by the following words (cf. Fubini [1]): «...quando dissi ad un altro grande matematico italiano, il Bianchi, che l'insieme dei numeri razionali ha misura nulla, egli mi rispose canzonandomi e dicendo che studiavo solo i paradossi dell'infinito».

3. Let us recall that during 1902 Vitali taught at the secondary technical school of Sassari (Sardinia) and during 1903, 1904 Vitali taught at the Liceo at Voghera. Recall also that Vitali's habilitation dissertation was that of a researcher experienced in a specialized chapter of Complex Analysis.

Now, if we look at Vitali's papers on Real Analysis reprinted in *Opere* we see, with astonishment!, that in those years Vitali discovers solitarily the following results, where the number of page is that of *Opere*.

a) The so-called Lebesgue - Vitali theorem on the Riemann integrability of functions (p. 125-128, dated Voghera, 25th April, 1903, p. 133-137).

b) Lebesgue measure (p. 139-149, dated Voghera, 5th November, 1903) (The construction relies upon exterior measure; it is independent of the construction by Lebesgue; it may be clarifying to quote here Vitali's foot-note on pag. 149, loc. cit. «Il presente lavoro era già in corso di stampa quando mi fu segnalata dal il$.^{mo}$ Prof. S. Pincherle la Nota *Sur une généralisation de l'intégrale définie*, par M.H. Lebesgue, pubblicata nei Comptes Rendus des séances de l'Académie de Sciences de Paris, t. 132, p. 1025 (a. 1901), nella quale si accenna ai concetti qui trattati. Il sig. Lebesgue fa uso di questi concetti per la costruzione delle primitive delle funzioni derivate che non sono integrabili»).

c) A measurable function is the sum of a function of Baire class ≤ 2 and of a function which takes the value O almost everywhere, and the so-called Luzin theorem on the almost continuity of measurable functions (p. 183-187) (This paper was surely written in 1903 or 1904: it is quoted in the above - mentioned letter of Fubini dated, Catania, 1 giugno 1905. Luzin's theorem appears on C.R. Ac. Sc. Paris the year 1912).

d) Equivalence of the concepts of Baire function and Borel - measurable function (p. 189-192) (A first draft of this paper was written in 1903. This follows from the contents of the letter of Pincherle to Vitali on p. 435 of *Opere* dated 22th November 1903. This letter desserves several comments. I restrain myself to note: 1) the difference of dates between that of the letter (1903) and that of the printed text (1905); 2) Pincherle valuates one of the important results of measure theory as follows «Ho avuto la Sua nota, che contiene una osservazione interessante.»!

The implication «Baire function $\Rightarrow$ Borel-measurable function» is already in Lebsegue's *Leçons sur l'intégration*, a note of Vitali).

e) The fundamental concept of absolutely continuous function and its characterization as a primitive of a summable function (p. 207-220) (The concept and the characterization is a creation of Vitali; the paper contains other results, now classical, on functions of bounded variation and on derived numbers. The paper was written probably by the last months of 1904 or beginning 1905; it was presented at the R. Ac. Sc. di Torino the 25th June, 1905).

I will not go on to catalogue all Vitali's discoveries in Real Analysis (cf. Pepe [3], Viola [1]). Let me only remember: p. 231-235, 1905, first example of a Lebesgue non - measurable set; p. 237-255, 1907, Vitali - Hahn - Nikodym - Saks theorem which became a central result of modern Functional Analysis; p. 257-276, 1908, covering theorem which is a fundamental tool in analysis and geometric measure theory; p. 347-360, 1922, Banach - Vitali theorem.

All this points out Vitali as one of the founders of modern analysis. In our historical context I must also mention Vitali [6] on the integration of series. This paper is closely related to Vitali's papers on sequences of holomorphic functions which I will analyse later on. Let me emphasize moreover that Vitali could add to the rich harvest of results achieved during a so short time the papers on complex analysis Vitali [7], Vitali [8], Vitali [9], *all three from the year* 1903, where Vitali gives the celebrated theorems on relative compacity of families of holomorphic functions; these theorems alone would grant to Vitali a place in the history of Analysis.

4. At this point we reach the central part of my talk and it will be convenient first to give very sketchily some information about mathematics at Bologna University in that enthusiastic period of history of Italy known as Risorgimento and its successive phases of development.

Let me begin with a quotation from E. Bortolotti [1], a document written for the International Congress of Mathematicians that met at Bologna in September 1928. The quotation is from p. 73, ch. III: «La période qui va de la restauration de 1814 à l'avénement de l'État National est une des plus obscures pour notre École de Mathématique.

Toute trace d'autonomie a disparu, ... Le gouvernement de l'instruction supérieure resta uniquement confié à la Cour Pontificale. Les chaires de Mathématiques furent réduites à quatre[8]...

Malgré quelques bons éléments qui restèrent encore dans le Corps Académique... notre école mathématique était réduite aux conditions les plus misérables, et s'acheminait sans doute vers sa fin».

The revival of a prestigious mathematical tradition at Bologna University was mainly the work of two eminent scholars: Luigi Cremona (1830-1903) e Eugenio Beltrami (1835-1900). Cremona taught at Bologna from 1860 till 1866; Beltrami in 1862 and next from 1866 till 1873 after his stay at Pisa where he could take advantage of the presence of Riemann.

In the year 1882 was created at Bologna University a course in Mathematics (Corso di Laurea in Matematica).

It is worthy to note that by the end of the century the staff of professors of Mathematics of Bologna University was composed by men open to the new trends of research and actively inserted in the world community of mathematicians.

The stars of this staff were Pincherle, Arzelà and Enriques; in a way or another, all three will influence the life's course of Vitali.

From what I have said about the results of Vitali obtained (in a so short period) in first years of our century, it would be quite natural to imagine that Vitali would obtain soon a position at some university, perhaps not far from Bologna due to the close scientific collaboration established between Arzelà and Vitali (as we will see later on). Actually, Vitali's fellows, who had already obtained their position, thought so. Let me quote from a postcard (1904) of Fubini, professor at Catania, to Vitali (cf. *Opere*, p. 446): «...mio così caro amico, onore della scienza italiana e certo destinato ai più alti destini...». Also form another postcard (1904) of Ugo Amaldi (1875-1957) professor at Cagliari (cf. *Opere*, p. 444): «...e non ho neppure *una* persona con cui conversare di argomenti matematici. Quanto sarei felice di avere qui un amico bravo e valente quale sei tu! Nota che qui non sarebbe difficile fargli avere un posticino, naturalmente retribuito, di assistente. Ma a te auguro cose ben migliori...».

---

[8] That is, cf. loc. cit., Introduction au Calcul, Calcul Sublime, Optique et Astronomie, Mécanique et Hydraulique. On p. 71, loc. cit. Bortolotti informs that «La faculté Physico-Mathématique, dans l'Université napoléonienne, fut constituée par le décret de 31 octobre 1803, avec les Chaire suivantes: Éléments de Géométrie et d'Algèbre, Introduction au Calcul Sublime, Mathématiques Appliquées, Architecture Civile et Militaire, Principes de dessin, Astronomie (two chairs), Physique Générale, Physique Expérimentale, Agronomie.

We have moreover two postcards (1904) of Arzelà (cf. *Opere*, p. 449, 450) that seem to suggest Arzelà's attention for Vitali's career. That is all! What we can find more in the collection of letters in *Opere* is a reference to a free course (Dini's letter, 1907, p. 464) and a letter of Mario Pieri (1860-1913), p. 471, with an offer of an «incarico (provvisorio)» at Parma University...

Indeed, the development of the story will be another...

In 1906 Leonida Tonelli (1885-1946) completes his university studies at Bologna and fascinates his professors Pincherle and Arzelà who disputed him as assistant. Tonelli was endowed with a marked personality. He was too much conscious of his value and possessed also a strong practical sense. His family belonged to the middle class. His dissertation to obtain the degree (Laurea in Matematica) is an excellent original memoir on Čebyšev polynomials which reveals already the power of the founder of modern Calculus of Variations. He published quickly some beautiful papers and by the year 1911 he was ready, with the support of Arzelà, to partecipate to a competition for a chair at Parma University. The death of Arzelà in 1912 complicated the situation and actually Tonelli was called to Parma only in 1917 (On all this cf. S. Cinquini [1]). Tonelli fought during the World War I and to him were awarded several medals. Tonelli could reach the chair of Analysis at Bologna University only in 1922.

Vitali was another type of man (and his progress a very different one!)[9].

Meanwhile Enriques left Bologna (1922) for Rome, replaced by Enrico Bompiani (1889-1975), and Bologna University could count also Beppo Levi (1875-1860) in its staff.

Pincherle was pensioned off by 1928. He was chosen in 1924, during the Toronto Congress, as a President of the International Mathematical Union. By that time Pincherle was a universally esteemed personality in the world community of mathematicians. His wisdom and sagacity could help Pincherle to overcome grudges and he could organize the successful 1928 Congress at Bologna with the participation of more than 1000 guests. Mathematicians of all belligerent countries could gather themselves for the first time after the World War I[10].

After 1928 come to Bologna Vitali, Luigi Fantappiè (1901-1956), Beniamino Segre (1903-1977). Bologna was then one among the good european mathematical centers until the race Law (1938), and the atrocious World War II had closed an historical period...

5. Now, let me come some decades back to spend some words about Enriques, Pincherle, Arzelà and the young Vitali.

As it is well - known Enriques created together with Castelnuovo (1865-1952) the theory of algebraic surfaces during their famous walks along the streets of Rome (cf. G. Castelnuovo [1]). (The word «created» is the correct one because we may say that this theory was invented ex nihilo: the methods of Max Noether were insuitable and his memoir on surfaces, notwithstanding its importance, is obscure and also some of its basic assertions revealed themselves wrong or led out of the way.)

Federigo Enriques was then constructing his projective models and setting up the geometric approach that led him to the complete classification of algebraic surfaces, an achievement

---

[9] Let us quote from Tonolo [1], in reference to the tardy recognition of Vitali by the academic *milieu*: «A ciò ha indubbiamente contribuito il suo carattere riservato e alieno nonchè del farsi largo...».

[10] Let us recall that from the two international Congresses after World War I, Strasbourg 1920, Toronto 1924, were excluded the mathematicians with german, austrian, hungarian and bulgarian citizenship. At Toronto Congress the American Section of the International Union (supported by Italy, Denmark, Holland, Sweden, Norway and Great Britain) proposed the removal of restrictions. The grudge was comprehensibly deep and Pincherle must overcame heavy difficulties.

that history of science will consacrate as one of the boldest conquests of human mind in those times. Enriques had widespread cultural interests. He was endowed with an astonishing capacity of intuition and his personality was a rich, fascinating one. As Felix Klein (1849-1925) in Germany, Enriques, with similar cultural motivations, but perhaps with deeper philosophical views, considered an important task the popularization of mathematical thinking and the preparation of skilled teachers of mathematics for secondary schools.

The young professor discovered at a glance his young pupil Vitali. We have a testimony of this already in a letter from 1897 (cf. L. Pepe [3], p. 200) to a certain Prof. Zannoni in order to obtain a school - grant for Vitali: «Alla Sua domanda di informazioni relative al Sig. Giuseppe Vitali già allievo del nostro primo biennio di Matematiche, posso rispondere nel modo più favorevole.

Attesto... che egli abbia ingegno promettente... sono sicuro che non avrà mai a pentirsene della Sua proposta, perchè ritengo sia questo *uno dei pochi casi in cui è veramente utile concedere un sussidio*».

A touching testimony are also two letters, cf. *Opere*, pag. 417, 418, 419, where Enriques gives to the second - year student Vitali precious advice concerning his entrance the Scuola Normale Superiore.

On account of what I have said about the seriousity of Enriques' endeavour for mathematical education, it is of much interest to notice the proposal of collaboration to the «Questioni riguardanti la geometria elementare» made by Enriques to Vitali already in 1899 cf. *Opere* letters on p. 419, 420, 421, 422, 423. On this type of collaboration we will have, later on, letters form 1903, 1904, 1922, cf. *Opere*, p. 429-430, 447, 521, 522.

There is evidence that Vitali immediately after the obtention of his degree (Laurea in Matematica) (1899) tried to study some interesting questions of Algebraic Geometry, surely under the influence of Enriques.

I could not found written notes by Vitali or something else that is sufficient to trace back the ideas of Vitali. The best I can make is to call attention for a long quotation from a letter of Enriques to Vitali dated «Saltino (Firenze), the 27th July 1899» which has a real interest and is characteristic of Enriques style (cf. *Opere*, p. 423):

«... Per quanto si riferisce al Suo lavoro di Geometria Le dirò che l'argomento di cui Ella mi parla può condurre a questioni interessanti, almeno mi sembra. Bisogna però non limitarsi a dare quelle facili estensioni che appunto perchè previste sono da riguardarsi quasi come note: piuttosto è il caso di proporsi lo studio di ciò che l'argomento può offrire di essenzialmente *nuovo*, e quindi non più così facile.

Mi permetterò d'indicarle la mia nota dei Math. Ann. 95 o 96 «Sui sistemi lineari di superficie...» ove troverà considerate le varietà a sezioni razionali e ellittiche rappresentate da 2 sistemi di quadriche. E giacchè ho parlato di quella nota Le farò notare che la discussione delle possibili varietà a curve sezioni ellittiche con più di 3 dimensioni, offre un argomento nuovo (in parte connesso con quello di cui Ella mi parla) e, se non sbaglio, promettente. Io credo che andando colle dimensioni ($n-1$) al di sopra d'un certo limite non si abbiano altre varietà a curve sezioni ellittiche che la varietà cubica (di S_n) e l'intersezione di due quadriche (di S_{n+1}). E' un'ipotesi che mi è suggerita dal fatto che proiettando la varietà (rapportata in S_{n+2} cioè d'ordine 5) da due punti si ha in S_n una varietà cubica con due S_{n-2} che devono avere punti comuni...

Ma anche se l'ipotesi predetta non sarà vera la classificazione completa di quelle varietà dovrà condurre ad un risultato semplice».

Perhaps Vitali at Pisa, without the possibility of frequent talks with Enriques about such a program, could not progress in this way. In those times the knowledge of an enormous amount of models was necessary and the techniques of those times of approach to such problems were yet, for a large part, in the mind of their creators. On the other hand Vitali was then inserted in a different school and must there preparate his dissertation for habilitation degree. It is thus comprehensible that Vitali took naturally another way... Although the correspondence between Enriques and Vitali continued for many years (the last letter in *Opere* is from 1922), and there was surely an intellectual affinity between these two men, I could not discover more references to subjects belonging to Algebraic Geometry. This may be also explained, at a certain extent, by the endeavour and the success of Vitali in other fields and, later on, in his professional and social activities. If in the last part of his life Vitali's interests turn anew to Geometry, specifically to Differential Geometry, as we shall see, it is as an analyst that Vitali gave his major contribution to Science.

Analysis at Bologna was then represented by two eminent scholars: Pincherle and Arzelà (who died in 1912). The personalities of Pincherle and Arzelà were rather different, and very different was their approach to Analysis and their research subjects; it seems also that the relationships between these two men were not too much hearty.

Although Pincherle obtained his diplom at Scuola Normale Superiore (Pisa) and frequented at Pavia the courses of F. Casorati (1835-1890) to whom (together with E. Betti 1823-1892) is due the introduction and propagation of Riemann's ideas on function theory in Italy[11], Pincherle considered himself as a pupil of Weierstrass whose lectures he attended in Berlin (1877-78). The textes of his courses on complex function theory leaved to us, and his conception of complex analysis were rigidly weierstrassian (as that of Mittag - Leffler). In 1886 appears his memoir «Studi sopra alcune operazioni funzionali»[12] and in 1897 his «Mémoire sur le Calcul fonctionnel distributif»[13]. It is well known that Functional Analysis has its roots in the pioneering work of Pincherle and (the rather different) work of Volterra. Pincherle's goal and viewpoint is illustrated by the beginning words of the first memoir refered to, and from «Appunti di Calcolo funzionale distributivo»[14] appeared also in 1897: «chiamo *operazione funzionale* qualunque operazione che eseguita sopra una funzione analitica dà per risultato una funzione analitica. Sono tali, per esempio, le operazioni aritmetiche in numero finito, e per classi numerose di casi, anche in numero infinito, la derivazione e l'integrazione, la risoluzione di equazioni finite o differenziali, la sostituzione, ecc.»; «Un'operazione funzionale distributiva si può definire in due modi: o mediante le funzioni che essa fa corrispondere alle potenze intere positive e nulla della variabile x, o mediante il suo sviluppo, sempre possibile, secondo le potenze intere positive del simbolo di derivazione D».

As it is well - known, in those times there was a widespread interest in the inversion of

---

[11] In 1858 Casorati and Betti in company of Brioschi (1824-1897) visited France and Germany; Volterra (Bull. Am. Math. Soc., 7 (1900), p. 60-62) pretends that the scientific existence of Italy as a nation can be dated from this travel. Cf. also Volterra, *Opere Mat.* v. 3, Roma, 1957, p. 1-13: Betti, Brioschi, Casorati, trois analystes italiens et trois manières d'envisager les questions d'analyse. For information about Casorati's life and work cf. E. Bertini, *Della Vita e delle Opere di Felice Casorati*, Rend. del Reale Ist. Lombardo, serie II, vol. 25, (1892), p. 1206-1236, or *Opere* (cf. Casorati [1]) p. 3-30. Also articles by U. Bottazzini, E. Neuenschwander and A. Weil in vol. 18, 20, 21 of Archive for History of exact Sciences (1977-78-79-80).

[12] Memorie della R. Acc. delle Sc. dell'Ist. di Bologna (4) 7 (1886), p. 393-442; or Pincherle [1], vol. I, p. 92-141.

[13] Math. Ann. 49 (1897), p. 325-382, or Pincherle [1], vol. II, p. 1-70.

[14] Reale Ist. Lomb. di Sc. e Lett., Rend. Cl. di Sc. Mat. e Nat. (2) 30 (1897), p. 1031-1039, or Pincherle [1], vol. I, p. 388-396.

integrals in the complex domain connected with several questions of Analysis and Geometry, and also in the representation and extension of holomorphic functions by finite or infinite algorithms. In most cases the output appears to depend linearly on the Weierstrass element, or as we would say today, on the germ of the given functions, that is to say, on the sequence of Taylor coefficients, which involves powers of the differential operator D. Abel's paper Abel [1] is a prototype, and it was deeply meditated by Pincherle. It must also be said that Pincherle knew the formal improvements bringed by Murphy to the old work of Gregory and Newton on difference equations and that to this subject Casorati gave a less formalistic and more substantial contribution in a memoir Casorati [1] of 1880 by using holomorphic function theory, and this paper undoubtly influenced Pincherle's conceptions. In Pincherle's work analytic continuation is sometimes considered as an «operazione distributiva» in such a context that suggests unavoidably to the reader of our days the theory of resurgent functions. The adherence of Pincherle to weierstrassian viewpoint has thus a motivation and gives not the (modern) reader that feeling of parti pris that issues from certain Mittag - Leffler's writings[15]. It must also be stressed that the analysis of inner derivations of an algebra of linear operators on a space of holomorphic functions plays an important rôle in Pincherle's work. From Pincherle we have an edition in two volumes of Opere Scelte (cf. Pincherle [1]). A reader of our days can discover here several lines of thought that lead far in the time to analytic functionals and hyperfunctions (Fantappiè, Silva, Köthe, Grothendieck, Martineau, Sato,...) and to inverse problems and Ecalle's theory. However, although Pincherle's papers contain a lot of pearls, we must historically constatate that interest for Pincherle's work lessens quickly from the twenties on, and today it is difficult to find in the literature references to Pincherle's results or to Pincherle's work[16]. Such is the course of history! Today a renewal of interest for some

---

[15] Here we may recall the discussion between Borel and Mittag - Leffler, cf. in particular E. Borel, *Leçons sur les fonctions monogènes uniformes d'une variable complexe*, Paris, 1917.

[16] The starting point of Fantappiè was the study of the functional that associates to a holomorphic function the value of its derivative at a given point (a problem suggested to him by Severi). The fist paper published by Fantappiè (1901-1956) on functional analysis in 1925 (Rend. Acc. Lincei, s. 6, *1*, 1° sem. 1925) has the eloquent title: *Sulla riduzione delle operazioni distributive di Pincherle alle funzionali lineari di Volterra*. Pincherle's work is mentioned by Fantappiè in several of his papers. Curiously enough in the systematic exposition of his theory *Teoría de los funcionales analíticos y sus aplicaciones*, Barcelona, 1943, (cf. *Opere Scelte* di Luigi Fantappiè, Unione Matematica Italiana 1973, vol. II, p. 893-1058), there is *no mention* of Pincherle. I can not refrain myself from quoting from p. 11 (p. 895 of Opere Scelte) because this is also interesting for the knowledge of Fantappiè's historical appraisal of (a part) of Arzelà and Ascoli's work: «3. *Diversas direcciones en el estudio de los funcionales*. Pasemos ahora a señalar las principales direcciones en que se ha orientado la teoría de los funcionales, aunque, naturalmente, sólo podemos hacer una exposición muy general, para hacer patente el lugar que ocupa entre las diversas corrientes o direcciones el estudio de los funcionales analíticos. Itália. – Poco después de los clásicos trabajos de Volterra, continuados por numerosos discípulos que hacen que la teoría de los funcionales pueda considerarse una gloria italiana, fué Arzelà el primero que marcó un nuevo camino. Así como la ordinaria teoría de funciones se inicia con el estudio de la teoría de conjuntos de números, así él comenzó por estudiar los conjuntos de funciones. Especial interés tiene la consideración de una función que sea elemento de *acumulación* de otras funciones, sobre la cual dió Arzelà su célebre teorema (análogo al de Bolzano - Weierstrass para los conjuntos de puntos) y cuya demostración perfeccionó Ascoli».

Forward Fantappiè mentions shortly Tonelli and goes on to refer on France and Germany.

Actually the interest for Pincherle's results on difference equations and integrals of Laplace type seems to grow today. Let me mention the paper by M.A.B. Deakin, G.J. Troup, K.P. Dabke, *Pincherle on the functional equation* $\sum h_\nu \phi(x + x_\nu) = f(x)$, Department of Mathematics Monash University (Clayton

subjects of Pincherle's research (in a broader context!) is present and thus a critical rewiew of Pincherle's work will be strongly appealing.

Pincherle was a well - informed and accurate mathematician. As we have seen, Vitali's début was that of a specialist of complex analysis of great force (1899-1904), the main field of interest for Pincherle. So it remains a mystery for me the fact that Pincherle seems completely ignore Vitali; we will return on this later.

The work of Arzelà (and the work of Volterra on pure real analysis) can be considered as a direct continuation of Dini's research on real function theory. The writings of Arzelà do not contain radical discoveries[17], but are characterized by a clear explanatory style and a subtle discussion of several questions of Analysis essential for the work of the next generation, in particular for Vitali's papers on Analysis. Today Arzelà is perhaps best known for his reformulation of Ascoli's compacity theorem (first exposed in Ascoli [1] in a rather entangled language), his necessary and sufficient condition for the continuity of the limit of continuous functions («convergenza uniforme a tratti» or «quasi - uniform convergence», a terminology introduced by the french school), and for his attempts to justify the Dirichlet principle. Actually the papers of Arzelà cover other topics in real analysis and are rich of subtle contrivance; as collected papers of Arzelà with critical comments are now being edited by the Unione Matematica Italiana I will not be more specific on the general contents of Arzelà's work.

Arzelà represents (as Enriques in the terms I have indicated) an intellectual reference mark for the young Vitali at Bologna. *Opere* gives notice of 13 mail communications of Arzelà to Vitali and surely Vitali met Arzelà during his visits to Bologna and perhaps aslo during the frequent stays of Arzelà at S. Stefano di Magra (Liguria). Incidentally it is interesting to remark that in a postcard (dated S. Stefano di Magra, 25th April 1908), cf. *Opere* pag. 470, Arzelà writes «Ho una nota del mio allievo Tonelli relativa alla lunghezza, e che si riattacca all'ultima sua nota.

«Io sono occupatissimo: potrebbe leggerla lei? è breve». It is question of the note Tonelli [2], or perhaps a first draft of it, where the theorem, now classical, on the rectification of curves is given; Tonelli begins his note with the words «Il prof. Vitali, in una recente nota[1] , ha dato una proposizione relativa ai gruppi di punti...» (« [1] G. Vitali, Sui gruppi di punti e sulle funzioni di variabili reali...» (cf. *Opere*, pag. 257)) and indeed Vitali's results and Vitali's concept of an absolutely continuous function are essential for Tonelli's theorem.

Vitali possessed a detailed knowledge of Arzelà's work and it influenced some important papers of Vitali (cf. *Opere*, p. 471-472 postcard of Arzelà), namely the papers on series or sequences of holomorphic functions to which will be dedicated next section: however, the influence of Arzelà's work on Vitali must not be overestimated because Vitali's contribution to analysis (as that of Lebesgue) was conceptually a novelty and truly original (original at a point

---

Victoria Australia) 1982; it contains an english translation of two papers of Pincherle. (To place the work of Pincherle on difference equations in its classical context see also N.E. Nörlund, Vorlesungen über Differenzenrechnung, Berlin, 1923). Cf. also last issues of Arch. for H. of exact Sc.

In recent textbooks reference to Pincherle's work is almost a rarity. (The more recent reference I can remember now is that of a result of Pincherle on Dirichlet series on p. 118 of A.F. Leont'ev *Riady Eksponent*, Moskva 1976. Indicative are also the references to Pincherle in the authoritative monograph of M. Kuczma, Functional equations, PWN, Warszawa, 1968 and its bibliography). Let me close this too long note by remember that Pincherle himself wrote an account of his work published by Acta Math., 1925, cf. Pincherle [1] vol. I, p. 45-63.

[17] The concept of equicontinuity and the relative compacity theorem are indeed due to Ascoli (1883-84), cf. Ascoli [1], Parte seconda. Yet Prof. L. Pepe informs me that the important idea of majorated convergence is already present in Arzelà's work.

that gave place to some misunderstanding testified by two letters of Lebesgue from 1907 (cf. *Opere*, pag. 457, 462), that fortunately could be soon cleared). We may say that Arzelà's work is an impulse towards the hatching of the new world of the ideas of Vitali and Lebesgue.

6. In this section I will consider the papers Vitali [7, 8, 9] in more detail than we can find in the other writings on Vitali (that cover the whole, or other aspects of Vitali's work). All this deals with one complex variable.

In [7], cf. *Opere*, p. 129, Vitali establishes that a sequence of holomorphic functions, uniformly bounded on a (connected) open set Δ and convergent at each point of a subset of Δ with a cluster point in Δ, will converge uniformly on compact subsets of Δ to a holomorphic function. This assertion, that I will call Th. I, extends a theorem by Stieltjes (1856-1894) (cf. Stieltjes [1], where it is supposed that the sequence converges on a subset of Δ with some interior point), and a theorem by Osgood (1864-1943) (cf. Osgood [1], where it is supposed that the convergence of the sequence takes place at each point of a dense subset of Δ).

Vitali [8], dated Bologna, the 16th July 1903, is a fundamental paper as I shall explain. Here Vitali takes again the contents of [7] and states that for a uniformly bounded sequence of holomorphic functions on Δ there is some holomorphic function on Δ to which a subsequence will converge (uniformly on compact subsets); cf. [8], p. 8, or p. 158 of *Opere*. This theorem is usually attributed to Montel (1876-1967). To the eye of charity, it is somehow implicit in Arzelà's Nota that is at the origin of Vitali [7], cf. Arzelà [1]; although Montel shows a detailed knowledge of Arzelà's work, and quotations from it abound in Montel's papers before World War I, Montel will *never* mention Vitali (nor Arzelà) for what concerns this (non obvious part of the) theorem of compacity (th. II).

The most important result in Vitali [8] is the following one (cf. p. 17, or *Opere*, p. 167) that, in agreement with the present times, I will call *tautness theorem* (th. III): «Se

$$u_1, u_2, \ldots$$

è una successione di funzioni analitiche finite e monodromi convergenti in ogni punto di un campo semplicemente connesso C, e se inoltre le funzioni suddette non assumono mai i valori 0 ed 1 la successione converge verso una funzione analitica finita e monodroma in C».

Moreover Vitali adds: «§4. Si può facilmente togliere la condizione che il campo C sia semplicemente connesso. Se C è più volte connesso si può arrivare alle stesse conclusioni spezzandolo in parti semplicemente connesse».

I will return on this result later.

Arzelà [1] and Vitali [7, 8] are conceptually simpler than Osgood [1]. In Vitali [8], Vitali takes again the study of convergent (uniformly or not) sequences of holomorphic functions (Capitolo Terzo), pp. 10-15, by Osgood [1], §2; Vitali and Osgood show essentially that the set Δ^* constituted by the points z of Δ such that a given sequence converges uniformly in a neighbourhood of z is a dense open subset of Δ .

This question interested very much Montel who completed this result of Osgood 1901 with several new essential properties. Montel considered the properties of what he called «fonctions de première classe» of a complex variable (they are of first Baire class, but the two concepts are not the same!) in several of his first mathematical writings (cf. Montel [1, 2, 3, 4]); the difficulties where not all overcame and are illustrated by the fact established by Lebesgue that the limit of a (uniformly convergent) sequence of functions of that class may not belong to the given class (cf. Lebesgue [1][18]).

---

[18] Let us quote from Montel [1], «*l'ensemble des points exceptionnels d'une série convergeante de fonc-*

Vitali [9] dated Bologna, september 1903, is actually a continuation of the study of convergence of a sequence of holomorphic functions and is motivated by a result of Arzelà [2]. Vitali proves among other:
(th. IV'), cf. p. 9 (*Opere*, p. 177) «Se la serie di funzioni analitiche

$$w_1(z) + w_2(z) + \ldots + w_n(z) + \ldots$$

ha la convergenza uniforme a strati[1] in un campo connesso C e se inoltre sopra ogni retta x = cost e y = cost è numerabile o finito il gruppo di punti di convergenza singolare, allora affinchè la serie converga verso una funzione analitica è necessario e sufficiente che esista una funzione $\Omega(z)$ continua in tutto il campo verso la quale la serie delle derivate converge nei punti di convergenza uniforme»; (th. IV"), cf. p. 13 (*Opere*, p. 181) «Se una serie di funzioni analitiche ha la convergenza uniforme a strati in un campo connesso C e se inoltre sopra ogni retta x = cost e y = cost è rinchiudibile il gruppo dei punti di convergenza singolare, allora affinchè la serie converga verso una funzione analitica è necessario e sufficiente che esista una funzione continua $\Omega(z)$ verso cui la serie delle derivate converge nei punti di convergenza uniforme e che in ogni campo interno a C i rapporti incrementali della funzione verso cui converge la serie restino in valore assoluto tutti minori di una quantità finita fissa».

Th. I and similar results are consequence of compacity theorem (th. II) and some other principle stating that a holomorphic function which satisfies it must vanish identically, such as the principle if isolated zeros, Blaschke condition, boundary value bahaviour, and so on[19].

Th. IV', IV" and similars rest on the continuity of the limit of a quasi-uniformly convergent sequence of continuous functions (Arzelà's theorem), th. II, and a condition to be satisfied by a continuous function F holomorphic on an open dense subset Δ^* of Δ to be holomorphic on Δ. The study of such conditions, and the analogues for harmonic or subharmonic functions (a subject also considered by Vitali [10][20]), was also treated by Montel in his thesis,

---

tions analytiques, régulières dans D , est parfait, non dense et d'un seul tenant avec la frontière C . On voit qu'une série convergeante de fonctions analytiques possède à l'intérieur de tout domaine un nouveau domaine où elle converge uniformément: sous cette forme la proposition a été aussi démontrée par M. Lebesgue».

This last result is thus in Osgood [1], 1901; Montel does not ascribe the statement to Osgood in [1] (and [3]) but only in [4], p. 108.

[19] Blaschke condition was stated in W. Blaschke, *Eine Erweiterung des Satzes von Vitali über Folgen analytischer Funktionen*, Ber. der Math. - Phys. Kl. der Sächsischen G. der Wiss. zu Leipzig, LXVII (1915), p. 194-200, cf. also E. Landau, *Über die Blaschkesche Verallgemeinerung des Vitalischen Satzes*, ib. LXX (1918), p. 156-159. Later on 1922, F. and R. Nevanlinna gave their well - known condition and we will have I. Privalov's Eine Erweiterung des Satzes von Vitali über Folgen analytischer Funktionen, Math. Zeit., 1924, p. 149.

[20] In [10] from 1912 Vitali refines a characterization of harmonic functions stated by E.E. Levi in 1909 (cf. E.E. Levi, Opere, Cremonese, ed. Roma 1959, p. 180-186) that had also been taken up by Tonelli [1], vol. I, p. 148-155. For such a subject and subharmonic functions cf. also T. Radò, *Subharmonic functions*, Berlin, 1937.

Here I want chiefly to call your attention to a postcard of Fubini to Vitali, dated 8th June 1912, on p. 477-478 of Opere. This postcard is very interesting because testify by 1912 scientific relationships between Vitali and the true genius that was Eugenio Elia Levi (killed at the front (Bainsizza, today Bate, Slovenia, 1917) during the World War I). Perhaps this is at the origin of Vitali [10]. Fubini's words show that Vitali began to interest himself to integro-differential equations, in particular to existence theorems. It is possible that the death of Levi precluded such a development of Vitali's research.

in Montel [5] (whose main statement Montel never proved), and blossomed between the two World Wars, so that Vitali's results may be today considerably improved. A plenty of such results is known so that the best to do is to refer to Menchoff [1] or Saks [1] (Taking account of Weyl's lemma [21] today we may add to all of this conditions of functional analytic type such as «F as a distribution or a hyperfunction satisfies Cauchy-Riemann equation».)

Th. III and Picard's theorems on exceptional values are much deeper results; their *raison d'être* is a complex differential-geometric property of $\mathbb{C}\setminus\{0,1\}$, i.e. $\mathbb{C}\setminus\{0,1\}$ is a complete hyperbolic manifold, cf. Grauert - Reckziegel [1] (for what concerns hyperbolic manifolds, taut manifolds and related topics we refer you to Wu [1], Kobayashi [1,2]).

The discovery of th. III is characteristic of the capacity of Vitali to grasp the essence of things and of his innovator mind.

Let us observe that in his thesis (the results of which were published in Montel [3] 1907), and also in his papers published before it, Montel never gives a statement like th. III; the first result of Montel of the type of th. III appears in 1907 (cf. Montel [6]) (indeed in the mentioned papers the conditions considered by Montel (such us $|f_n - a| \geq k$) are not conformally different from the boundedness of sequences).

Later on Montel established a certain number of results in the logical flow of th. III that play an important rôle in his influential classical treatise Montel [7], relying ultimately on Landau's papers on Picard's theorems.

The precious experience gained by Vitali during the preparation of his dissertations Vitali [3, 4, 5] has certainly contributed to the discovery of th. III; his familiarity with the geometry of complex analysis has, on the other hand, surely contributed to the hastiness of Vitali's proof that indeed is incomplete: the use of some metrical consideration or some estimate, e.g. of Schottky type, that ensures the tautness is lacking. This had not been observed by Montel, and others, until it was noted by Carathéodory and Landau [22] in their paper Beiträge zur Konvergenz von Funktionenfolgen (1911), cf. Carathéodory - Landau [1]. Carathéodory and Landau point out the gap in Vitali's proof in a footnote on p. 599 of their paper («Der Beweis von Hrn. Vitali... für dieses Resultat scheinen aber die Hilfsmittel, welche Hrn. Vitali zur Verfügung standen, nicht auszureichen»).

For the «little history» it is perhaps interesting to quote some other specification from Carathéodory - Landau [1]. In these quotations the name of Severini (1872-1951) is mentioned; later on I will say something more about Severini's papers. Montel's writings (that come after Vitali [8]) refered to in Carathéodory - Landau [1] are Montel [3, 4, 6]. We have:

– from p. 589, footnote 2: «Hrn. Montels Literaturangaben daselbst ist obiger Hinweis auf Vitali und Porter hinzuzufügen»;

– from p. 596, footnote 1: «Hr. Montel schreibt auf S. 124 seines Buches[23] irrtümlich die Schottkyschen Sätze III, IV Landau zu und formuliert IV ausserdem unrichtig. Er gibt

---

[21] About 25 years ago G. Fichera put to J.S. Silva the following question: to characterize (up to isomorphism), if it exists, the maximal functional space where Weyl's Lemma holds. J.S. Silva has considered this problem in *O Lema de Weyl no quadro das Ultradistribuições* (Bol. da Ac. das C. de Lisboa, XXXVII (1965), cf. J.S. Silva [1], vol. III, p. 293-302). This is an interesting question within the theory of differential operators that deserves still further research.

[22] Let me point out here a postcard of Landau to Vitali dated Göttingen, 21th October, 1910, cf. *Opere*, p. 472: «Sehr geehrte Herr Kollege! Indem ich Ihnen meine letzte Arbeit sende, bitte ich um Separata Ihrer 3 Arbeiten «Sopra le serie di funzioni analitiche» aus den jahren 1903, 1904, sowie etwas neuerer über dies Thema...»

[23] Montel [4].

nämlich die ϵ — Bedingung (die seinerzeit noch erforderlich war) richtig an, lässt jedoch die ω — Bedingung fort; dadurch entsteht ein offenkundig falscher Satz, wie schon das triviale Beispiel $F(x) = a_0$ lehrt. Wenngleich Hrn. Montels Behandlung der vorliegenden Probleme uns zu verschiedenen Beanstandungen historischer und sachlicher Art veranlasst, so wollen wir doch nicht unterlassen, besonders hervorzuheben, dass wir sowohl seine Thèse als auch sein Buch sehr hoch schätzen und viel Neues daraus gelernt haben. (...)».

— From p. 597: «Die HH. Severini und Montel waren nicht im Besitze des Satzes V, sondern bedienten sich bei ihren Untersuchungen des Satzes IV mit dem erschwerenden Ballast der ϵ — Bedingung» (cf. also footnote 3, p. 597);

— From p. 599, footnote: «Vgl. die in der Einleitung zitierten Stellen bei den HH. Vitali, Montel und Severini. Dass auch der im Text genannte Wortlaut, der weniger als VI besagt, von den HH. Montel und Severini nicht einwandfrei bewiesen wurde, haben wir schon erwähnt... Anderseits müssen wir konstatieren, dass Hr. Montel (auf. S. 912 seiner in der Einleitung zitierten Note) und Hr. Severini (auf. S. 188 seiner ebenda genannte Abhandlung) die Stelle bei Hrn. Vitali erwähnen, ohne irgendein Bedenken geltend zu machen».

Let me complete this picture of minute notices. Montel published on the Comptes Rendus de l'Académie des Sciences de Paris (and on Bulletin les Sciences Mathématiques, $2^{\text{èm}}$ série, XXX, juin 1906, p. 189-192, but this paper is not of interest for our discussion) a summary of the results of his thesis Montel [3], 1907. The first paper on C.R. was presented by Painlevé (1863-1933) may, 25th, 1903 and concerns the integration of differential forms of degree one. The second paper on C.R. presented also by Painlevé is from February, 22th, 1904 and it is just dedicated to sequences of holomorphic functions. We must observe that this paper has thus appeared after Vitali [7] (th. I) and was presented after Vitali [8, 9] (th. II, III, IV', IV"). On a copy of Montel's book *Leçons sur les Séries de Polynomes* (Montel [4], printed 1910) in our library (leaved to it by Pincherle), Montel states th. I on p. 20 without mention of Vitali or Porter, and th. II on p. 21 without mention of Vitali; handwritten by Pincherle we have a note to the statement of th. I saying «Questo teorema è di *Vitali*. V. Landau Carathédory, Sitzungsbericht Ak. Berlin 1911, 18 Mai, p. 988» and a note to the statement of th. II saying «théor. fondamental».

Actually, th. I, to which Carathéodory and Laudau refer in loc. cit. by writing (p. 588) «nennen wir jetzt folgenden schönen Satz, den man Hrn. Vitali verdankt: ...», was stated and proved, clearly independently of Vitali, by M.B. Porter in the paper Porter [1] appeared on the Annals of Mathematics in the issue of July 1905 (this fact is also stressed by Carathéodory - Landau [1]).

Finally, th. III is ascribed by Montel to Vitali in Montel [6], 1907, presented also by Painlevé, and th. I is also ascribed by Montel to Vitali in his treatise Montel [7], 1927 (§116 of ch. I has the title: Théorème de M. Vitali).

It will be also interesting to examine a sample of textbooks of different periods dealing with complex analysis in one dimension, published in the first third of our century, for what concerns the more elementary theorems I, II.

These theorems are not included in Pincherle's textbook «Gli elementi della teoria delle funzioni analitiche (Parte Prima)», Nicola Zanichelli, ed., Bologna, of 1922 because Pincherle perhaps considered (?) them a matter of a more advanced level than that of the text.

The german translation of Vivanti's treatise «Theorie der eindeutigen analytischen funktionen», «Umarbeitung unter Mitwirkung des Verfassers» by A. Gutzmer, B.G. Teubner, Leipzig 1906, includes in its bibliographical references Vitali [2, 3, 7, 8, 9] and other two papers, Montel [1], a paper of M.B. Porter from 1904, but not Porter [1], Osgood [1] nor Stieltjes [1]. However, the book does not contain the statement of th. I, nor th. II.

Bieberbach's textbook «Lehrbuch der Funktionentheorie, Band I», 2nd ed., Teubner, Leipzig, Berlin, 1923, contains th. I on p. 165: «§13. Der Vitalische Doppelreihensatz».

The beautiful «Vorlesungen über allgemeine Funktionentheorie und elliptische Funktionen», 3 ed., Springer, Berlin 1929, of A. Hurwitz[24] and R. Courant contains th. I, II; th. I is named «Satz von Vitali», cf. p. 319 loc. cit.

It goes without saying that Vitali's results are quoted in Peano's Formulaire (cf. letters of Peano to Vitali in *Opere*).

It remains to say some words about Severini's papers on sequences of holomorphic functions. We have 7 papers from 1903 to 1912. There is no truly new result in these papers! Severini restrains himself essentially to bring some (rather obvious) modifications to the hypothesis of Vitali's theorems or methods of proof. It is astonishing that Pincherle had presented to the Reale Accademia dei Lincei (August, 2nd, 1903) one of such papers (that mentions Vitali [7]!), and that Severini presented himself another paper to the International Congress at Rome 1908, cf. Atti del IV Congresso Internazionale dei Matematici, 1909, vol. II, p. 183-193. (Although this does not belong to my subject let me note that Tonelli pretended that the celebrated Egorov theorem was discovered by Severini (cf. L. Tonelli, Opere Scelte, ed. Cremonese, Roma 1950, vol. I, p. 421), this is truly questionable Severini became professor at Catania University in 1910 with the support of Pincherle, I suppose).

Th. I was considered in those times an important improvement of Weierstrass theorem at a point that, much later, Lindelöf in 1914 (Bull. p. 171) considered it useful to dedicate a paper to this theorem and give of it a direct proof deducing next from it th. II (In Lindelöf's own words: Jusqu' à présent, on a toujours commencé par établir le théorème du n. 6, puis on en a déduit, par voie indirecte, le théorème du n. 1. Il nous semble que la marche que nous avons suivie ci-dessus est plus simple et plus naturelle).

Also Jentzsch in 1917 includes a proof of th. I in his long paper Jentzsch [1]. Theorems I, II, III, and other important results from the beginning of our century, could be discovered due to the pioneering work of Volterra, Ascoli and Arzelà (who first understood well what *now* we call the pre-compacity of a set of functions). It is remarkable that none of these theorems could spring out of some hundred papers dedicated before to sequences of holomorphic functions (see Vivanti's references in Vivanti l.c.); it will be surely instructive for the understanding of the historical flow of the ideas to close this section by quoting the final paragraph of Carathéodory - Landau [1]: «Die ganzen Resultate dieser Untersuchungen dürfen wir wohl als recht merkwürdig bezeichnen; hatte doch bereits Stieltjes, der eine gewisse Zwischen - station zwischen Satz I und Satz II (historisch die erste über Weierstrass hinausgehende) erreicht hatte, in einem Briefe an Hermite[25] (vom 14.2.1894) seiner Verwunderung über sein eigenes Ergebnis in folgenden Worten Ausdruck verliehen: ayant longuement réfléchi sur cette démonstration, je suis sûr qu'elle est bonne, solide et valable. J'ai dû l'examiner avec autant plus de soin qu' *a priori* il me semblait que le théorème énoncé *ne pouvait pas exister* et *devait être faux*. Je vous avouerai cependant que je serais heureux si quelqu'un voulait examiner la démonstration; peut- être M. Picard qui a le coup d'oeil si facile et si juste...« Mit ähnlichen Empfindungen hatte Landau im Jahre 1904 den Beweis seiner Verallgemeinerung des Picardschen Satzes[26] betrachtet und lange mit der Publikation gezögert, da auch der Beweis richtig,

---

[24] Hurwitz and Pincherle attended both the lectures of Weierstrass in Berlin and had leaved to us two very different versions of Weirstrass' course.

[25] Correspondance d'Hermite et de Stieltjes, vol. II (1905), p. 370.

[26] F. Schottky *Über den Picard' schen Satz und die Borel' schen Ungleichungen.* Sitzungsber. der Kön. Preuss. Ak. der Wiss. (1904) p. 1244-1262.

aber der Satz zu unwahrscheinlich erschien. Und nun findet das merkwürdige Zusammentreffen statt, dass der Stieltjessche Satz (in der Vitalischen Verschärfung) und der Landausche Satz (in der Schottky - Landauschen Verschärfung) vor dasselbe Problem mit Erfolg gespannt wurden».

7. In October 1922 took place a competition for a chair of mathematical Analysis at the University of Modena. The board of examiners was constitued by Fubini, Levi-Civita, Pincherle, Tonelli and G. Torelli. The list of the three winners was (by order of classification): 1st Gustavo Sannia[27], 2nd Giuseppe Vitali, 3^d Pia Nalli[28] with the following voting result: for the 1st position Sannia 3, Vitali 2; for the 2nd position Vitali 5; for the 3^d position Nalli 5. It is known that Pincherle, Tonelli, Torelli voted for Sannia for the first position, and Fubini, Levi-Civita voted for Vitali for the first place. It is really surprising! Even astonishing is the vote of Tonelli in whose work the results of Vitali are everywhere present. Fortunately Sannia renounced to the chair at Modena because to him meanwhile was offered a chair at Torino, and so Vitali could occupy his chair at Modena University. He will be at Modena till 1925 when he substituted in Padova Ricci-Curbastro (1853-1925), the eminent geometer, who died that year, and who tributed to Vitali a great esteem.

Vitali taught at Padova University for five years, and here, although in the meantime he was stricken with paralysis of the right part of his body, he developed an intensive mathematical activity surrounded by an agreeable atmosphere in which was born the Seminario Matematico dell'Università di Padova that had Vitali as its first director.

For that period of Vitali's life cf. Pepe [1], and Borgato - Vaz Ferreira [1] in which is focused the research of Vitali in Differential Geometry that became the main subject of interest of Vitali (the calculus of connections on higher-order tangent bundles to manifolds (that Vitali immersed in Hilbert space as the concept of bundle was not yet appeared in those times)[29]). One of the interesting achievements of Vitali is the introduction of a parallel transport, referred to by Struik (cf. Struik [2]) as Weitzenböck-Vitali (cf. also R. Weitzenböck, Invariantentheorie, P. Noordhoff (1923), Groningen, §7, 11, ch. XIII). Einstein used (without mention of Vitali) in a paper of 1928 (cf. A. Einstein, Riemann - Geometrie mit Aufrechterhaltung des Begriffes des Fernparallelismus, Sitzungsber. preuss. Akad. Wiss. 1928, pp. 217-221) Vitali's covariant derivative. The generalized Vitali's covariant derivatives are essential in the work of Enea Bortolotti (1866-1942).

In 1928 Vitali was invited to Pisa to occupy the chair of Bianchi who died this year. Vitali did not accept the invitation, perhaps on account of his health conditions and his family interests

---

[27] Sannia (1875-1930) could present at the competition a lot of honest, but modest, papers from which one appeared on Math. Ann.; two notes on C.R. Ac. Sc. Paris; the other papers were published by Ann. Mat. p. ed appl., Giorn. di Mat. (Battaglini), Rend. Acc. Lincei, Mem. and Atti Acc. Torino, Rend. Acc. Napoli. They deal with Borel (and other) methods of summation, Differential Geometry and partial differential equations.

[28] Nalli (1982-1964) was an excellent mathematician. By the time of competition she had published 18 papers dedicated mainly to real analysis and integral equations. One of this papers is a beautiful exposition of Denjoy integral with some original results of her. Her papers on integral equations are piercing and announce some much posterior results on spectral theory (recall that abstract Hilbert spaces were introduced by von Neumann much later). Cf. P. Nalli, *Opere scelte*, Unione Matematica Italiana, 1976.

[29] Cf. G. Vitali, *Geometria nello spazio hilbertiano*. N. Zanichelli, ed. Bologna, 1929, and G. Vitali, *Nuovi contributi alla nozione di derivazione covariante*, Rend. Sem. Mat. Padova, anno I (1930), p. 46-72, G. Vitali, *Sulle derivazioni covarianti*, ib. *3* (1932), p. 1-2.

centered at Bologna. This chair will be occupied by Tonelli and Vitali will take the place of Tonelli at Bologna University in the year 1930.

Vitali was replaced at Padova by Renato Caccioppoli (1904-1959), a 26 years-old gifted boy grandson, from his mother side, of the russian revolutionary Bakunin, who was in possess of a powerful mathematical mind [30].

Giuseppe Vitali died untimely the 29th February, 1932 in the afternoon. This day, after his lectures, he left University in company with Ettore Bortolotti (1896-1947) speaking about some mathematical papers. The death took him suddenly in S. Vitali street. The funeral honours were organized by the Town Council of Bologna. Funeral orations were made by his brother Vitichindo Vitali and by Renato Caccioppoli.

Vitali taught at Bologna University for the short period 1930-1932, and here he had as his last pupil Tullio Viola, a subtle analyst endowed with a gentle and refined personality, recently deceased.

In spite of the fact that Vitali belonged to the board of university professors only during the last ten years of his life when he was stricken by his disease, he was not isolated among the mathematical community as testify his correspondence with some of the better mathematicians of his time (100 textes recorded in Opere, and other documents in the Archivio Giuseppe Vitali bequeathed to the Unione Matematica Italiana by Vitali's daughter Miss Luisa Vitali before her recent death). Particulary hearty were Vitali's relationship with Otton Nikodym (1889-1974) and the Polish School (Vitali contributed with a short communication to the I Congrès des Mathématiciens des Pays Slaves held at Warszawa 1929, Vitali was a member of the Société Polonaise de Mathématique, and he published on Fundamenta Math. and on Ann. de la Soc. Pol. de Math.). On the occasion of Vitali's death, N. Luzin (1883-1950) sent to the Seminario Matematico di Padova a touching letter that you can find recordered in Pepe [3].

Let me finish my speach by quoting some words from the obituary notice of Seminario Mat. Univ. di Padova written by a Vitali's friend and colleague at Padova University (cf. Tonolo [1]) that illuminate Vitali's personality: Il Nostro era veramente un matematico di classe. Dotato di penetrante intuizione, Egli sentiva la verità d'una proposizione anche se Gli mancavano gli elementi logici per una tale persuasione. ... d'un colpo d'occhio vedeva il piano generale del problema, partiva deciso verso la soluzione, avanzando senza timore d'ingannarsi. L'originalità era una Sua caratteristica mentale, nel senso che non desiderava servirsi di quanto prima di Lui era stato fatto, ed anche per cose note, sentiva il bisogno di dare al loro studio una personale impronta. Maestro nel senso più elevato e più ampio della parola, aveva fervido l'amore per la Scuola. Agli allievi insegnava con l'esempio il metodo della ricerca matematica... Come lo Scienziato fu prodigo distributore di sapere, così l'Uomo fu largo di generosi affetti. Sotto il sembiante riservato, Egli custodiva tesori di delicati sentimenti... e sento che non darò mai abbastanza simpatia ad un'Anima così bella e così pura, ad un Uomo così mite e così buono.

REFERENCES

N.H. ABEL [1], Sur les fonctions génératrices et leurs déterminantes, *Oeuvres.* Nouvelle édition, tome II, p. 67-81, Christiania, 1881.

P. APPELL [1], Sur les intégrales de fonctions à multiplicateurs et leur application au dévelop-

---

[30] Cf. R. Caccioppoli, *Opere*, Cremonese ed., Roma 1963. In the second volume there are deep papers on complex analysis (several variables), quasi-analytic functions, etc.

pement des fonctions abéliennes en séries trigonométriques, *Acta Math.* XIII (1890), p. 3-174.

C. ARZELÁ [1], Sulle serie di funzioni analitiche, *R. Acc. delle Sc. dell'Ist. di Bologna*, 30 Novembre 1902.

C. ARZELÁ [2], Sulle serie di funzioni, *R. Acc. delle Sc. di Bologna*, serie 5, VIII (1900).

G. ASCOLI [1], Sulle curve limite di una varietà data di curve, *Mem. Acc. Lincei* (III), XVIII (1883-84), p. 521-586.

M.T. BORGATO and A. VAZ FERREIRA [1] Giuseppe Vitali: Ricerca matematica e attività accademica dopo il 1918, Atti del Convegno *«La matematica italiana tra le due guerre mondiali»*, Milano, Gargano del Garda, 8-11 ottobre 1986, p. 43-58.

Et. BORTOLOTTI [1], *L'école mathématique de Bologne*, Congrès International des Mathématiciens, Bologne - Settembre 1928, N. Zanichelli, ed.

C. CARATHÉODORY and E. LANDAU [1] Beiträge zur Konvergenz von Funktionenfolgen. *Sitzungsb. der Kön. Preuss. Ak. der Wiss.* XXVI (1911), p. 587-613.

F. CASORATI [1], Il calcolo delle differenze finite interpretato ed accresciuto di nuovi teoremi a sussidio principalmente delle odierne ricerche basate sulla variabilità complessa. *Ann. di Mat. pura ed appl.* II, **10** (1880), p. 10-45. Or *Opere*, Roma 1951-52.

G. CASTELNUOVO [1], Federigo Enriques. *Rend. Acc. Naz. Lincei*, serie VIII, vol. II (1947).

S. CINQUINI [1], Della vita e delle Opere di Leonida Tonelli. *Ann. Scuola Norm. Sup. Pisa*, serie II, XV (1946) pp. 1-37 and Tonelli [1], p. 1-35, vol. I.

P. DELIGNE [1], *Équations Différentielles à points singuliers réguliers*. Lecture Notes in Mathematics n. 163, Springer Verlag, 1970.

G. FUBINI [1], Il teorema di riduzione per gli integrali doppi, *Rend. Sem. Mat. Torino*, IX (1949), p. 125-133 or *Opere Scelte*, Cremonese ed. Roma 1962, vol. III, p. 402.

H. GRAUERT - H. RECKZIEGEL [1] Hermitesche Metriken und normale Familien holomorpher Abbildungen. *Math. Zeit.* **89** (1965), p. 108-125.

R.C. GUNNING [1], *Lectures on vector bundles on Riemann surfaces*, Princeton Un. Press.

R. JENTZSCH [1], Untersuchungen zur theorie der Folgen analytischer Funktionen, *Acta Math.* **41** (1917), 219-70.

K. KOBAYASHI [1], *Hyperbolic Manifolds and Holomorphic Mappings*, Marcel Dekker, Inc., New York, 1970.

K. KOBAYASHI [2], Intrinsic distances, measures and geometric function theory. *Bull. Am. Math. Soc.*, **82** (1976), p. 357-416.

H. LEBESGUE [1], Sur la représentation analytique des fonctions continues, *Bull. des Sc. Math.*, $2^{\text{ème}}$ série, XXVII (1903), p. 82.

D. MENCHOFF [1], *Les conditions de monogénéité*. Hermann et c^{ie}, éd. Paris 1936.

P. MONTEL [1], Sur les suites de fonctions analytiques, *C.R. Ac. des Sc.* Paris **138** (1904), p. 469-471.

P. MONTEL [2], Sur le séries de fonctions analytiques, *Bull. des Sc. Math.*, $2^{\text{ème}}$ série, XXX (1906), p. 189-192.

P. MONTEL [3], Sur les suites infinies de fonctions, Thèse, Paris, 1907. Or *Ann. Sc. de l'Éc. Norm. Sup.*, Sér. III, XXIV (1907).

P. MONTEL [4], *Leçons sur les séries de polynomes à une variable complexe.* Paris 1910.

P. MONTEL [5], Sur les différentielles totales et les fonctions monogènes. *C.R. Ac. Sc. Paris,* **156** (1913), p. 1820-1822.

P. MONTEL [6], Sur les points irréguliers des séries convergeantes de fonctions analytiques, *C.R. Ac. Sc. Paris* **145** (1907), p. 910-913.

P. MONTEL [7], *Leçons sur les familles normales de fonctions analytiques et leurs applications,* Gauthier - Villars et c^{ie} , éd. Paris, 1927.

W.F. OSGOOD [1], Note on the functions defined by infinite series whose terms are analytic functions of a complex variable; with corresponding theorems for definite integrals. *Ann. Math.* 2nd series, **3** (1901), p. 25-34.

L. PEPE [1], *Una biografia di Giuseppe Vitali,* cf. Vitali [0], p. 1-24.

L. PEPE [2], Giuseppe Vitali e la didattica della matematica, *Archimede,* 1983, p. 163-176.

L. PEPE [3], Giuseppe Vitali e l'analisi reale, *Rend. Sem. Mat. e Fisica di Milano,* LIV (1984), p. 187-201.

S. PINCHERLE [1], *Opere scelte,* Cremonese ed., Roma 1954.

M.B. PORTER [1], Concerning series of analytic functions. *Ann. Math.* 2nd series, VI (1904-1905), p. 190-192.

S. SAKS [1], *Theory of the integral,* 2nd ed., Dover publ., New York, 1964.

J.S. SILVA [1], *Obras,* INIC, Lisboa, 1985.

T.J. STIELTJES [1], Recherches sur les fractions continues, *Ann. de la Faculté des Sc. de Toulouse,* **8** (1894), p. 56, or *Oeuvres,* p. 457.

D.J. STRUIK [1], *A concise history of Mathematics,* 1987.

D.J. STRUIK [2], *The theory of linear connections,* Springer Verlag. Berlin 1934.

L. TONELLI [1] *Opere Scelte,* Cremonese ed. Roma, 1960.

L. TONELLI [2] Sulla rettificazione delle curve, *Atti R. Accad. delle Sc. di Torino,* XLIII (1908), p. 789-800. Or [1], vol. I, p. 52-68.

A. TONOLO [1] Commemorazine di Giuseppe Vitali. *Rend. Sem. Mat. Univ. Padova,* **3** (1932), p. 67-81.

T. VIOLA [1], Ricordo di Giuseppe Vitali, a 50 anni dalla Sua scomparsa, *Atti del Convegno «La Storia delle Matematiche in Italia»,* Cagliari, 29-30 settembre e 1 ottobre 1982, p. 535-544.

VITALI [0], *Opere sull'analisi reale e complessa. Carteggio.,* Edizioni Cremonese, 1984.

VITALI [1], Sulle applicazioni del Postulato della continuità nella geometria elementare. *Questioni riguardanti la geometria elementare,* Bologna, Zanichelli, 1900.

VITALI [2], Sui limiti per $n = \infty$ delle derivate n^{me} delle funzioni analitiche. *Rend Circ. Mat. Palermo,* **14** (1900) p. 209-216.

VITALI [3], Sulle funzioni analitiche sopra le superficie di Riemann. *Rend. Circ. Mat. Palermo,* **14** (1900) p. 202-208.

VITALI [4], Sopra le equazioni differenziali lineari omogenee a coefficienti algebrici. *Rend. Circ. Mat. Palermo,* **16** (1902) p. 57-69.

VITALI [5], Sopra le equazioni differenziali lineari omogenee a coefficienti algebrici. *Ann. Scuola Norm. Sup. Pisa,* **9** (1903) p. 3-57.

VITALI [6], Sopra l'integrazione di serie di funzioni di una variabile reale *Boll. dell'Acc. Gioenia di Sc. Nat. in Catania*, fasc. LXXXVI (Maggio 1905), p. 1-6.

VITALI [7], Sopra le serie di funzioni analitiche. *Rend. Ist. Lombardo*, serie 2^a **36** (1903) pp. 772-774.

VITALI [8], Sopra le serie di funzioni analitiche. *Ann. Mat. Pura e Appl.*, serie 3^a **10** (1904) p. 65-82.

VITALI [9], Sopra le serie di funzioni analitiche. *Atti Acc. Sc. Torino*, Cl. Sc. Fis. Mat. Natur. **39** (1903-04), p. 22-32.

VITALI [10], Sopra una proprietà caratteristica delle funzioni armoniche. *Atti Acc. Naz. Lincei, Rend. Cl. Sc. Fis. Mat. Natur.* 5^a **21** (1912) p.315-320.

H. WU [1], Normal families of holomorphic mappings. *Acta Math.* **119** (1967) p.193-233.

Contact Curves of Two Kummer Surfaces

ALESSANDRO VERRA, Dipartimento di Matematica, Università di Genova
via L. B. Alberti, 16100 Genova, Italy

Introduction. The aim of these notes is to display once more some classical projective constructions related to Kummer surfaces, in particular we consider contact curves of two Kummer surfaces in $\mathbf{P}^3$. Let C be one of these curves, H a plane section of C, K the canonical divisor : C has degree 8 and arithmetic genus 5 ; moreover $L_C = O_C(H - K)$ is a non trivial line bundle of degree zero. We are mainly interested to the following question :

(1.1) describe the family T_n of curves C such that L^n is trivial.

The study of these contact curves has various motivations at least in the known cases $n = 2, 3$. Each element C of T_2 is a Prym canonical curve, constructing from C its associated Prym variety $P(C)$ one obtains a 4-dimensional principally polarized abelian variety. It turns out ([V]) that $P(C)$ does not depend on C and it is the unique 4-dimensional principally polarized abelian variety with 10 theta nulls (i.e. the singular points of the theta divisor are ten nodes and they are elements of order 2 in $P(C)$). Again, for $n = 3$, a unique geometric object comes up from the family T_3: it is the Horrocks-Mumford bundle H on $\mathbf{P}^4$. This is shown in [B]: each element of T_3 parametrizes the jumping lines of H passing through a given point of $\mathbf{P}^4$; moreover the existence of a smooth, irreducible element of T_3 implies the existence of H. Perhaps there is still some interesting geometry behind the families $T_n, n > 3$.

Throughout the paper we will give some more details on all this, together with the construction of contact curves; then we will explain the main result of $[B - V]$ where problem (1.1) is discussed for more general families of $K3$ surfaces. The result is the following: let X be a $K3$ surface, C and H two polarizations on X such that $C^2 = C.H = 2g - 2, H^2 = 2g - 6$ and C is ample; consider again the family $T_n = \{C\epsilon \mid C \mid /O_C(H - C)^n = O_C\}$ then

(1.2) T_n is a finite set of cardinality $\binom{2n^2-2}{g}$

for a sufficiently general X as above.

Aknowledgements I wish to thank the organizers of the meeting "Geometria e variabile complessa a Bologna" for having given me the opportunity of taking part to it. Specially I wish to thank Prof. S. Coen for his patient work of editing the proceedings.

Contact curves. Let X be a Kummer surface in $\mathbf{P}^3$ i.e. a quartic surface with 16 nodes as its only singularities. The construction of X is well known: consider $A =$ jacobian of E ($E =$ smooth, irreducible genus 2 curve) or, equivalently, a 2-dimensional complex torus with a fixed, non product, principal polarization Θ. Up to translations on A, Θ is the image of E by the Abel map. One can choose Θ symmetric with respect to -1 multiplication on A and consider the linear system $\mid 2\Theta \mid$. It turns out that its associated map $f : X \to \mathbf{P}^3$ is a degree 2 morphism and $X = f(A)$. Actually $X = A/<-1>$ and its 16 nodes are the images of the 16 order 2 elements of A.

It is useful to recall at this point that Kummer did not construct X in this way; in fact the relation of X to complex tori was later recognized by Cayley (see A. Weil's introduction to Kummer's collected works) and Kummer came on X through his investigations on the geometry of lines of $\mathbf{P}^3$.

Since we will need it later, we want to recall this beautiful construction of X as the focal surface of a quadratic complex ($[G - H]$ch.6). Let

(2.1) $$G \subset \mathbf{P}^5$$

be the Plucker embedding of the grassmannian of lines of $\mathbf{P}^3$, G is a smooth quadric. A quadratic complex

(2.2) $$Q = G \cap F$$

is the intersection of G with another quadric F. We assume Q is smooth; to Q we can naturally associate a conic bundle

(2.3) $$\hat{a} : \hat{Q} \to \mathbf{P}^3.$$

Indeed let $I = \{(x,r) \in G \times \mathbf{P}^3 / x \in r\}$ be the incidence correspondence of points and lines in $\mathbf{P}^3$ together with its natural projections

(2.4) $$\mathbf{P}^3 \xleftarrow{a} I \xrightarrow{b} G$$

observe that $a^{-1}(x)$, ($x \in \mathbf{P}^3$), is the $\mathbf{P}^2$ of lines through x and $\mathbf{P}_x^2 = b(a^{-1}(x))$ is a plane in G. Since $\text{Pic}(Q) = \mathbf{Z}$ (Lefschetz thm.) Q cannot contain a plane and

(2.5) $$C_x = \mathbf{P}_x^2 . Q$$

is always a conic. Now consider $b^*(Q) = \hat{Q}$, the projection $\hat{a} = a/\hat{Q}$ realizes $\hat{Q}$ as a conic bundle over $\mathbf{P}^3$.

The fibre of $\hat{a}$ over x is C_x; Kummer showed that the discriminant locus of $\hat{a}$

$$\{x \in \mathbf{P}^3 / C_x \text{ is singular}\}$$

is exactly the Kummer surface X. In particular x is a node of X if and only if C_x is a double line and the surface

(2.6) $$\hat{X} = \{(x,r) \in \hat{Q}/r \in \operatorname{Sing}(C_x)\}$$

is the desingularization of X by blowing up its nodes.

The previous construction yelds a linear system of curves on $\hat{X}$ (and X) which is the most interesting to us: consider the surface

(2.7) $$Y = b(\hat{X}) \subset \hat{Q} ,$$

is turns out that $b/\hat{X}$ is an embedding on $\hat{X}$ in $\mathbf{P}^5$ so that Y is a smooth $K3$ surface in Q. By adjunction formula and $\operatorname{Pic}(Q) = \mathbf{Z}$ it follows $Y \in \mid O_Q(2) \mid$. Hence Y is a smooth complete intersection of 3 quadrics in $\mathbf{P}^5$. This defines the linear system

(2.8) $$\mid C \mid = \mid O_Y(1) \mid$$

on $\hat{X}$. Mapping the elements of $\mid C \mid$ in $\mathbf{P}^3$ by a one obtains the contact curves of X and another Kummer surface.

We want to show this and some more properties of $\mid C \mid$; let

(2.9) $$\mid H \mid = \mid (\hat{a}/\hat{X})^* O_X(1) \mid ,$$
(2.10) $$R_i = \text{desingularization of the node } o_i \in X.$$

It is well known and not difficult to be shown that a basis for $\operatorname{Pic}(\hat{X}) \otimes \mathbf{Q}$ is given by the classes of $H, R_1, \ldots, R_{16}$: Let $C \sim mH + \sum n_i R_i, (m, n_i \in \mathbf{Q})$; note that, by construction, $b(R_i)$ is a line so that $C.R_i = 1$. Since $H.R_i = 0, R_i^2 = -2$ one computes $n_i = -\frac{1}{2}$ and, from $C.C = 8, m = 2$. Therefore

(2.11) $$C \sim 2H - \frac{1}{2}\sum R_i .$$

Finally consider $a(C) : a/C$ is a morphism of degree 1; the degree of $a(C)$ in $\mathbf{P}^3$ is 8 ($H.C = 8$) and the arithmetic genus is 5 ($C.C = 8$).

PROPOSITION 2.1   There exists another Kummer surface X' in $\mathbf{P}^3$ such that the intersection cycle of X and X' is $2a(C)$.

PROOF : (this is again classically well known, a modern proof is given in [B]); let P be the hyperplane cutting C on Y, assume for simplicity P does not contain any plane $\mathbf{P}_x^2$ as above. Then consider in G the divisor $2(P \cap G) \doteq Q_\infty$ and the pencil of quadratic complexes

$$S = \{Q_t, t \in \mathbf{P}^1\}$$

with $Q_t = Q + tQ_\infty$. For general t Q_t is smooth and one can reconstruct as above its associated Kummer surface X_t in $\mathbf{P}^3$. Let $x \in \mathbf{P}^3, S_x$ the restriction of S to $\mathbf{P}_x^2 : S_x$ is a pencil of conics which contains the double line $2L = Q_\infty.\mathbf{P}_x^2$. Then, by definition, x belongs to each Kummer surface X_t if and only if S_x is a pencil of singular conics. This happens if and only if a reduced singular conic of S_x has its node on L. Therefore the intersection of all Kummer surfaces X_t is the curve $a(C)$. To complete the proof it remains to show that X_t moves in a pencil of quartic surfaces: let $B \subset \mathbf{P} = \mathbf{P}H°(O_{\mathbf{P}^3}(4))$ be the non complete curve parametrizing X_t for Q_t smooth, $\overline{B}$ its Zariski closure. Take a point $x \in \mathbf{P}^3$ which is not on a quartic surface parametrized by a point of $\overline{B} - B$ nor on $a(C)$. Then the general conic of $\mathcal{S}_x$ is smooth and the only singular ones are: the double line Q_∞. $\mathbf{P}_x^2$ and a reduced singular conic C_t. Hence the only quartic surface of $\overline{B}$ passing through x is X_t. This implies that $\overline{B}$ is a pencil whose base locus is the curve $a(C)$ endowed with a double structure. If P contains some $\mathbf{P}_x^2$ the proof is similar.

Since it satisfies the previous proposition $a(C)$ is by definition a contact curve of two Kummer surfaces in $\mathbf{P}^3$.

Observe that $a(C)$ is the image of C in $\mathbf{P}^3$ by $O_C(H)$; let $\mid L \mid = \mid H - C \mid$, just writing in a different way we have

(2.12) $$O_C(H) = \omega_C \otimes L_C$$

with $\omega_C = O_C(C) = $ dualizing sheaf of C, $L = O_{\hat{X}}(H - C)$, $L_C = L \otimes O_C$. Obviously $\deg(L_C) = O$, consider the cohomology exact sequence

$$0 \to H°(O_{\hat{X}}(-C + L)) \to H°(O_{\hat{X}}(L)) \to H°(L_C) \to H^1(O_{\hat{X}}(-C + L))\ ;$$

since $L.H = -4 < O$, $h°(O_{\hat{X}}(L)) = O$. On the other hand, computing the class of $C - L$ in $\text{Pic}(\hat{X}) \otimes \mathbf{Q}$, one obtains

(2.13) $$C - L \sim 3H - \sum R_i$$

which is the linear system mapping X onto its dual surface. This implies the vanishing of $H^1(O_{\hat{X}}(-C + L))$ because $(C - L)^2$ is positive $(= 4)$ and the general element of $\mid C - L \mid$ is reduced irreducible. Therefore $H°(L_C) = O$ and we have shown:

PROPOSITION 2.2 $\forall C \in \mid C \mid$, the line bundle L_C is not trivial.

In other words for $n = 1$ the family T_n considered in (1.1) is empty.

Prym canonical contact curves. We want to consider now the family T_2. Its elements are curves C such that $a(C)$ is Prym canonically embedded in $\mathbf{P}^3$, to describe them we introduce some more properties of Kummer surfaces.

Recall ([G−H], ch.6) that $\hat{X}$ contains a classical (16,6) configuration of 32 smooth, irreducible rational curves $R_i, T_j (i, j = 1, \ldots, 16)$ such that:

(3.1) (i) $R_i.R_j = T_i.T_j = 0$ for $i \neq j$;
(ii) each row of the matrix $(R_i.T_j)$ has six terms equal to 1 the remaining ones being zero;
(iii) let $T_j.R_k = 1 (k \in K \subset \{1, \ldots, 16\}, \# K = 6)$ then $2T_j + \sum R_k \sim H$;
(iv) let $R_i.T_k = 1 (k \in K \sim \{1, \ldots, 16\}, \# K = 6)$ then $2R_i + \sum T_k \sim 3H - \sum_{j=1}^{16} R_j$.

Note that, by (i) and (iii), $H.R_i = 0$ so that, as in (2.10), $R_i = $ desingularization of a node of X. On other hand, by (ii), $a(T_j)$ is one of the 16 "double conics" of X i.e. $a(T_j) = X \cap \pi_j$ where π_j is a plane everywhere tangent to X.

Since $C.R_i = C.T_j = 1$ each element of the configuration is a line in Y, let us fix one of them: e.g. $L = R_i$ and assume $T_1, \ldots, T_6$ are the 6 lines of Y intersecting L in the (16,6) configuration. Let N be the net of quadrics containing Y; for each point $y \in Y - (L + T_1 + \ldots + T_6)$ there exists a unique quadric q of N which contains the plane L_y spanned by y and L. This defines a rational map

(3.2) $$h : Y \to N$$

sending y to q. It is easy to show that h is a morphism. Assume q is smooth, then q contains two families q', q'' of rationally equivalent planes and L_y is in one of them. On the other hand consider the finite double covering $s : \hat{N} \to N$ parametrizing elements q', q'', then $s(q') = s(q'') = q$ and s is branched over the discriminant curve of the net

(3.3) $$\Delta = \{q \in N / q \text{ is singular}\}.$$

Therefore, by the previous remarks, we have also a rational map $\hat{h} : Y \to \hat{N}$ sending y to q' or q'' and such that

(3.4) $$s \cdot \hat{h} = h$$

Since there exists a unique plane of the family $q'(q'')$ which contains L it follows that $\hat{h}$ is birational. A more precise result we do not show is that (3.4) is the Stein factorization of h.

PROPOSITION 3.1 (i) h is the morphism associated to the linear system $| H - L |$;
(ii) the discriminant curve Δ of N is the union of six distinct lines touching the same smooth conic B.

PROOF : (i) let P be a general line in N : P is a general pencil of quadrics containing Y; since Y is smooth we can assume the same for the base locus S of P. Then $s^{-1}(P)$ is a smooth irreducible genus 2 curve ([D]). Let $D = h^{-1}(P)$, since $\hat{h}$ is birational D is isomorphic to $s^{-1}(P)$; this implies $D^2 = 2 \dim | D | = 2$. The following is well known ([D], [G - H]ch.6): the family of lines E of $S(E \neq L)$ which intersect L is parametrized by $s^{-1}(P)$ and the union of them is the intersection of S with a rank 4 quadric F singular along L. Observe that $D \subset F$: let $y \in D(y \notin L)$, the plane spanned by L and y contains a (unique) line E of S passing through y; hence $y \in E \subset F$. Obviously F contains the lines $T_1, \ldots, T_6$ of Y which intersect L. Since F does not contain Y we have

$$\dim | 2C - 2L - \sum_{j=1}^{6} T_j - D | \geq 0 \ .$$

By (2.11) and (3.1) $2C - 2L - \sum T_j \sim H$ so that D is a plane quartic curve with a double point in the Kummer model X. This implies $| D | = | H - R_i |$ for some $i = 0 \ldots 16$. Finally it is not hard to see that $D.T_j = 1$ so that $R_i.T_j = 1$ and $R_i = L$.
(ii) Let $o \in X$ be the node whose blowing up is L, the morphism defined by $| H - L |$ is just the projection $p : X \to \mathbf{P}^2$ of X from o composed with the desingularization

$\hat{a} : \hat{X} \to X$. Since X contains 15 nodes besides o the branch locus of $p \cdot \hat{a}$ is a sextic curve with 15 nodes i.e. the union of six distinct lines. They touch the same conic $B = p \cdot \hat{a}(L)$.

PROPOSITION 3.2 (i) N contains six pencils of quadrics P_i passing through a fixed singular base point $U_i (i = 0, \ldots, 5)$;
(ii) after a proper choice of coordinates $(u_o : \ldots : u_5)$ on $\mathbf{P}^5$ one can reduce the equations of Y to the normal form

(3.5) $$\sum \mu_i^2 u_i^2 = \sum \lambda_i \mu_i u_i^2 = \sum \lambda_i^2 u_i^2 = 0 \quad (i = 0, \ldots, 5) \ ;$$

(iii) Y carries the six projective involutions $p_i : Y \to Y$, where p_i is changing the sign of the i-th coordinate of $(u_o : \ldots : u_5)$.

PROOF: (i) let P_i be one of the six lines in the discriminant curve Δ of N. The general quadric $q \in P_i$ is a smooth point of Δ, hence its rank is 5. Let $U(q)$ be the unique singular point of q: by Bertini thm. $U(q)$ is in the base locus S of P_i, in particular in $\text{Sing}(S)$. If $U(q)$ is not fixed $\dim(\text{Sing}(S)) \geq 1$ and $Y \cap \text{Sing}(S) \neq \emptyset$; hence Y is singular: contradiction.

(ii) Let U_i be the fixed singular base point of $P_i (i = 0, \ldots, 5)$, clearly $U_i \neq U_j (i \neq j)$. Fix coordinates $(u_o : \ldots : u_5)$ such that $U_o, \ldots, U_5$ are the 6 fundamental points: a standard exercise in linear algebra shows that all the quadrics of N have a diagonal equation. Fix coordinates $(z_o : z_1 : z_2)$ on N, then $N = \sum A_i u_i^2 = 0$, with $A_i = $ linear form in z_o, z_1, z_2. The equation of P_i is $A_i = 0$; since $P_o, \ldots, P_5$ are tangent to the same smooth conic B, in the dual plane $\hat{N} P_o, \ldots, P_5$ are points of the dual conic $\hat{B}$. Let $(\hat{z}_o : \hat{z}_1 : \hat{z}_2)$ be the dual coordinates on $\hat{N}$, we can assume $\hat{B} = \{\hat{z}_1^2 - \hat{z}_o \hat{z}_1\} = 0$; then $A_i = z_o \mu_i^2 + z_1 \mu_i \lambda_i + \lambda_i^2 z_2$ and the equations of Y are as in (3.5).
(iii) it follows immediately from (3.5).

Proposition 3.2 enables us to describe the family T_2:

PROPOSITION 3.3 The elements of T_2 are the six curves of $|C|$ which are fixed by all the projective involutions p_i. In other words: $C \in T_2$ if and only if $C = Y \cap \{u_i = 0\}$ for some $i = 0 \ldots 5$.

PROOF : as in (2.12) let $|L| = |H - C|, L_C = O_C(L), \hat{L}_C = O_C(-L); C \in T_2$ if and only if $L_C \cong \hat{L}_C$ i.e. if and only if the linear systems $|C + L| = |H|$ and $|C - L| = |3H - \sum R_i|$ cut linearly equivalent divisors on C. By Riemann Roch $h^0(O_{\hat{X}}(\pm L)) = h^1(O_{\hat{X}}(\pm L)) = O$, then, using standard exact sequences, one obtains that the restrictions

(3.5) $\quad f_\pm : H^0(O_{\hat{X}}(C \pm L)) \to H^0(O_C(C \pm L)), r : H^0(O_{\hat{X}}(C)) \to H^0(O_H(C))$

($H \in |H|$) are isomorphisms. Fix a general curve H, i.e. a smooth plane quartic in X. Let h be the intersection divisor of H and $C \in |C|$; since C is a contact curve of two quartic surfaces we have $2h \in |O_H(4H)|$. Hence $h \in |2K + \epsilon|$, with $K = $ canonical divisor of $H, \epsilon = $ non trivial order 2 element of $\text{Pic}(H)$. On the other hand let W be the restriction of $|C - L|$ to H; recall that $|C - L|$ is cut on X by the cubic surfaces passing through the nodes of X. Hence there is a web of plane cubic curves cutting W on H and $W \subset |3K|$.

Assume $C \in T_2$, so that $O_C(C+L) = O_C(C-L)$; by the previous remarks there exists a unique $\hat{H} \in |\, C - L\, |$ cutting on C the divisor h. Equivalently there exists a unique $w \in W$ such that

(3.6)
$$w = h + b$$

with b = effective divisor of degree 4 on H. In particular $|\, b\, | = |\, K + \varepsilon\, |$ which is one dimensional. Since there is a unique C cutting h on H (3.5) the elements of W satisfying (3.6) correspond bijectively to the elements of T_2. To check these special elements of W consider the natural map $\Phi : |\, K + \epsilon\, | \times |\, 2K + \epsilon\, | \to |\, 3K\, |$. Φ is a morphism of degree 1 and its image Z is some linear projection in $\mathbf{P}^9 = |\, 3K\, |$ of the Segre embedding of $\mathbf{P}^1 \times \mathbf{P}^5 = |\, K + \epsilon\, | \times |\, 2K + \varepsilon\, |$. Hence $\deg(Z) = 6$ and, since $W = \mathbf{P}^3$, one expects 6 divisors $w \in Z \cap W$ and 6 curves $C \in T_2$. This is true if Z and W are transversal: it is a not difficult but quite long exercise to show that $Z \cap W$ is a finite set; therefore we omit it. Then, to complete the proof, it suffices to produce 6 elements of T_2; namely the curves $\{u_i = O\} \cap Y$. For this observe that

(3.7)
$$p \ast (C+L) \simeq C - L$$

for each involution $p = p_i$. Indeed by (3.1): $2T + R_1 + \ldots + R_6 \sim H \sim C + L$; where $T \in \{T_j\}$ and we assume $R_1.T = \ldots = R_6.T = 1$. The set of fixed points of p is a smooth hyperplane section of $Y(\{u_i = O\}$ if $p = p_i)$, hence, for each line $L \subset Y, L \neq p(L)$ and $L \cap p(L)$ = one point. Since Y contains exactly 32 lines it follows $p^\ast T_j = R_i (i, j = 1 \ldots 16)$. Let $p(T) = R, p(R_i) = T_i (i = 1 \ldots 6)$ then $p^\ast(C+L) \sim 2R+T_1+\ldots T_6$. Since $(p^\ast(C+L))^2 = 4$ it follows $R.T_1 = \ldots = R.T_6 = 1$. Hence, by (3.1), $2R + T_1 + \ldots + T_6 \sim 3H - \sum R_i \sim C - L$.

Finally let $C = Y \cap \{u_i = O\}$; C is the set of fixed points of p_i. Hence H and $\hat{H} = p^\ast(H)$ cut on C the same divisor. Since $\hat{H} \in |\, C - L\, |$ we have $O_C(C+L) \cong O_C(C-L)$ i.e. $C \in T_2$.

There is a lot of nice classical and modern geometry concerning Prym canonical contact curves of Kummer surfaces; we limit ourselves to mention freely some of their properties. Let $C = \{u_i = O\}$ as above; C is the base locus of a net of quadrics of $\mathbf{P}^4$. By prop.3.2 all the quadrics of this net have a diagonal equation and the discriminant curve

(3.8)
$$\Delta_i = \prod_j (\mu_j^2 z_o + \mu_j \lambda_j z_1 + \lambda_j^2 z_2) \quad j \neq i \quad j = 0, \ldots, 5.$$

splits in the union of 5 lines. This last condition characterizes in $\mathbf{P}^4$ nets of quadrics which can be simultaneously diagonalized. Conversely, using prop.3.2 , it follows that, for a sufficiently general such a net M, the base locus of M is a curve $C \in T_2$ for some Kummer surface X. We can see from this that Prym canonical contact curves depend on 2 moduli.

Let C be any non trigonal genus five canonical curve, M the net of quadrics through C then C is a double cover of an elliptic curve if and only if the discriminant of M contains a line ([ACGH] p.272); therefore the family of curves we are considering is the family of genus 5 curves having 5 elliptic involutions. These curves were first studied by Humbert and they are sometimes called now Humbert curves.

A generalization We want to consider now a somehow more general situation: let X be any $K3$ surface, $|C|, |H|$ linear systems on X. Assume $|C|$ is ample and $C^2 = C.H = 2g - 2, H^2 = 2g - 6$; the question is again:

(4.1) \qquad describe the family $T_n = \{C \in |C| / L_C^n = O_C\}$

with $|L| = |H - C|, L_C = O_C(L)$. An answer to this problem is given in $[B - V]$ under the assumption X sufficiently general, that is

(4.2) \qquad $\mathrm{Pic}(X) \cong \mathbf{Z} \oplus \mathbf{Z}$ and generated by C and L:

PROPOSITION 4.1 Assume X satisfies (4.2) and $2n^2 - 2 \geq g$, then T_n is a finite set and its cardinality (counting multiplicities) is

$$\binom{2n^2 - 2}{g}.$$

We want to sketch the proof of this and to give some examples. Let $C \in |C|$, consider the exact sequence

(4.3) \qquad $0 \to O_X(-C) \to O_X \to O_C \to 0$

and tensor it by $O_X(-nL)$. Computing the cohomology groups of the associated long exact sequence we have: $h^0(O_X(nL)) = h^2(O_X(nL)) = 0$ because $C.L = 0$; hence, by Riemann-Roch,

(4.4) \qquad $h^1(O_X(nL)) = 2n^2 - 2.$

Since $C.(nL - C) < 0$, $h^0(O_X(nL - C)) = 0$ so that

(4.5) \qquad $h^1(O_X(nL - C)) = 2n^2 - (g+1) + d, d = \dim H^2(O_X(nL - C)).$

The long exact sequence of (4.3) is

$0 \to H^0(L_C^n) \to H^1(O_X(nL - C)) \xrightarrow{f_C} H^1(O_X(nL)) \to$
$\to H^1(L_C^n) \to H^2(O_X(nL - C)) \to 0.$

One shows easily that, since $\mathrm{Pic}(X)$ is generated by C and L, each $C \in |C|$ is irreducible. Under this assumption $C \in T_n$ if and only if $h^0(L_C^n) = 1$ and this is also equivalent to $\dim(\mathrm{Ker}(f_C)) = 1$.

Consider now the map

(4.6) \qquad $m : W = H(O_X(C)) \otimes H^1(O_X(nL - C)) \to H^1(O_X(nL))$

sending $c \otimes s$ in $f_C(s)$. Let $K = \mathrm{Ker}(m)$: the intersection of K with the subset of indecomposable vectors of W gives the sections $c \in H^0(O_X(C))$ such that $\dim \mathrm{Ker}(f_C) = 1, (C = \mathrm{div}(c))$. After projectivization this means that

(4.7) \qquad $T_n = \mathbf{P}(K) \cap S$

where S is the Segre embedding of $\mathbf{P}H^{\circ}(O_X(C)) \times \mathbf{P}H^1(O_X(nL-C))$ in $\mathbf{P}(W)$.
Computing dimensions we have:
codim $\mathbf{P}(K) \leq 2n^2 - 2$, dim $S = H^{\circ}(O_X(C)) - 1 + h^{\circ}(O_X(nL-C)) - 1 = 2n^2 - 2 + d$,
hence

(4.8) $$T_n \neq \emptyset, \ \dim T_n \geq d.$$

Let us show $d = h^2(O_X(nL-C)) = 0$: assume not, then there exists a complete family $T \subset T_n$ of irreducible curves C_t, $t \in T$, such that $L_{C_t}^n$ is trivial. This contradicts the following

(4.9) LEMMA ([B − V]) Let L be a line bundle on any surface X such that: L_{C_t} is trivial for some complete and 1-dimensional family $\{C_t\}$ of linearly equivalent irreducible ample curves of X. Then L is trivial.

Since $O_X(nL)$ is not trivial $d = 0$. Hence, by (4.7),

(4.10) $$\text{cardinality of } T_n = \deg(S) = \binom{2n^2-2}{g}$$

(counting multiplicities) and the proof is complete.

Note that in the Kummer surface case (which does not satisfy the previous generality assumptions) we already obtained this number for $n = 2, g = 5$.
As an example we discuss more in detail some of the cases with $n = 2$. Let X, C, L be as in prop. 4.1 and assume first $g = 2$. Then $|C|$ defines a double covering $f : X \to \mathbf{P}^2$ branched on a smooth plane sextic Δ. In this case $|C + L| (|C - L|)$ contains a unique element $H(\hat{H})$ which is a smooth irreducible rational curve. Since $H + \hat{H} \in |2C|$ we have $f(H) = f(\hat{H}) = B =$ smooth conic everywhere tangent to Δ. Let $p_1, \ldots, p_6$ be the six tangency points of Δ and B: for X sufficiently general we can assume that they are distinct and that any line joining two of them is not tangent to Δ. Let p, p' be two of these points, M the line joining them. Then $f^{-1}(M) = C$ is smooth and $w = f^{-1}(p), w' = f^{-1}(p')$ are two Weierstrass points of C. Therefore $w_C(-w - w') = O_C(C - H) = L_C$ is a non trivial 2-torsion element of Pic (C) and $C \in T_2$. Since there are 15 lines joining two by two the previous six points, the number 15 we obtain in prop. 4.1 for $n = 2, g = 2$ is explained.

Let $g = 3$, assume C is very ample and $X =$ smooth quartic surface in $\mathbf{P}^3(O_X(1) = O_X(C))$. In this case $|C \pm L|$ is a pencil of quartic elliptic curves. Clearly, for each pair $(H, \hat{H}) \in |C + L| \times |C - L|$ there exists a unique quadric Q such that $Q \cap X = H + \hat{H}$. Hence $|C + L| \times |C - L|$ defines a web W of quadrics containing the family of quadrics Q as above as the Segre embedding of $\mathbf{P}^1 \times \mathbf{P}^1$. Note that the restriction map r from $H^{\circ}(O_X(C+L)) \otimes H^{\circ}(O_X(C-L))$ to $H^{\circ}(O_C(C+L)) \otimes H^{\circ}(O_C(C-L))$ is an isomorphism. Moreover the multiplication map m_C from $H^{\circ}(O_C(C+L)) \otimes H^{\circ}(O_C(C-L))$ to $H^{\circ}(O_C(2C))$ is not surjective if and only if $O_C(C+L) = O_C(C-L)$ i.e. iff $C \in T_2$. Now m_C is not surjective if and only if the restriction of W to the plane π containing C is not 3-dimensional; i.e. if and only if π is contained in a quadric of W. A general web of quadrics contains ten reducible quadrics; hence we aspect 20 planes as above: this explains the number 20 for $n = 2, g = 3$.

Actually we can apply the same argument for $g = 3, 4$: consider (for any g) the commutative diagram

$$H^\circ(O_X(C+L)) \otimes H^\circ(O_X(C-L)) \xrightarrow{r} H^\circ(O_C(C+L)) \otimes H^\circ(O_C(C-L))$$

$$\downarrow m \qquad\qquad\qquad\qquad\qquad \downarrow m_c$$

$$H^\circ(O_X(2C)) \xrightarrow{r_c} H^\circ(O_C(2C))$$

and the multiplication

$$H^\circ(O_X(C)) \otimes H^\circ(O_X(C)) \xrightarrow{a} H^\circ(O_X(2C))$$

r is an isomorphism. Assume $g = 3, 4$. In this case m is injective. Therefore $\mathrm{Ker}(m_C)$ injects in $\mathrm{Ker}(r_C)$ by $r^{-1} \cdot m$. Note also that $\mathrm{Ker}\,(r_C) = a(H^\circ(O_X(C)) \otimes <c>)$, $(\mathrm{div}(c) = C)$, and that its dimension is $g+1$. Then observe that $O_C(C+l) \cong O_C(C-L)$ (i.e. $C \in T_2$) if and only if $\mathrm{rank}\,(m_C) \le \frac{1}{2}g(g-1)$. Computing dimensions it follows that $C \in T_2$ if and only if

$$\dim(\mathrm{Ker}(r_C) \cap \cdot mr^{-1}(\mathrm{Ker}(m_C)) \ge \tfrac{1}{2}g(g-3) + 1.$$

Finally consider $a^{-1}(Im(m)) = K$: it is a subspace of codimension $6g - 3 - g^2$ in $H^\circ(O_X(C)) \otimes H^\circ(O_X(C))$. Considering its intersection with the set of indecomposable vectors and projectivizing we obtain

$$S = \mathbf{P}(K) \cap \mathbf{P}^g \times \mathbf{P}^g.$$

S is the zero locus in $\mathbf{P}^g \times \mathbf{P}^g$ of $6g-3-g^2$ symmetric bilinear forms. Consider the projection $p: S \to \mathbf{P}^g$ on one factor, by our previous remarks the elements of T_2 corresponds to the fibres of p having dimension $\ge \frac{1}{2}g(g-3)$:
$g = 3$, S is zero dimensional; its lenght as a zero cycle equals the degree of the Segre embedding of $\mathbf{P}^3 \times \mathbf{P}^3$ which is 20;
$g = 4$, $p(S)$ is a symmetric-determinantal quintic 3-fold in $\mathbf{P}^4$; there are 15 planes in S which are contracted to 15 points by p or, equivalently, S contains 15 planes (which are contracted by the projection on the other factor).

Horrocks-Mumford curves The interest for the family T_3 on a Kummer surface X is motivated by the following result of W. Barth:
(5.1)   Let C be a smooth, irreducible, linearly normal curve in $\mathbf{P}^3$. assume $\deg(C) = m^2/2$ and C = contact curve of two surfaces X, X' of degree m. Let N be the normal bundle of C and $L \subset N$ the line subbundle defined by the tangent planes to X (and X'). Assume $N/L = L(-1)$, then one can reconstruct from C a stable rank 2 vector bundle on $\mathbf{P}^4$ having Chern clases $c_1 = 2m-1, c_2 = m^2$.

For $m = 4$ C is a contact curve of two Kummer surfaces and belongs to T_3, H is the Horrocks-Mumford bundle twisted by 1 . Therefore producing a smooth element of T_3 is an alternative way of showing the existence of the Horrocks-Mumford bundle. This has been done recently by O. Schreier and Decker with the help of a computational method (Maccauley). Though the assumptions of prop. 4.1 are not satisfied by X because Pic (X) has rank 17, nevertheless, applying the proof of 4.1, we can say at least that the family

(5.2) $$S_3 = \{C \in |C| / h^\circ(L_C) \geq 1\}$$

is not empty; unfortunately this is not enough to produce smooth elements of S_3 (hence of T_3). Note also that, in contrast with the general situation described in 4.1, S_3 is not a finite set. Indeed S_3 contains 3-dimensional linear systems of reducible curves. To see this fix for instance $T \in \{T_j\}$ as in (2.1) and consider any curve $C = T + D$ with D smooth. One computes $D.T = 3$ and $h^\circ(O_D \otimes L^3) = O$ ($L = O_X(H - C)$); moreover $O_T \otimes L^3 = O_{\mathbf{P}^1}(3)$. Therefore a section of L_C^3 is just a section of $O_T \otimes L^3$ vanishing on the intersection of D and T. Of course this implies $h^\circ(L_C^3) = 1$; hence the linear system $T+ \mid D \mid$ is in S_3.

Besides these degenerate components one has to expect a two dimensional family of smooth curves in S_3. The reason is the following ([B]): fix a point p of $\mathbf{P}^4$ and consider the family of jumping lines of the Horrocks-Mumford bundle which are passing through p. This is a contact curve C for a given pencil of Kummer surfaces and an element of T_3 for all of them. Therefore the family of pairs (C, X)(C as above, $X =$ Kummer surface through C) is 5 dimensional. Since Kummer surfaces depend on 3 moduli, one expects a 2 dimensional family of these curves on each Kummer surface.

REFERENCES

ACGH. E. Arbarello, M. Cornalba, P.A. Griffiths, J. Harris, *Geometry of Algebraic Curves I*, Springer-Verlag, Berlin, (1985).

B. W. Barth, *Kummer surfaces associated with the Horrocks-Mumford bundle*, Proceedings Angers conference on Algebraic Geometry (1979).

BV. W. Barth, A. Verra, *Torsion on K3 sections*, to appear in Proceedings of Cortona conference on Algebraic Surfaces (1988).

D. R. Donagi, *Group law on the intersection of two quadrics*, Ann. Scuola Norm. Sup. Pisa, v. 7 (1980), 217–239.

GH. P.A. Griffiths, J. Harris, *Principles of Algebraic Geometry*, Wiley Interscience, New York (1978).

H. R. Hudson, *Kummer's quartic surface*, Cambridge Univ. Press, (1905).

V. R. Varley, *Weddle's surfaces, Humbert curves and a certain 4-dimensional abelian variety*, Am. J. of Math. v. 108-2 (1986), 931–951.

Kreĭn Spaces and Holomorphic Isometries of Cartan Domains

EDOARDO VESENTINI Scuola Normale Superiore, I-56100 Pisa, Italy

1.   Let $\mathcal{H}_1$ and $\mathcal{H}_2$ be complex Hilbert spaces and let B be the open unit ball of the complex Banach space $L(\mathcal{H}_2,\mathcal{H}_1)$ of all continuous linear maps of $\mathcal{H}_2$ into $\mathcal{H}_1$. The group Aut B of all holomorphic automorphisms of B, which acts transitively on B [2], can be described in terms of the Kreĭn space $(\mathcal{H}, \mathfrak{a})$ [3]. This space is, by definition, the pair consisting of the Hilbert space direct sum, $\mathcal{H} = \mathcal{H}_1 \oplus \mathcal{H}_2$, of $\mathcal{H}_1$ and $\mathcal{H}_2$, and of the non-degenerate, hermitian, sesquilinear form $\mathfrak{a}$ associated to the matrix

$$J = \begin{pmatrix} I_1 & 0 \\ 0 & -I_2 \end{pmatrix},$$

where I_j is the identity operator in $\mathcal{H}_j$ (and where, to avoid trivialities, $\mathcal{H}_j$ will be assumed $\neq \{0\}$ (j=1,2)). To be more specific, let $\Gamma(J)$ be the group consisting of all invertible elements $S \in L(\mathcal{H}) = L(\mathcal{H},\mathcal{H})$, such that

(1) $\qquad\qquad\qquad S^*JS = J.$

Extending a result established by H. Klingen in [4,5] when $\dim_C \mathcal{H}_1 < \infty$, $\dim_C \mathcal{H}_2 < \infty$, a homomorphism $\phi_0: \Gamma(J) \to$ Aut B was described by T. Franzoni in [1], which is surjective if $\mathcal{H}_1$ and $\mathcal{H}_2$ have different (transfinite) dimensions and is surjective up to conjugation if $\dim_C \mathcal{H}_1 = \dim_C \mathcal{H}_2$.

In the infinite dimensional case a new phenomenon arises. The Carathéodory differential metric $Z \to |\ |_Z$ ($Z \in B$) (which coincides with the Kobayashi differential metric because B is a homogeneous open ball) is a norm which, for every $Z \in B$, is equivalent to the uniform norm in $L(\mathcal{H}_2,\mathcal{H}_1)$. For any fixed $U \in L(\mathcal{H}_2,\mathcal{H}_1)$, the function $Z \to |U|_Z$ is a locally Lipschitz function on B.

Let Iso B be the semigroup of all holomorphic isometries for the Carathéodory differential metric of B. If at least one of the Hilbert spaces $\mathcal{H}_1$ and $\mathcal{H}_2$ has infinite dimension, Aut B is a proper subgroup of Iso B. If both $\mathcal{H}_1$ and $\mathcal{H}_2$ are infinite dimensional, the structure of the semigroup Iso B turns out to be much more complicated than that of Aut B, and still largely unknown. In fact, H. Cartan's theorem whereby every $f \in$ Aut B fixing the origin is necessarily linear does not hold when $f \in$ Iso B, as an example constructed in [7] shows. However, if one of the spaces $\mathcal{H}_1$ and $\mathcal{H}_2$, $\mathcal{H}_2$ say, is finite dimensional, then the homomorphism ϕ_0 can be extended to a homomorphism $\phi: \Lambda(J) \to$ Iso B, where $\Lambda(J)$ is the semigroup consisting of all linear maps $S \in L(\mathcal{H})$ satisfying (1).

If $n = \dim_C \mathcal{H}_2 < \infty$, for all $S \in \Lambda(J)$, $\phi(S)$ can be analytically extended to a neighbourhood of the closure $\overline{B}$ of B. The restriction of this extension to $\overline{B}$ turns out to be continuous for the weak topology on $L(\mathcal{H}_2,\mathcal{H}_1)$. The Schauder-Tychonoff fixed point theorem implies then that (the continuous extension of) $\phi(S)$ has at least one fixed point in $\overline{B}$. This entails that every $S \in \Lambda(J)$ has at least one non-vanishing eigenvalue. The cardinality of the point-spectrum of $S \in \Lambda(J)$ is estimated by the following lemma [8].

LEMMA. *If* $n = \dim_C \mathcal{H}_2 < \infty$, *the number of distinct eigenvalues of* S, *no pairs of which are equivalent under the map* $\zeta \to \bar{\zeta}^{-1}$, *does not exceed* n. *The dimension of the eigenspace corresponding to any eigenvalue not contained in the unit circle is less than or equal to* n.

2.   Let $T: \mathbf{R}_+ \to L(\mathcal{H})$ be a strongly continuous semigroup, and let X be its infinitesimal generator. Under which conditions on X is $T(\mathbf{R}_+) \subset \Lambda(J)$, i.e., is

(2) $$T(t)^* J T(t) = J$$

for all $t \geq 0$?

By Theorem I of [6], (2) holds for all $t \geq 0$ if, and only if, iJX is a symmetric operator. If iJX is symmetric, then T is the restriction to $\mathbf{R}_+$ of a strongly continuous group $\mathbf{R} \to L(\mathcal{H})$ if, and only if, iJX is self-adjoint. In this case the image of $\mathbf{R}$ is contained in $\Gamma(J)$. This result singles out, among all one-parameter strongly continuous semigroups or groups, the ones belonging to $\Lambda(J)$ or to $\Gamma(J)$. If $\dim_C \mathcal{H}_2 < \infty$, the following theorem characterizes, within the family of all linear operators on $\mathcal{H}$, the ones which generate strongly continuous semigroups satisfying the above condition.

THEOREM I. *Let* $n = \dim_C \mathcal{H}_2 < \infty$. *A linear operator* X *with domain* $D(X) \subset \mathcal{H}_1 \oplus \mathcal{H}_2$ *is the infinitesimal generator of a strongly continuous semigroup* $T: \mathbf{R}_+ \to L(\mathcal{H}_1 \oplus \mathcal{H}_2)$ *satisfying* (2) *for all* $t \geq 0$ *if, and only if, there exists a dense linear subspace* $\mathcal{D}_1 \subset \mathcal{H}_1$ *such that*

$$\mathcal{D}(X) = \mathcal{D}_1 \oplus \mathcal{H}_2,$$

and X is expressed by the matrix

$$X = \begin{pmatrix} X_{11} & X_{12} \\ X_{21} & X_{22} \end{pmatrix}$$

where: iX_{22} *is a hermitian operator on* $\mathcal{H}_2$; $X_{12} \in L(\mathcal{H}_2, \mathcal{H}_1)$; $X_{21} = X_{12}^*|_{\mathcal{D}_1}$; iX_{11} *is a closed symmetric operator with domain* $\mathcal{D}(X_{11}) = \mathcal{D}_1$, *and the spectrum* $\sigma(X_{11}) \subset \{\zeta \in \mathbf{C} : \operatorname{Re}\zeta \leq 0\}$.

Furthermore, X generates a strongly continuous group $T: \mathbf{R} \to L(\mathcal{H}_1 \oplus \mathcal{H}_2)$ *satisfying* (2) *for all* $t \in \mathbf{R}$ *if, and only if,* iX_{11} *is self-adjoint.*

The proof of this theorem [8] depends strongly on the fact that $\dim_{\mathbf{C}} \mathcal{H}_2 < \infty$. The following result, also established in [8], shows that nothing is gained if the latter condition is replaced by the analyticity of the semigroup.

THEOREM II. *If the semigroup* $T: \mathbf{R}_+ \to \Lambda(J)$ *is non-trivial, the function T cannot be extended to a holomorphic map* $\tilde{T}: D \to L(\mathcal{H}_1 \oplus \mathcal{H}_2)$ *of any open connected neighbourhood D of* $\mathbf{R}_+^*$ *in* $\mathbf{C}$ *into* $L(\mathcal{H}_1 \oplus \mathcal{H}_2)$ *such that*

(3) $$\tilde{T}(z)^* J \tilde{T}(z) = J$$

for all $z \in D$.

This theorem raises the question whether there exist non-constant $\Lambda(J)$-valued holomorphic functions on D (whose restrictions to $\mathbf{R}_+^*$ are not semigroups). The answer turns out to be affirmative if both $\mathcal{H}_1$ and $\mathcal{H}_2$ are infinite dimensional (in which case non-constant affine maps $\mathbf{C} \to \Lambda(J)$ can be constructed), but negative otherwise [8].

3. The hypothesis $n = \dim_{\mathbf{C}} \mathcal{H}_2 < \infty$ leads to a description of the structure of the spectrum $\sigma(X)$ of X or, more exactly, of the relationship between $\sigma(X)$ and $\sigma(X_{11})$.

If the closed operator iX_{11} is symmetric but not self-adjoint, there exists, in the open right half-plane $\{\zeta \in \mathbf{C}: \text{Re}\zeta > 0\}$, a finite set C with cardinality Card $C \le n$ whose points are eigenvalues of X, such that:

$$\sigma(X) = C \cup \{\zeta \in \mathbf{C}: \text{Re}\zeta \le 0\};$$

if C' is the image of C by the reflection $\zeta \to -\bar{\zeta}$ around the imaginary axis, then $\{\zeta \in \mathbf{C}: \text{Re}\zeta \le 0\} \setminus C'$ is contained in the residual spectrum of X.

If iX_{11} is self-adjoint, $\sigma(X)$ is symmetric with respect to the imaginary axis. Furthermore $\sigma(X) \cap \{\zeta \in \mathbf{C}: \text{Re}\zeta > 0\}$ consists of no more than n points, each of which is an eigenvalue of X.

4. Throughout this section $\mathcal{H}_2$ will be assumed to be finite dimensional. Any one-parameter semigroup in $\Lambda(J)$ (or any one-parameter group in $\Gamma(J)$) defines, through the homomorphism ϕ a semigroup in Iso B (or, through ϕ_0, a subgroup of Aut B).

If T satisfies (2), the semigroup $\hat{T}$ defined on $L(\mathcal{H}_2, \mathcal{H})$ by

$$\hat{T}(t)W = T(t) \circ W \qquad (t \ge 0, W \in L(\mathcal{H}_2, \mathcal{H}))$$

is strongly continuous. Its infinitesimal generator $\hat{X}$ is defined by

$$\hat{X}W = X \circ W$$

on the domain

$$D(\hat{X}) = \{W \in L(\mathcal{H}_2, \mathcal{H}): Wx_2 \in D(X) \text{ for all } x_2 \in \mathcal{H}_2\}.$$

In a similar way, let

$$\hat{D}_1 = \{Z \in L(\mathcal{H}_2, \mathcal{H}_1): Zx_2 \in D(X_{11}) \text{ for all } x_2 \in \mathcal{H}_2\}.$$

For every $Z_0 \in B \cap \hat{D}_1$ and for $t \ge 0$, let

$$Z(t) = (\phi \circ T(t))Z_0.$$

Then $Z(t) \in \hat{D}_1$ for all $t \ge 0$, and $Z(t)$ satisfies the differential equation

(4) $$\dot{Z}(t) = X_{11} \circ Z(t) - Z(t) \circ X_{22} - Z(t) \circ X_{12}^* \circ Z(t) + X_{12}$$

with the initial condition

(5) $$Z(0) = Z_0.$$

Introducing on $\hat{D}_1$ the graph-norm

(6) $$Z \to \|Z\| + \|X_{11} \circ Z\|,$$

the following theorem holds [8].

THEOREM III. *If* $\dim_C \mathcal{H}_2 < \infty$, *for every* $\gamma > 0$ *and every* $Z_0 \in B \cup \hat{D}_1$, *the function* $t \to Z(t)$ *is the unique solution of* (4) *with the initial condition* (5) *which is contained in* $C^1([0,\gamma], L(\mathcal{H}_2, \mathcal{H}_1))$ *and is continuous for the graph-norm* (6).

REFERENCES

[1] T. FRANZONI, The group of holomorphic automorphisms in certain J*-algebras, *Ann. Mat. Pura Appl.*, (4) CXVII (1981), 51-66.

[2] L. A. HARRIS, *Bounded symmetric homogeneous domains in infinite dimensional spaces*, Lecture notes in mathematics, # 369, Springer-Verlag, Berlin/Heidelberg/New York, 1973, 13-40.

[3] J. W. HELTON, Operator theory, analytic functions, matrices and electrical engineering, *Regional conference series in mathematics*, # 68, Amer. Math. Soc., Providence, R.I., 1987.

[4] H. KLINGEN, Diskontinuierliche Gruppen in symmetrischen Räumen. I, *Math. Annalen*, 129 (1955), 345-369.

[5] H. KLINGEN, Über die analytischen Abbildungen verallgemeinerter Einheitkreise auf sich, *Math. Annalen*, 132 (1956), 134-144.

[6] E. VESENTINI, Semigroups of holomorphic isometries, *Advances in Math.*, 65 (1987), 272-306.

[7] E. VESENTINI, *Holomorphic families of holomorphic isometries*, Lecture notes in Mathematics, # 1277, Springer-Verlag, Berlin/Heidelberg/New York, 1987, 291-302.

[8] E. VESENTINI, Semigroups on Kreĭn spaces, to appear.

L'insegnamento della Matematica nell'Università di Bologna dal 1860 al 1940

STEFANO FRANCESCONI

SUMMARY

Which were the professors of mathematics at the University of Bologna? How long did they teach there? How changed the curricola for the students in mathematics from 1860 to 1940?

The aim of the paper is to provide a tentative answer to these questions.

The political Unification of Italy in 1860 marked a turning point for the history of Bologna University, too. At that time a new academic policy provided new regulations.

During the seventies, however, students could not complete their mathematical studies in consequence of the lack of professors for the higher courses.

Since the beginning of the eighties onwards the situation was restablished and students could get again their degree. At that time the leading professors at Bologna University were Cesare Arzelà, Federigo Enriques and Salvatore Pincherle.

After World War I, the Unione Matematica Italiana (U.M.I.) was founded on the initiative of Salvatore Pincherle. Pincherle also organized the International Congress of Mathematicians held in Bologna in 1928.

Ten years later, the fascist dictatorship introduced the racial laws in consequence of which all jewish mathematicians and scientists were expelled from Universities and Scientific Institutions.

Lavoro compiuto con il contributo della borsa di studio A. Arnese.

INTRODUZIONE

Nel 1947 Ettore Bortolotti pubblicava la *Storia della matematica nella Università di Bologna*, un volume in cui era delineata l'attività scientifica e didattica dei matematici attivi presso l'Ateneo bolognese dalle origini fino alla ricostituzione del corso di laurea in matematica avvenuta nel 1881.

Questo lavoro intende essere un primo tentativo di proseguire lo studio compiuto da Bortolotti per quanto riguarda, in particolare, l'organizzazione del corso di laurea in matematica e la composizione del corpo docente nel periodo che va dal 1860 al 1940.

L'Unità d'Italia segnò infatti una svolta non solo nella vita politica e civile del nostro paese, ma anche sul terreno dell'istruzione secondaria e universitaria. Negli anni dell'Unificazione si presentò infatti, insieme al problema dell'organizzazione del nuovo stato, anche quello di una profonda riforma dell'istruzione. Il dibattito che ne seguì impegnò direttamente la comunità scientifica, e quella matematica in particolare, non solo nelle aule parlamentari e nelle sedi istituzionali, ma anche nelle pagine di quotidiani e riviste.

1. GLI ANNI DELL'UNITÀ

Nel 1862 la rivista «Il Politecnico» pubblicava alcune riflessioni di Carlo Cattaneo (1801-1869) «sul riordinamento degli studii scientifici in Italia» (1). Avendo presente la necessità della preparazione di tecnici ed ingegneri per il nuovo Stato, Cattaneo auspicava una maggiore «specializzazione» degli studi. Infatti, rivolgendosi al ministro della pubblica istruzione Carlo Matteucci (1811-1868), affermava:

> «Sì, è la suddivisione delle facultà, è la loro specificazione ch'io vi dimando per gli ingegneri italiani; voi potete farvene l'istitutore. Voi darete all'Italia ingegneri architetti, ingegneri idraulici, ingegneri agronomi, ingegneri censuarii, ingegneri delle miniere, ingegneri militari, navali, geografi, ferroviari, e uomini nati col genio mecanico» (2).

Cattaneo, dopo aver osservato che tali «facultà» avrebbero dovuto presentare una comune formazione matematica, aggiungeva:

> «se intendete conferire ad alcuno il titolo alquanto arduo di dottore in matematica, e non volete che sia una vanità, è necessario che dimandiate ad una facultà speciale di prepararvi i professori di matematica pei cento licei d'Italia, per tutti li altri stabilimenti di primo e secondo ordine, per le scôle militari e maritime, per le specule astronomiche, per l'insegnamento privato, e in fine, di educare degnamente quei rari intelletti, che cercano la scienza per la scienza, e hanno diritto d'attingere immantinente e solamente alle fonti più sublimi» (3).

Il direttore del «Politecnico» osservava inoltre che si sarebbero potute

> «ripartire in luoghi diversi le varie sezioni degli studi, [in modo tale da] collocare [...] la scôla degli ingegneri navali nel più gran centro di costruzione, o almeno di navigazione; [...] collocare gli ingegneri fluviali in Padova o in Bologna, o in ambo le città sino a che il genio dei professori, o l'incremento della scienza, o i bisogni della nazione, addittassero qualche ulteriore specificazione; di che non è a dubitarsi [...] collocare gli ingegneri delle miniere in luogo dove non possono girare li occhi senza vedersi affacciare da monti e valli le evoluzioni del globo» (4).

Le idee di Cattaneo trovavano realizzazione solo in parte in un regio decreto emanato il 14 settembre 1862 per iniziativa del Matteucci. Il «decreto Matteucci», stabilendo il nuovo «regolamento per la facoltà di scienze matematiche, fisiche e naturali», modificava profondamente anche la situazione degli studi scientifici nell'Ateneo bolognese.

Sino all'approvazione del «nuovo regolamento», il piano degli studi della «facoltà matematica» era costituito da diciotto insegnamenti (compresi i periodi di pratica) suddivisi in cinque anni. L'esame di laurea era posto al termine del quarto anno, mentre l'esame di libera pratica, che permetteva d'esercitare la professione d'ingegnere civile, era posto al termine del quinto anno.

La seguente tabella illustra il piano degli studi della «facoltà matematica» durante l'anno accademico 1859-'60.

anno I
- Storia naturale
- Fisica sperimentale
- Architettura civile
- Introduzione al Calcolo

esame di passaggio

anno II
- Meccanica ed Idraulica
- Ottica ed Astronomia
- Calcolo sublime
- Agronomia teorico-pratica

esame di baccellierato

anno III
- Meccanica ed Idraulica
- Ottica ed Astronomia
- Istituzioni civili
- Agronomia teorico-pratica

esame di licenza

anno IV
- Pratica nello studio di un ingegnere approvato
- Economia pubblica
- Agronomia teorico-pratica

esame di laurea

anno V
- Pratica nello studio di un ingegnere approvato
- Pratica nello studio di un architetto o di un idraulico approvato
- Agronomia teorico-pratica

esame di libera pratica.

Gli insegnamenti di matematica pura erano quindi soltanto due: Introduzione al Calcolo, Calcolo sublime, come anche quelli di matematica applicata: Meccanica ed Idraulica, Ottica ed Astronomia (i quali però, diversamente dai precedenti, avevano durata biennale). Gli altri corsi spaziavano dalle scienze fisiche a quelle biologiche, dal diritto alla architettura e dalle scienze agrarie a quelle geologiche. Si è dunque tentati di concludere che la «facoltà matematica», anzichè fornire ai laureandi una specifica cultura tecnico-scientifica, forniva una cultura generale, di «base», che avrebbe dovuto essere valida per ogni disciplina scientifica e per ogni branca dell'ingegneria. Inoltre, se tale organizzazione degli studi poteva essere funzionale per uno stato caratterizzato da bassi indici di sviluppo industriale e da una relativa arretratezza economica come quello pontificio, non rappresentava certamente la situazione ideale per uno stato che ambiva essere incluso nel novero delle nazioni più progredite.

2. IL PERIODO DI CREMONA E DI BELTRAMI

A Bologna, un primo tentativo di fornire la specializzazione degli studi fu realizzato mediante un decreto-legge emanato da Farini nel febbraio del 1860 (5) col quale veniva istituita, fra le altre, la cattedra di Chimica mineralogica ed analitica. Il mese successivo un secondo decreto legge (6) istituiva tre nuove cattedre: Geodesia, Meccanica applicata e Geometria superiore; in luogo della cattedra di Storia naturale venivano create quelle di Mineralogia, Geologia e Zoologia, e infine quella di Ottica ed Astronomia veniva sdoppiata in due: Ottica, Astronomia.

Le nuove cattedre così istituite consentivano certo di colmare alcune lacune, ma non modificavano sostanzialmente la struttura degli studi scientifici. La «facoltà matematica» continuava a proporre lo stesso piano degli studi a tutti i suoi iscritti senza effettuare alcuna distinzione tra studi di carattere matematico, fisico, chimico, biologico o ingegneristico. Da questo punto di vista l'introduzione dei nuovi corsi appare un rimedio temporaneo, un palliativo in attesa d'un radicale rinnovamento. A riformare l'ordinamento degli studi scientifici doveva provvedere, come si è accennato, il «decreto Matteucci».

Tale decreto prevedeva, nell'ambito della «facoltà matematica», quattro corsi di laurea: in scienze matematiche pure, in scienze fisico-matematiche, in scienze fisico-chimiche, in storia naturale.

I primi due corsi di laurea citati erano costituiti rispettivamente da quindici e sedici insegnamenti suddivisi in quattro anni. La tabella seguente ne illustra i piani di studio.

	Corso di laurea in scienze matematiche pure	Corso di laurea in scienze fisico – matematiche
anno I	Algebra complementare Geometria analitica Chimica inorganica Disegno	Algebra complementare Fisica Chimica inorganica Esercizi di chimica
anno II	Calcolo differenziale ed integrale Geometria descrittiva Fisica Disegno	Calcolo differenziale ed integrale Fisica Chimica organica Esercizi di chimica
anno III	Meccanica razionale Geodesia teorica Fisica Disegno	Meccanica razionale Analisi e Geometria superiore Mineralogia e Geologia Esercizi pratici di Fisica
anno IV	Analisi e Geometria superiore Astronomia e meccanica celeste Fisica matematica	Astronomia e Meccanica celeste Fisica matematica Esercizi pratici di Fisica Esercizi pratici di Astronomia e Geodesia

Il confronto tra i piani degli studi vigenti negli anni accademici precedenti l'Unità con quelli riportati in tabella, può fornire una prima indicazione del rinnovamento introdotto dal «decreto Matteucci». Inoltre, sebbene ciò non appaia esplicitamente, le norme elaborate da Matteucci incidevano anche sulla formazione degli ingegneri. Quest'ultima infatti veniva affidata alla facoltà di scienze per il solo «biennio preparatorio», al termine del quale si otteneva la licenza fisico-matematica. Per il successivo triennio, non essendo prevista nell'ordinamento universitario italiano la facoltà d'ingegneria, si provvedeva ad istituire le R. Scuole d'Applicazione per gli Ingegneri. A Bologna, peraltro, vigeva una situazione particolare poichè, sino al 1875, per ottenere il diploma d'ingegnere era necessario anzitutto conseguire la licenza fisico-matematica, quindi frequentare i Corsi Pratici per gli Ingegneri sostenendo i relativi esami e fare pratica presso lo studio d'un ingegnere approvato. I Corsi Pratici per gli Ingegneri erano costituiti da tre insegnamenti universitari: Meccanica applicata, Agronomia teorico-pratica, Mineralogia e Geologia, ognuno dei quali durava due anni. Questi ultimi, sin dalla loro istituzione, erano stati però criticati perchè ritenuti incompleti ed insufficienti. Di conseguenza, molti aspiranti ingegneri si vedevano costretti a trasferirsi a Milano o a Torino, città i cui Istituti si caratterizzavano per un maggior numero di specializzazioni e per una preparazione più adeguata all'attività di ingegnere (7).

Due decreti emanati dal Governo nazionale fra il 1861 ed il 1862 (8) modificavano ulteriormente l'ordinamento universitario bolognese. Infatti, sino al 1862, era rimasta in vigore la «stantia istituzione dei Collegi di esaminatori» (9). Questi ultimi avevano lo scopo di valutare gli studenti agli esami di baccellierato, licenza e laurea ed erano costituiti da alcuni docenti e da alcuni «titolari emeriti o estranei nominati dal papa...[che] non erano affatto titolari di cattedra» (10). In seguito alla applicazione dei decreti detti «la distinzione fra Collegio esaminatore e Corpo insegnante scomparve, perchè nel primo entrarono, secondo l'anzianità, solo gli insegnanti e così essi si fusero» (11). Lo spirito che aveva animato i decreti emanati dal Governatore delle R. Provincie dell'Emilia a partire dal 1859, era quello di uniformare l'ordinamento universitario bolognese alle norme del Regno Sardo.

«In pendenza delle riforme che, sugli studi, si stanno elaborando in Piemonte, e per uniformarsi intanto a ciò che è stato sancito per le provincie modenesi e parmensi...»,

dichiarava infatti la premessa del decreto-legge del 1859 col quale veniva abolita la facoltà di teologia (12), mentre la premessa del decreto del 1860 (13) che stabiliva l'appartenenza dell'Università di Bologna alle Università di primo ordine affermava:

«Considerando essere necessità che tutte le istituzioni esistenti nelle R. Provincie dell'Emilia siano in modo uniforme ordinate, prendendo per base, per quanto sia possibile, le leggi e le istituzioni del R. Sardo; Veduto che le Università degli Studi nelle provincie dell'Emilia sono in modo disformi governate; e che, fatta ragione della loro importanza relativa, possono e devono ordinarsi conformemente a quelle del Regno;...».

In questo modo alcuni importanti aspetti dell'ordinamento universitario bolognese quali, ad esempio, gli organi di governo dell'Università, le tasse scolastiche, le retribuzioni agli insegnanti, l'ordine degli studi (in particolare quello degli studi scientifici esclusa la formazione degli ingegneri) e l'identificazione del Corpo insegnante coi Collegi esaminatori, erano stati uniformati agli ordinamenti delle altre Università. D'altra parte negli anni della creazione dello Stato unitario la legislazione universitaria era estremamente intricata. La «legge Casati» era stata inizialmente emanata nel 1859 per il Regno Sardo e la Lombardia e ad essa, nella situazione creatasi con la progressiva unificazione politica del paese, avrebbe dovuto far riferimento l'ordinamento d'ogni Università. Tale legge tuttavia

«non fu mai estesa nè all'Università di Bologna, nè alle altre governate dalle rispettive leggi locali... [e non poteva dunque essere applicata alcuna] limitazione del numero dei professori ordinari di ciascuna Facoltà, propria della legge Casati» (14).

L'organico della facoltà di scienze pertanto veniva deciso in base al «regolamento Matteucci» e ad alcune «leggi locali» (15).

Come si è già accennato nel marzo del 1860 il decreto-legge di Farini istituiva alcune nuove cattedre. Nello stesso anno, relativamente a quelle cattedre, il ministro Terenzio Mamiani (1799-1885) nominava per la Geodesia Matteo Fiorini (1827-1901) (16), per la Meccanica applicata Quirico Filopanti (1812-1894) (17) e per la Geometria superiore Luigi Cremona (1830-1903) (18).

Cremona rimaneva a Bologna sino al 1866, anno in cui veniva comandato presso l'Istituto Tecnico Superiore di Milano. Durante il periodo trascorso nell'Ateneo bolognese egli compiva le ricerche che più caratterizzano la sua opera scientifica e, inoltre, veniva incaricato di tenere i corsi di Geometria analitica, Geometria descrittiva e Meccanica razionale. L'ultimo dei corsi citati era tenuto da Cremona nell'a.a. 1864-'65, in seguito al rifiuto di p. Domenico Chelini (1802-1878) (19), titolare della cattedra sino all'anno precedente, di giurare fedeltà al nuovo Stato. Per i suoi studi di Meccanica razionale Chelini si era conquistato la stima del mondo scientifico, e in primo luogo di Cremona, ma a nulla valsero i tentativi d'aggirare la norma secondo la quale ogni funzionario statale era tenuto a prestare giuramento. Di conseguenza veniva collocato a riposo insieme ad altri due docenti che avevano compiuto la stessa scelta: Giuseppe Bianconi (1809-1878) e Lorenzo Respighi (1824-1899) (20). Due anni dopo era chiamato sulla cattedra di Meccanica razionale Eugenio Beltrami (1835-1900) (21). Questo ultimo, che aveva già insegnato a Bologna durante l'a.a. 1862-'63, riceveva anche l'incarico di tenere il corso di Geometria descrittiva. Come per Cremona, anche per Beltrami il periodo bolognese, che terminò nel 1873, fu di intensa attività scientifica.

3. IL PERIODO DAL 1873 AL 1877

Con il collocamento a riposo di Respighi e la partenza di Cremona, rimanevano scoperte le cattedre di Astronomia e di Geometria superiore e, dal 1874, anche quella di Meccanica razionale. In seguito alla partenza di Beltrami quest'ultimo corso era infatti tenuto per un solo anno da Cesare Razzaboni (1827-1893) (22), il quale veniva poi comandato presso la Scuola d'Applicazione di Roma. Rimaneva infine vacante la cattedra di Fisica matematica, per la quale non era stata compiuta alcuna nomina sin dalla sua istituzione. In quel periodo rimanevano quindi senza docenti alcuni corsi del terzo anno e i corsi del quarto, sicchè la facoltà poteva conferire la sola licenza fisico matematica.

Inoltre, in seguito alle critiche cui si è fatto riferimento in precedenza, nel 1875 i Corsi Pratici per gli Ingegneri venivano soppressi. Tuttavia, due anni più tardi, si fondava la Scuola d'Applicazione per gli Ingegneri in Bologna, la cui gestione però veniva garantita dallo Stato per il solo primo anno. A questo proposito, nel discorso di apertura dell'a.a. 1876-'77, il rettore Luigi Calori (1807-1897), criticando non senza enfasi la situazione creatasi, ricordava che:

> «or fa un anno, [...] un lutto oppressò cotesto Ateneo, [...] e la cagione del grave lutto fu [...] il colpo mortale dato alle matematiche in questa Università. [...] Ma che avvenne? Udite: quando per ragioni a tutti ben note rimasero vuote le cattedre di Astronomia, di Geometria superiore, di Meccanica applicata e Idraulica i Proposti [sic] della Pubblica Istruzione non vi pensarono punto, e pregandoli noi che provvedessero, scusaronsi dicendo non esserci soggetti idonei. Non ci sono soggetti idonei! Ah si, perchè vi siete chiodato nella testa che buon insegnante tanto valga quanto celebrità [...]. Ma le ragioni ed i rammarichi erano niente: il colpo era dato e la vittima a terra. Non più corso di ingegneria pratica, non più laurea matematica e per conseguenza non più Studio generale od Università» (23).

Calori procedeva nella sua prolusione descrivendo i tentativi compiuti allo scopo di ristabilire le condizioni precedenti, quindi ringraziava il prefetto di Bologna, senatore Gravina, il quale

> «piegava S.E. il signor Ministro della Pubblica Istruzione, e conducevalo nella determinazione di completarci la sopra detta Facoltà rispetto alle Matematiche Pure, acciò che si potesse novellamente conferire la Laurea [...] [Il signor Ministro, pertanto,] ci ha concedute due Cattedre nuove, la Meccanica superiore e l'Analisi superiore. Lo che ci sia arra che non andrà molto che avremo anche il restante» (24).

Il rettore concludeva affermando che:

> «completa nella suddetta maniera la Facoltà matematica, il reintegramento è ben lontano dall'essere pieno, mancando noi onde sopperire al bisogno di que' giovani che debbano fare i due ultimi anni del corso pratico d'Ingegneria. E qui la verità vuol che si dica non essere d'accagionare di tal difetto il signor Ministro, il quale è dispostissimo anche a questa concessione ognorachè e Municipio e Provincia vogliano provvedere a tant'uopo, e reggere la spesa» (25).

In seguito alla proposta del ministro, si costituiva un consorzio composto dal Comune e dalla Provincia di Bologna, dalle aziende Aldini e Valeriani e dai collegi Bertocchi e Comelli. Il consorzio stipulava quindi una convenzione con lo Stato in base alla quale gli enti consociati avrebbero finanziato gli ultimi due anni della scuola d'Applicazione (26). L'organizzazione dell'Istituto veniva affidata a Razzaboni il quale, tornato a Bologna dopo tre anni di permanenza alla Scuola d'Applicazione di Roma, oltre ad assumere la carica di direttore otteneva la cattedra di Idraulica.

La seguente tabella illustra l'ordine degli studi e l'organico della Scuola d'Applicazione nell'a.a. 1877-'78 (con «ord» vengono indicati i professori ordinari, con «straord» quelli straordinari e con «inc» gli incaricati):

anno I

Applicazioni di Geometria descrittiva	P. Boschi (inc)
Disegno di applicazioni di Geom. descritt.	P. Boschi (inc)
Statica grafica	A. Fais (inc)
Disegno di Statica grafica	A. Fais (inc)
Geodesia teoretica	M. Fiorini (ord)
Meccanica razionale	F. Ruffini (inc)
Chimica Docimastica	D. Santagata (inc)
Manipolazioni chimiche	D. Santagata (inc)

anno II

Mineralogia applicata alle costruzioni	L. Bombicci (inc)
Geologia applicata alle costruzioni	G. Capellini (inc)
Fisica tecnica	L. Donati (straord)
Meccanica applicata alle costruzioni	S. Canevazzi (inc)
Materie giuridiche	O. Regnoli (inc)
Geometria pratica	P. Riccardi (ord)
Disegno di Meccanica applicata alle costruz.	S. Canevazzi (inc)
Meccanica applicata alle macchine	G. Silvani (inc)
Disegno di Meccanica applicata alle macchine	G. Silvani (inc)
Disegno topografico	P. Riccardi (inc)

anno III

Materiali da costruzioni	L. Venturi (straord)
Economia ed estimo rurale	L.F. Botter (ord)
Macchine	J. Benetti (ord)
Strade ordinarie	C. Stabilini (inc)
Idraulica	C. Razzaboni (ord)
Ponti	S. Canevazzi (inc)
Architettura tecnica	F. Lodi (inc)
Disegno di macchine	J. Benetti (inc)
Disegno di ponti	S. Canevazzi (inc)
Disegno di strade	G. Stabilini (inc)
Progetti di Architettura	F. Lodi (inc)
Ferrovie	J. Benetti (inc)

Nel 1877, oltre alla istituzione della Scuola d'Applicazione, venivano nuovamente tenuti gli insegnamenti del terzo anno del corso di laurea in matematica. Il corso di Geodesia, che durante gli aa.aa. 1875-'76 e 1876-'77 non era stato tenuto, veniva nuovamente affidato a Fiorini, quello di Meccanica razionale a Paolo Ferdinando Ruffini (1823-1908) (27), mentre Antonio Saporetti (1821-1901) (28) dalla cattedra di Calcolo infinitesimale si trasferiva su quella di Astronomia. In quel periodo insegnarono presso la facoltà di scienze e la Scuola d'Applicazione matematici come Francesco d'Arcais (1849-1927) (29), Antonio Fais (1841-1925) (30) e Riccardo de Paolis (1854-1829) (31).

4. LA RICOSTITUZIONE DEL CORSO DI LAUREA

Nel discorso d'apertura dell'a.a. 1876-'77, il rettore Calori, riferendosi al concorso per le cattedre di Analisi superiore e di Meccanica superiore, osservava che:

> «per mala ventura il concorso d'Analisi è andato deserto; l'altro non pare; ma non ne sappiamo niente per ancora» (32).

Tuttavia dal 1881 gli studenti potevano completare il corso degli studi in matematica a Bologna. Infatti dal 1879 Luigi Donati (1846-1932) (33) iniziava a tenere, presso la facoltà di scienze, il corso di Fisica matematica e, presso la Scuola d'Applicazione, quello di Fisica tecnica. Dal 1881 Cesare Arzelà (1847-1912) (34), professore straordinario di calcolo infinitesimale, e Salvatore Pincherle (1853-1936) (35), professore straordinario di Algebra e Geometria analitica, venivano rispettivamente incaricati di tenere i corsi di Analisi superiore e di Geometria superiore.

Nel 1885 veniva varato un nuovo regolamento universitario che però non intruduceva alcuna novità sostanziale relativamente all'ordine degli studi in matematica.

Nel 1894 Federigo Enriques (1871-1946) (36) iniziava a Bologna la carriera scientifica e, due anni dopo, era chiamato sulla cattedra di Geometria descrittiva e proiettiva.

Nel gennaio del 1897 il ministro della pubblica istruzione Emanuele Gianturco (1857-1907), visitando l'Università di Bologna, veniva violentemente contestato dagli studenti. I disordini che seguirono furono in breve sedati dall'intervento dell'esercito, ma la

> «scintilla, che a torto si disse partita da Bologna, divampò incendio in tutte le Università del

Regno... [e la rimanente parte dell'a.a. 1896-'97] fu quasi del tutto speso nella cura molesta, del pari che infruttuosa, di agitazioni di ogni genere e di tumulti» (37).

Fra le molteplici cause dei moti studenteschi di quel periodo alcune erano riconducibili ai provvedimenti restrittivi proposti dal Gianturco quale, ad esempio, l'abolizione della sessione straordinaria di esami. Altre erano connesse alle particolari situazioni nelle quali si trovavano le singole Università. Nel discorso d'apertura dell'a.a. 1898-'99 il rettore dell'Ateneo bolognese, riferendosi alla copertura finanziaria per il «mantenimento» della Scuola d'Applicazione, affermava:

«E intanto, i bisogni che si moltiplicavano o si facevano più gravi, le speranze cadute, le promesse di parziali provvedimenti fatte e non mantenute, lo stesso trattamento migliore reso dallo Stato a tutte le altre Università e i larghi aiuti a molte altre forniti, produssero in questa una condizione veramente insoffribile di vita; si creò, anzi, uno stato permanente, bene a tutti noto, di inquietudine e di agitazione, così nei docenti come nei discepoli; di guisa che la maggior parte del tempo e la migliore attività di tutti fu ed è tuttora intesa a prevenire o a far cessare disordini, tanto più temibili, in quanto sostanzialmente fondati sopra giusta ragione» (38).

Altri motivi, infine, erano imputabili alle tensioni sociali che caratterizzavano l'Italia della fine del secolo. Tensioni acuite dalla grave crisi economica verificatasi alla fine degli anni '80 e dal clima di repressione instaurato dai governi crispini e postcrispini.

Coi primi del '900, in seguito alla svolta politica attuata da Giolitti, il clima sociale mutò radicalmente. La crisi economica parve essere superata, in particolare, afferma V. Castronovo:

«il sistema industriale riuscì in complesso a reggere l'urto d'una crisi così grave e trovò la forza di riprendersi nel giro di pochi anni, ancor prima che la tempesta abbattutasi sull'economia italiana si fosse del tutto dileguata [...]. Insomma, [...] il sistema industriale costituiva oramai una realtà tangibile e in vario modo consolidata» (39).

Sin dai primi anni della ripresa si manifestava in maniera sempre più evidente, almeno nella componente «industrialista» della borghesia italiana, l'esigenza d'una maggiore interazione tra il mondo produttivo e quello dell'istruzione tecnico-scientifica. Allo scopo di soddisfare alcune delle richieste avanzate dagli imprenditori si fondavano Istituti tecnici, si consolidavano le Scuole d'Applicazione e, all'interno di queste ultime, si istituivano nuovi indirizzi. Ciò avveniva però non senza suscitare le resistenze di alcuni settori della classe dirigente e, inoltre, non si può dimenticare che da questo processo di sviluppo veniva esclusa una larga parte del paese. Nel settentrione si assisteva comunque ad un generale incremento degli studi tecnico-scientifici e delle istituzioni ad essi collegate. In particolare a Bologna, oltre alla realizzazione del nuovo «polo» universitario di via Irnerio, un decreto emanato nel 1899 ed approvato nonostante alcune «deplorevoli circostanze» (40), dichiarava Regia la Scuola d'Applicazione. Lo Stato si assumeva perciò l'onere dell'intero corso degli studi, anche se il consorzio avrebbe continuato a contribuire al finanziamento della Scuola ancora a lungo.

La seguente tabella illustra l'ordine degli studi della sezione di ingegneria civile per l'a.a. 1899-1900.

anno I
$\begin{cases} \text{Geodesia teoretica} \\ \text{Meccanica razionale} \\ \text{Statica grafica} \\ \text{Applicazioni alla Geometria descrittiva} \\ \text{Chimica docimastica} \\ \text{Geologia applicata} \\ \text{Costruzioni civili e fondazioni} \end{cases}$

anno II
- Geometria pratica e celerimensura
- Fisica tecnica
- Costruzioni civili e fondazioni
- Meccanica applicata alle costruzioni
- Meccanica applicata alle macchine
- Economia rurale ed estimo
- Idraulica

anno III
- Idraulica
- Macchine termiche ed idrauliche
- Materie giuridiche
- Architettura tecnica
- Ponti e costruzioni idrauliche
- Costruzioni stradali e ferroviarie
- Ferrovie
- Fisica tecnica (Elettrotecnica)
- Igiene applicata all'ingegneria (corso facoltativo)

Va rilevato che durante il periodo compreso tra il 1877 ed il 1899, nella Scuola d'Applicazione di Bologna venivano realizzate due sole specializzazioni: ingegneria civile ed architettura, e che, diversamene da quanto era stato fatto a Milano o a Torino sin dal 1863, continuava a mancare una sezione di ingegneria «industriale».

Un primo incremento della specializzazione degli studi di «ingegneria» si registra, a Bologna, all'inizio del '900. Nel 1901 infatti, un finanziamento della Cassa di Risparmio di Bologna permette di fondare la R. Scuola Superiore di Agraria, mentre le prime indicazioni atte ad istituire una sezione di ingegneria «industriale» appariranno solo nell'annuario del 1913 (41).

Le ragioni del ritardo col quale venne introdotta una maggiore specializzazione degli studi di carattere ingegneristico risiedevano in buona parte nella struttura economica delle regioni di provenienza degli iscritti alla Scuola d'Applicazione. In effetti lo sviluppo economico di regioni quali l'Emilia Romagna, le Marche, l'Umbria, parte del Veneto e della Toscana e, infine, di alcune zone del meridione, per la assenza di grandi industrie meccaniche e siderurgiche, tessili e chimiche, rimaneva strettamente connesso all'agricoltura. Pertanto gli ingegneri venivano preparati ad una attività legata quasi esclusivamente all'amministrazione statale ed alla costruzioni di edifici, ponti, strade e linee ferroviarie.

Nel 1901 venne emanato un nuovo regolamento universitario. Esso introduceva alcune novità: ogni facoltà doveva infatti indicare, anno per anno, i corsi fondamentali, i corsi liberi e quelli complementari da tenersi nella facoltà stessa e, inoltre, i corsi ritenuti di «coltura generale [...] professati a qualunque titolo nell'Università» (42). Per uno studente che avesse voluto laurearsi in matematica il piano degli studi doveva essere composto dai seguenti corsi obbligatori: Fisica sperimentale, Chimica organica ed inorganica, Mineralogia, Analisi algebrica, Analisi infinitesimale, Geometria analitica, Geometria proiettiva con disegno, Geometria descrittiva con disegno, Meccanica razionale; lo studente doveva quindi scegliere quattro fra i seguenti corsi del secondo biennio: Matematiche superiori (corso di Pincherle) (*), Matematiche superiori (corso di Arzelà) (*), Fisica matematica, Astronomia, Chimica fisica, Geodesia (i corsi contraddistinti dal simbolo (*) erano quelli consigliati dalla facoltà); e infine, fra i corsi liberi lo studente poteva scegliere, oltre ai corsi liberi della facoltà di scienze, i seguenti corsi tenuti nelle facoltà di medicina, giurisprudenza, lettere e filosofia:

Facoltà di medicina:	1. Fisiologia
	2. Patologia generale
	3. Igiene
	4. Embriologia
Facoltà di giurisprudenza:	1. Economia politica
	2. Statistica
	3. Filosofia del Diritto
Facoltà di lettere e filosofia:	1. Letteratura italiana
	2. Glottologia
	3. Filosofia teoretica
	4. Storia della Filosofia
	5. Storia della Pedagogia

(il regolamento non specifica però il numero dei corsi liberi che lo studente poteva scegliere).

E' significativo che Enriques e Carlo Emery (1848-1925) sottolineassero la rilevanza sul piano culturale della scelta dei corsi liberi. Questa infatti doveva avere lo scopo di

«indicare ai giovani quegli studi di coltura generale che sotto vari aspetti possono integrarne la coltura speciale; e così, da una parte, aprire al loro spirito la visione più larga della scienza, che va al di là e al di sopra di ogni ordine limitato di conoscenze; per l'altra parte, suggerire alle loro attività più vari mezzi per esercitarsi, segnando, ove occorra, scopi nuovi, in armonia colle esigenze della vita sociale [...] [Inoltre tale procedura] viene ad assumere una importanza notevole, in un regime di studi rigorosamente stabilito secondo una netta divisione delle carriere scientifiche; onde è lecito trarne l'augurio che esso sia principio d'una nuova riforma, la quale consacri l'unità della istituzione universitaria e la libertà degli studi» (43).

Il nuovo regolamento istituiva, infine, la sezione matematica della Scuola di Magistero. Scopo della scuola era di formare gli insegnanti di matematica per la scuola secondaria. I docenti appartenevano al corpo insegnante della facoltà di scienze o a quello della scuola secondaria ed erano diretti, in generale, dal preside d'un Istituto tecnico o d'un Liceo. I corsi, ai quali potevano iscriversi gli studenti del quarto anno e i laureati, avevano una durata biennale. Oltre a cicli di conferenze di matematica, storia delle istituzioni scolastiche, legislazione scolastica comparata e didattica generale, essi comprendevano anche un periodo di tirocinio presso una scuola secondaria.

Nel 1906, cioè a soli quattro anni dall'entrata in vigore di queste norme, veniva emanato un nuovo regolamento che, sebbene mantenesse invariati i corsi fondamentali, negava agli studenti la possibilità di frequentare corsi tenuti in facoltà o scuole diverse da quella di appartenenza. Enriques, relativamente all'inversione di tendenza sanzionata dal regolamento del 1906, affermava:

«la maltolta libertà ha vietato ai nostri studenti di scegliere fra gli insegnamenti universitari quelli che convengono ai più diversi scopi della vita sociale, pei quali si richiedono ognora più ricche varietà di combinazioni e coordinazioni possibili, all'infuori dei programmi tradizionali» (44).

5. GLI ANNI DELL'INIZIO DEL SECOLO E DELLA «GRANDE GUERRA»

Nel primo decennio del '900 la facoltà di scienze vedeva l'avvicendamento di alcuni insegnanti: Michele Rajna (1854-1920) (45) e Federigo Guarducci (1851-1930) (46) ottenevano rispettivamente le cattedre di Astronomia e di Geodesia, mentre dal 1908 Pietro Burgatti (1868-1938) (47) era chiamato a tenere il corso di Meccanica razionale. Inoltre, in seguito alla scomparsa di Arzelà, avvenuta nel 1912, il corso di Calcolo infinitesimale veniva affidato a Pincherle mentre Enriques veniva incaricato di tenere il corso di Analisi superiore.

Nel 1915 l'Italia entrava nel conflitto europeo iniziato l'anno precedente. Per l'inaugurazione di quell'anno accademico, Pincherle teneva un discorso in cui si augurava che a «signoreggiare» nel «nuovo periodo storico che si aprirà al chiudersi delle porte del tempio di Giano» fosse la «forza del Diritto» e non il «Diritto della forza» (48), ma il «Diritto» che avrebbe «signoreggiato» di lì a qualche anno non era probabilmente quello auspicato da Pincherle.

Per ciò che riguarda la composizione del corpo docente, negli anni immediatamente seguenti la fine del conflitto avvenivano alcuni cambiamenti: Ettore Bortolotti (1866-1947) (49) era chiamato sulla cattedra di Algebra complementare e Leonida Tonelli (1885-1946) (50) su quella di Analisi superiore, mentre Enriques nel 1921, dopo quasi trent'anni di insegnamento a Bologna, veniva comandato all'Università di Roma.

Nel 1921 veniva fondata la Scuola Superiore di Chimica industriale e inoltre, presso la facoltà di scienze, veniva istituito il corso di laurea mista in scienze fisico-matematiche il cui piano degli studi era il seguente:

anno I $\begin{cases} \text{i corsi del primo anno erano gli stessi} \\ \text{del corso di laurea in matematica} \end{cases}$

anno II $\begin{cases} \text{i corsi del secondo anno erano gli stessi} \\ \text{del corso di laurea in matematica} \end{cases}$

anno III $\begin{cases} \text{Meccanica razionale} \\ \text{Elementi di Teoria delle funzioni} \\ \text{Matematiche complementari I} \\ \text{Fisica complementare I} \end{cases}$

anno IV $\begin{cases} \text{Matematiche complementari II} \\ \text{Fisica complementare II} \\ \text{Fisica matematica} \\ \text{un corso complementare} \end{cases}$

(i corsi complementari erano i seguenti: Meccanica superiore, Geometria superiore, Fisica superiore, Oscillazioni elettriche, Calcolo delle probabilità e Statistica matematica, Astronomia, Geodesia).

L'anno seguente i matematici italiani venivano coinvolti in una iniziativa di Pincherle, il quale proponeva di dar vita ad una organizzazione in grado di costituire un punto di riferimento per chi svolgeva attività di ricerca e, contemporaneamente, di rappresentare la comunità matematica italiana a livello internazionale. In seguito alle numerose adesioni raccolte, nel 1922, cioè nello stesso anno in cui Pincherle aveva illustrato il suo progetto, veniva fondata l'Unione Matematica Italiana (U.M.I.) (51). Due anni più tardi si svolgeva, a Toronto, il primo congresso dell'Unione Matematica Internazionale. Il congresso designava Pincherle presidente dell'Unione Matematica Internazionale e lo incaricava di preparare il congresso successivo.

L'insegnamento della matematica dal 1860 al 1940

Pincherle, dopo aver comunicato tale decisione all'U.M.I., costituiva un comitato il quale riusciva a superare quelle difficoltà di natura politica che avevano impedito ai matematici tedeschi ed austriaci di partecipare al congresso di Toronto. Sicchè, al secondo congresso dell'Unione Matematica Internazionale, svoltosi nel 1928 a Bologna e nella sua parte finale a Firenze, per la prima volta dopo la fine del conflitto potevano aderire le associazioni matematiche di tutte le nazioni.

6. DALLA «RIFORMA GENTILE» AL 1940

Nello stesso anno in cui l'U.M.I. iniziava la propria attività, l'interesse della comunità scientifica veniva attratto anche da altri eventi. Infatti, a solo un anno dalla «marcia su Roma», il governo fascista promuoveva la «riforma Gentile». Se, da un lato, quest'ultima pareva concedere alle Università una maggiore autonomia didattica ed economica, dall'altro imponeva un rigido controllo esercitato dal ministro della pubblica istruzione che comportava, fra l'altro, la nomina ministeriale del rettore e dei presidi di facoltà. Una disposizione contenuta nella riforma proposta da Gentile sanciva inoltre la classificazione delle Università in tre categorie. L'Ateneo bolognese apparteneva alla prima categoria che comprendeva gli Istituti totalmente a carico dello Stato. La Scuola d'Applicazione, la Scuola d'Agraria e la Scuola di Chimica industriale di Bologna appartenevano, invece, alla seconda categoria, composta da quegli Istituti parzialmente finanziati dallo Stato. L'ostilità con la quale alcuni membri della comunità scientifica accoglievano la «riforma Gentile», era motivata in parte dalla svolta autoritaria impressa dal nuovo ordinamento universitario, in parte da riserve di principio e in parte da aspetti più particolari quali, ad esempio, l'istituzione di una cattedra unica di Fisica e Matematica nei Licei. Ma erano soprattutto motivi di carattere politico quelli per cui, a pochi anni dalla sua emanazione, la «riforma Gentile» iniziava a subire gli effetti di un processo di erosione che l'avrebbe in larga misura snaturata.

Nel 1926, in seguito all'approvazione di un nuovo regolamento, a Bologna veniva proposto, per la laurea in matematica, il seguente piano degli studi:

anno I
- Analisi algebrica (con esercitazioni)
- Geometria analitica (con esercitazioni)
- Istituzioni di Geometria proiettiva e descrittiva
- Fisica sperimentale (biennale)
- Chimica generale inorganica con elementi di Chimica organica (facoltativo)
- Disegno di Ornato e Architettura (facoltativo)

anno II
- Analisi infinitesimale (con esercitazioni)
- Fisica sperimentale (con esercitazioni)
- Complementi di Analisi e di Geometria

anno III
- Meccanica razionale (con esercitazioni)
- Elementi di Teoria delle funzioni

anno IV
- Fisica matematica
- Analisi superiore
- Geometria superiore
- un corso complementare

(i corsi complementari erano i seguenti: Meccanica superiore, Analisi superiore (biennale), Geometria superiore (biennale), Fisica superiore (biennale), Oscillazioni elettriche (biennale), Matematiche complementari (biennale), Fisica complementare e misure fisiche (biennale), Calcolo delle probabilità e Statistica matematica, Astronomia, Geodesia, Storia della matematica).

Nello stesso anno si svolgeva, a Bologna, la XV riunione della Società Italiana per il Progresso delle Scienze. Alla riunione partecipava anche Mussolini, il cui intervento è stato pubblicato sull'Annuario dell'Università (52). Nel suo discorso, oltre ad esprimere alcune riflessioni di carattere «epistemologico», il capo del governo rilevava la necessità di destinare maggiori finanziamenti alla ricerca scientifica. E' da notare tuttavia che, nonostante la deferenza mostrata dal rettore Sfameni, l'intervento del duce, anzichè apparire nella parte ufficiale dell'Annuario, vi era riportato soltanto in una nota apposta al resoconto della riunione.

Dal 1931, i docenti universitari venivano obbligati a giurare fedeltà «al re, ai suoi reali successori e al regime fascista» e, dal 1934, erano istituiti i corsi di Cultura militare il cui attestato di frequenza veniva richiesto in sede di laurea.

Nel 1936 veniva emanato lo statuto dell'Università di Bologna. In base all'articolo 1 di tale statuto, l'Università bolognese veniva divisa in nove facoltà: giurisprudenza, lettere e filosofia, medicina e chirurgia, scienze matematiche, fisiche e naturali, chimica industriale, farmacia, ingegneria, agraria, medicina veterinaria. Le varie «scuole superiori» assumevano quindi la qualifica di facoltà. Ad una prima lettura si è indotti a pensare che ciò significhi una maggiore considerazione per la cultura scientifica. In effetti, in quel periodo, il regime pareva affidare alla cultura tecnico-scientifica un ruolo ben diverso da quello nel quale lo aveva relegato la «riforma Gentile». A questo nuovo atteggiamento non era certamente estranea l'approvazione delle sanzioni economiche decretate dalla Società delle Nazioni nel 1935 e la conseguente adozione di una politica economica autarchica. L'attenzione del regime si concentrava, infatti, quasi esclusivamente su quelle ricerche potenzialmente destinate ad avere, in tempi brevi, delle applicazioni nei settori dell'industria civile e militare.

Per ciò che riguarda il corso di laurea in matematica, lo statuto consigliava il seguente piano degli studi:

anno I
- Analisi matematica I
- Geometria analitica con elementi di proiettiva
- Fisica sperimentale I
- Chimica generale ed inorganica con elementi di organica

anno II
- Analisi matematica II
- Geometria descrittiva con disegno
- Fisica sperimentale II
- Meccanica razionale
- un corso complementare

anno III
- Analisi Superiore
- Geometria Superiore
- un corso complementare

anno IV
- Fisica matematica
- un corso complementare

(i corsi complementari erano i seguenti: Matematiche complementari, Teoria delle funzioni, Geometria differenziale, Fisica teorica, Meccanica superiore, Astronomia, Geodesia). Confrontando questo piano degli studi con quello contenuto nel regolamento del 1926, si possono rilevare diverse novità: i corsi di Geometria del primo biennio sono stati riorganizzati, il corso di Meccanica razionale è stato anticipato al secondo anno, il corso di Teoria delle funzioni è stato inserito fra gli insegnamenti complementari, e infine, è stata resa obbligatoria la frequenza ai corsi di Analisi superiore e di Geometria superiore.

Osservando l'appendice A si può inoltre notare come, durante gli anni '30, il corpo docente del corso di laurea in matematica a Bologna sia stato frequentemente rinnovato. I corsi d'Analisi del primo biennio, ad esempio, vennero tenuti, dopo la chiamata di Tonelli a Pisa, da Beppo Levi (1875-1961) (53) e Giuseppe Vitali (1875-1932) (54), quindi, dopo la scomparsa di quest'ultimo, da B. Levi e Luigi Fantappiè (1901-1956) (55). Nel 1934 Fantappiè lasciava l'Italia per il Brasile, per cui gli insegnamenti d'Analisi vennero affidati a B. Levi e a Beniamino Segre (1903-1977) (56) e, dall'anno successivo, al solo Levi. I corsi di Geometria vedevano alternarsi meno docenti ma risentivano dei cambiamenti introdotti col regolamento del '36. Relativamente ai corsi del secondo biennio, potevano svolgere attività didattica con continuità Burgatti, docente di Meccanica razionale e di Fisica matematica, e Segre, docente di Geometria Superiore. Levi e Segre venivano però espulsi dall'Università nel 1938 in seguito alle leggi razziali.

Queste ultime, allontanando dall'Università un cospicuo numero di docenti e compromettendo ulteriormente col regime chi rimaneva, creavano una profonda lacerazione anche all'interno della comunità scientifica. In particolare, matematici quali Ascoli, Castelnuovo, Enriques, Fubini, B. Levi, Levi-Civita, Segre, Terracini, ..., oltre ad essere espulsi dalle Università, venivano radiati anche dalle varie associazioni scientifiche. La matematica italiana perdeva quindi il contributo di numerose fra le sue personalità più importanti. In tutt'altro modo, ovviamente, il regime valutava le conseguenze delle «leggi sull'integrità della razza». Infatti, durante il II Congresso della U.M.I., tenutosi a Bologna nell'aprile del 1940, e cioè a due mesi dal coinvolgimento dell'Italia nella seconda guerra mondiale, Bottai affermava:

> «la matematica italiana, non più monopolio di Geometri di altre razze, ritrova la genialità e la poliedricità tutta sua propria, per cui furono grandi, nel clima dell'unità della Patria, i Casorati, i Brioschi, i Betti, i Cremona, i Beltrami, e riprende, con la potenza della razza purificata e liberata, il suo cammino ascensionale» (57).

Ma con gli avvenimenti che seguirono, il «cammino ascensionale» della matematica e, più in generale, della cultura italiana, doveva presto assumere tutt'altra direzione da quella auspicata da Bottai.

NOTE

(1) C. Cattaneo, Sul riordinamento degli studii scientifici in Italia, in *Il Politecnico*, XII, 1862, fasc. LXVII, pp. 61-75, ora in C. Cattaneo, *Scritti politici*, a cura di M. Boneschi, vol. III, 1965, pp. 111-129.
(2) Ibid., p. 115.
(3) Ibid., p. 115.
(4) Ibid., p. 116.
(5) Decreto-legge del governatore delle R. Provincie dell'Emilia (L.C. Farini) (ministro A. Montanari), 9 febbraio 1860.
(6) Decreto-legge del governatore delle R. Provincie dell'Emilia (L.C. Farini) (ministro A. Montanari), 8 marzo 1860.
(7) Si veda C.G. Lacaita, *Istruzione e sviluppo industriale in Italia 1859-1914*, Firenze 1973.

(8) Regio decreto 14 febbraio 1861, N. 4662 (ministro T. Mamiani), e regio decreto 21 aprile 1862, N. 602 (ministro C. Matteucci).
(9) L. Simeoni, *Storia della Università di Bologna*, vol. II, Bologna, 1940, p. 210.
(10) Ibid., p. 212.
(11) Ibid., p. 214.
(12) Decreto-legge del governatore generale delle Romagne (L. Cipriani) (ministro C. Albicini), 2 novembre 1859.
(13) Decreto-legge del governatore delle R. Provincie dell'Emilia (L.C. Farini) (ministro A. Montanari), 22 gennaio 1860.
(14) D.M. Orsetti, Le Leggi dell'Università di Bologna, in *Annuario della R. Università di Bologna*, a.a. 1897-98, Bologna, 1898, p. 365.
(15) Si tratta dei Decreti-legge citati alle note n. 5 e 6.
(16) Si veda F. Cavani, Commemorazione di Matteo Fiorini, in *Annuario della R. Scuola d'Applicazione per gli ingegneri in Bologna*, a.a. 1901-02 e 1902-03, Bologna, 1903, pp. 87-99.
(17) Si veda *Discorsi e scritti in onore di Quirico Filopanti*, Pubblicazione del comitato esecutivo per un ricordo monumentale a Quirico Filopanti, Bologna, 1898.
(18) Si veda G. Loria, Luigi Cremona et son oeuvre mathématique, in *Bibliotheca Mathematica*, ser. 3, vol. V., 1904, pp. 125-195; si veda anche L. Cremona, *Opere Matematiche*, 3 voll., Milano, 1914-15-17.
(19) Si veda E. Beltrami, Della vita e delle opere di Domenico Chelini, in (a cura di) E. Beltrami e L. Cremona, *Collectanea mathematica inedita in memoriam Dominici Chelini*, Milano, 1881, pp. I-XXXII.
(20) Alcune note biografiche di L. Respighi sono contenute in G. Armellini, Il primo centenario dell'Osservatorio del Campidoglio, *Calendario del R. Osservatorio di Roma*, Roma, 1927. Un necrologio di L. Respighi si trova in: P. Tacchini, cenno necrologico di L. Respighi, in *Atti R. Accademia dei Lincei, Rendiconti*, ser. 4, vol. VI, sem. 1, Roma, 1890, pp. 106-110.
(21) Si veda G. Loria, Eugenio Beltrami e le sue opere mathematiche, *Bibliotheca Mathematica*, ser. 3, vol. II, 1901, pp. 392-440; si veda anche E. Beltrami, *Opere Matematiche*, 4 voll., Milano, 1902-04-11-20.
(22) Si veda F Cavani, *Elogio storico di Cesare Razzaboni*, Bologna, 1899.
(23) L. Calori, Discorso del rettore, in *Annuario della R. Università di Bologna*, a.a. 1876-77, Bologna, 1877, p. 5.
(24) Ibid., p. 10.
(25) Ibid., p. 10.
(26) Si veda Vicende storiche, dati statistici complessivi ed altre generalità, in *Annuario della R. Scuola d'Applicazione per gli Ingegneri in Bologna*, a.a. 1899-1900, Bologna, 1900, pp. 3-17.
(27) Si veda F. Cavani, *Della vita e delle opere del prof. ing. Paolo F. Ruffini*, Bologna, 1908, pp. 41.
(28) L'unico riferimento biografico pare essere contenuto in F.G. Tricomi, Matematici italiani del primo secolo dello Stato unitario, in *Memorie dell'Accademia delle Scienze di Torino, classe di scienze fis. mat. e nat.*, ser. 4, vol. I, Torino, 1962, p. 98.
(29) Un breve cenno commemorativo (anonimo) di F. D'Arcais si trova in *B.U.M.I.*, vol. VII, 1928, p. 64.
(30) Si veda A. Fais, *Pagine autobiografiche*, Sassari, 1923.
(31) Si veda C. Segre, Riccardo de Paolis cenni biografici, in *Rendiconti del Circolo Matematico di Palermo*, vol. VI, Palermo, 1892, pp. 208-224.

(32) L. Calori, Relazione delle cose universitarie più notabili accadute nell'Anno Scolastico 1876-1877, in *Annuario della R. Università di Bologna*, a.a. 1877-78, Bologna, 1878, p. 4.

(33) Si veda E. Foà, Luigi Donati, in *B.U.M.I.*, vol. XI, 1932, pp. 127-128; si veda anche L. Donati, *Memorie e note scientifiche*, Bologna, 1925.

(34) Si veda S. Pincherle, Commemorazione del prof. Arzelà, in *Rendiconto della R. Accademia delle Scienze dell'Istituto di Bologna*, classe di scienze fisiche, nuova serie, vol. XVI, 1911-12, Bologna, 1912, pp. 159-179.

(35) Si veda U. Amaldi, Della vita e delle opere di Salvatore Pincherle, in S. Pincherle, *Opere scelte*, vol. I, Roma, 1954, pp. 3-16; si veda anche S. Pincherle, Notice sur les travaux, in *Acta Mathematica*, vol. XLVI, pp. 341-362, Stoccolma, 1925, ora in S. Pincherle, *Opere scelte*, vol. I, Roma, 1954, pp. 45-63.

(36) Si veda G. Castelnuovo, Federigo Enriques, in *Rendiconti dell'Accademia Nazionale dei Lincei*, ser. VIII, vol. II, 1947, ora in F. Enriques, *Memorie scelte di Geometria*, vol. I, Bologna, 1959, p.p. IX-XXII.

(37) V. Puntoni, Per la solenne apertura dell'Università. Parole del rettore Vittorio Puntoni, in *Annuario della R. Università di Bologna*, a.a. 1898-99, Bologna, 1899, p. 11.

(38) Ibid., p. 10.

(39) V. Castronovo, *L'industria italiana dall'ottocento ad oggi*, Milano, 1980, p. 53.

(40) V. Puntoni, Per la solenne apertura dell'Università. Parole del rettore Vittorio Puntoni, in *Annuario della R. Università di Bologna*, a.a. 1898-99, Bologna, 1899, p. 11.

(41) Si veda l'art. 12 del Regolamento interno della R. Scuola d'Applicazione di Bologna, in *Annuario della R. Scuola d'Applicazione per gli Ingegneri in Bologna*, aa.aa. dal 1906-07 al 1914-15, Bologna.

(42) Si veda il Regolamento speciale per la facoltà di scienze fisiche, matematiche e naturali, in *Annuario della R. Università di Bologna*, a.a. 1901-02, Bologna, 1902, pp. 331-348,; si veda anche il Regolamento generale del 1901 che è pubblicato sullo stesso annuario.

(43) C. Emery, F. Enriques, Relazione dei professori Emery ed Enriques intorno all'applicazione dell'articolo 85 del Regolamento generale, in *L'Università italiana*, vol. II, 1903, p. 4.

(44) F. Enriques, Sull'ordinamento dell'Università in rapporto alla Filosofia, in *L'Università italiana*, vol. V, 1906, p. 177.

(45) Si veda F. Guarducci, Michele Rajna, in *Annuario della R. Università di Bologna*, a.a. 1920-21, Bologna, 1921, pp. 126-128.

(46) Si veda P.D. (probabilmente Paolo Dore), Federigo Guarducci, in *Annuario della R. Università di Bologna*, a.a. 1930-31, Bologna, 1931, pp. 357-359.

(47) Si veda D. Graffi, Pietro Burgatti, in *B.U.M.I.*, vol. XVII, 1938, pp. 145-156; si veda anche P. Burgatti, *Memorie scelte*, Bologna, 1951.

(48) Si veda S. Pincherle, La Matematica e il Futuro, in *Annuario della R. Università di Bologna*, a.a. 1915-16, Bologna, 1916, pp. 61-84.

(49) Si veda E. Carruccio, Commemorazione di Ettore Bortolotti, in *Periodico di Matematica*, ser. 4, vol. XXVI, 1948, pp. 1-13.

(50) Si veda S. Cinquini, Della vita e delle opere di Leonida Tonelli, in L. Tonelli, *Opere scelte*, vol. I, Roma, 1960, pp. 1-35.

(51) Si veda C. Pucci, L'Unione Matematica Italiana dal 1922 al 1944: documenti e riflessioni, in *Symposia Mathematica*, vol. XXVII, pp. 187-212.

(52) Si veda l'*Annuario della R. Università di Bologna*, a.a. 1926-27, Bologna, 1927, pp. 26-28.

(53) Si veda A. Terracini, Commemorazione del corrispondente Beppo Levi, in *Rendiconti dell'Accademia dei Lincei*, ser. 8, vol. XXXIV, 1963, pp. 590-606.

(54) Si veda L. Pepe, Una biografia di Giuseppe Vitali, in G. Vitali, *Opere sull'Analisi reale e complessa-Carteggio*, Firenze, 1984, pp. 1-24.

(55) Si veda G. Fichera, La vita matematica di Luigi Fantappiè, in *Rendiconti di Matematica*, ser. 5, vol. XVI, Roma, 1957, pp. 140-160; si veda anche L. Fantappiè, *Opere scelte*, 1973.

(56) Si veda E. Marchionna, Ricordo di Beniamino Segre, in *Atti della Accademia delle Scienze di Torino*, n. 113, Torino, 1979, pp. 513-547, ora in B. Segre, *Opere scelte*, vol. I, 1987, pp. XXI-XXXI; si veda anche E. Vesentini, Beniamino Segre (1903-1977), in *B.U.M.I.*, vol. XV, 1978, pp. 699-714.

(57) Giuseppe Bottai, Discorso dell'Ecc. Giuseppe Bottai al II Congresso Nazionale dell'Unione Matematica Italiana, in *Annuario della R. Università di Bologna*, a.a. 1940-41, Bologna, 1941, p. 108.

APPENDICE A

La presente appendice illustra gli insegnamenti del corso di laurea in matematica tenuti presso la Facoltà di Scienze dell'Università di Bologna dal 1860 al 1940. Di ogni insegnamento è specificato il docente e, fra parentesi, il tipo di qualifica di quest'ultimo. Allo scopo di facilitare la consultazione, si tengano presenti le seguenti norme:

1) se di un corso le tabelle non riportano il nome del docente, significa che, pur essendo nel piano degli studi, quel corso non veniva tenuto;

2) gli annuari dell'Università, in numerose occasioni, specificano che i corsi di esercitazioni o di disegno relativi ad un determinato insegnamento erano stati tenuti da assistenti. Ove non si è certi della qualifica dell'insegnante che ha tenuto il corso di esercitazioni o di disegno nelle tabelle appare il simbolo «(?)»;

3) poichè dall'a.a. 1925-26 i docenti dei corsi di Analisi del primo biennio si alternavano, la qualifica viene indicata solo quando l'insegnante teneva il corso di cui era titolare.

La compilazione della presente appendice è stata effettuata consultando essenzialmente gli annuari della Università di Bologna. Per indicare la qualifica dei docenti nei diversi periodi è necessario, nella lettura delle tabelle, utilizzare le seguenti legende:

legenda n. 1 (vale dall'a.a. 1859-60 all'a.a. 1873-74)

p.i.c. = professore insegnante collegiato
p.i. = professore insegnante (non collegiato)
straord. = professore straordinario
inc. = professore incaricato
ass. = assistente

legenda n. 2 (vale dall'a.a. 1874-75 all'a.a. 1923-24)

ord. coll. = professore ordinario collegiato
ord. = professore ordinario (non collegiato)
straord. = professore straordinario
inc. = professore incaricato
ass. = assistente

legenda n. 3 (vale dall'a.a. 1924-25 all'a.a. 1930-31)

p.s.c. = professore stabile collegiato
p.s. = professore stabile (non collegiato)
p.n.s. = professore non stabile
inc. = professore incaricato
ass. = assistente

legenda n. 4 (vale dall'a.a. 1931-32 all'a.a. 1939-40)

ord. coll. = professore ordinario collegiato
ord. = professore ordinario (non collegiato)
straord. = professore straordinario
inc. = professore incaricato
ass. = assistente.

Si ricorda, inoltre, che gli annuari compilati durante la prima guerra mondiale e negli anni immediatamente successivi non contengono le informazioni necessarie per la stesura delle tabelle relative agli aa. aa. compresi fra il 1916 ed il 1920.

	1859-'60
Storia naturale Fisica sperimentale Archittettura civile Introduzione al Calcolo	*Bianconi G.G. (p.i.c.)* *Della Casa L. (pic.)* *Lodi F. (?)* *Ramenghi S.(p.i.c.)*
Meccanica ed Idraulica Ottica ed Astronomia Calcolo sublime Agronomia teorico-pratica	*Chelini D. (p.i.c.)* *Respighi L. (p.i.c.)* *Saporetti A. (p.i.c.)* *Botter L.F. (p.i.c.)*
Meccanica ed Idraulica Ottica ed Astronomia Istituzioni civili Agronomia teorico-pratica	*Chelini D. (p.i.c.)* *Respighi L. (p.i.c.)* *Giusti E. (p.i.)* *Botter L.F. (p.i.c.)*
Economia pubblica Agronomia teorico-pratica Pratica nello studio di un ingegnere approvato	*Mariscotti A. (p.i.)* *Botter L.F. (p.i.c.)*
Agronomia teorico-pratica Pratica nello studio di un ingegnere approvato Pratica nello studio di un architetto o di un idraulico approvato	*Botter L.F. (p.i.c.)*

	1860-'61
Introduzione al Calcolo Fisica sperimentale Chimica generale Architettura civile	*Ramenghi S. (p.i.c.)* *Della Casa L. (pic.)* *Santagata D. (p.i.)* *Lodi F. (p.i.c.)*
Calcolo sublime Chimica metallurgica Mineralogia Architettura civile	*Saporetti A. (p.i.c.)* *Tassinari P. (?)* *Lodi F. (p.i.c.)*
Meccanica ed Idraulica Astronomia Geometria Superiore Geologia Agronomia teorico-pratica	*Chelini D. (p.i.c.)* *Cremona L. (p.i.)* *Capellini G. (p.i.)* *Botter L.F. (p.i.c.)*
Ottica Meccanica applicata Agronomia teorico-pratica Pratica presso un ingegnere approvato	*Respighi L. (p.i.c.)* *Filopanti Q. (p.i.)* *Botter L.F. (p.i.c.)*
Economia pubblica Geodesia teorico-pratica Agronomia teorico-pratica Pratica presso un ingegnere approvato	*Mariscotti A. (p.i.)* *Fiorini M. (p.i.)* *Botter L.F. (p.i.c.)*

	1861-'62
Introduzione al Calcolo Fisica sperimentale Chimica generale Architettura civile	*Ramenghi S. (p.i.c.)* *Della Casa L. (p.i.c.)* *Santagata D. (p.i.)* *Lodi F. (p.i.c.)*
Calcolo sublime Meccanica razionale Mineralogia Architettura civile	*Saporetti A. (p.i.c.)* *Chelini D. (p.i.c.)* *Bombicci L. (p.i.)* *Lodi F. (p.i.c.)*
Meccanica razionale Astronomia Geometria superiore e Geom. descrittiva Geologia	*Chelini D. (p.i.c.)* *Respighi L. (p.i.c.)* *Cremona L. (p.i.)* *Capellini G. (p.i.)*
Meccanica applicata Agronomia teorico-pratica Chimica mineralogica ed analitica Pratica presso un ingegnere architetto approvato	*Filopanti Q. (p.i.)* *Botter L.F. (p.i.c.)*
Geodesia teorico-pratica Agronomia teorico-pratica Pratica presso un ingegnere architetto approvato	*Fiorini M. (p.i.)* *Botter L.F. (p.i.c.)*

	1862-'63	1863-'64	1864-'65	1865-'66
Algebra complementare (1)	Beltrami E. (straord.)	Boschi P. (straord.)	Boschi P. (straord.)	Boschi P. (straord.)
Geometria analitica (1)	Cremona L. (inc.)	Boschi P. (straord.)	Boschi P. (straord.)	Boschi P. (straord.)
Chimica inorganica	Santagata D. (p.i.)	Santagata D.. (p.i.)	Santagata D. (p.i.)	Santagata D. (p.i.)
Disegno	Lodi F. (p.i.c.)	Lodi F. (p.i.c.)	Lodi F. (p.i.c.)	Lodi F. (p.i.c.)
Calcolo differenziale e integrale	Saporetti A. (p.i.c.)	Saporetti A. (p.i.c.)	Saporetti A. (p.i.c.)	Saporetti A. (p.i.c.)
Geometria descrittiva	Cremona L. (inc.)	Cremona L. (p.i.)	Cremona L. (inc.)	Cremona L. (inc.)
Fisica (2)	Della Casa L. (p.i.c.)	Della Casa L. (p.i.c.)	Della Casa L. (p.i.c.)	Della Casa L. (p.i.c.)
Disegno	Lodi F. (p.i.c.)	Lodi F. (p.i.c.)	Lodi F. (p.i.c.)	Lodi F. (p.i.c.)
Meccanica razionale	Chelini D. (p.i.c.)	Chelini D. (p.i.c.)	Cremona L. (inc.)	Venturi L. (straord.)
Geodesia teoretica	Fiorini M. (p.i.)	Fiorini M. (p.i.)	Fiorini M. (p.i.)	Fiorini M. (p.i.)
Fisica (2)	Della Casa L. (p.i.c.)	Della Casa L. (p.i.c.)	Della Casa L. (p.i.c.)	Della Casa L. (p.i.c.)
Disegno	Lodi F. (p.i.c.)	Lodi F. (p.i.c.)	Lodi F. (p.i.c.)	Lodi F. (p.i.c.)
Analisi e Geometria Superiore	Cremona L. (p.i.)	Cremona L. (p.i.)	Cremona L. (p.i.c.)	Cremona L. (p.i.c.)
Astronomia e Meccanica celeste	Respighi L. (p.i.c.)	Respighi L. (p.i.c.)		
Fisica matematica				

(1) Pur essendo distinti, i corsi di Algebra complementare e di Geometria analitica corrispondevano ad una unica cattedra.
(2) Dal 1865 gli studenti che intendevano proseguire gli studi alle Scuole d'Applicazione potevano frequentare i corsi di Fisica al primo e al secondo anno sostenendone l'esame al termine del secondo.

	1866-'67	1867-'68	1868-'69
Algebra complementare (1)	Boschi P. (straord.)	Boschi P. (straord.)	Boschi P. (straord.)
Geometria analitica (1)	Boschi P. (straord.)	Boschi P. (straord.)	Boschi P. (straord.)
Chimica inorganica	Santagata D. (p.i.)	Santagata D. (p.i.)	Santagata D. (p.i.)
Disegno	Lodi F. (p.i.c.)	Lodi F. (p.i.c.)	Lodi F. (p.i.c.)
Calcolo differenziale e integrale	Saporetti A. (p.i.c.)	Saporetti A. (p.i.c.)	Saporetti A. (p.i.c.)
Geometria descrittiva	Beltrami E. (inc.)	Beltrami E. (inc.)	Boschi P. (inc.)
Fisica (2)	Della Casa L. (p.i.c.)	Della Casa L. (p.i.c.)	Della Casa L. (p.i.c.)
Disegno	Lodi F. (p.i.c.)	Lodi F. (p.i.c.)	Lodi F. (p.i.c.)
Meccanica razionale	Beltrami E. (p.i.)	Beltrami E. (p.i.)	Beltrami E. (p.i.c.)
Geodesia teoretica	Fiorini M. (p.i.)	Fiorini M. (p.i.)	Fiorini M. (p.i.)
Fisica (2)	Della Casa L. (p.i.c.)	Della Casa L. (p.i.c.)	Della Casa L. (p.i.c.)
Disegno	Lodi F. (p.i.c.)	Lodi F. (p.i.c.)	Lodi F. (p.i.c.)
Analisi e Geometria Superiore (3)	Cremona L. (p.i.)	Cremona L. (p.i.)	Cremona L. (p.i.)
Astronomia e Meccanica celeste			
Fisica matematica			

(1) Pur essendo distinti, i corsi di Algebra complementare e di Geometria analitica corrispondevano ad una unica cattedra.
(2) Gli studenti che intendevano proseguire gli studi alla Scuola d'Applicazione potevano frequentare i corsi di Fisica al primo e al secondo anno sostenendone l'esame al termine del secondo.
(3) Dal 1866 L. Cremona veniva comandato presso l'Istituto Tecnico Superiore di Milano.

	1869-'70	1870-'71	1871-'72
Algebra complementare (1)	*Boschi P.* *(straord.)*	*Boschi P.* *(straord.)*	*Boschi P.* *(straord.)*
Geometria analitica (1)	*Boschi P.* *(straord.)*	*Boschi P.* *(straord.)*	*Boschi P.* *(straord.)*
Chimica inorganica	*Santagata D.* *(p.i.)*	*Santagata D.* *(p.i.)*	*Santagata D.* *(p.i.)*
Disegno	*Lodi F.* *(p.i.c.)*	*Lodi F.* *(p.i.c.)*	*Lodi F.* *(p.i.c.)*
Calcolo differenziale e integrale	*Saporetti A.* *(p.i.c.)*	*Saporetti A.* *(p.i.c.)*	*Saporetti A.* *(p.i.c.)*
Geometria descrittiva	*Boschi P.* *(inc.)*	*Boschi P.* *(inc.)*	*Boschi P.* *(inc.)*
Fisica (2)	*Della Casa L.* *(p.i.c.)*	*Della Casa L.* *(p.i.c.)*	*Villari E.* *(p.i.)*
Disegno	*Lodi F.* *(p.i.c.)*	*Lodi F.* *(p.i.c.)*	*Lodi F.* *(p.i.c.)*
Meccanica razionale	*Beltrami E.* *(p.i.c.)*	*Beltrami E.* *(p.i.c.)*	*Beltrami E.* *(p.i.c.)*
Geodesia teoretica	*Fiorini M.* *(p.i.)*	*Fiorini M.* *(p.i.)*	*Fiorini M.* *(p.i.)*
Fisica (2)	*Della Casa L.* *(p.i.c.)*	*Della Casa L.* *(p.i.c.)*	*Villari E.* *(p.i.)*
Disegno	*Lodi F.* *(p.i.c.)*	*Lodi F.* *(p.i.c.)*	*Lodi F.* *(p.i.c.)*
Analisi e Geometria superiore (3)	*Cremona L.* *(p.i.c.)*	*Cremona L.* *(p.i.c.)*	*Cremona L.* *(p.i.c.)*
Astronomia e Meccanica celeste			
Fisica matematica			

(1) Pur essendo distinti, i corsi di Algebra complementare e di Geometria analitica corrispondevano ad una unica cattedra.
(2) Gli studenti che intendevano proseguire gli studi alla Scuola d'Applicazione potevano frequentare i corsi di Fisica al primo e al secondo anno sostenendone l'esame al termine del secondo.
(3) Sino al 1872 L. Cremona veniva comandato presso l'Istituto Tecnico Superiore di Milano.

	1872-'73	1873-'74	1874-'75
Algebra complementare (1)	*Boschi P.* *(straord.)*	*Boschi P.* *(straord.)*	*Boschi P.* *(straord.)*
Geometria analitica (1)	*Boschi P.* *(straord.)*	*Boschi P.* *(straord.)*	*Boschi P.* *(straord.)*
Chimica inorganica	*Santagata D.* *(p.i.)*	*Santagata D.* *(p.i.)*	*Santagata D.* *(ord.)*
Disegno	*Lodi F.* *(p.i.c.)*	*Lodi F.* *(p.i.c.)*	*Lodi F.* *(ord. coll.)*
Calcolo differenziale e integrale	*Saporetti A.* *(p.i.c.)*	*Saporetti A.* *(p.i.c.)*	*Saporetti A.* *(ord. coll.)*
Geometria descrittiva	*Boschi P.* *(inc.)*	*Boschi P.* *(inc.)*	*Boschi P.* *(inc.)*
Fisica (2)	*Villari E.* *(p.i.)*	*Villari E.* *(p.i.)*	*Villari E.* *(ord. coll.)*
Disegno	*Lodi F.* *(p.i.c.)*	*Lodi F.* *(p.i.c.)*	*Lodi F.* *(ord. coll.)*
Meccanica razionale	*Beltrami E.* *(p.i.c.)*	*Razzaboni C.* *(p.i.)*	
Geodesia teoretica	*Fiorini M.* *(p.i.)*	*Fiorini M.* *(p.i.)*	*Fiorini M.* *(ord. coll.)*
Fisica (2)	*Villari E.* *(p.i.)*	*Villari E.* *(p.i.)*	*Villari E.* *(ord. coll)*
Disegno	*Lodi F.* *(p.i.c.)*	*Lodi F.* *(p.i.c.)*	*Lodi F.* *(ord. coll.)*
Analisi e Geometria superiore			
Astronomia e Meccanica celeste			
Fisica matematica			

(1) Pur essendo distinti, i corsi di Algebra complementare e di Geometria analitica corrispondevano ad una unica cattedra.
(2) Gli studenti che intendevano proseguire gli studi alla Scuola d'Applicazione potevano frequentare i corsi di Fisica al primo e al secondo anno sostenendone l'esame al termine del secondo.

	1875-'76
Analisi algebrica (1) (2)	*D'Arcais F. (straord.)*
Geometria analitica (1) (2)	*D'Arcais F. (straord.)*
Geometria proiettiva e descrittiva (2)	*Boschi P. (straord.)*
Disegno di Geometria proiettiva e descrittiva	*Boschi P.*
Disegno di Ornato e di Architettura elementare	*Lodi F. (ord. coll.)*
Analisi infinitesimale (2)	*Saporetti A. (ord. coll.)*
Geometria proiettiva e descrittiva (2)	*Boschi P. (straord.)*
Fisica sperimentale (2)	*Villari E. (ord. coll.)*
Chimica inorganica (2)	*Santagata D. (ord.)*
Disegno di Geometria proiettiva e descrittiva	*Boschi P.*
Disegno di Ornato e di Architettura elementare	*Lodi F. (ord. coll.)*
Geodesia teoretica	
Meccanica razionale	
Astronomia	
Fisica matematica	

(1) Pur essendo distinti i corsi di Algebra e di Geometria analitica corrispondevano ad una unica cattedra.
(2) Questi corsi erano necessari per conseguire la licenza fisico-matematica. Quest'ultima permetteva di proseguire gli studi di Fisica e di Matematica all'interno della Facoltà, oppure, «insieme coi certificati di diligenza» degli altri corsi permetteva l'iscrizione alla Scuola d'Applicazione.

	1876-'77
Algebra (1) (2)	*Fais A. (straord.)*
Geometria analitica (1) (2)	*Fais A. (straord.)*
Geometria proiettiva	*Boschi P. (straord.)*
Disegno di Geometria proiettiva	*Lodi F. (?)*
Disegno di Ornato	*Lodi F. (ord. coll.)*
Mineralogia	*Bombicci L. (ord.)*
Calcolo infinitesimale (2)	*Saporetti A. (ord. coll.)*
Geometria descrittiva (2)	*Boschi P. (straord.)*
Fisica (2)	*Villari E. (ord. coll.)*
Chimica inorganica (2)	*Santagata D. (ord.)*
Disegno di Geometria descrittiva	*Lodi F. (?)*
Disegno di Architettura	*Lodi F. (ord. coll.)*
Mineralogia	*Bombicci L. (ord.)*
Geologia	*Capellini G. (ord.)*
Disegno di Ornato	*Lodi F. (ord. coll.)*
Geodesia teoretica Meccanica razionale Astronomia Fisica matematica	

(1) Pur essendo distinti i corsi di Algebra e di Geometria analitica corrispondevano ad una unica cattedra.
(2) Questi corsi erano necessari per conseguire la licenza fisico-matematica. Quest'ultima permetteva di proseguire gli studi di Fisica e di Matematica all'interno della Facoltà, oppure, «insieme coi certificati di diligenza» degli altri corsi permetteva l'iscrizione alla Scuola d'Applicazione.

	1877-'78
Algebra (1) (2)	*Donati L. (inc.)*
Geometria analitica (1) (2)	*Ruffini F. (inc.)*
Geometria proiettiva (2)	*Boschi P. (ord.)*
Disegno di Geometria proiettiva	*Boschi P.*
Disegno di Ornato	*Lodi F. (ord. coll.)*
Mineralogia	*Bombicci L. (ord.)*
Chimica inorganica (2)	*Santagata D. (ord.)*
Fisica (2)	*Villari E. (ord. coll.)*
Calcolo infinitesimale (2)	*Fais A. (straord.)*
Geometria descrittiva (2)	*Boschi P. (ord.)*
Fisica (2)	*Villari E. (ord. coll.)*
Disegno di Geometria descrittiva	*Boschi P.*
Mineralogia	*Bombicci L. (ord.)*
Geologia	*Capellini G. (ord.)*
Disegno di Ornato	*Lodi F. (ord. coll.)*
Geodesia teorica	*Fiorini M. (ord. coll.)*
Meccanica razionale	*Ruffini F. (inc.)*
Astronomia	*Saporetti A. (ord. coll.)*
Fisica matematica	

(1) Pur essendo distinti i corsi di Algebra e di Geometria analitica corrispondevano ad una unica cattedra.

(2) Questi corsi erano necessari per conseguire la licenza fisico-matematica. Quest'ultima permetteva di proseguire gli studi di Fisica e di Matematica all'interno della Facoltà, oppure, «coi certificati di diligenza» degli altri corsi permetteva l'iscrizione alla Scuola d'Applicazione.

	1878-'79
Algebra (1) (2)	*De Paolis R. (straord.)*
Geometria analitica (1) (2)	*De Paolis R. (straord.)*
Geometria proiettiva (2)	*Boschi P. (ord.)*
Disegno di Geometria proiettiva	*Boschi P.*
Disegno di Ornato	*Lodi F. (ord. coll.)*
Mineralogia	*Bombicci L. (ord.)*
Chimica inorganica (2)	*Santagata D. (ord.)*
Fisica (2)	*Villari E. (ord. coll.)*
Calcolo infinitesimale (2)	*Fais A. (straord.)*
Geometria descrittiva (2)	*Boschi P. (ord.)*
Disegno di Geometria descrittiva	*Boschi P.*
Geologia	*Capellini G. (ord.)*
Disegno di Ornato	*Lodi F. (ord. coll.)*
Geodesia teoretica	*Fiorini M. (ord. coll.)*
Meccanica razionale	*Ruffini F. (inc.)*
Astronomia	*Saporetti A. (ord. coll.)*
Fisica matematica	

(1) Pur essendo distinti i corsi di Algebra e di Geometria analitica corrispondevano ad una unica cattedra.
(2) Questi corsi erano necessari per conseguire la licenza fisico-matematica. Quest'ultima permetteva di proseguire gli studi di Fisica e di Matematica all'interno della Facoltà, oppure, «coi certificati di diligenza» degli altri corsi permetteva l'iscrizione alla Scuola d'Applicazione.

	1879-'80	1880-'81
Algebra e Geometria analitica (1)	De Paolis R. (straord.)	Arzelà C. (inc.)
Geometria proiettiva (1)	Boschi P. (ord.)	Boschi P. (ord.)
Disegno di Geometria proiettiva	Boschi P.	Boschi P.
Disegno di Ornato e di Architettura	Lodi F. (ord. coll.)	Lodi F. (ord. coll.)
Mineralogia		
Chimica inorganica (1)	Santagata D. (ord.)	Santagata D. (ord.)
Fisica (1)	Villari E. (ord. coll.)	Villari E. (ord. coll.)
Calcolo infinitesimale (1)	Fais A. (straord.)	Arzelà C. (straord.)
Geometria descrittiva (1)	Boschi P. (ord.)	Boschi P. (ord.)
Disegno di Geometria descrittiva	Boschi P.	Boschi P.
Disegno di Architettura	Lodi F. (ord. coll.)	Lodi F. (ord. coll.)
Geologia	Capellini G. (ord.)	Capellini G. (ord.)
Geodesia teoretica	Fiorini M. (ord. coll.)	Fiorini M. (ord. coll.)
Meccanica razionale	Ruffini F. (ord.)	Ruffini F. (ord.)
Astronomia	Saporetti A. (ord. coll.)	Saporetti A. (ord. coll.)
Fisica matematica	Donati L. (straord.)	Donati L. (straord.)

(1) Questi corsi erano necessari per conseguire la licenza fisico-matematica. Quest'ultima permetteva di proseguire gli studi in Fisica o in Matematica all'interno della Facoltà, oppure «insieme coi certificati di diligenza» degli altri corsi permetteva la iscrizione alla Scuola di Applicazione.

	1881-'82
Algebra (1) (2)	*Pincherle S. (straord.)*
Geometria analitica (1) (2)	*Pincherle S. (straord.)*
Geometria proiettiva e Disegno (2)	*Boschi P. (ord.)*
Chimica inorganica (2)	*Santagata D. (ord.)*
Fisica (2)	*Villari E. (ord. coll.)*
Mineralogia	*Bombicci L. (ord. coll.)*
Disegno di Ornato	*Lodi F. (ord. coll.)*
Geometria descrittiva e Disegno (2)	*Boschi P. (ord.)*
Disegno di Architettura	*Lodi F. (ord. coll.)*
Geologia	*Capellini G. (ord. coll.)*
Calcolo infinitesimale (2)	*Arzelà C. (straord.)*
Geodesia teoretica	*Fiorini M. (ord. coll.)*
Meccanica razionale	*Ruffini M. (ord.)*
Geometria superiore	*Pincherle S. (inc.)*
Fisica matematica	*Donati L. (straord.)*
Astronomia	*Saporetti A. (ord. coll.)*
Analisi superiore	*Arzelà C. (inc.)*

(1) Pur essendo distinti i corsi di Algebra e di Geometria analitica corrispondevano ad una unica cattedra.
(2) Questi corsi erano necessari per conseguire la licenza fisico-matematica. Quest'ultima permetteva di proseguire gli studi in Fisica o in Matematica all'interno della Facoltà, oppure, «insieme coi certificati di diligenza» degli altri corsi permetteva l'iscrizione alla Scuola d'Applicazione.

	1882-'83	1883-'84
Algebra (1) (2)	*Pincherle S.* (straord.)	*Pincherle S.* (straord.)
Geometria analitica (1) (2)	*Pincherle S.* (straord.)	*Pincherle S.* (straord.)
Geometria proiettiva (2) e Disegno	*Boschi P.* (ord.)	*Boschi P.* (ord. coll.)
Chimica inorganica (2)	*Santagata D.* (ord.)	*Santagata D.* (ord.)
Fisica (2)	*Villari E.* (ord. coll.)	*Villari E.* (ord. coll.)
Mineralogia	*Bombicci L.* (ord. coll.)	*Bombicci L.* (ord. coll.)
Disegno di Ornato	*Luminasi G.* (inc.)	*Zannoni A.* (inc.)
Esercizi di Algebra e Geometria analitica	*Pesci G.* (ass.)	*Pesci G.* (ass.)
Geometria descrittiva e Disegno (2)	*Boschi P.* (ord.)	*Boschi P.* (ord. coll.)
Disegno di Archittettura	*Luminasi G.* (inc.)	*Zannoni A.* (inc.)
Geologia	*Capellini G.* (ord. coll.)	*Capellini G.* (ord. coll.)
Calcolo infinitesimale (2)	*Arzelà C.* (straord.)	*Arzelà C.* (straord.)
Esercizi di Calcolo infinitesimale	*Pesci G.* (ass.)	*Pesci G.* (ass.)
Geodesia teoretica	*Fiorini M.* (ord. coll.)	*Fiorini M.* (ord. coll.)
Meccanica razionale	*Ruffini F.* (ord.)	*Ruffini F.* (ord. coll.)
Geometria superiore	*Pincherle S.* (inc.)	*Pincherle S.* (inc.)
Fisica matematica	*Donati L.* (straord.)	*Donati L.* (straord.)
Astronomia	*Saporetti A.* (ord. coll.)	*Saporetti A.* (ord. coll.)
Analisi superiore	*Arzelà C..* (inc.)	*Arzelà C.* (inc.)

(1) Pur essendo distinti i corsi di Algebra e di Geometria analitica corrispondevano ad una unica cattedra.
(2) Questi corsi erano necessari per conseguire la licenza fisico-matematica. Quest'ultima permetteva di proseguire gli studi in Fisica o in Matematica all'interno della Facoltà, oppure, «insieme coi certificati di diligenza» degli altri corsi permetteva la iscrizione alla Scuola d'Applicazione.

	1884-'85	1885-'86	1886-'87
Algebra (1) (2)	Pincherle S. (straord.)	Pincherle S. (straord.)	Pincherle S. (straord.)
Geometria analitica (1) (2)	Pincherle S. (straord.)	Pincherle S. (straord.)	Pincherle S. (straord.)
Geometria proiettiva e disegno (2)	Boschi P. (ord. coll.)	Boschi P. (ord. coll.)	Boschi P. (ord. coll.)
Chimica inorganica (2)	Santagata D. (ord.)	Santagata D. (ord.)	Cavazzi A. (inc.)
Fisica (2)	Villari E. (ord. coll.)	Villari E. (ord. coll.)	Villari E. (ord. coll.)
Disegno di Ornato	Zannoni A. (inc.)	Zannoni A. (inc.)	Zannoni A. (inc.)
Esercizi di Algebra e Geometria analitica	Pesci G. (ass.)	Pesci G. (ass.)	Pesci G. (ass.)
Geometria descrittiva e Disegno (2)	Boschi P. (ord. coll.)	Boschi P. (ord. coll.)	Boschi P. (ord. coll.)
Disegno di Architettura	Zannoni A. (inc.)	Zannoni A. (inc.)	Zannoni A. (inc.)
Mineralogia	Bombicci L. (ord. coll.)	Bombicci L. (ord. coll.)	Bombicci L. (ord. coll.)
Calcolo infinitesimale (2)	Arzelà C. (straord.)	Arzelà C. (straord.)	Arzelà C. (straord.)
Esercizi di Calcolo infinitesimale	Pesci G. (ass.)	Pesci G. (ass.)	Pesci G. (ass.)
Geodesia teorica	Fiorini M. (ord. coll.)	Fiorini M. (ord. coll.)	Fiorini M. (ord. coll.)
Meccanica razionale	Ruffini F. (ord. coll.)	Ruffini F. (ord. coll.)	Ruffini F. (ord. coll)
Geometria superiore	Pincherle S. (inc.)	Pincherle S. (inc.)	Pincherle S. (inc.)
Fisica matematica	Donati L. (straord.)	Donati L. (straord.)	Donati L. (straord.)
Astronomia	Saporetti A. (ord. coll.)	Saporetti A. (ord. coll.)	Saporetti A. (ord. coll.)
Analisi superiore	Arzelà C. (inc.)	Arzelà C. (inc.)	Arzelà C. (inc.)

(1) Pur essendo distinti, i corsi di Algebra e di Geometria analitica corrispondevano ad una unica cattedra.
(2) Questi corsi erano necessari per conseguire la licenza fisico-matematica. Quest'ultima permetteva di proseguire gli studi in Fisica o in Matematica all'interno della Facoltà, oppure, «insieme coi certificati di diligenza» degli altri corsi permetteva la iscrizione alla Scuola d'Applicazione.

	1887-'88	1888-'89
Algebra (1) (2)	Pincherle S. (straord.)	Pincherle S. (ord.)
Geometria analitica (1) (2)	Pincherle S. (straord.)	Pincherle S. (ord.)
Geometria proiettiva e Disegno (2)	Razzaboni A. (inc.)	Montesano D. (?)
Chimica inorganica (2)	Santagata D. (ord.)	Cavazzi A. (inc.)
Fisica (2)	Villari E. (ord. coll.)	Villari E. (ord. coll.)
Disegno di Ornato	Zannoni A. (inc.)	Zannoni A. (inc.)
Esercizi di Algebra e Geometria analitica		Bortolotti E. (ass.)
Geometria descrittiva e Disegno (2)	Razzaboni A. (inc.)	Montesano D. (?)
Disegno di Architettura	Zannoni A. (inc.)	Zannoni A. (inc.)
Mineralogia	Bombicci L. (ord. coll.)	Bombicci L. (ord. coll.)
Calcolo infinitesimale (2)	Arzelà C. (ord. coll.)	Arzelà C. (ord. coll.)
Esercizi di Calcolo infinitesimale		Ingrami G. (ass.)
Geodesia teoretica	Fiorini M. (ord. coll.)	Fiorini M. (ord. coll.)
Meccanica razionale	Ruffini F. (ord. coll.)	Ruffini F. (ord. coll)
Geometria superiore	Pincherle S. (inc.)	Pincherle S. (inc.)
Fisica matematica	Donati L. (straord.)	Donati L. (straord.)
Astronomia	Saporetti A. (ord. coll.)	Saporetti A. (ord. coll.)
Analisi superiore	Arzelà C. (inc.)	Arzelà C. (inc.)

(1) Pur essendo distinti, i corsi di Algebra e di Geometria analitica corrispondevano ad una unica cattedra.
(2) Questi corsi erano necessari per conseguire la licenza fisico-matematica. Quest'ultima permetteva di proseguire gli studi in Fisica o in Matematica all'interno della Facoltà, oppure, «insieme coi certificati di diligenza» degli altri corsi permetteva la iscrizione alla Scuola d'Applicazione.

	1889-'90
Algebra (1) (2)	*Pincherle S. (ord.)*
Geometria analitica (1) (2)	*Pincherle S. (ord.)*
Geometria proiettiva lezione (2)	*Montesano D. (straord.)*
Geometria proiettiva disegno (2)	*Montesano D.*
Chimica generale (2)	*Ciamician G. (ord.)*
Fisica (2)	*Righi A. (ord.)*
Disegno di Ornato e d'Archittettura	*Zannoni A. (inc.)*
Esercizi di Algebra e di Geometria	*Bortolotti Ett. (ass.)*
Geometria descrittiva lezione (2)	*Montesano D. (straord.)*
Geometria descrittiva Disegno (2)	*Montesano D.*
Disegno d'Ornato e d'Archittettura (2)	*Zannoni A. (inc.)*
Calcolo infinitesimale (2)	*Arzelà C. (ord. coll.)*
Esercizi di Calcolo infinitesimale	*Cicognani L. (ass.)*
Mineralogia (2)	*Bombicci L. (ord. coll.)*
Geodesia teoretica	*Fiorini M. (ord. coll.)*
Meccanica razionale	*Ruffini F. (ord. coll.)*
Matematiche superiori (corso di)	*Arzelà C. (inc.)*
" " (corso di)	*Pincherle S. (inc.)*
Fisica matematica	*Donati L. (straord.)*
Astronomia	*Saporetti A. (ord. coll.)*
Matematiche superiori (corso di)	*Arzelà C. (inc.)*
" " (corso di)	*Pincherle S. (inc.)*

(1) I corsi di Algebra e Geometria analitica, pur essendo distinti, corrispondevano ad una unica cattedra.
(2) Questi corsi erano quelli necessari per ottenere la licenza fisico-matematica. Quest'ultima permetteva di proseguire gli studi di fisica e di matematica presso l'Università, oppure, «insieme coi certificati di diligenza» degli altri corsi, permetteva l'iscrizione alle Scuole Superiori d'Applicazione.

	1890-'91	1891-'92	1892-'93
Algebra (1)	*Pincherle S.* *(ord.)*	*Pincherle S.* *(ord.)*	*Pincherle S.* *(ord.)*
Geometria analitica (1)	*Pincherle S.* *(ord.)*	*Pincherle S.* *(ord.)*	*Pincherle S.* *(ord.)*
Geometria proiettiva	*Montesano D.* *(straord.)*	*Montesano D.* *(straord.)*	*Montesano D.* *(straord.)*
Chimica generale	*Ciamician G.* *(ord. coll.)*	*Ciamician G.* *(ord. coll.)*	*Ciamician G.* *(ord. coll.)*
Fisica sperimentale	*Righi A.* *(ord. coll.)*	*Righi A.* *(ord. coll.)*	*Righi A.* *(ord. coll.)*
Disegno di Ornato e d'Architettura (3)	*Zannoni A.* *(inc.)*	*Zannoni A.* *(inc.)*	*Zannoni A.* *(inc.)*
Disegno di Geometria proiettiva	*Montesano D.*	*Montesano D.*	*Montesano D.*
Geometria descrittiva	*Montesano D.* *(straord.)*	*Montesano D.* *(straord.)*	*Montesano D.* *(straord.)*
Disegno di Ornato e d'Architettura (3)	*Zannoni A.* *(inc.)*	*Zannoni A.* *(inc.)*	*Zannoni A.* *(inc.)*
Calcolo infinitesimale	*Arzelà C.* *(ord. coll.)*	*Arzelà C.* *(ord. coll.)*	*Arzelà C.* *(ord. coll.)*
Disegno di Geometria descrittiva	*Montesano D.*	*Montesano D.*	*Montesano D.*
Mineralogia (3)	*Bombicci L.* *(ord. coll.)*	*Bombicci L.* *(ord. coll.)*	*Bombicci L.* *(ord. coll.)*
Fisica sperimentale (2)	*Righi A.* *(ord. coll.)*	*Righi A.* *(ord. coll.)*	*Righi A.* *(ord. coll.)*
Geodesia teoretica	*Fiorini M.* *(ord. coll.)*	*Fiorini M.* *(ord. coll.)*	*Fiorini M.* *(ord. coll.)*
Meccanica razionale	*Ruffini F.* *(ord. coll.)*	*Ruffini F.* *(ord. coll.)*	*Ruffini F.* *(ord. coll)*
Matematiche superiori	*Arzelà C. (inc.)*	*Arzelà C. (inc.)*	*Arzelà C. (inc.)*
Matematiche superiori	*Pincherle S. (inc.)*	*Pincherle S. (inc.)*	*Pincherle S. (inc.)*
Fisica matematica	*Donati L.* *(straord.)*	*Donati L.* *(straord.)*	*Donati L.* *(straord.)*
Astronomia	*Saporetti A.* *(ord. coll.)*	*Saporetti A.* *(ord. coll.)*	*Saporetti A.* *(ord. coll.)*
Matematiche superiori	*Arzelà C. (inc.)*	*Arzelà C. (inc.)*	*Arzelà C. (inc.)*
Matematiche superiori	*Pincherle S. (inc.)*	*Pincherle S. (inc.)*	*Pincherle S. (inc.)*

(1) Pur essendo distinti, i corsi di Algebra e di Geometria analitica corrispondevano ad una unica cattedra.
(2) Dal 1890 il corso di Fisica II diventava facoltativo e senza esame.
(3) Dal 1891 i corsi di Mineralogia e di Disegno di Ornato diventavano facoltativi e senza esame.

	1893-'94	1894-'95	1895-'96
Algebra (1)	Pincherle S. (ord. coll.)	Pincherle S. (ord. coll.)	Pincherle S. (ord. coll.)
Geometria analitica (1)	Pincherle S. (ord. coll.)	Pincherle S. (ord. coll.)	Pincherle S. (ord. coll.)
Geometria proiettiva	Montesano D. (straord.)	Enriques F. (inc.)	Enriques F. (inc.)
Chimica generale	Ciamician G. (ord. coll.)	Ciamician G. (ord. coll.)	Ciamician G. (ord. coll.)
Fisica sperimentale	Righi A. (ord. coll.)	Righi A. (ord. coll.)	Righi A. (ord. coll.)
Disegno di Ornato e d'Architettura (2)	Zannoni A. (inc.)	Zannoni A. (inc.)	Zannoni A. (inc.)
Disegno di Geometria proiettiva	Montesano D.	Maccaferri E. (ass.)	Maccaferri E. (ass.)
Geometria descrittiva	Montesano D. (straord.)	Enriques F. (inc.)	Enriques F. (inc.)
Disegno di Ornato e d'Architettura (2)	Zannoni A. (inc.)	Zannoni A. (inc.)	Zannoni A. (inc.)
Calcolo infinitesimale (2)	Arzelà C. (ord. coll.)	Arzelà C. (ord. coll.)	Arzelà C. (ord. coll.)
Disegno di Geometria descrittiva	Montesano D.	Maccaferri E. (ass.)	Maccaferri E. (ass.)
Mineralogia (2)	Bombicci L. (ord. coll.)	Bombicci L. (ord. coll.)	Bombicci L. (ord. coll.)
Fisica sperimentale (2)	Righi A. (ord. coll.)	Righi A. (ord. coll.)	Righi A. (ord. coll.)
Geodesia teoretica	Fiorini M. (ord. coll.)	Fiorini M. (ord. coll.)	Fiorini M. (ord. coll.)
Meccanica razionale	Ruffini F. (ord. coll.)	Ruffini F. (ord. coll.)	Ruffini F. (ord. coll)
Matematiche superiori	Arzelà C. (inc.)	Arzelà C. (inc.)	Arzelà C. (inc.)
Matematiche superiori	Pincherle S. (inc.)	Pincherle S. (inc.)	Pincherle S. (inc.)
Fisica matematica	Donati L. (straord.)	Donati L. (straord.)	Donati L. (straord.)
Astronomia	Saporetti A. (ord. coll.)	Saporetti A. (ord. coll.)	Saporetti A. (ord. coll.)
Matematiche superiori	Arzelà C. (inc.)	Arzelà C. (inc.)	Arzelà C. (inc.)
Matematiche superiori	Pincherle S. (inc.)	Pincherle S. (inc.)	Pincherle S. (inc.)

(1) Pur essendo distinti, i corsi di Algebra e di Geometria analitica corrispondevano ad una unica cattedra.
(2) I corsi di Fisica II, Mineralogia, Disegno di Ornato erano facoltativi e senza esame.

	1896-'97	1897-'98	1898-'99
Algebra (1)	*Pincherle S.* *(ord. coll.)*	*Pincherle S.* *(ord. coll.)*	*Pincherle S.* *(ord. coll.)*
Geometria analitica (1)	*Pincherle S.* *(ord. coll.)*	*Pincherle S.* *(ord. coll.)*	*Pincherle S.* *(ord. coll.)*
Geometria proiettiva	*Enriques F.* *(straord.)*	*Enriques F.* *(straord.)*	*Enriques F.* *(straord.)*
Chimica generale	*Ciamician G.* *(ord. coll.)*	*Ciamician G.* *(ord. coll.)*	*Ciamician G.* *(ord. coll.)*
Fisica sperimentale	*Righi A.* *(ord. coll.)*	*Righi A.* *(ord. coll.)*	*Righi A.* *(ord. coll.)*
Disegno di Ornato e d'Architettura (2)	*Zannoni A.* *(inc.)*	*Zannoni A.* *(inc.)*	*Zannoni A.* *(inc.)*
Disegno di Geometria proiettiva	*Enriques F.*	*Enriques F.*	*Enriques F.*
Geometria descrittiva	*Enriques F.* *(straord.)*	*Enriques F.* *(straord.)*	*Enriques F.* *(straord.)*
Disegno di Ornato e d'Architettura (2)	*Zannoni A.* *(inc.)*	*Zannoni A.* *(inc.)*	*Zannoni A.* *(inc.)*
Calcolo infinitesimale (2)	*Arzelà C.* *(ord. coll.)*	*Arzelà C.* *(ord. coll.)*	*Arzelà C.* *(ord. coll.)*
Disegno di Geometria descrittiva	*Enriques F.*	*Enriques F.*	*Enriques F.*
Mineralogia (2)	*Bombicci L.* *(ord. coll.)*	*Bombicci L.* *(ord. coll.)*	*Bombicci L.* *(ord. coll.)*
Fisica sperimentale (2)	*Righi A.* *(ord. coll.)*	*Righi A.* *(ord. coll.)*	*Righi A.* *(ord. coll.)*
Geodesia teoretica	*Fiorini M.* *(ord. coll.)*	*Fiorini M.* *(ord. coll.)*	*Fiorini M.* *(ord. coll.)*
Meccanica razionale	*Ruffini F.* *(ord. coll.)*	*Ruffini F.* *(ord. coll.)*	*Ruffini F.* *(ord. coll)*
Matematiche superiori	*Arzelà C. (inc.)*	*Arzelà C. (inc.)*	*Arzelà C. (inc.)*
Matematiche superiori	*Pincherle S. (inc.)*	*Pincherle S. (inc.)*	*Pincherle S. (inc.)*
Fisica matematica	*Donati L.* *(straord.)*	*Donati L.* *(straord.)*	*Donati L.* *(straord.)*
Astronomia	*Saporetti A.* *(ord. coll.)*	*Saporetti A.* *(ord. coll.)*	*Saporetti A.* *(ord. coll.)*
Matematiche superiori	*Arzelà C. (inc.)*	*Arzelà C. (inc.)*	*Arzelà C. (inc.)*
Matematiche superiori	*Pincherle S. (inc.)*	*Pincherle S. (inc.)*	*Pincherle S. (inc.)*

(1) Pur essendo distinti, i corsi di Algebra e di Geometria analitica corrispondevano ad una unica cattedra.
(2) I corsi di Fisica II, Mineralogia, Disegno di Ornato erano facoltativi e senza esame.

	1899-'900	1900-'01	1901-'02
Algebra (1)	Pincherle S. (ord. coll.)	Pincherle S. (ord. coll.)	Pincherle S. (ord. coll.)
Geometria analitica (1)	Pincherle S. (ord. coll.)	Pincherle S. (ord. coll.)	Pincherle S. (ord. coll.)
Geometria proiettiva	Enriques F. (straord.)	Enriques F. (ord. coll.)	Enriques F. (ord. coll.)
Chimica generale	Ciamician G. (ord. coll.)	Ciamician G. (ord. coll.)	Ciamician G. (ord. coll.)
Fisica sperimentale	Righi A. (ord. coll.)	Righi A. (ord. coll.)	Righi A. (ord. coll.)
Disegno di Ornato e d'Architettura (2)	Zannoni A. (inc.)	Zannoni A. (inc.)	Zannoni A. (inc.)
Disegno di Geometria proiettiva	Enriques F.	Enriques F.	Enriques F.
Geometria descrittiva	Enriques F. (straord.)	Enriques F. (ord. coll.)	Enriques F. (ord. coll.)
Disegno di Ornato e d'Architettura (2)	Zannoni A. (inc.)	Zannoni A. (inc.)	Zannoni A. (inc.)
Calcolo infinitesimale	Arzelà C. (ord. coll.)	Arzelà C. (ord. coll.)	Arzelà C. (ord. coll.)
Disegno di Geometria descrittiva (2)	Enriques F.	Enriques F.	Enriques F.
Mineralogia (2)	Bombicci L. (ord. coll.)	Bombicci L. (ord. coll.)	Bombicci L. (ord. coll.)
Fisica sperimentale (2)	Righi A. (ord. coll.)	Righi A. (ord. coll.)	Righi A. (ord. coll.)
Geodesia teoretica	Fiorini M. (ord. coll.)	Fiorini M. (ord. coll.)	Sacchetti
Meccanica razionale	Ruffini F. (ord. coll.)	Ruffini F. (ord. coll.)	Ruffini F. (ord. coll.)
Matematiche superiori	Arzelà C. (inc.)	Arzelà C. (inc.)	Arzelà C. (inc.)
Matematiche superiori	Pincherle S. (inc.)	Pincherle S. (inc.)	Pincherle S. (inc.)
Fisica matematica	Donati L. (straord.)	Donati L. (straord.)	Donati L. (straord.)
Astronomia	Saporetti A. (ord. coll.)	Saporetti A. (ord. coll.)	Saporetti A. (ord. coll.)
Matematiche superiori	Arzelà C. (inc.)	Arzelà C. (inc.)	Arzelà C. (inc.)
Matematiche superiori	Pincherle S. (inc.)	Pincherle S. (inc.)	Pincherle S. (inc.)

(1) Pur essendo distinti, i corsi di Algebra e di Geometria analitica corrispondevano ad una unica cattedra.
(2) I corsi di Fisica II, Mineralogia, Disegno di Ornato, Disegno di Geometria proiettiva e descrittiva erano facoltativi e senza esame.

	1902-'03	1903-'04
Algebra (1)	*Pincherle S.* (ord. coll.)	*Pincherle S.* (ord. coll.)
Geometria analitica (1)	*Pincherle S.* (ord. coll.)	*Pincherle S.* (ord. coll.)
Geometria proiettiva	*Enriques F.* (ord. coll.)	*Enriques F.* (ord. coll.)
Chimica generale	*Ciamician G.* (ord. coll.)	*Ciamician G.* (ord. coll.)
Fisica sperimentale	*Righi A.* (ord. coll.)	*Righi A.* (ord. coll.)
Disegno di Ornato e d'Architettura (2)	*Zannoni A.* (inc.)	*Zannoni A.* (inc.)
Disegno di Geometria proiettiva (2)	*Enriques F.*	*Vecchi M.* (ass.)
Geometria descrittiva	*Enriques F.* (ord. coll.)	*Enriques F.* (ord. coll.)
Disegno di Ornato e d'Architettura (2)	*Zannoni A.* (inc.)	*Zannoni A.* (inc.)
Calcolo infinitesimale	*Arzelà C.* (ord. coll.)	*Arzelà C.* (ord. coll.)
Disegno di Geometria descrittiva (2)	*Enriques F.*	*Enriques F.*
Mineralogia (2)	*Bombicci L.* (ord. coll.)	*Bombicci L.* (ord. coll.)
Fisica sperimentale (2)	*Righi A.* (ord. coll.)	*Righi A.* (ord. coll.)
Geodesia teoretica	*Guarducci F.* (ord. coll.)	*Guarducci F.* (ord. coll.)
Meccanica razionale	*Ruffini F.* (ord. coll.)	*Ruffini F.* (ord. coll)
Matematiche superiori	*Arzelà C. (inc.)*	*Arzelà C. (inc.)*
Matematiche superiori	*Pincherle S. (inc.)*	*Pincherle S. (inc.)*
Fisica matematica	*Donati L.* (straord.)	*Donati L.* (straord.)
Astronomia	*Rajna M.* (ord. coll.)	*Rajna M.* (ord. coll.)
Matematiche superiori	*Arzelà C. (inc.)*	*Arzelà C. (inc.)*
Matematiche superiori	*Pincherle S. (inc.)*	*Pincherle S. (inc.)*

(1) Pur essendo distinti, i corsi di Algebra e di Geometria analitica corrispondevano ad una unica cattedra.
(2) I corsi di Fisica II, Mineralogia, Disegno di Ornato, Disegno di Geometria proiettiva e descrittiva erano facoltativi e senza esame.

	1904-'05	1905-'06
Analisi algebrica (1)	*Pincherle S.* (ord. coll.)	*Pincherle S.* (ord. coll.)
Geometria analitica (1)	*Pincherle S.* (ord. coll.)	*Pincherle S.* (ord. coll.)
Geometria proiettiva	*Enriques F.* (ord. coll.)	*Enriques F.* (ord. coll.)
Fisica sperimentale	*Righi A.* (ord. coll.)	*Righi A.* (ord. coll.)
Chimica inorganica	*Ciamician G.* (ord. coll.)	*Ciamician G.* (ord. coll.)
Esercitazioni di Analisi algebrica e Geometria analitica	*Galvani L.* (ass.)	*Galvani L.* (ass.)
Disegno di Geometria proiettiva	*Pasquali C.* (ass.)	*Pasquali C.* (ass.)
Calcolo infinitesimale	*Arzelà C.* (ord. coll.)	*Arzelà C.* (ord. coll.)
Geometria descrittiva	*Enriques F.* (ord. coll.)	*Enriques F.* (ord. coll.)
Fisica sperimentale	*Righi A.* (ord. coll.)	*Righi A.* (ord. coll.)
Esercitazioni di Calcolo infinitesimale	*Sibirani F.* (ass.)	*Sibirani F.* (ass.)
Disegno di Geometria descrittiva	*Pasquali C.* (ass.)	*Pasquali C.* (ass.)
Meccanica razionale	*Ruffini F.* (ord. coll.)	*Rajna M.* (inc.)
Geodesia teorica	*Guarducci F.* (ord. coll.)	*Guarducci F.* (ord. coll.)
Matematiche superiori	*Pincherle S. (inc.)*	*Pincherle S. (inc.)*
Analisi superiore	*Arzelà C. (inc.)*	*Arzelà C. (inc.)*
Astronomia	*Rajna M.* (ord. coll.)	*Rajna M.* (ord. coll.)
Fisica matematica	*Donati L.* (straord.)	*Donati L.* (straord.)
Matematiche superiori	*Pincherle S. (inc.)*	*Pincherle S. (inc.)*
Analisi superiore	*Arzelà C. (inc.)*	*Arzelà C. (inc.)*

(1) Pur essendo distinti, i corsi di Algebra e di Geometria analitica corrispondevano ad una unica cattedra.

	1906-'07	1907-'08	1908-'09	1909-'10
Algebra complementare (1)	*Pincherle S.* *(ord. coll.)*	*Pincherle S.* *(ord. coll.)*	*Pincherle S.* *(ord. coll.)*	*Pincherle S.* *(ord. coll.)*
Geometria analitica (1)	*Pincherle S.* *(ord. coll.)*	*Pincherle S.* *(ord. coll.)*	*Pincherle S.* *(ord. coll.)*	*Pincherle S.* *(ord. coll.)*
Geometria proiettiva	*Enriques F.* *(ord. coll.)*	*Enriques F.* *(ord. coll.)*	*Enriques F.* *(ord. coll.)*	*Enriques F.* *(ord. coll.)*
Fisica sperimentale	*Righi A.* *(ord. coll.)*	*Righi A.* *(ord. coll.)*	*Righi A.* *(ord. coll.)*	*Righi A.* *(ord. coll.)*
Chimica inorganica	*Ciamician G.* *(ord. coll.)*	*Ciamician G.* *(ord. coll.)*	*Ciamician G.* *(ord. coll.)*	*Ciamician G.* *(ord. coll.)*
Esercitazioni di Algebra complementare e Geometria analitica	*Galvani L.* *(ass.)*	*Tonelli L.* *(ass.)*	*Tonelli L.* *(ass.)*	*Tonelli L.* *(ass.)*
Disegno di Geometria proiettiva	*Enriques F.*	*Enriques F.*	*Enriques F.*	*Enriques F.*
Calcolo infinitesimale	*Arzelà C.* *(ord. coll.)*	*Arzelà C.* *(ord. coll.)*	*Arzelà C.* *(ord. coll.)*	*Arzelà C.* *(ord. coll.)*
Geometria descrittiva	*Enriques F.* *(ord. coll.)*	*Enriques F.* *(ord. coll.)*	*Enriques F.* *(ord. coll.)*	*Enriques F.* *(ord. coll.)*
Fisica sperimentale	*Righi A.* *(ord. coll.)*	*Righi A.* *(ord. coll.)*	*Righi A.* *(ord. coll.)*	*Righi A.* *(ord. coll.)*
Chimica organica	*Ciamician G.* *(ord. coll.)*	*Ciamician G.* *(ord. coll.)*	*Ciamician G.* *(ord. coll.)*	*Ciamician G.* *(ord. coll.)*
Esercitazioni di Calcolo infinitesimale	*Sibirani F.* *(ass.)*	*Sibirani F.* *(ass.)*	*Tardini L.* *(ass.)*	*Tardini L.* *(ass.)*
Disegno di Geometria descrittiva	*Enriques F.*	*Enriques F.*	*Enriques F.*	*Enriques F.*
Meccanica razionale	*Rajna M.* *(inc.)*	*Picciati G.* *(straord.)*	*Burgatti P.* *(straord.)*	*Burgatti P.* *(straord.)*
Geodesia teoretica	*Guarducci F.* *(ord. coll.)*	*Guarducci F.* *(ord. coll.)*	*Guarducci F.* *(ord. coll.)*	*Guarducci F.* *(ord. coll.)*
Matematiche superiori	*Pincherle S.* *(inc.)*	*Pincherle S.* *(inc.)*	*Pincherle S.* *(inc.)*	*Pincherle S.* *(inc.)*
Analisi superiore	*Arzelà C.* *(inc.)*	*Arzelà C.* *(inc.)*	*Arzelà C.* *(inc.)*	*Arzelà C.* *(inc.)*
Astronomia	*Rajna M.* *(ord. coll.)*	*Rajna M.* *(ord. coll.)*	*Rajna M.* *(ord. coll.)*	*Rajna M.* *(ord. coll.)*
Fisica matematica	*Donati L.* *(straord.)*	*Donati L.* *(straord.)*	*Donati L.* *(straord.)*	*Donati L.* *(straord.)*
Matematiche superiori	*Pincherle S.* *(inc.)*	*Pincherle S.* *(inc.)*	*Pincherle S.* *(inc.)*	*Pincherle S.* *(inc.)*
Analisi superiore	*Arzelà C.* *(inc.)*	*Arzelà C.* *(inc.)*	*Arzelà C.* *(inc.)*	*Arzelà C.* *(inc.)*

(1) Pur essendo distinti, i corsi di Algebra e di Geometria analitica corrispondevano ad una unica cattedra.

	1910-'11	1911-'12	1912-'13
Algebra complementare (1)	*Pincherle S.* (ord. coll.)	*Pincherle S.* (ord. coll.)	*Tonelli L.* (inc.)
Geometria analitica (1)	*Pincherle S.* (ord. coll.)	*Pincherle S.* (ord. coll.)	*Sibirani F.* (inc.)
Geometria proiettiva	*Enriques F.* (ord. coll.)	*Enriques F.* (ord. coll.)	*Enriques F.* (ord. coll.)
Fisica sperimentale	*Righi A.* (ord. coll.)	*Righi A.* (ord. coll.)	*Righi A.* (ord. coll.)
Chimica inorganica	*Ciamician G.* (ord. coll.)	*Ciamician G.* (ord. coll.)	*Ciamician G.* (ord. coll.)
Chimica organica	*Ciamician G.* (ord. coll.)	*Ciamician G.* (ord. coll.)	*Ciamician G.* (ord. coll.)
Esercitazioni di Algebra complementare e Geometria analitica	*Pincherle S.*	*Pincherle S.*	*Tonelli L.*
Disegno di Geometria proiettiva	*Enriques F.*	*Enriques F.*	*Enriques F.*
Calcolo infinitesimale	*Arzelà C.* (ord. coll.)	*Arzelà C.* (ord. coll.)	*Pincherle S.* (ord. coll.)
Geometria descrittiva	*Enriques F.* (ord. coll.)	*Enriques F.* (ord. coll.)	*Enriques F.* (ord. coll.)
Fisica sperimentale	*Righi A.* (ord. coll.)	*Righi A.* (ord. coll.)	*Righi A.* (ord. coll.)
Esercitazioni di Calcolo infinitesimale	*Arzelà C.*	*Arzelà C.*	*Pincherle S.*
Disegno di Geometria descrittiva	*Enriques F.*	*Enriques F.*	*Enriques F.*
Meccanica razionale	*Burgatti P.* (ord. coll.)	*Burgatti P.* (ord. coll.)	*Burgatti P.* (ord. coll.)
Geodesia teoretica	*Guarducci F.* (ord. coll.)	*Guarducci F.* (ord. coll.)	*Guarducci F.* (ord. coll.)
Matematiche superiori	*Pincherle S.* (inc.)	*Pincherle S.* (inc.)	*Pincherle S.* (inc.)
Analisi superiore	*Arzelà C.* (inc.)	*Arzelà C.* (inc.)	*Pincherle S.* (inc.)
Astronomia	*Rajna M.* (ord. coll.)	*Rajna M.* (ord. coll.)	*Rajna M.* (ord. coll.)
Fisica matematica	*Donati L.* (straord.)	*Donati L.* (straord.)	*Donati L.* (straord.)
Matematiche superiori	*Pincherle S.* (inc.)	*Pincherle S.* (inc.)	*Pincherle S.* (inc.)
Analisi superiore	*Arzelà C.* (inc.)	*Arzelà C.* (inc.)	*Pincherle S.* (inc.)
Meccanica superiore	*Burgatti P.* (inc.)	*Burgatti P.* (inc.)	*Burgatti P.* (inc.)

(1) Pur essendo distinti, i corsi di Algebra e di Geometria analitica corrispondevano ad una unica cattedra.

	1913-'14	1914-'15	1915-'16
Algebra complementare (1)	Sibirani F. (inc.)	Sibirani F. (inc.)	Sibirani F. (inc.)
Geometria analitica (1)	Razzaboni A. (inc.)	Razzaboni A. (inc.)	Razzaboni A. (inc.)
Geometria proiettiva	Enriques F. (ord. coll.)	Enriques F. (ord. coll.)	Enriques F. (ord. coll.)
Fisica sperimentale	Righi A. (ord. coll.)	Righi A. (ord. coll.)	Righi A. (ord. coll.)
Chimica inorganica	Ciamician G. (ord. coll.)	Ciamician G. (ord. coll.)	Ciamician G. (ord. coll.)
Chimica organica	Ciamician G. (ord. coll.)	Ciamician G. (ord. coll.)	Ciamician G. (ord. coll.)
Esercitazioni di Algebra complementare e Geometria analitica	Sibirani F.	Sibirani F.	Sibirani F.
Disegno di Geometria proiettiva	Enriques F.	Enriques F.	Enriques F.
Calcolo infinitesimale	Pincherle S. (ord. coll.)	Pincherle S. (ord. coll.)	Pincherle S. (ord. coll.)
Geometria descrittiva	Enriques F. (ord. coll.)	Enriques F. (ord. coll.)	Enriques F. (ord. coll.)
Fisica sperimentale	Righi A. (ord. coll.)	Righi A. (ord. coll.)	Righi A. (ord. coll.)
Esercitazioni di Calcolo infinitesimale	Pincherle S.	Pincherle S.	Pincherle S.
Disegno di Geometria descrittiva	Enriques F.	Enriques F.	Enriques F.
Meccanica razionale	Burgatti P. (ord. coll.)	Burgatti P. (ord. coll.)	Burgatti P. (ord. coll.)
Geodesia teoretica	Guarducci F. (ord. coll.)	Guarducci F. (ord. coll.)	Guarducci F. (ord. coll.)
Matematiche superiori	Pincherle S. (inc.)	Pincherle S. (inc.)	Pincherle S. (inc.)
Analisi superiore	Enriques F. (inc.)	Enriques F. (inc.)	Enriques F. (inc.)
Astronomia	Rajna M. (ord. coll.)	Rajna M. (ord. coll.)	Rajna M. (ord. coll.)
Fisica matematica	Donati L. (straord.)	Donati L. (straord.)	Donati L. (straord.)
Matematiche superiori	Pincherle S. (inc.)	Pincherle S. (inc.)	Pincherle S. (inc.)
Analisi superiore	Enriques F. (inc.)	Enriques F. (inc.)	Enriques F. (inc.)
Meccanica superiore	Burgatti P. (inc.)	Burgatti P. (inc.)	Burgatti P. (inc.)

(1) Pur essendo distinti, i corsi di Algebra e di Geometria analitica corrispondevano ad una unica cattedra.

	1921-'22
Algebra complementare Geometria analitica ed esercizi Geometria proiettiva e Disegno (1) Fisica sperimentale I Chimica inorganica (2) Chimica organica (2)	*Bortolotti Ett. (ord.)* *Bortolotti Ett. (ord.)* *Enriques F. (ord. coll.) (1)* *Majorana Q. (ord.)* *Ciusa R. (inc.)* *Scagliarini G. (inc.)*
Calcolo infinitesimale ed esercizi Fisica sperimentale II Geometria descrittiva e Disegno (1)	*Pincherle S. (ord. coll.)* *Majorana Q. (ord.)* *Enriques F. (ord. coll.)*
Analisi superiore ed esercizi Astronomia ed esercizi Geodesia teoretica ed esercizi Matematiche superiori Meccanica razionale ed esercizi	*Tonelli L. (ord.)* *Horn d'Arturo G. (inc.)* *Guarducci F. (ord. coll.)* *Pincherle S. (inc.)* *Burgatti P. (ord.)*
Meccanica superiore ed esercizi Fisica matematica ed esercizi Matematiche superiori ed esercizi Analisi superiore ed esercizi	*Burgatti P. (inc.)* *Burgatti P. (inc.)* *Pincherle S. (inc.)* *Tonelli L. (ord.)*

(1) Negli aa.aa. 1921-22 e 1922-23 Enriques veniva comandato presso l'Università di Roma.
(2) Nell'a.a. 1921-22 gli studenti potevano sostenere in una unica seduta l'esame di Chimica inorganica e di Chimica organica.

	1922-'23	1923-'24	1924-'25
Analisi algebrica ed esercizi	*Bortolotti Ett.* *(ord.)*	*Bortolotti Ett.* *(ord. coll.)*	*Bortolotti Ett.* *(p.s.c.)*
Geometria analitica ed esercizi	*Bortolotti Ett.*	*Agostini A.* *(inc.)*	*Tonelli L.* *inc.)*
Geometria proiettiva e Disegno (1)	*Enriques F.* *(ord. coll.)*	*Bompiani E.* *(straord.)*	*Bompiani E.* *(p.n.s.)*
Fisica sperimentale I	*Majorana Q.* *(ord.)*	*Majorana Q.* *(ord. coll.)*	*Majorana Q.* *(p.s.c.)*
Chimica generale ed inorganica con elementi di Chimica organica	*Ciusa R.* *(inc.)*	*Betti M.* *(ord.)*	*Betti M.* *(p.s.c.)*
Analisi infinitesimale ed esercizi	*Pincherle S.* *(ord. coll.)*	*Pincherle S.* *(ord. coll.)*	*Pincherle S.* *(p.s.c.)*
Fisica sperimentale II	*Majorana Q.* *(ord.)*	*Majorana Q.* *(ord. coll.)*	*Majorana Q.* *(p.s.c.)*
Geometria descrittiva e Disegno (1)	*Enriques F.* *(ord. coll.)*	*Bompiani E.* *(straord.)*	*Bompiani E.* *(p.n.s.)*
Analisi superiore ed esercizi	*Tonelli L.* *(ord.)*	*Tonelli L.* *(ord. coll.)*	*Tonelli L.* *(p.s.c.)*
Astronomia ed esercizi	*Horn d'Arturo G.* *(inc.)*	*Horn d'Arturo G.* *(inc.)*	*Horn d'Arturo G.* *(p.s.c.)*
Geodesia teoretica ed esercizi	*Guarducci F.* *(ord. coll.)*	*Guarducci F.* *(ord. coll.)*	*Guarducci F.* *(p.s.c.)*
Geometria superiore ed esercizi	*Tonelli L.* *(inc.)*	*Bortolotti En.* *(inc.)*	*Bompiani E.* *(inc.)*
Meccanica razionale ed esercizi	*Burgatti P.* *(ord. coll.)*	*Burgatti P.* *(ord. coll.)*	*Burgatti P.* *(p.s.c.)*
Fisica matematica ed esercizi	*Roghi R.* *(inc.)*	*Burgatti P.* *(inc.)*	*Burgatti P.* *(inc.)*
Geometria superiore ed esercizi	*Tonelli L.* *(inc.)*	*Bortolotti En.* *(inc.)*	*Bortolotti En.* *(inc.)*
Analisi superiore ed esercizi	*Tonelli L.* *(ord.).*	*Tonelli L.* *(ord. coll.)*	*Tonelli L.* *(p.s.c.)*

(1) Negli aa.aa. 1921-22 e 1922-23 Enriques veniva comandato presso l'Università di Roma.

L'insegnamento della matematica dal 1860 al 1940

	1925-'26
Analisi algebrica	*Tonelli L. (inc.)*
Esercitazioni di Analisi algebrica	*Lelli M. (?)*
Geometria analitica	*Bortolotti Ett. (inc.)*
Esercitazioni di Geometria analitica	*Onofri L. (?)*
Istituzioni di Geometria proiettiva e descrittiva	*Bompiani E. (p.s.)*
Esercitazioni di Geometria proiettiva e descrittiva	*Supino G. (?)*
Fisica sperimentale (biennale)	*Majorana Q. (p.s.c.)*
Chimica generale e inorganica con elementi di Chimica organica	*Betti M. (p.s.c.)*
Disegno di Ornato e di Architettura	*Collamarini E. (inc.)*
Analisi infinitesimale	*Pincherle S. (p.s.c.)*
Esercitazioni di Analisi infinitesimale	*Onofri L. (?)*
Fisica sperimentale (biennale)	*Majorana Q. (p.s.c.)*
Esercitazioni di Fisica (facoltativo)	*Majorana Q.*
Complementi di Analisi e di Geometria	*Bortolotti En. (inc.)*
Meccanica razionale	*Burgatti P. (p.s.c.)*
Esercitazioni di Meccanica razionale	*Manarini M. (?)*
Elementi di Teoria delle funzioni	*Pincherle S. (inc.)*
Fisica matematica	*Burgatti P. (inc.)*
Analisi superiore II	*Tonelli L. (p.s.c.)*
Geometria superiore II	*Bompiani E. (inc.)*
Un corso complementare	

	1926-'27	1927-'28
Analisi algebrica	*Pincherle S.*	*Tonelli L.* *(p.s.c.)*
Esercitazioni di Analisi algebrica	*Onofri L.* *(?)*	*Mambriani A.* *(?)*
Geometria analitica	*Bortolotti Ett.* *(?)*	*Bortolotti Ett.* *(p.s.c.)*
Esercitazioni di Geometria analitica	*Onofri L.* *(?)*	*Onofri L.* *(?)*
Istituzioni di Geometria proiettiva e descrittiva	*Bompiani E.* *(p.s.)*	*Supino G.* *(inc.)*
Esercitazioni di Geometria proiettiva e descrittiva	*Supino G.* *(?)*	*Supino G.*
Fisica sperimentale (biennale)	*Majorana Q.* *(p.s.c.)*	*Majorana Q.* *(p.s.c.)*
Chimica generale e inorganica con elementi di Chimica organica	*Betti M.* *(p.s.c.)*	*Betti M.* *(p.s.c.)*
Disegno di Ornato e di Architettura (facoltativo)	*Collamarini E.* *(inc.)*	*Collamarini E.* *(inc.)*
Analisi infinitesimale	*Tonelli L.*	*Pincherle S.* *(p.s.c.)*
Esercitazioni di Analisi infinitesimale	*Lelli M.* *(?)*	*Onofri L.* *(?)*
Fisica sperimentale (biennale)	*Majorana Q.* *(p.s.c.)*	*Majorana Q.* *(p.s.c.)*
Esercitazioni di Fisica sperimentale (facoltativo)	*Majorana Q.*	*Majorana Q.*
Meccanica razionale	*Burgatti P.* *(p.s.c.)*	*Burgatti P.* *(p.s.c.)*
Esercitazioni di Meccanica razionale	*Manarini M.* *(?)*	*Manarini M.* *(?)*
Elementi di Teoria delle funzioni	*Pincherle S.* *(inc.)*	*Pincherle S.* *(inc.)*
Fisica matematica	*Burgatti P.* *(inc.)*	*Burgatti P.* *(inc.)*
Analisi superiore II	*Tonelli L.* *(p.s.c.)*	*Tonelli L.* *(inc.)*
Geometria superiore II	*Bompiani E.* *(inc.)*	*Bortolotti En.* *(inc.)*
Un corso complementare		

L'insegnamento della matematica dal 1860 al 1940 463

	1928-'29	1929-'30	1930-'31
Analisi algebrica	*Tonelli L.*	*Tonelli L.*	*Levi B.* (p.s.)
Esercitazioni di Analisi algebrica	*Mambriani A.* (?)	*Mambriani A.* (?)	*Levi B.*
Geometria analitica	*Bortolotti Ett.* (p.s.c.)	*Bortolotti Ett.* (p.s.c.)	*Bortolotti Ett.* (p.s.c.)
Esercitazioni di Geometria analitica	*Onofri L.* (?)	*Onofri L.* (?)	*Onofri L.* (?)
Istituzioni di Geometria proiettiva e descrittiva	*Supino G.* (inc.)	*Supino G.* (inc.)	*Supino G.* (inc.)
Fisica sperimentale (biennale)	*Majorana Q.* (p.s.c.)	*Majorana Q.* (p.s.c.)	*Majorana Q.* (p.s.c.)
Chimica generale inorganica con elementi di Chimica organica	*Betti M.* (p.s.c.)	*Betti M.* (p.s.c.)	*Betti M.* (p.s.c.)
Disegno di Ornato e di Architettura (facoltativo)	*Ricci P.* (inc.)	*Ricci P.* (inc.)	*Zucchini G.* (inc.)
Analisi infinitesimale	*Tonelli L.* (p.s.c.)	*Tonelli L.* (p.s.c.)	*Vitali G.* (p.s.)
Esercitazioni di Analisi infinitesimale	*Mambriani A.* (?)	*Mambriani A.* (?)	*Mambriani A.* (?)
Fisica sperimentale (biennale)	*Majorana Q.* (p.s.c.)	*Majorana Q.* (p.s.c.)	*Majorana Q.* (p.s.c.)
Esercitazioni di Fisica sperimentale (facoltativo)	*Majorana Q.*	*Majorana Q.*	*Majorana Q.*
Meccanica razionale	*Burgatti P.* (p.s.c.)	*Burgatti P.* (p.s.c.)	*Burgatti P.* (p.s.c.)
Esercitazioni di Meccanica razionale	*Manarini M.* (?)	*Manarini M.* (?)	*Manarini M.* (?)
Elementi di Teoria delle funzioni	*Levi B.* (p.s.)	*Levi B.* (p.s.)	*Levi B.* (?)
Esercitazioni di Teoria delle funzioni	*Levi B.*	*Levi B.*	*Levi B.*
Fisica matematica	*Burgatti P.* (inc.)	*Burgatti P.* (inc.)	*Burgatti P.* (inc.)
Analisi superiore II	*Levi B.* (inc.)	*Levi B.* (inc.)	*Levi B.* (inc.)
Geometria superiore II	*Levi B.* (inc.)	*Levi B.* (inc.)	*Levi B.* (inc.)
Un corso complementare			

	1931-'32
Analisi algebrica	*Vitali G.*
Esercitazioni di Analisi algebrica	*Mambriani A. (?)*
Geometria analitica	*Bortolotti Ett. (ord. coll.)*
Esercitazioni di Geometria analitica	*Onofri L. (?)*
Istituzioni di Geometria proiettiva e descrittiva	*Segre B. (straord.)*
Esercitazioni di Geometria proiettiva e descrittiva	*Pini E. (?)*
Fisica sperimentale (biennale)	*Majorana Q. (ord. coll.)*
Chimica generale e inorganica con elementi di Chimica organica	*Betti M. (ord. coll.)*
Disegno di Ornato e di Architettura (facoltativo)	*Zucchini G. (inc.)*
Analisi infinitesimale	*Levi B.*
Esercitazioni di Analisi infinitesimale	*Viola T. (?)*
Fisica sperimentale (biennale)	*Majorana Q. (ord. coll.)*
Esercitazioni di Fisica sperimentale (facoltativo)	*Majorana Q.*
Meccanica razionale	*Burgatti P. (ord. coll.)*
Esercitazioni di Meccanica razionale	*Manarini M. (?)*
Elementi di Teoria delle funzioni	*Levi B. (inc.)*
Fisica matematica	*Burgatti P. (inc.)*
Analisi superiore II	*Manarini M. (inc.)*
Geometria superiore II	*Segre B. (inc.)*
Un corso complementare	

	1932-'33
Analisi algebrica	*Levi B. (ord.)*
Esercitazioni di Analisi algebrica	*Mambriani A. (?)*
Geometria analitica	*Bortolotti Ett. (ord. coll.)*
Esercitazioni di Geometria analitica	*Onofri L. (?)*
Istituzioni di Geometria proiettiva e descrittiva	*Segre B. (straord.)*
Esercitazioni di Geometria proiettiva e descrittiva	*Pini E. (?)*
Fisica sperimentale (biennale)	*Majorana Q. (ord. coll.)*
Chimica generale ed inorganica	*Betti M. (ord. coll.)*
Elementi di Chimica organica	*Betti M. (inc.)*
Analisi infinitesimale	*Fantappiè L. (ord.)*
Esercitazioni di Analisi infinitesimale	*Mambriani A. (?)*
Fisica sperimentale (biennale)	*Majorana Q. (ord. coll.)*
Esercitazioni di Fisica sperimentale (facoltativo)	*Majorana Q.*
Geometria proiettiva	*Segre B. (straord.)*
Meccanica razionale	*Burgatti P. (ord. coll.)*
Esercitazioni di Meccanica razionale	*Manarini M. (?)*
Elementi di Teoria delle funzioni	*Levi B. (inc.)*
Un corso complementare	
Fisica matematica	*Burgatti P. (inc.)*
Analisi superiore	*Fantappiè L. (inc.)*
Geometria superiore II	*Segre B. (inc.)*
Un corso complementare	

	1933-'34
Analisi algebrica	*Fantappiè L.*
Esercitazioni di Analisi algebrica	*Fantappiè L.*
Geometria analitica	*Bortolotti Ett. (ord. coll.)*
Esercitazioni di Geometria analitica	*Bortolotti Ett.*
Istituzioni di Geometria proiettiva e descrittiva	*Segre B. (straord.)*
Esercitazioni di Geometria proiettiva e descrittiva	*Segre B.*
Fisica sperimentale (biennale)	*Majorana Q. (ord. coll.)*
Chimica generale ed inorganica	*Betti M. (ord. coll.)*
Elementi di Chimica organica	*Betti M. (inc.)*
Analisi infinitesimale	*Levi B.*
Esercitazioni di Analisi infinitesimale	*Levi B.*
Fisica sperimentale (biennale)	*Majorana Q. (ord. coll.)*
Esercizi di Fisica sperimentale (facoltativo)	*Majorana Q.*
Complementi di Geometria (conferenze)	*Segre B. (straord.)*
Meccanica razionale	*Burgatti P. (ord. coll.)*
Esercitazioni di Meccanica razionale	*Burgatti P.*
Elementi di Teoria delle funzioni	*Levi B. (inc.)*
Analisi superiore (biennale)	*Fantappiè L. (inc.)*
Geometria superiore (biennale)	*Segre B. (inc.)*
Un corso complementare	
Fisica matematica	*Burgatti P. (inc.)*
Analisi superiore	*Fantappiè L. (inc.)*
Geometria superiore	*Segre B. (inc.)*
Un corso complementare	

	1934-'35
Analisi algebrica	*Levi B. (ord.)*
Esercitazioni di Analisi algebrica	*Levi B.*
Geometria analitica	*Bortolotti Ett. (ord. coll.)*
Esercitazioni di Geometria analitica	*Bortolotti Ett.*
Istituzioni di Geometria proiettiva e descrittiva	*Segre B. (straord.)*
Esercitazioni di Geometria proiettiva e descrittiva	*Segre B.*
Fisica sperimentale (biennale)	*Majorana Q. (ord. coll.)*
Chimica generale ed inorganica	*Betti M. (ord. coll.)*
Elementi di Chimica organica	*Betti M. (inc.)*
Analisi infinitesimale	*Segre B. (?)*
Esercitazioni di Analisi infinitesimale	*Segre B.*
Fisica sperimentale (biennale)	*Majorana Q. (ord. coll.)*
Esercizi di Fisica sperimentale	*Majorana Q.*
Applicazioni di Geometria descrittiva	*Onofri L. (inc.)*
Meccanica razionale	*Burgatti P. (ord. coll.)*
Esercitazioni di Meccanica razionale	*Burgatti P.*
Elementi di Teoria delle funzioni	*Mambriani A. (inc.)*
Analisi superiore (biennale)	*Levi B. (inc.)*
Geometria superiore (biennale)	*Segre B. (inc.)*
Un corso complementare	
Fisica matematica	*Burgatti P. (inc.)*
Analisi superiore	*Levi B. (inc.)*
Geometria superiore	*Segre B. (inc.)*
Un corso complementare	

	1935-'36
Analisi matematica I	Levi B. (ord.)
Esercitazioni di Analisi mat.	Levi B.
Geometria analitica	Bortolotti E. (ord. coll.)
Esercitazioni di Geom. anal.	Bortolotti E.
Fisica sperimentale I	Majorana Q. (ord. coll.)
Chimica generale ed inorganica	Betti M. (ord. coll.)
Analisi matematica II	Levi B.
Esercitazioni di Analisi mat.	Levi B.
Geometria proiettiva e descritt.	Segre B. (ord.)
Esercitazioni e Disegno di Geometria proiett. e descritt.	Segre B.
Fisica sperimentale II	Majorana Q. (ord. coll.)
Esercizi di laboratorio di Fisica sperimentale	Majorana Q.
Meccanica razionale con elementi di Statica grafica	Burgatti P. (ord. coll.)
Esercizi di Mecc. raz. e Disegno di Statica grafica	Burgatti P.
Un corso complementare	
Analisi superiore I	Levi B. (inc.)
Geometria superiore I	Segre B. (inc.)
Fisica matematica I	Burgatti P. (inc.)
Un corso complementare	
Analisi superiore II	Levi B. (inc.)
Geometria superiore II	Segre B. (inc.)
Fisica matematica II	Burgatti P. (inc.)
Un corso complementare	

	1936-'37	1937-'38	1938-'39
Analisi matematica I	Levi B. (ord.)	Levi B. (ord.)	Scorza G. (inc.)
Esercitazioni di Analisi matematica	Levi B.	Levi B.	Scorza G.
Geometria analitica con elementi di Geometria	Segre B. (ord.)	Segre B. (ord.)	Onofri L. (inc.)
Esercitazioni di Geometria analitica e proiettiva	Segre B.	Segre B.	Onofri L.
Fisica sperimentale	Majorana Q. (ord. coll.)	Majorana Q. (ord. coll.)	Majorana Q. (ord. coll.)
Esercizi di laboratorio di Fisica sperimentale	Majorana Q.	Majorana Q.	Majorana Q.
Chimica generale ed inorganica	Betti M. (ord. coll.)	Betti M. (ord. coll.)	Betti M. (ord. coll.)
Elementi di Chimica organica	Betti M. (inc.)	Betti M. (inc.)	Betti M. (inc.)
Cultura militare	Approsio L. (inc.)	Approsio L. (inc.)	Approsio L. (inc.)
Analisi matematica II	Levi B.	Levi B.	Mambriani A. (inc.)
Esercitazioni di Analisi matematica	Levi B.	Levi B.	Mambriani A.
Geometria proiettiva e descrittiva	Segre B. (ord.)	Segre B. (ord.)	Onofri L. (inc.)
Disegno di Geometria proiettiva e descrittiva	Segre B.	Segre B.	Onofri L.
Fisica sperimentale	Majorana Q. (ord. coll.)	Majorana Q. (ord. coll.)	Majorana Q. (ord. coll.)
Esercizi di laboratorio di Fisica sperimentale	Majorana Q.	Majorana Q.	Majorana Q.
Meccanica razionale con elementi di Statica grafica	Burgatti P. (ord. coll.)	Burgatti P. (ord. coll.)	Graffi D. (straord.)
Esercitazioni di Meccanica razionale e Disegno di Statica grafica	Burgatti P.	Burgatti P.	Graffi D.
Cultura militare	Approsio L. (inc.)	Approsio L. (inc.)	Approsio L. (inc.)
Un corso complementare			
Analisi superiore	Comessatti A. (inc.)	Levi B. (inc.)	Scorza G. (inc.)
Geometria superiore	Segre B. (inc.)	Segre B. (inc.)	Bassi A. (inc.)
Un corso complementare			
Fisica matematica	Burgatti P. (inc.)	Burgatti P. (inc.)	Manarini M. (inc.)
Un corso complementare			

	1939-'40
Analisi matematica I	*Cimmino G. (straord.)*
Esercitazioni di Analisi matematica	*Cimmino G.*
Geometria analitica con elementi di Geometria proiettiva	*Villa M. (straord.)*
Esercitazioni di Geometria analitica e proiettiva	*Villa M.*
Fisica sperimentale	*Majorana Q. (ord. coll.)*
Chimica generale ed inorganica	*Betti M. (ord. coll.)*
Elementi di Chimica inorganica	*Betti M. (inc.)*
Cultura militare	*Approsio L. (inc.)*
Analisi matematica II	*Zagar F. (inc.)*
Esercitazioni di Analisi matematica	*Zagar F.*
Geometria proiettiva e descrittiva	*Onofri L. (inc.)*
Esercitazioni di Geometria proiettiva e descrittiva	*Onofri L.*
Disegno di Geometria proiettiva e descrittiva	*Onofri L.*
Fisica sperimentale	*Majorana Q. (ord. coll.)*
Esercizi di laboratorio di Fisica sperimentale	*Majorana Q.*
Meccanica razionale con elementi di Statica grafica	*Graffi D. (ord.)*
Esercitazioni di Meccanica razionale e Disegno di Statica grafica	*Graffi D. (ord.)*
Un corso complementare	
Cultura militare	*Approsio L. (inc.)*
Analisi superiore	*Cimmino G. (inc.)*
Geometria superiore	*Villa M. (inc.)*
Esercitazioni di Fisica sperimentale	*Majorana Q. (ord. coll.)*
Un corso complementare	
Fisica matematica	*Manarini M. (inc.)*
Un corso complementare	

Insegnamenti complementari

	1925-'26	1926-'27	1927-'28	1928-'29	1929-'30
Analisi superiore (biennale)	Tonelli L. (inc.)	Tonelli L. (p.s.c.)	Tonelli L. (inc.)	Levi B. (inc.)	Levi B. (inc.)
Geometria superiore (biennale)	Bompiani E. (inc.)	Bompiani E. (inc.)	Bortolotti En. (inc.)	Levi B. (inc.)	Levi B. (inc.)
Fisica superiore (biennale)	Dalla Noce G. (inc.)	Dalla Noce G. (inc.)	Brunetti R. (inc.)	Specchia O. (inc.)	Specchia O. (inc.)
Oscillazioni elettriche (biennale)	Todesco G. (inc.)	Todesco G. (inc.)	Todesco G. (inc.)	Todesco G. (inc.)	Todesco G. (inc.)
Matematiche compl. (biennale)	Bortolotti Ett. (inc.)	Bortolotti Ett. (inc.)	Bortolotti Ett. (inc.)	Bortolotti Ett. (inc.)	Bortolotti Ett. (inc.)
Fisica complementare e misure fisiche (biennale)	Piola F. (inc.)	Brunetti R. (inc.)	XXX	XXX	XXX
Calcolo della probabilità e Statistica matematica	Horn d'Arturo G. (inc.)	Horn d'Arturo G. (inc.)	XXX	XXX	XXX
Astronomia	Horn d'Arturo G. (p.n.s.)	Horn d'Arturo G. (p.n.s.)	Horn d'Arturo G. (p.s.)	Horn d'Arturo G. (p.s.)	Horn d'Arturo G. (p.s.)
Geodesia	Guarducci F. (p.s.c.)	Guarducci F. (p.s.c.)	Dore P. (inc.)	Dore P. (inc.)	Dore P. (inc.)
Conferenze di Storia della matematica	Bortolotti Ett. (inc.)	Bortolotti Ett. (inc.)	XXX	XXX	XXX
Fisica teorica	XXX	XXX	XXX	Dalla Noce G. (inc.)	Dalla Noce G. (inc.)

	1930-'31	1931-'32	1932-'33	1933-'34	1934-'35
Analisi superiore (biennale)	Vitali G. (inc.)	Manarini M. (inc.)	Fantappiè L. (inc.)	XXX	XXX
Geometria superiore (biennale)	Levi B. (inc.)	Segre B. (inc.)	Segre B. (inc.)	XXX	XXX
Fisica superiore (biennale)	Specchia O. (inc.)	Specchia O. (inc.)	Todesco G. (inc.)	Todesco G. (inc.)	Todesco G. (inc.)
Oscillazioni elettriche (biennale)	Graffi D. (inc.)	Graffi D. (inc.)	Graffi D. (inc.)	Graffi D. (inc.)	Graffi D. (inc.)
Matematiche compl. (biennale)	Bortolotti Ett. (inc.)	Bortolotti Ett. (inc.)	Bortolotti Ett. (inc.)	Bortolotti Ett. (inc.)	Bortolotti Ett. (inc.)
Astronomia	Horn d'Arturo G. (p.s.c.)	Horn d'Arturo G. (ord. coll.)	Horn d'Arturo G. (ord. coll.)	Horn d'Arturo G. (ord. coll.)	Horn d'Arturo G. (ord. coll.)
Geodesia	Dore P. (inc.)	Dore P. (inc.)	Dore P. (inc.)	Dore P. (inc.)	Dore P. (inc.)
Fisica teorica	Dalla Noce G. (inc.)	Dalla Noce G. (inc.)	Dalla Noce G. (inc.)	Dalla Noce G. (inc.)	Dalla Noce G. (inc.)
Meccanica superiore (corso pareggiato)	XXX	XXX	XXX	Manarini M. (inc.)	Manarini M. (inc.)

Il simbolo «XXX» indica i corsi esclusi dal piano degli studi in un determinato anno accademico.

Insegnamenti complementari

	1935-'36	1936-'37	1937-'38	1938-'39	1939-'40
Matematiche complementari	Bortolotti Ett. (inc.)	Comessatti A. (inc.)	Comessatti A. (inc.)	Comessatti A. (inc.)	Villa M. (inc.)
Teoria delle funzioni	Mambriani A. (inc.)		Comessatti A. (inc.)	Comessatti A. (inc.)	
Geometria differenziale	Onofri L. (inc.)		Onofri L. (inc.)	Onofri L. (inc.)	Onofri L. (inc.)
Fisica superiore	Ranzi I. (inc.)	Ranzi I. (inc.)	Todesco G. (inc.)	Bernardini G. (straord.)	Bernardini G. (straord.)
Fisica teorica	Dalla Noce G. (inc.)	Dalla Noce G. (inc.)	Dalla Noce G. (inc.)	Dalla Noce G. (inc.)	Dalla Noce G. (inc.)
Meccanica superiore (corso pareggiato)	Manarini M.	Manarini M.	XXX	XXX	Zagar F. (inc.)
Astronomia	Horn d'Arturo G. (ord. coll.)	Horn d'Arturo G. (ord. coll.)	Horn d'Arturo G. (ord. coll.)	Zagar F. (straord.)	Zagar F. (ord.)
Geodesia	Dore P. (inc.)	Dore P. (inc.)	Dore P. (inc.)	Dore P. (inc.)	Dore P. (inc.)

Il simbolo «XXX» indica i corsi esclusi dal piano degli studi in un determinato anno accademico.

APPENDICE B

ELENCHI DEI PROFESSORI DI RUOLO DEL CORSO DI LAUREA IN MATEMATICA DELLA UNIVERSITÁ DI BOLOGNA DAL 1860 AL 1940.

Professori insegnanti collegiati, insegnanti, ordinari collegiati, ordinari, stabili collegiati, stabili.

Arzelà C., *Calcolo infinitesimale*, 1885-911;
Beltrami E., *Meccanica razionale*, 1866-72;
Betti M., *Chimica generale e inorganica*, 1923;
Bianconi G., *Storia naturale*, 1859;
Bombicci L., *Mineralogia*, 1861 e 1876-903;
Bompiani E., *Geometria proiettiva e descrittiva*, 1925-27;
Bortolotti Ett., *Algebra complementare*, 1921-26, *Geometria analitica*, 1927-35;
Boschi P., *Geometria proiettiva e descrittiva*, 1877-86;
Botter L.F., *Agronomia*, 1859-61;
Burgatti P., *Meccanica razionale*, 1911-37;
Capellini G., *Geologia*, 1860-61 e 1876-83;
Chelini D., *Meccanica e Idraulica*, 1859-60, *Meccanica razionale*, 1861-63;
Ciamician G., *Chimica generale*, 1889-903, *Chimica inorganica*, 1904-1915 (1), *Chimica organica*, 1906-1915 (1);
Cremona L., *Geometria superiore*, 1860-1871;
Della Casa L., *Fisica sperimentale*, 1859-70;
Enriques F., *Geometria proiettiva e descrittiva*, 1900-22;
Fantappiè L., *Analisi infinitesimale*, 1932-33;
Filopanti Q., *Meccanica applicata*, 1860-900;
Fiorini M., *Geodesia*, 1860-900;
Giusti E., *Istituzioni civili*, 1859;
Graffi D., *Meccanica razionale*, 1939-;
Guarducci F., *Geodesia*, 1902-27;
Horn d'Arturo G., *Astronomia*, 1927-37;
Levi B., *Teoria delle funzioni*, 1928-29, *Analisi algebrica*, 1930-34, *Analisi matematica I*, 35-37;
Lodi F., *Disegno d'ornato e d'Architettura*, 1860-81;
Majorana Q., *Fisica sperimentale*, 1921- (1);
Mariscotti A., *Economia pubblica*, 1859-60;
Pincherle S., *Algebra e Geometria analitica*, 1888-911, *Calcolo infinitesimale*, 1912-27;
Rajna M., *Astronomia*, 1902-1915 (1);
Ramenghi S., *Introduzione al Calcolo*, 1859-61;
Razzaboni C., *Meccanica razionale*, 1873;
Respighi L., *Ottica ed Astronomia*, 1859, *Ottica*, 1860, *Astronomia*, 1861-63;
Righi A., *Fisica*, 1889-1915 (1);
Ruffini P.F., *Meccanica razionale*, 1879-904;
Santagata D., *Chimica generale*, 1860-61, *Chimica inorganica*, 1862-87;
Saporetti A., *Calcolo sublime*, 1859-61, *Calcolo differenziale ed integrale*, 1862-76, *Astronomia*, 1877-901;
Segre B., *Geometria proiettiva e descrittiva*, 1935-37;

Tonelli L., *Analisi superiore*, 1921-27, *Analisi algebrica*, 1927, *Analisi infinitesimale*, 1928-30;
Villari E., *Fisica* 1871-88;
Vitali G., *Analisi infinitesimale*, 1930-32;
Zagar F., *Astronomia*, 1939-.

Professori straordinari, non stabili.

Arzelà C., *Analisi infinitesimale*, 1880-84;
Beltrami E., *Algebra complementare*, 1862;
Bernardini G., *Fisica superiore*, 1938- ;
Bompiani E., *Geometria proiettiva e descrittiva*, 1923-24;
Boschi P., *Algebra e Geometria analitica*, 1863-76;
Burgatti P., *Meccanica razionale*, 1908-11;
Cimmino G., *Analisi matematica I*, 1939-;
D'Arcais F., *Algebra e Geometria analitica*, 1875;
De Paolis R., *Algebra e Geometria analitica*, 1878-79;
Donati L., *Fisica matematica*, 1879-1915 (1);
Enriques F., *Geometria proiettiva e descrittiva*, 1896-99;
Fais A., *Algebra e Geometria analitica*, 1876, *Analisi infinitesimale*, 1877-79;
Graffi D., *Meccanica razionale*, 1938-;
Horn d'Arturo G., *Astronomia*, 1924-26;
Montesano D., *Geometria proiettiva e descrittiva*, 1889-93;
Picciati G., *Meccanica razionale*, 1907;
Pincherle S., *Algebra e Geometria analitica*, 1881-87;
Segre B., *Geometria proiettiva e descrittiva*, 1931-34;
Venturi L., *Meccanica razionale*, 1865;
Villa M., *Geometria analitica*, 1939-;
Zagar F., *Astronomia*, 1938.

(1) Tra gli anni 1915 e 1921 gli Annuari della R. Università di Bologna non riportano nè i docenti nè i corsi da loro tenuti.

Index

Abel, N. 384, 392
 Abel's identity 243
 Abel logarithmic sequence 268
 Abel map 398
 Abel polynomial 243, 268
 Abel-Jacobi map 100-101
 Abel surface 69, 75
 Abel variety 84-85
Adamajan-Arov-Kreĭn parametrization 153
Agnesi, G. 324
Agnesi, M. G. v, 323-327
Agnesi, M. T. 323-324
Agnesi, don P. 323, 326
 Agnesi witch 323
Agostini, A. 460
Aizenberg, L. 153
Albicini, C. 430
Akahori, T. 231, 237
Akhiezer, N. I. 227
Albanese, G. 63, 131
 Albanese dimension 66, 69, 83
 Albanese general type fibrations 72-74, 85
 Albanese general type manifolds 69-70
 Albanese map 66, 80, 83, 100-102
 Albanese morphism 67

Albanese variety 63-66
Aldini, A. 304
Aldini, G. 304
Alegria, P. 153
Algarotti, F. 296
Amaldi, U. vi, 380, 431
Amasaki, M. 186
Amodeo, F. 135
Ancona, V. 1-4
Andersen, K. 201
André, J. 8
Angeli, S. 196, 199, 201
Aoki, T. 355
Appel, P. 376, 392
 Appel equations, 376
Approsio, L. 469
Arakelov, S. J. 77, 86
Arbarello, E. 91-92, 105, 406
Archimedes 201
Arnese, A. 415
Arocena parametrization 153
Artin, M. 64, 145, 168-169, 174
 Artin-Schreyer covering 64
Arzelà, C. v, 30, 292, 375-395, 415, 422, 424, 426, 431, 444-457, 473-474
Ascoli, G. 377, 384-385, 390, 393, 429
Askey, R. 269, 275

475

Atiyah, M. F. 86
Avanissian-Gay transform 22
Averbukh, B. 168-169, 174

Babbage, D. W. 92, 94, 96, 105-106, 176, 186
Baire, R. 378
  Baire class 379, 386
  Baire function 379
Baker, H. F. 138
Ballico, E. 186
Banach-Vitali theorem 379
Barber, E. 105
Barlotti, A. 5-9
Barnabei, M. 275
Barrow, I. 213
Bartalesi, A. 93, 105
Barth, W. 86, 93, 105, 163, 171, 174, 406
Bassi, A. 469
Battaglini, G. 27, 135
Beatrous, F. 150, 153
Beauville, A. 59, 69, 79-82, 84, 86-87, 93-95, 99, 105
Beccari, J. B. 323
Belgrado, J. 299
Belloni conte 324
Beltrami, Ed. 11-17
Beltrami, Eu. v, 11, 279, 380, 418, 420, 429-430, 436-439, 473-474
  (Eu.) Beltrami equation 155
  (Eu.) Beltrami operator 12, 14-15, 17
Benedetto XIV (Pope) v, 292-293, 295-296, 299, 323-327
Benetti, J. 422
Bengel, G. 330, 355

Bengel P-functional 354-355
Bengel-Kawai duality 355
Berenstein, C. A. 19-23, 350, 356
Berg, C. 227
Bergman, S. 238
  Bergman kernel 238
  Bergman space 152
  Bergman-Shiffer theorem 148, 150
Bernardini, G. 472, 474
Bernoulli, J. 302
  Bernoulli numbers 255, 276
  Bernoulli operator 247
  Bernoulli lemniscata 300
Bernstein, S. N. 223, 227
  Bernstein theorem 221-227
  Bernstein-Hausdorff theorem 223
Bertini, E. 130, 376, 383
  Bertini's theorem 46, 56, 402
Bertrand, J. 26
Bessel functions 31-32, 34
Betti, E. 28-29, 35, 39, 383, 429
Betti, M. 460-470, 473
  (E.) Betti numbers 64
Beutelspacher, A. 8
Bézout, E. 43
  Bézout's theorem 53, 56-57
Bianchi, L. 130, 376-379, 391
Bianconi, G. 420, 434, 473
Bieberbach, L. 390
Biggeri, C. 114, 117
Black, C. W. M., 129
Blaschke, W. 387
  Blaschke condition 387
  Blaschke product 217-218
Bleeker, D. 163
Bliss, G. A. 128
Blokhuis, A. 6, 8

Boas Jr., R. P. 118

Boas, H. P. 231, 237

Bochner, S. 148, 227, 311, 320
  Bochner (generalized) theorem 149-151
  Bochner-Martinelli integral formula 311
  Bochner-Martinelli kernel 313
  Bochner-Martinelli-Koppelman formula 313

Bockstein homomorphisms 63

Bogomolov, F. 86, 98, 105

Bohr, H. A. 113

Bolondi, G. iii

Bolzano-Weierstrass theorem 26, 384

Bombelli, R. iv

Bombicci, L. 421, 435, 441-454, 473

Bombieri, E. 61, 64, 75, 86, 98, 102, 105, 174

Bompiani, E. viii, xiv, 277, 279, 381, 460-462, 471, 473-474

Bonasoni, Al. 197

Bonasoni, An. 197

Bonasoni, An.-Maria 197

Bonasoni, Ga. 197

Bonasoni, Giov. 197

Bonasoni, Giu. 197

Bonasoni, P. 197

Boneschi, M. 429

Bonfioli Malvezzi, A. 291, 295, 299-300, 303-304

Boratynski, M. 55

Borel, E. 33, 378, 384
  Borel-measurable functions 379

Borgato, M. T. 375, 391, 393

Bortolotti, En. 162, 391, 460-461, 471

Bortolotti, Et. vi, vii, 295, 304, 375, 380, 392-393, 416, 426, 431, 448-449, 459-468, 471-473

Boschi, P. 421, 436-447, 473-474

Bose, R. C. 7, 8

Bossut, C. 300

Bottai, G. 429, 432

Bottazzini, U. iii, 25-40, 383

Botter, F. L. 422, 434-435, 473

Bouligand, C. 116

Bourbaki, N. 275

Boutet de Movel, L. 230, 237

Bremermann, H. 316

Brill, A. 90, 92

Brini, A. 275

Brioschi, F. 383, 429

Briot, Ch. A. A. 376

Brivio, A. 324

Bruck, R. H. 8

Bruen, A. A. 6, 8

Brunetti, R. 471

Bruzual, R. 153

Buchsbaum, D. 355
  Buchsbaum curves 182, 186-187

Burbea, J. 150, 153

Burgatti, P. vi, 426, 429, 431, 456-469, 473-474

Burns, D. 230, 237

Bussey, W. H. 136

Caccioppoli, R. 392

Caire, L. 54

Caldani, P. 299, 302

Calori, L. 420-422, 430-431

Cambij, G. M. 197

Campana, F. 163

Campedelli, L. 93, 105, 167

Canevazzi, S. 421-422

Canterzani, S. 291-292, 299-304
Cantoni, R. 125
Cantor, G. 36, 291
Capellini, G. vii, xiii, 421, 434-435, 441-446, 473
Carafa, M. 330
Carathéodory, C. 388-390, 393
 Carathéodory metric 116, 409-410
Cardano, G. iv, 292
Cardinali, S. 300, 304
Carducci, G. v
Carleman, T. 284
Carruccio, E. 431
Cartan, E. 369, 373
(H.) Cartan theorem 410
Cartier divisor 44, 48
Cartier operator 64
Casaglia, I. 8
Casali, G. 291, 299-302
Casas-Alvero, E. 361, 366
Casini, P. 304
Casorati, F. 25-40, 383-384, 393, 429
Casse, L. R. A. 8
Cassini, G. v, 292
 Cassini curve 34
 Cassini ovals 300
Castelnuovo, G. 42, 55, 60-62, 64, 75, 86, 89-109, 165-168, 174, 186-187, 381, 393, 429, 431
 Castelnuovo's criterion of rationality 141
 Castelnuovo-De Franchis theorem 67-69, 80
 Castelnuovo-Mumford lemma 59, 181, 183-185
 Castelnuovo's theorem 44, 47
Castro, F. 366

Castronovo, V. 424, 431
Casulleras, J. 283, 289
Cataldi, P. A. iv
Catalisano, M. V. 41-57
Catanese, F. 59-88, 89, 93-94, 96, 99, 101-102, 105-106
Catlin, D. 235, 237, 319, 320
Cattaneo, C. 416, 429
Cauchy, A. L. 26-27, 35, 291
 Cauchy integral formula 38, 311, 321, 337-338
 Cauchy-Kowaleskaja theorem 231, 286, 288
 Cauchy-Fantappiè formula 307-321
 Cauchy-Fantappiè forms 308, 313-318, 321
 Cauchy-Fantappiè kernel 320
 Cauchy-Fantappiè-Leray formula 148, 222
 Cauchy problem 281-82, 310, 321, 330
 Cauchy-Riemann (CR) equations 237, 308, 321, 333
 Cauchy-Riemann (CR) functions 229
 Cauchy-Riemann (CR) hypersurface 370-371
 Cauchy-Riemann (CR) isometry 371
 Cauchy-Riemann (CR) manifolds 229-238
 Cauchy-Riemann (CR) map 370
 Cauchy-Riemann (CR) operator 237
 Cauchy-Riemann (CR) structure 229, 369
 Cauchy-Riemann (CR) system 337, 348, 354
 Cauchy-Riemann (CR) vector field 372
Cavalieri, B. iv, 195-213, 292

Index 479

Cavani, F. 430
Cavazza, M. 304
Cavazzi, A. 447-448
Cayley, A. 398
Cebisev, P. 225
 Cebisev polynomials 381
Chasles-Cremona principle 125
Chelini, D. 420, 430, 434-436, 473
Chern, S. S. 369, 373
 Chern class 75, 77, 156, 180
Chiellini, A. 125
Chisini, O. vi, 56, 106, 130-131, 139
Christ, M. 235-237
Ciamician, G 449-458, 473
Cicognani, L. 449
Ciliberto, Ci. 60, 85-86, 89-109
Cimmino, G. 470, 474
Cinquini, S. 381, 393, 431
Cipriani, L. 430
Ciusa, R. 459-460
Clebsch, A. 90, 92, 165-166, 174
Clemens, H. 189
Clemente XIV, (Pope) 299
Clifford inequality 78
Coen, S. iii, vii, 111-139
Cohen-Macaulay curves 44, 177, 181
Cohen-Macaulay subscheme 44
Collamarini, E. 461-462
Collins, J. 213
Comessatti, A. vi, 469, 472
Condorcet, M.-J.-N. Caritat, Marquis de 299-300
Conte, A. 141-145
Copernicus, N. iv, 295-296
Corazzi, E. 293
Cornalba, M. 406
Cossec, F. 172, 174

Cotlar, M. 114, 117-118, 147-153, 170
Courant, R. 289, 390
Cremante, R. 304
Cremona, L. v, vii, 380, 418, 420, 429-430, 434-38, 473

Dabke, K. P. 384
D'Alambert, J. Le Rond 292, 300
Dal Ferro, S. iv
Dalla Noce, G. 471-472
D'Almeida, J. 185-186
Damiani, E. 275
Dandolo, V. 303
Danti, E. 292
D'Antona, O. 275
Darboux, G. 40, 169, 175, 279, 377
 Darboux system 278
D'Arcais, F. 422, 430, 440, 474
Dattari, S. 197
Davis, E. 44, 50, 56
Deakin, M. A. B. 384
Debarre, O. 99, 106
De Bartolomeis, P. 155-163
De Brosses, C. 324-325
De Bruijn, N. G. 21-22
Dedekind, J. W. R. 38, 92
De Franchis, M. 67-69, 86
Deligne, P. 376, 393
Della Casa, L. 434-438, 473
Del Re, A. 130
De Maria, C. 126
De Moivre equality 116
Denisjuk, A. 227
Denjoy integral 391

De Paolis, R. v, 135, 422, 430, 443- 444, 474
Del Pezzo, P. 127- 129, 134
  Del Pezzo fiber space 142
  Del Pezzo surface 142, 169
De Rham, G. 64
De Rolandis, G. B. 302
Derwidué, L. 132, 361, 366
Desargues spaces 7
Descartes, R. 200, 291
Deschamps, M. 86
Deuring, M. 132
Deutsch, J. C. 137
Di Martino, N. 296
Dienes, P. 22, 116
Dieudonné, J. 129, 139, 335
Dini, U. 28, 31- 33, 36, 39, 376- 378, 381, 385
Dirac's delta function 282
Dirichlet-Lejeune, P. G. L. 36
  Dirichlet's principle 27, 28, 31, 113, 116, 378, 385
Dolgachev, I. V. 60, 89, 165- 176
Domìnguez, M. 153
Donagi, D. R. 407
Donati, L. vi, 421- 422, 431, 442, 444- 458, 474
Dore, P. 431, 471- 472
D' Souza, H. 142, 145
Du Bois Reymond, P. 31
Du Cloux, F. 151, 153
Dugac, P. 31, 34, 39, 377
Duhamel's principle 288
Dupin hypersurface 279
Du Val, P. 366

Ecalle, J. 19- 21, 23
  Ecalle's resurgent functions 19- 23
Egorov theorem 390
Ehrenpreis, L. 330, 337, 345, 355
Einstein, A. 391
Ellia, Ph. 177- 188
Ellinsgrud, G. 181, 186
Emery, C. 425, 431
Enriques, F. vi, vii, 41, 54, 56, 59- 109, 130, 135, 137, 139, 165- 177, 186, 292, 376- 377, 380- 383, 385, 393, 415, 422, 425- 426, 429, 431, 451- 460, 473- 474
  Enriques double plane 168- 169
  Enriques double quintic 94, 96
  Enriques sextic 168- 170
  Enriques surfaces 91, 165- 176
Euclid 126, 133
  Euclid's geometry 136
Euler, L. 292, 300, 326
  Euler class 156
  Euler function 225
  Euler sequence 177

Faber, G. 20
Fabry- Hadamard gap theorem 20
Faedo, S. 129- 130
Fais, A. 421- 422, 430, 441- 444, 474
Fano, G. 124, 135, 169- 171, 175
  Fano model 172
  Fano variety 142
Fantappiè, L. vi, vii, 115, 147- 148, 153, 224, 227, 281- 283, 289, 307-

Index

321, 329-356, 381, 384, 429, 432, 465-466, 471, 473
Fantappiè- Fichera- Martineau- Komatsu duality theorem 348
Fantappiè functionals 330, 333
Fantappiè indicatrix 147, 152, 221-227, 341, 343-344
Fantappiè- Martineau theorem 222
Fantuzzi, G. 195-196
Farini, L. C. 418, 420, 429-430
Fatou, P. 216, 219
Fatou limit 215-219
Favaro, A. 196
Fefferman, C. L. 235, 237, 318-320
Fermat, P. de 323
Ferrari, L. iv
Fichera, G. 118, 139, 330, 337, 346, 355, 388, 432
Filopanti, Q. 420, 430, 434-435, 473
Fiorentini, M. 179, 186
Fiorini, M. 420-422, 430, 434-439, 442-453, 473
Flenner, H. 87
Foà, E. 431
Fock spaces 148, 152-153
Foias, C. 148
Folland, G. B. 153
Fontana, G. 303
Forni, G. iii
Fossum, R. 87
Fourier- Borel transforms 356
Fourier transforms 150, 223, 232, 283-284, 349, 351
Fourier series 35
Francesconi, S. 31, 39, 415-474
Franchetta, A. 93, 96-98, 107
Francia, P. 98, 102, 107

Franzoni, T. 409, 413
Freedman, M. H. 74, 87
Frisi, A. 324, 326
Frisi, P. 326
Fritzsche, B. 153
Frobenius, G. 31, 35, 38-39
Fubini, G. 139, 375-395, 429
Fujita, T. 76, 87

Gaffney, T. 366
Galiani, C. 294, 296, 304
Galilei, G. iv, 196, 291
Galois, E. 291
Galois planes 6
Galois spaces 9
Galois theory 5-9
Galvani, L. 304, 455-456
Gario, P. 129
Gaveau, B. 3, 4
Gauss, K. F. 255
Gauss ψ (psi) function 256
Gauss series 36
Gay, R. 22
Gemelli, N. 324
Gentile, G. 427
Geramita, A. 50, 52, 56, 186
Gherardelli, F. 189-194
Ghione, F. 106
Gianturco, E. 422-423
Gigante, G. 369, 373
Gilden, P. 23
Gimigliano, A. 49, 53, 56
Giolitti, G. 423
Giuffridda, S. 50, 56
Giuntini, S. 304

Giusti, E. iv, vii, 195- 213, 434, 473
Gizatullin, M. Gb. 6, 144
Glynn, D. G. 8
Godeaux, L. 93, 107, 167, 170, 175
Godement, R. 355
Gordan, P. 174
Gould polynomials 270
Goursat, E. 376
Graffi, D. 431, 469- 471, 473- 474
Grandi, G. 323
Grassmann varieties 73
Grauert, H. 1, 4, 314, 317, 320, 355, 388, 393
Gravina, L. 421
Greco, S. 41- 57
Green, M. 59, 69, 79- 80, 82- 83, 87, 91, 107, 177, 183, 187- 188
Gregory, J. 384
Griffiths, P. A. 279, 406- 407
Grothendieck, A. 63, 79, 99, 148, 340, 355, 384
Grothendieck duality theorem 355
Grugnetti, L. 39, 138
Gruson, L. 56, 185
Guarducci, F. 426, 431, 454- 460, 471, 473
Guglielmini, D. v, 292- 293, 295, 299, 303- 304
Gunning, R. C. 376, 393
Gutzmer, A. 389

Hadamard, J. 20, 116, 281, 378
Hahn, H. 379
Hall planes 8
Halphen, G.- H. 186

Hankel forms 147- 153
Harbourne, B. 50, 56
Harder- Narasimhan filtration 76
Hardy, G. H. 275
Hardy field 248
Harris, J. 107, 279, 406- 407
Harris, L.- A. 413
Hartogs, F. 337, 353
Hartogs' phenomena 337
Hartshorne, R. 87, 187
Harvey, R. 317, 320, 345- 346
Heft, S. M. 7, 8
Heine, H. 31, 33
Heins, M. iii, 215- 219
Heisenberg group 152
Helson- Szegö theorem 149- 150
Helton, J. W. 413
Henkin, G. M. 221- 227, 314- 315, 317, 320
Henkin- Ramirez kernel 315
Henry, J. P. G. 366
Hensel, K. 92, 129
Herglotz, G. 148, 281, 283
Hermite, C. 34, 390
Hilbert, D. 115, 135, 139, 289, 349
Hilbert function 41- 57, 105, 179
Hilbert modular group 108
Hilbert operator 148
Hilbert polynomial 179
Hilbert scheme 82, 101- 102, 181, 183 185, 187
Hilbert- Schmidt operator 151
Hilbert syzygy theorem 349
Hilbert transforms 147- 153
Hill, R. 7, 8
Hironaka, H. 363
Hirschfeld, J. W. P. 8

Index

Hirschowitz, A. 177, 183, 186-187
Hitchin, N. 163
Hobson, E. W. 118
Hodge, W. 64
  Hodge decomposition 66
  Hodge-De Rham theory 65
  Hodge-De Rham theorem 64
    Hodge-De Rham theorem (twisted version) 65
  Hodge operator 158
  Hodge theory 80
Hölder, O. 230
  Hölder class 236
  Hölder estimates 230, 235, 237, 318-321
  Hölder norms 235
Holmgrem's theorem 288
Horikawa, E. 87, 96, 99, 107, 175
Hörmander, L. 237, 287, 289, 345
  Hörmander condition 233
  Hörmander operator 235
Horn D'Arturo, G. 459-460, 471-474
Horrocks criterion 179
Horrocks-Mumford bundle 105, 397
Houthakker, H. 226
Hudson, H. R. 130, 407
Hughes planes 8
Humbert, G. 403
  Humbert curves 403, 407
Hurwitz, A. 19, 21-23, 38, 390
  Hurwitz series 245
Huygens, C. 11
  Huygens' principle 11, 17

Idà, M. 177-188

Igusa, J. I. 62, 87
Illusie, L. 87
Ingrami, G. 448
Iskovskikh, V. A. 144-145
  Iskovskikh's conjecture 142
Itaka, S. 61
Iujakov, A. 153

Jacobi, C. 49, 56-57
  Jacobi theta-functions 117
Jacobowitz, H. 373
Janson, S. 148, 153
Jentzsch, M. 390, 393
John, F. 282-283, 287, 289
Johansen, L. 226-227
Johnson, R. A. 138
Jongmans, F. 99, 107
Jouanolou, J. P. 56

Kahana, D. 276
Kähler 1
  Kähler-Hodge theory 59, 61
  Kähler manifolds 1-4, 59-88, 108, 158-159
  Kähler metric 159
  Kaluza-Klein theories 160
Kantor, S. 144-145
Kashiwara, M. 289, 330, 355
Kawai, T. 287, 289, 330, 349, 355-356
Kawamata, Y. 61, 66, 87
Kelleher, J. 356
Kerzman, N. 153, 317, 320
Kimura, T. 330, 355

Kirchoff, G.- B. 281

Kirstein, B. 153

Klein, F. 26, 38- 39, 137, 382

Klingen, H. 409, 413

Kobayashi, S. 163, 388, 393
  Kobayashi metric 409

Kobb, G. 127, 129

Kodaira, K. 4, 61, 74, 87, 98, 107, 167, 171
  Kodaira dimension 91, 141, 167
  Kodaira vanishing theorem 107

Kohn, J. J. iii, 163, 229- 238, 318- 320

Kollar, J. 85, 87

Komatsu, H. 330- 331, 344- 348, 353, 356

Kondo, S. 171, 175

Koppelman, W. 312- 314, 321
  Koppelman formulas 312, 317
  Koppelman's second formula 314

Korevaar, J. 227

Kössler, M. 116

Koszul, J.- L. 76
  Koszul cohomology 107, 187- 188
  Koszul complex 182, 345, 350
  Koszul sequence 76

Köthe, G. 148, 330, 340, 356, 384

Kowaleskaja, S. 29

Kreĭn space 409- 413

Kronecker, L. 26- 27, 38, 92

Kuczma, M. 385

Kuiper, N. 163

Kummer, E. 134, 398
  Kummer surface 170, 174, 397- 407

Kuranishi, M. 74, 87, 231, 238
  Kuranishi family 75

Kustaanheimo, P. 5, 6
  Kustaanheimo conjecture 5

Lacaita, C. G. 429

Lagrange, J. L. 292
  Lagrange inversion formula 261, 265
  Lagrange variety 360

Laguerre logarithmic sequence 267, 274

Laguerre operator 245

Laguerre polynomials 267

Lambert series 34

Lambertini, P. (see Benedetto XIV)

Lamé, G. 32
  Lamé functions 32, 34

Lampariello, G. 375

Landau, E. 387- 391, 393

Lang, S. 87

Lang, W. 87

Laplace, P. S. 4
  Laplace- Beltrami operator 4
  Laplace operator 13
  Laplace transform 223, 225, 356

Lascu, A. 179, 186

Laufer, H. 364, 366

Laurent series 252

Lauricella, S. 377

Lax, P. 11, 17

Lazarsfeld, R. 56, 59, 69, 79- 83, 87, 185, 187

Lê Dũng Tráng 366

Leau, L. 20
  Leau- Wigert- Faber theorem 22

Lebesgue, H. 137, 377- 379, 385- 386, 393
  Lebesgue density 118
  Lebesgue measure 379
  Lebesgue- Vitali theorem 379

Le Brun, C. 163

Lefschetz, S. 398

Index

Lefschetz theorem 398
Legendre, A. M. 117
 Legendre polynomials 31
 Legendre transform 226
Leibniz, G. W. 209, 249, 291, 293, 304-305
Leiterer, J. 320
Lejeune-Jalabert, M. 363-366
Lelli, M. 461-462
Lelong, P. 227
Leont'ev, A. F. 385
Leont'ev, V. 226
Leprotti, A. 295
Leray, J. 308, 310-312, 320-321, 343
 Leray's Cauchy integral formula 312-313
 Leray covering 343
Levi Civita, T. 391, 429
 Levi Civita connection 157
Levi, B. vi, 111-139, 357, 366, 381, 429, 432, 463-469, 471, 473
Levi, E. E. 113, 315, 377, 387
 (E. E.) *Levi form* 189-194, 316, 318
 (E. E.) *Levi problem* 113, 308, 316-317
Levy, P. 153
Lewy, H. 19, 21
 Lewy operator 237
 Lewy solution 21, 22
Lie algebra 233
Lieb, I. 314, 320-321
Lindelöf, E. 390
 Lindelöf sectorial limit theorem 216
Lindemann, C. L. F. 34-35
Lodi, F. 422, 434-445, 473
Lipschitz class 236
Lobachevsky, N. 125

Lobachevsky geometry 136
Loeb, D. 239-276
Loeser, F. 365-366
Logan, B. F. 227
Lojasiewicz, S.
 Lojasiewicz inequalities 365
Longo, C. 134
Loria, G. 304, 430
Luminasi, G. 446
Lüroth, J. 141
 Lüroth's problem 141
Luzin, N. 379, 392

Maccaferri, E. 451
Maccauley, F. S. 406
Macdonald, I. G. 276
Mac Rae, R. E. 103, 107
Maggioni, R. 53, 56
Magini, G. A. v, 292
Majorana, Q. 459-470, 473
Malaguti, F. 299
Malfatti, G. 291-305
Malgrange, B. 345
Malliavin, P. 4
Malpighi, M. 295
Malvezzi, V. 299
Mambriani, A. 462-465, 467-468
Mamiani della Rovere, T. v, 420, 430
Manaresi, M. iii
Manarini, M. 61-65, 69-72
Mandelbrojt, S. 116
Manfredi, E. 293, 295-296, 299
Manfredi, G. v, 291-294, 296, 304
Marchionna, E. 432

Maria Teresa d' Austria (Maria- Theresia, Kaiserin) 323- 327
Mariscotti, A. 434, 473
Markov, A. 225
Maroscia, P. 56
Marsigli, L. F. v, 292- 293, 295, 299, 303
Martineau, A. 224, 282- 283, 289, 330- 331, 339- 344, 348, 356, 384
Martineau- Aizenberg functional representation 148
Martineau- Aizenberg formula 152
Martinelli, E. 311, 312
Martinelli integral formula 211
Martinet, J. 21
Mascheroni, L. 303
Mateo, J. 138
Matteucci, C. 416- 417, 419, 430
Maupertius, P. 300
May, D. C. 311
Mazzetti, S. 195- 196
Mc Neal, J. 235, 238
Menchoff, D. 388, 393
Menegaux, R. 87
Mengoli, P. iv, vii, 195- 213
Mengoli, S. 195
Menichetti, G. iii
Meril, A. 356
Merle, M. 363- 364, 366
Miani, L. 304
Michel, F. 366
Migliore, J. 186
Migliorini, L. 163
Millevoi, T. 144- 145
Millevoi's counterexample 144
Milne, J. S. 87
Milnor numbers 364

Miranda, R. 78
Mirò- Roig, R. M. 187
Mittag- Leffler, G. 20, 34- 35, 38, 383- 384
Miyaoka, I. 61, 98, 107, 141, 279
Möbius, A. 278
Möbius group 159
Möbius transformation 217
Moishezon, B. 66, 88
Monge- Ampère equations 108
Montalbani, O. 196
Montaldo, O. 39, 138
Montanari, A. 429- 430
Montanari, G. 292, 295
Montel, P. 115- 116, 375- 395
Montesano, D. 448- 451, 474
Monti, F. M. 299
Moore, G. H. 139
Morán, N. 153
Morgagni, G. 296
Mori, S. 61, 141
Morrow, J. 87
Morse critical point 365
Moser, J. 369, 373
Mullin, R. 276
Mumford, D. 61, 64, 88, 98- 99, 107, 145, 174- 175, 187
Muracchini, L. 277- 279
Murphy, R. 384
Murray, J. 17
Murray model 14
Mussolini, B. 428
Musti, R. iii

Naftalevich, A. 19, 21, 23

Index

Nagel, A. 235, 238, 318, 321
Nagy, B. Sz. 153
Nagy-Foias theorem 148, 150
Nakano, S. 4
Nakayama's lemma 52
Naldi, G. 275
Nalli, P. 391
Nannicini, A. 163
Napoleon, 292, 302
Nash, J. F. 358, 367
Natali, G. 304
Nehari problem 149, 153
Nehari theorem 150, 152
Neuenschwander, E. 27-28, 39, 383
Neumann, C. 31-32, 39, 376, 391
von Neumann, J. 263
Neumann problem 28
Nevanlinna, F. 387
Nevanlinna, R. 387
Newlander, A. 163
  Newlander-Nirenberg theorem 157
Newton, I. 28, 116, 133, 291-292, 305, 384
  Newton's formula 241, 257
  Newton polygon 363-366
Nicoletti, G. 275
Nijenhuis tensor 157
Nikodym, O. 392
Nikulin, V. 171-172, 176
Nirenberg, L. 105, 163, 231, 238
Noether, M. 61, 74, 88, 90-93, 107, 128-129, 133, 166, 168, 176, 381
  Noether Restsatz 43, 55
  Noether's formula 77
Nomizu, K. 163
Norguet, F. 316
Nörlund, N. E. 385

Novara, D. M. iv
Nygaard, N. O. 88

Odlyzko, A. 276
Ohtsuka, M. 216, 219
Oka, K. 309, 316
Onofri, L. 461-465, 467, 469-470, 472
Orecchia, F. 52, 56
Orlandi, P. 196
Orsetti-Mantovani, D. 430
Osgood, W. F. 129, 386-389, 393
  Osgood space 309

Pacioli, L. iv
Padé approximation 225
Painlevé, P. 389
Palamodov, V. P. 88, 227, 281-289, 345, 349, 356
Paley, R. 227
  Paley-Wiener inversion formula 225
Palladino, F. 304
Pallavicini, G. L. 326
Panella, G. 8
Paranjape, K. 187
Pareschi, G. 187
Pascal, B. 297
Pasch, M. 135, 139
Pasquali, C. 455
Pavarino, L. 275
Paxia, G. 47, 56
Peano, G. 123, 135-136, 139, 390
Peetre, J. 148, 153
Pellegrino, F. 153, 356

Pellegrino, G. 7, 9

Pepe, L. 291- 305, 375, 379, 382, 385, 391- 393, 432

Perrin, D. 187

Perron, O. 377

Persson, U. 107

Pesci, G. 446- 447

Peskine, C. 56, 185, 187

Peters, C. 69, 86, 93, 105, 163, 171, 174

Petracca, A. 114, 117

Petrovski, I. G. 287

Petri, K. 91- 92, 102, 107, 177, 187

Pham, D. 20

Phillips, R. 11, 17

Phong, D. 237

Picard, E. 35, 62, 64, 99, 128- 129, 139, 378, 388, 390

  Picard group 170- 172, 174

  Picard's principle 116

  Picard variety 62, 65

Picciati, G. 456, 474

Pieri, M. 130, 136, 139, 381

Pincherle, S. v, vi, vii, 19- 23, 25- 40, 375- 395, 415, 422, 424, 426- 427, 431, 445- 462, 473- 474

  Pincherle derivative 261, 266

Pini, E. M. 464- 465

Pinkall, U. 279

Piola, F. 471

Plancherel theorem 151

Plücker, J.

  Plücker formulas 361

Poincaré, H. 20, 34, 36- 39, 62, 64, 88, 129, 369, 373

  Poincaré bundle 63

  Poincaré conjecture 21

  Poincaré problem 189, 333

Poisson, S. 281

  Poisson integral 219

  Poisson- Lebesgue integral 216

Polking, J. 317, 320

Pompilj, G. 93, 107

Porter, M. B. 389, 393

Pringsheim, N. 116

Privalov, I. 387

Prym, E. 143, 397

  Prym canonical curve 397- 407

  Prym variety 143, 397

Pucci, C. 431

Puiseux, V. 364

  Puiseux exponent 364

Puntoni, V. 431

Qvist, B. 7, 9

Radò, T. 387

Radon, J. 289

  Radon's inversion formula 282

  Radon transform 221, 225- 227

Ragusa, A. 53, 56

Rajna, M. 426, 431, 454- 458, 473

Ramanan, S. 187

Ramanujan, C. P. 97, 107

Rameau, J. 325

Ramenghi, S. 434- 435, 473

Ramey logarithmic sequence 269

Ramirez, E. 314, 321

Ramis, J.- P. 21

Rampinelli, R. 324- 326

Ran, Z. 69, 88

Index

Range, R. M. 307-321
Ranzi, I. 472
Rao, A. P. 187
  Rao module 182
Razzaboni, A. 448, 458
Razzaboni, C. 420-422, 430, 439, 473
Reckziegel, H. 388, 393
Regnoli, O. 421
Reid, M. 99
Reider, I. 59, 75-76, 88, 98, 102, 107
Reilly, R. 11, 17
Reiss, M. 57
Respighi, L. 420, 430, 434-436, 473
Reye, T. 169, 176
  Reye congruences 169, 171, 174
Riccardi, P. 196, 421
Riccati, G. 299
Riccati, J. 297
Riccati, V. v, 291, 296-299
  Riccati functional 309-310
Ricci, G. 196
Ricci, P. 463
Ricci-Curbastro, G. 377, 391
Riemann, B. 26-29, 31, 36, 90, 107, 165-166, 281, 291, 380, 383
  Riemann integral 378
  Riemann geometry 136, 391
  Riemann-Roch theorem 90, 109, 172, 184, 402, 404
  Riemann surface 20, 68, 73, 115, 335, 376
Riemenschneider, O. 1, 4
Riesz, F. 216-217, 219
Riesz, M. 216-217, 219
  Riesz theorem 150
Rigaud, S. J. 213
Righi, A. 299, 449-458, 473

Roberts, L. G. 56
Robinet, A. 305
Rocca, G. A. 195-196
Rochberg, R. 148, 153
Rodrigues, O. 263
  Rodrigues formula 267
Roghi, R. 460
Rohn, K. 134
Roman, S. 249, 251, 276
  Roman coefficient 253
  Roman factorial 249-250
  Roman module 251
  Roman shift 262, 265, 270
Rondelli, G. 296
Rosay, J. P. 235, 238, 321
Rossi, H. 230, 238
Rota, G.-C. iii, 239-276
Rotschild, L. P. 234-235, 238
Ruffini, P. 303
Ruffini, F. P. 421-422, 430, 442-455, 473

Sabbah, C. 365-366
Sacchetti, A. 453
Sadosky, C. 153
Saks, S. 388, 394
Saladini, G. 297-299
Salamon, S. 163
Salmon, P. iii, 163
Sampson, J.-H. 323-327
Sannia, G. 391
Sansone, G. 377
Santagata, D. 421, 434-448, 473
Santaló, L. A. 116, 125-126, 147
Saporetti, A. 422, 434-453, 473

Sarkisov, V. G. 145
Sato, M. 289, 330, 348, 356, 384
 Sato indicatrix 342
Savioli, L. 299
Scagliarini, G. 459
Schapp, A. 139
Schauder- Tychonoff fixed point theorem 410
Schläfli, L. 28
Schoen, R. 163
Schottky, G. 338, 390
Schreyer, F. O. 64, 187, 406
Schrödinger representation 151
Schubert, H. 358
 Schubert cycle 358
Schur, F. 125, 136
Schwartz, L. 283, 330, 340, 355
 Schwartz spaces 151
Schwarz, H. A. 27- 28, 37, 40
 Schwarz lemma 222
Scorza, G. 469
Sebbar, A. 19, 22
Segre, B. vi, vii, 5- 6, 8- 9, 41, 43, 56- 57, 93, 108, 132, 189, 277- 279, 357- 367, 369, 373, 381, 429, 432, 464- 469, 471, 473- 474
Segre, C. 123- 139, 430,
 (B.) *Segre class* 357
 (C.) *Segre embedding* 402, 404, 406
 (C.) *Segre surface* 278
 (C.) *Segre variety* 338, 340
Semin, F. 279
Sernesi, E. 91- 92, 102, 107
Serre, J. P. 88, 107, 330, 356
 Serre duality lemma 75, 81, 104, 341

Severi, F. vi, 54, 57, 62, 75, 78, 88, 108, 116, 124, 128, 130- 132, 135, 139, 170, 307, 358, 384
 Severi conjecture 76, 99
 Severi's inequality 99
Severini, C. 377, 388- 390
Shafarevich, I. R. 60, 88, 98, 108, 167, 171, 174
Shananin, A. A. 221- 227
Shaw, M. C. 231, 237
Sheffer sequence 240, 275
Sfameni, P. 428
Shokurov, V. V. 143- 145
Sibirani, F. 455- 458
Silva, J. S. 148, 330, 340, 356, 384, 388, 394
Silvani, G. 421
Simart, G. 128- 129, 139
Simeoni, L. 430
Siu, Y. T. 59, 69, 88, 321
Skoda, H. 319, 321
Slupinski, M. 163
Smith, S. H. J. 361, 366
Sobolev, S. 283
 Sobolev estimates 237
 Sobolev spaces 231- 232
Somigliana, C. 377
Spaventa, B. v
Specchia, O. 471
Spencer, D. C. 74
 Spencer- Quillen condition 286
Spivak, M. 369, 373
Spivakovsky, M. 367
Stabilini, G. 422
Stancari, V. F. 293, 295- 296
Stein, E. M. 234- 235, 238, 317- 318, 320- 321

Stein factorization 68, 82-83, 401
(K.) Stein manifold 230-231
Steiner surface 278-279
Steinhaus, H. 118
 Steinhaus lemma 118
Stiefel-Whitney class 156
Stieltjes, T. 225, 386, 389-391, 394
 Stieltjes transform 225
Stirling number 249-250, 273, 275
Stolz, D. 215
 Stolz angle, 215, 218
Struik, D. J. 378, 391, 394
Struppa, D. C. 329-356
Supino, G. 461-463
Szegö, G. 238
 Szegö kernel 238, 320-321
 Szegö operator 229, 235
Szpiro, L. 88, 187

Tacchini, P. 430
Tanaka, N. 369, 373
Tardini, L. 456
Tartaglia, N. iv
Tassinari, P. 434
Taylor, A. E. 377
Taylor, B. 295
Taylor, B. A. 350, 356
Taylor's formula 242
Taylor's logarithmic theorem 254
Taylor's theorem 260, 268, 272
Tchakaloff, V. 227
Tega, W. 304-305
Teissier, B. 357-367
Terracini, A. 113, 124, 133, 139, 429, 432
Thas, J. A. 6, 8, 9

Thom, R. 360
Thomé, L. W. 31
Tinaglia, C. iii
Todd, J.-A. 358
 Todd polar class 357
Todesco, G. 71-72
Toeplitz forms 141-153
Toeplitz kernels 153
Toeplitz operator 153
Togliatti, E. 124
Tomassini, G. 369-377
Tonelli, L. vi, 381, 384, 391, 393-394, 426, 429, 431, 456-457, 459-463, 471, 474
Tonolo, A. 375, 381, 392, 394
Torelli, G. 100, 385, 387, 391-392
Torricelli, E. 196
Touchard, J. 273, 276
Tricomi, F. 375, 430
Trionfetti, G. B. 299
Trivulzi, P. 326
Troup, G. J. 384
Tsen's theorem 143
Tsuji, M. 216, 219
Turing, A. 17
 Turing model 14
Turrini, C. 356

Uccelli, L. 195
Ueno, K. 61, 66, 88
Uspenski, J. V. 116

Van der Geer, G. 94, 108

Vandermonde, C. 241

Vandermonde's identity 241

Van de Ven, A. 86, 93, 105, 163, 172

Varley, V. R. 407

Vaz Ferreira, A. iii, 375- 395

Veblen, O. 136

Vecchi, I. 8, 9

Vecchi, M. 454

Ventura, I. 304

Venturi, L. 422, 436, 474

Veronese, G. 54, 57, 135, 139

Veronese surface 95

Verra, A. 170, 176, 397- 407

Verzaglia, G. 293

Vesentini, E. iii, 409- 413, 432

Viète, F. 200

Villa, M. 470, 472, 474

Villari, E. 438- 448, 474

Viola, T. 113, 115- 116, 138, 375, 379, 392, 394, 464

Vitali, D. 376

Vitali, G. vi, 138, 375- 395, 429, 432, 463- 464, 471, 474

Vitali, L. 392

Vitali, V. 392

Vitali- Hahn- Nikodym- Saks theorem 379

Vivanti, G. 116, 389

Voisin, C. 188

Volterra, V. 147, 281, 307- 308, 377- 378, 383, 385, 390

Weber, C. 366

Weber, H. 27- 28, 38, 40

Webster, S. M. 369, 373

Webster metric 369, 371- 372

Weddle's surface 407

Weierstrass, K. 20, 25- 40, 133, 291, 383- 384, 390

Weierstrass operator 246- 247, 274

Weierstrass point 405

Weierstrass preparation theorem 121- 122

Weierstrass σ (sigma) function 114

Weierstrass ζ (zeta) function 114

Weil, A. 170, 383, 398

Weitzenböck, R. 391

Weitzenböck formula 3

Weyl, H. 151

Weyl's lemma 388

Weyl transform 151

Whitney, H. 359, 366- 367

Whitney equisingularity theorem 358

Whitney filtration 360

Whitney stratification 360

Widder inversion formula 225

Wiegenerinck, J. 227

Wiener, N. 227, 378

Wiener- Hopf equation, 22

Wigert, N. 20

Wu, H. 388, 395

Wu, W.T. 156

Wu, Wen Tsu 163

Wainger, S. 235, 238, 318, 321

Wallach, N. 151

Walker, R. 131

Xiao, G. 59, 76, 93, 108

Index

Yau, S. T. 108, 163
Young, W. H. 118

Zagar, F. 470, 472, 474
Zagier, D. 94, 108
Zamboni, L. 302
Zannoni, A. 382, 446-454
Zanotti, F. M. 291-292, 295-296, 299
Zariski, O. 60, 97-98, 105, 108, 124, 126, 129-132, 139, 361, 367
  Zariski closure 399
  Zariski tangent space 82
Zeuthen, H. 77
  Zeuthen-Segre formula 78
  Zeuthen-Segre invariant 77
Zermelo's axiom of choice 115, 132
Zucchini, G. 463-464
Zygmund, A. 118, 237